2019
서울도시건축비엔날레
SEOUL BIENNALE
OF ARCHITECTURE
AND URBANISM

KB154987

집합도시
COLLECTIVE
CITY

2019
서울도시건축비엔날레
SEOUL BIENNALE
OF ARCHITECTURE
AND URBANISM

집합도시
COLLE
CITY

CTIVE

차례

주 제 전

더불어 사는 일상 • SNS 집합도시 • 두레주택 • 건축의
공적 역할 • 난민 헤리티지 • 두 개의 변신 전략 •
돌무더기에서 나와 • 건물들과 그 영역 • 1부: 주택부족 •
3부: 새로운 주택 • 기후변화대응조치 2.0: 건축학적 연대
• 멕시코 주거 도시화 • 바우하우스 데사우와 건축(1부) •
디지털 광장 • 메카 군중 • 사계절 내내 낙엽은 떨어진다
• 멀티플라이도시 • 1,000명을 위한 사무실 • '데 드리
호벤', 현대식 실버타운 신축 • 고시원 엿보기

도시의 경기장 • 코펜하겐이라 부르는 도시 • 리버풀의
하루 • 집합도시: 인도 현대 건축의 집합적 수행의 형태 •
고물 놀이터 • 극소수부대 – 이야기의 변주 • 지상의 조직
• 타인과 어울리는 방법 • 탈주 미래주의자 • 중세시대
마을(렉스톤) • 공간적 가치의 창조 • 베이험거리: 장소의
서사를 매핑하기 • 작동방식 감당하기 • 집, 몽상 •
서구의 꿈 • 민중의 집 • 일곱가지 서적을 올린 제단 • 매니

Table of Contents

THEMATIC EXHIBITION

Collective Form of the Everyday • SNS
Collective City • Du-Re House • The Public
Role of Architecture • Refugee Heritage • Two
Transformation Stratgies • Out of The Rubble •
Buildings and their Territories • Part I. Housing
Shortage • Part III: The New House • Climate Actions
2.0: Architectural Solidarities • Housing Urbanism
Mexico • The Bauhaus Dessau and its Construction
(Part 1) • Digital Agora • Mecca Crowd • Leaves Fall
in All Seasons • MultipliCity • Office Building for
1000 People • Opening of "De Drie Hoven," Modern
Retirement Housing Complex • Out with a Bang

Urban Enclosure • A City Called Copenhagen • Day
in Liverpool • Collective Cities: Notes on Forms
Of Collective Practices in Contemporary Indian
Architecture • The Junk Playground • Microbrigades
– Variations of a Story • Institute Above-Ground
• Playing Well With Others • Fugitive Futurist •
Medieaval Village (Laxton) • Creation of Spatial Value

도 시 전

CITIES EXHIBITION

글 로 벌 스 튜 디 오

GLOBAL STUDIO

현 장 프 로 젝 트

LIVE PROJECTS

SEOUL MADANG

서울마당

시민참여프로그램

PUBLIC PROGRAM

인사말

2019 서울도시건축비엔날레는 올해 2회를 맞이하여 '집합도시'를 주제로 새로운 차원의 협업과 모델을 연구하고 정보를 교환하며 토론 할 수 있는 장을 마련하였습니다. 이러한 과정과 결과를 통해 도시의 새로운 전략과 비전을 제시하며 서울이 '함께 만들고 함께 누리는 도시'로 자리 잡는 밑거름이 될 것입니다.

서울비엔날레는 다양한 담론을 논의하는 글로벌 플랫폼으로 자리매김하며 동시에 세계 곳곳에서 발생하는 도시 불균형을 해소하기 위한 첨단기술과 진화하는 도시 시스템 등 다양한 연구 결과물들을 보여줍니다.

2019 서울비엔날레는 동대문디자인플라자(DDP), 돈의문박물관마을, 서울도시건축전시관, 세운상가 등 역사도심 곳곳에서 진행되었으며, 이는 향후 도시건축 정책 발굴의 주요거점이 될 것입니다. 또한 시민들과 함께 현대 도시에 관한 이슈를 공유하고 함께 미래도시에 대한 대안을 제안할 수 있는 다채로운 프로그램들도 마련하였습니다. 이로써 서울비엔날레가 진행되는 65일간, 서울 시민들이 함께 즐기고 참여하며 세계인들과 함께 도시와 건축에 대한 생각을 발전시켜 나갈 것으로 기대해 봅니다.

이번 도록에는 '집합도시'라는 주제에 대한 전문가들의 전시 내용과 함께 서울비엔날레 진행 기간 동안의 여러 가지 시민 참여 모습들을 담았습니다. 전 세계 다양한 참여자들의 도시건축에 대한 생각(연구), 작업 진행과정 및 그 결과물을 담은 이 책은 향후 미래 도시건축에 대한 자료로 적극 활용될 것으로 기대합니다.

"우리가 건축을 만들지만, 다시 그 건축이 우리를 만든다." 윈스턴 처칠이 한 말입니다. 시민이 도시를 만들고, 도시가 다시 시민을 만들어 가는 서울이 되기 위해 서울비엔날레가 시민과 도시 속에서 생동하며 지속적으로 발전해 나가기를 기원합니다.

감사합니다.

2019년 11월
서울특별시장 박원순

Greetings

The Seoul Biennale of Architecture and Urbanism 2019 celebrates its 2nd anniversary this year. Under the theme of "Collective City," the Seoul Biennale offers a venue to study new dimensions of collaboration and models to exchange information and engage in in-depth discussions. The Seoul Biennale offers new strategies and visions for the city through such processes, serving as the foundation for Seoul residents to reclaim the city and redefine architecture.

The Seoul Biennale has become a global platform of urban discourses. At the same time, the Biennale displays a wide range of research such as urban systems that evolve with cutting-edge technologies to alleviate urban imbalances in global cities.

The historical venues for the Seoul Biennale 2019—Dongdaemun Design Plaza (DDP), Donuimun Museum Village, Seoul Hall of Urbanism & Architecture and Sewoon Plaza—serve as hubs for developing urban and architecture policy. Moreover, the Biennale offers various programs for Seoul residents to share in the discussion of urban issues and suggest alternatives for cities in the future. I hope Seoul residents enjoy, participate in, and continue to develop ideas on cities and architecture with people around the world throughout the 65 days of the Seoul Biennale.

This year's catalogue provides not only the different exhibitions, but also the various scenes of participation in public programs. It indeed holds urban and architectural perspectives of various participants, processes of projects, and their outcomes. I hope the catalogue serves as a good example for future urbanism and architecture. Winston Churchill said "we shape our buildings, and afterwards, our buildings shape us." I hope the Seoul Biennale vibrantly interacts with its residents and the city and continues to develop to help bring about a city that is shaped by its residents, and which, in turn, will go on to shape them.

Thank you.

November 2019
Won-soon Park, Mayor of Seoul

2019 서울도시건축비엔날레의 여정: 새로운 집합 유형을 찾아서

임재용, 2019 서울도시건축비엔날레 총감독

The Journey of the Seoul Biennale of Architecture and Urbanism 2019: Finding New Forms of the Collective

Jaeyong Lim, Co-director, Seoul Biennale of Architecture and Urbanism 2019

원래 도시는 사람들이 모여 사는 인간중심의 집합체였다. 도시의 규모가 커지면서 도시를 효과적으로 관리할 시스템이 필요하게 되었고 효율과 기능을 중요시하다 보니 도시는 점점 시스템 중심으로 진화하게 되고 도시의 본질인 인간중심의 집합체는 파괴되고 있다. 이러한 도시의 본질인 집합성의 붕괴는 이제 어느 한 도시에 국한된 문제가 아니고 이미 세계 모든 도시들이 직면하고 있는 글로벌 이슈가 되었다. 서울시는 이러한 문제의 심각성을 인지하고 시스템 중심의 도시를 다시 인간 중심의 도시로 회복시키기 위하여 세계 각국의 도시들이 서로 토론하고 해법을 찾아갈 글로벌 플랫폼의 필요성을 느꼈고 2017년 제1회 서울도시건축비엔날레를 개최하게 되었다.

올해로 2회를 맞는 서울도시건축비엔날레는 세 가지 목표를 가지고 출발하였다. 첫째는 서울비엔날레가 유럽과 북미는 물론 남미, 아프리카, 아시아 그리고 오세아니아의 도시들을 대거 포함하는 진정한 의미의 글로벌 플랫폼이 되는 것이다. 둘째로 시민들이 도시의 문제를 쉽게 이해하고 도시를 만드는 과정에 자연스럽게 참여하게 하는 소통의 플랫폼이 되는 것이다. 마지막으로 서울도시건축비엔날레가 일회성 행사로 그치지 않고 지식과 정보가 쌓이는 지속 가능한 시스템이 되고 나아가 도시 및 건축의 정책을 입안하는 중요한 수단이 되게 하는 것이다.

이러한 목표는 소기의 성과를 거두었다. 이번 비엔날레에 전세계 80여개의 도시에서 160개 팀이 참여하였고 유럽과 북미 그리고 나머지 대륙의 참여자 비율도 거의 50:50에 이르러 명실 상부한 글로벌 플랫폼이 되었다. 비엔날레를 통하여 시민들과의 소통하는 방식에 관해서도 다양한 실험이 있었다. 관람객들이 피동적으로 전시를 관람하는 것이 아니고 전시에 참여시켜 도시 문제에 관심을 가지게 하고 더 나아가서는 시민들이 도시를 만드는 과정에 자연스럽게 참여하게 하려는 다양한 시도가 있었다. 대표적인 사례는 시민 사진 및 영상 공모전인데 시민들에게 가장 좋아하는 공공공간의 사진이나

In the past, cities were regarded as people-centered living spaces. As the cities began to grow, the need for an effective management system grew as well. The importance of such systems marked the shift of cities from a people-driven city to a system-driven city. Throughout this process, the once prominent people-centered identity of these urban living spaces began to diminish noticeably. The collapse of this city model as a collective entity is no longer a problem of any one given city, but rather an issue that all cities are facing similarly at the global level. The city of Seoul recognizes the severity of this problem and aims to shift the core identity of the cities to the people-focused paradigm of the past. To achieve this, experts and other representatives from Seoul have initiated dialogues with other cities all around the world in search of a solution to this overwhelming issue. In 2017, realizing the need for an official, global platform to facilitate such discussions, the first edition of the Seoul Biennale of Architecture and Urbanism was introduced.

This year marks the second edition of the Seoul Biennale of Architecture and Urbanism, and three main topics have been selected for discussion. First is the goal to establish a truly global platform, including major cities not only from Europe and North America, but also from South America, Asia, Africa, and Oceania. Second is to take the necessary steps in becoming a platform for communication that facilitates the public's understanding of problems that cities currently face, while also enabling public participation in the process of urban planning and design. The last is to solidify the Biennale's role as a systematic tool for the constant gathering and exchanging of information and knowledge, and paving the way for the implementation of new policies, as opposed to being an event that is relevant once every two years.

Thankfully, these objectives led to an achievement of our desired results. This year's Biennale hosted 160 teams from more than 80 cities around the world, with an evenly spread proportion in

동영상을 찍어서 응모하게 하였다. 사진 1,519장과 동영상 100편이 접수되었고 당선작도 시민들이 직접 투표로 선정하도록 하였다. 접수된 작품들을 분석해 보면 시민들이 규정하는 집합도시의 한 단면을 볼 수 있다. 시민들이 직접 뽑은 사진 당선작 중 대상과 금상을 받은 작품은 여러 가지로 시사하는 바가 크다. 대상을 차지한 작품은 서울역 앞 버스정류장의 풍경이다. 물론 사진의 작품성이 뛰어나 뽑혔을 수도 있으나 교통 인프라는 이미 우리 생활의 중요한 부분이 되었고 집합도시의 중요한 요소라는 것이 증명된 셈이다. 금상은 최근 문을 연 공공 헌책방 서울책보고에서 시민들이 책을 보고 있는 풍경이다. 시민들은 문화 인프라도 공공공간의 중요한 부분이라고 인식하고 있다. 이는 집합도시의 부제인 "함께 만들고 함께 누리는 도시"에서 함께 누리는 도시를 가능하게 하는 공공공간의 영역 중에 교통 및 문화 인프라가 중요한 부분을 차지한다는 것을 시민들의 목소리를 통해서 확인한 셈이다. 이번 시민 공모전을 통해서 서울도시건축비엔날레가 시민의 소리를 듣고 그것을 반영하는 중요한 정책적 수단이 될 수 있다는 가능성을 확인하였다.

2019 서울도시건축비엔날레는 "집합도시"라는 주제를 내걸고 세계 각국의 도시들에게 아래의 질문들을 던지면서 소통의 장으로 초대했다.

오늘날 도시를 인간중심의 집합체로 회복시키는 것은 무엇을 의미하는가?
도시를 만드는 과정에서 새로운 유형의 집합체는 가능한가? 그 집합체에서 정부/지자체, 학계와 전문가 집단, 그리고 시민의 역할은 각각 무엇인가?
시민이 도시를 공평하게 누리게 하는 전략은 무엇인가?
함께 만들고 함께 누리는 집합도시의 새로운 유형은 무엇인가?

세계 각국의 참여자들이 집합도시에 대한 다양한 해석과 해법을 내어 놓았다. 참여 작품들을 내용적으로 분류하여 보면 새로운 주거의 집합유형 및 전략, 디지털 시대의 도시 전략, 도시를 만드는 새로운 유형의 집합체, 난민, 이민자, 인종갈등의 해법들, 새로운 유형의 도시 개발 전략, 시민이 도시를 공평하게 누리는 새로운 방식, 다양한 기후 변화 및 환경 문제에 대한 대응책, 새로운 유형의 집합공간, 도시를 바라보는 인식의 전환, 도시에서 물질 과 생산의 문제, 새로운 유형의 교통 인프라 등으로 정리할 수 있다.

representation from European and North American cities and cities from other continents, thus confirming this event's identity as a global platform. We implemented several new experimental methods for communicating with the public. Rather than offering passive exhibits that visitors passively enjoy, we designed various projects that encourage active participation. Through these activities, an average visitor can learn about, take interest in, and even contribute to solving urban issues. Visitors can even take part in the actual development, planning, and designing of cities. An example of such participatory projects was an open call for pictures and videos taken of public spaces that people are fond of or otherwise enjoy. We received 1,519 pictures and 100 video submissions for this event, and the general public selected the winners. The collection of entries displayed one aspect of the collective city. In particular, the winners of the Grand and Gold Prizes have numerous, important implications in this context. The Grand Prize was awarded to an image of a bus stop in front of Seoul Station. While the picture could have been selected for its unique artistic flair and the artist's talent, it also demonstrates how public transportation infrastructure is an irreplaceable aspect of our everyday lives and a critical factor in what defines a collective city. The Gold Prize was awarded to an image of people reading books at a public, second-hand bookstore, also demonstrating the public's perception of cultural infrastructure as an important part of public spaces. Indeed, these two images are symbolic of the goal of collective cities as urban spaces that people can create and enjoy collectively; these submissions confirmed that transportation and cultural infrastructure play a vital role in creating equitable public spaces. Through this event, we were able to gather the thoughts of the general public and became more aware of the potential of reflecting these thoughts in future public policies.

Under the theme of "Collective Cities," the 2019 Seoul Biennale of Architecture and Urbanism invites global cities to a platform for communication and presents them with the following series of questions:

What does it mean today to return cities to their previous identities as people-driven collective systems?
Is it possible to implement new collective models when designing and creating new cities?
What are the individual roles of the central and local governments, academia, expert groups, and the public within these new

새로운 집합 유형	주제전	도시전	글로벌 스튜디오
새로운 주거 집합유형 및 전략	8	6	4
디지털 시대의 도시 전략	3	3	2
도시를 만드는 새로운 유형의 집합체	7	11	3
난민, 이민자 및 인종갈등 문제의 집합적 해법	3	1	1
새로운 유형의 도시개발전략	7	20	8
시민이 도시를 공평하게 누리는 새로운 방식	3	1	0
다양한 기후변화와 환경 문제에 대한 대응책	2	4	1
새로운 유형의 집합공간	6	12	6
도시를 바라보는 인식의 전환 (리서치 프로젝트 등)	5	15	7
도시에서 물질 및 생산에 관한 문제	3	5	1
새로운 유형의 교통 인프라	0	2	1

참여자들은 도시를 만드는 새로운 유형의 집합체에 관련해서 21개 작품을 출품하였는데 도시를 만드는 주체가 기존의 정부주도형 탑다운 방식이 아니라 정부, 학계와 전문가, 시민이 협력하는 새로운 유형의 집합체가 주체가 되어야 한다고 주장한다. 함께 만들고 함께 누리는 도시를 부제로 하는 비엔날레의 취지와도 일치하는 부분이다. 다양한 주거유형과 개발방식에 대해서 18팀이 작품을 제출하였고 급변하는 거주환경에 대처하는 참신하고 적용 가능한 아이디어가 많았다. 모든 도시가 처한 상황과 환경이 다르다. 35개 팀이 각 도시에 적합한 다양한 유형의 도시개발전략을 소개하여 주었다. 24개 팀이 새로운 유형의 공공공간 또는 집합공간을 제안해서 시민들이 공평하게 누리는 도시의 풍경을 상상하게 해 주었다. 이외에도 다양한 참여자들이 도시문제를 새로운 시각으로 접근하는 전략, 난민이 이민자를 위한 집합적 해법, 기후변화와 환경문제, 새로운 교통 인프라 구축, 디지털 시대의 도시전략 등 도시 문제를 입체적으로 바라볼 수 있는 훌륭한 작품들을 선보였다. 이제 우리는 이러한 소중한 정보들을 더욱 깊이 들여다보고 각각의 도시에 적합한 사례와 전략들을 발굴하여 각 도시의 상황에 맞는 새로운 집합적 전략으로 재탄생 시켜야 할 것이다.

서울도시건축비엔날레는 일회성 전시행사가 아니다. 주최한 서울시는 물론이고 세계각국의 참여자들은 긴 시간 작품을 구상하고 준비해왔으며 작품을 전시장에 설치한 후 참여자들간에 심포지엄을 통해서 서로의 의견을 발표하고 토론하는 기회를 가졌다. 또 두 달이 넘는 전시기간 동안에 많은 관람객들의 피드백도 받았다. 참여자들은 이러한 소통의 장을 통해 자신들의 전략들을 검증하고 다른 참여자들의 경험을 통해서 새로운 방향을 모색하게 될 것이다. 이것이 비엔날레를 개최하고 참여하는 진정한 의미이다. 일반적으로

collective models?
What strategies can ensure equitability among the public within an urban setting? What are some new, collective city models that the public can create and enjoy together?

Participants from nations around the world presented their responses to these questions. When analyzed closely, the content of the presentations prepared by participating cities could be categorized as follows: new collective housing models and strategies; urban strategies in the digital era; new collective models that create cities; collective solutions to issues surrounding refugees, migrants, and race; new models of urban development strategies, new methods for equitability in urban populations; response measures against climate change and other environmental issues; new models of collective spaces; shifts in the recognition and acknowledgement of cities; issues with materials and production in cities; new models of transportation infrastructure.

New Collective Model	Thematic Exhibition	Cities Exhibition	Global Studio
New collective housing models and strategies	8	6	4
Urban strategies in the digital era	3	3	2
New collective models that create cities	7	11	3
Collective solutions to issues surrounding refugees, migrants, and race	3	1	1
New models of urban development strategies	7	20	8
New methods for equitability in urban populations	3	1	0
Response measures against climate change and other environmental issues	2	4	1
New models of collective spaces	6	12	6
Shifts in the recognition and acknowledgment of cities (includes research projects)	5	15	7
Issues with materials and production in cities	3	5	1
New models of transportation infrastructure	0	2	1

The participants prepared 21 projects on new collective models that create cities, which were no longer characterized by the previously dominant, government-led, top-down system; instead, these models proposed a new collective model led by public collaboration. In other words, the core of these models was reliant on the public creating and enjoying cities as a collective body, aligning with the theme and message of this year's biennale. There were also 18 projects on various housing models and development strategies, which outlined innovative and detailed response strategies to the rapidly changing housing environment. It is important to note that each city is faced with its

전시도록은 개막식에 맞추어 발간하지만 2019 서울도시건축비엔날레는 이러한 일련의 소통의 과정을 담아내기 위해 폐막식에 맞추어 전시도록을 발간했다. 전시도록에는 가이드북에 담지 못했던 작품의 정보는 물론 개막식 이후의 다양한 피드백들과 전시장 풍경 등을 담았다.

올해의 주제인 집합도시는 2017년의 주제인 공유도시의 연장선상에 있다. 서울도시건축비엔날레는 계속해서 도시 문제들을 토론하고 해법을 찾는 글로벌 플랫폼의 역할을 다하며 시민이 도시를 만드는 과정에 참여하는 소통의 장이 될 것이며 도시정책을 입안하는 중요한 수단으로서의 역할을 지속적으로 해나갈 것이다. 2019 서울도시건축비엔날레에 참여해주신 세계의 각국의 참여자들에게 다시 한번 감사 말씀을 전한다.

own unique situations and environments. Thirty-five teams prepared projects on various models of urban development strategies. Twenty-four teams shared new models of public or collective spaces that, when implemented, would enable all members of the public to equitably enjoy and benefit what the city has to offer. Other participants also prepared impressive projects that take innovative approaches to urban issues; solutions for refugee and migrant communities; strategies for climate change and other environmental issues; strategies for establishing new transportation infrastructure; urban strategies in the digital era; and other strategies for problems commonly faced by cities around the world. Now, it is our turn to take these important ideas and information and incorporate appropriate strategies and case studies from each city to develop a new set of collective urban strategies.

The Seoul Biennale of Architecture and Urbanism is not a one-time event. Cities from all around the world dedicated an extensive amount of time and effort to preparing their projects, which were then discussed through a symposium that facilitated the exchange of opinions and ideas. Over the course of the two months, we also received an enormous amount of feedback from visitors and participants alike, who not only shared their own strategies, but also had the opportunity to listen to and closely consider ideas of others. This is exactly what the biennale aimed to accomplish. Most exhibition catalogues are released on the day of the opening ceremony. But this year, it will be ready on the day of the closing ceremony in order to not only document the exhibition but also include the discussions held throughout the event. The catalogue contains information on the projects and various feedback and images that were collected since the opening.

The theme, "Collective City," is an extension of the theme of the 2017 biennale, "Imminent Commons." We are committed to carrying forward the message and goals of the Seoul Biennale of Architecture and Urbanism to provide a platform for discussing and seeking solutions for urban issues and to provide the public with opportunities to have their voices heard and to contribute to the development of cities. This event is crucial in ongoing efforts to develop and implement meaningful urban policies in the future. On that note, I would like to take this opportunity to extend my gratitude to all the representatives around the world for their sincere dedication to and participation in the 2019 Seoul Biennale of Architecture and Urbanism.

집합도시:
함께 만들고 함께 누리는 도시

프란시스코 사닌, 2019 서울도시건축비엔날레
총감독

Collective City:
Reclaiming the City Redefining Architecture

Francisco Sanin, Co-director, Seoul Biennale of Architecture and Urbanism 2019

도시는 경쟁이 벌어지는 공간이다. 그러나 본질적으로 절충의 공간이자, 여러 복잡한 주제, 다양한 권리·가치관·요구·상황이 존재하는 곳이기도 하다. 따라서 도시를 제대로 알려면 경쟁·절충·다양성이라는 도시의 모순적 특성을 이해해야 한다. 아울러 도시가 주거, 천연·사회 자원에 대한 접근성, 교통, 정치적 자유·정의에 대한 권리와 연관되어 있으면서 그와 동시에 국내·국외 이주민, 기후 변화, 불평등 심화 문제에 영향을 받는 물리·정치적 구성체임을 인식할 필요가 있다.

오늘날의 도시는 효율과 이윤의 논리에 따라 형성된 투기의 수단으로 작동한다. 공공의 자원과 공간이 점점 사유화·상품화되고 있다. 최근 부상하는 여러 도시 모델의 바탕에는 도시의 집합적 본질을 약화시키는 기업, 기술관료들의 비전에 특권적 우위를 부여하는 경영 모델이 있으며, 이는 우리의 정치·환경적 생존을 위협하고 있다. 이 과정에서 건축은 도시의 형태와 구성에서 그 중요성을 계속 잃어 갔으며, 학문적 영역 또한 점점 축소되었다. 기업 도시, 자유무역 지대부터 지정학적 고립 지역에 이르기까지, 도시는 본질적으로 중대한 위기를 맞이하고 있으며 이는 이번 비엔날레에서 다루어 진다.

건축은 세계의 도시화와 왜곡된 도시 성장으로 도시 형태와 그 구성 과정에서 차츰 소외되었고, 기존의 학문적 영역에서도 점차 사라지게 되었다. 오늘날의 도시는 전례 없는 경제 성장과 점점 복잡해지는 사회·정치·경제 체계 속에서 사람·상품·데이터의 흐름에 좌우되고 통계와 수익으로 성공을 가늠하는 공간이 된 것이다.

현재 우리가 직면한 시급한 과제는 국내부터 국외까지, 공적 공간부터 사회기반시설 네트워크까지, 직접 행동부터 제도적 차원의 역할까지 다양한 층위에 걸쳐 참여할 수 있는 수단과 방식을 개발하고, 건축의 학문적 영역을 지식 생산과 행동 체계로 재정의하는 일이다. 이번 2019 서울도시건축비엔날레는 건축이 도시 형성에 능동적으로 참여하고, 정치·문화적 도시 프로젝트를 실현하며, 오늘날 한정된 도시의 대상과 상징을 넘어

Cities are contested territories, by their very nature spaces of negotiation, multiple and complex subjects, divergent rights, values, claims and conditions. To understand the city one needs to understand this contradictory nature and to recognize and frame it as a physical and political construct complicit, in the right to housing, access to natural and social resources, transportation, political freedoms and justice simultaneously pressured by the challenges of internal and global migration, climate change and increasing inequity.

The contemporary city operates as a speculative apparatus, a mechanism of capital, shaped by the logic of efficiency and profit; public resources and spaces are increasingly being privatized and conceived of as commodities. Emerging models are based on a managerial model that privileges corporate and/or technocratic visions that undermine the collective nature of the city, and challenge both our political and environmental survival. From corporate cities, free-trade zones to geopolitical enclaves, the city is in a deep crisis as to the very nature of its project.

In the wake of global urbanization and the distorted growth of urban centres—cities—architecture has become progressively marginalized from the city's form and processes, increasingly relinquishing the traditional definition of its disciplinary scope. In the context of this continued and unprecedented growth and progressively complex social, political and economic systems, the contemporary city is overwhelmingly conceived in terms of flows and data system with success measured by statistics and profit.

The urgent challenge is to develop new tools, methods and forms of intervention at multiple scales from the domestic to the territorial scale, from public space to infrastructural networks, from activism to institutional roles, redefining architecture's disciplinary territories as a system of knowledge production. The Seoul Architecture and Urbanism Biennale (SBAU) 2019 claims architecture as an active participant in the

새로운 도시를 생각하고 그려내는 주체임을 소개하는 자리이다.

우리는 2019 도시건축비엔날레에서 참신한 사고와 상상력을 바탕으로 도시가 가진 집합 프로젝트의 본질과 실현 가능성과 구조를 면밀히 살펴볼 것이다. 도시 문제 해결은 우리 시대의 가장 시급한 과제가 되었다. 70억을 돌파한 세계 인구의 80%가 도시에서 살아가는 오늘날, 도심의 대변화는 우리가 공간과 시간적으로 이해할 수 있는 수준을 넘어 종래의 방식으로는 대응할 수 없는 속도로 이뤄지고 있다—현대 도시의 실패가 우리 눈 앞에서 일어나고 있는 것이다.

우리는 2019 서울 도시건축비엔날레를 통해 집합 공간이 어떤 의미를 가질 수 있는지, 어떻게 하면 '새로운 도시를 그려내기 위해 사회·환경적 공통 주제를 정치적 행동으로 묶어내는 주체'로서 집합공간을 재구성할 수 있는지 보여주고 그에 대한 근거를 제시하는 무대를 구축하고자 한다. 이번 비엔날레는 도시가 당면한 여러 문제에 대한 해결책을 제시하고, 새로운 집합 행위가 시대적 역경 앞에서 힘과 의미를 가지기 위해 어떻게 구성·검증·혁신·활용되어야 하는가를 이해하는 데 반드시 필요한 학문적·문화적 대변화를 모색하는 행사이다.

또한 이번 비엔날레는 진정한 글로벌 관객에게 열려 있으며 이들과 소통하는 연구와 토론의 장을 형성하고자 한다. 도시의 집합성이—현재 도시의 자연적 상태가 아니라—우리가 다시 쟁취하고 구성해야 할 목표라는 관점으로, 목표를 이루기 위한 조건을 연구하며 이 집합성이 실제로 정치적 행동을 촉진하고 도시를 전면적으로 변화시키는 수단임을 시사할 것이다. 서울시 주최로 열리는 이번 2019 서울도시건축비엔날레에서는 세계 각지의 도시에서 온 참여자들과 함께 도시 문제에 관한 글로벌 차원의 대화를 바탕으로 혁신적 집합성에 공간적 구성체, 다양한 예측과 연구, 그리고 다양한 거버넌스 유형과 새로운 사회 실천의 성격이 있음을 확인하고자 한다. 또한 새롭게 부상하는 전세계 규모의 협업과 도시 건설의 모델을 구현하고 탐구할 예정이다. 아울러 향후 열릴 서울도시건축비엔날레에서도 계속해서 탐구·토론·반론·재고의 대상이 될 만한 연구 주제와 변화 전략을 설정함으로서 지난 비엔날레들의 성과를 이어 나가고자 한다.

construction of the city and as an agent in realising a political and cultural project for the city, reimagining reconsidering it beyond the object-icon dimension in which is trapped today.

The SBAU has the privilege to rethink, reimagine and claim back the city as a collective project and to truly interrogate its viability and its construct. These questions could not be more immediate, timely or urgent. The global population tips the balance at 7 billion with more than 80% of people living in our cities, the transformation and sublimation of our urban centres transgresses comprehensible notions of scale and time, outpacing our hitherto ability to respond and act – we are witness to the failure of the modern city.

The biennale is an incitement to demonstrate and evidence what collective space can mean, how we can imagine it as an agent of political action of both of social and environmental commons in order to speculate on the new city. It is a platform to grapple with essential and necessary disciplinary and cultural transformations that must take place to enable us to answer these challenges and to understand how new collective practices can be mapped, tested, innovated and leveraged to truly gain agency and relevancy in the face of most certain adversity.

SBAU 2019 aims to create a space of research and debate opened to and engaged with a truly global audience. The Biennale will explore the condition of the collective in the city, not as the natural state of the city today, but rather as a condition to be reclaimed and reframed, suggesting that the collective subject is in fact the instrument of political action and transformation of the city. With the city of Seoul as host, SBAU proposes to engage in this global dialogue with partners from multiple cities around the world to identify innovative forms of collectivity as both spatial constructs, forms of speculation and research, modes of governance and new social practices. It aims to map and explore emerging models of collaboration and city making at a global scale. SBAU aims to establish a continuity building on previous biennales by establishing research agendas and transformative strategies in the city to be taken over, debated, refused or reconsidered in future editions of SBAU.

주제전
THEMA
EXHIBI

TIC

TION

주제전

베스 휴즈, 주제전 큐레이터

Thematic Exhibition

Beth Hughes, Curator of Thematic Exhibition

2019 서울도시건축비엔날레의 주제전은 '집합도시'를 주제로 집합적 실천과 행위가 어떻게 도시의 개발 패러다임을 변화시키고 공간 생산의 지배적 시스템에 저항할 수 있는지 질문을 던진다. 이번 서울비엔날레는 건축과 도시, 환경의 대안적 개념을 제시하고 건축의 정치적 동력을 탐색하기 위해서 공존, 사회적 실천, 거버넌스, 연구 및 추측의 새로운 모델을 반추하고자 한다. 주제전은 현재의 도시 구성을 재해석하라는 권유이자, 우선순위를 재배열해보자는 자극제다. 이때 부동산 투기와 토지 상품화를 통한 개인 및 자본의 성공으로부터 집합적 권리와 도시가 공유 투자라는 논점으로 초점이 변화된다.

전세계에서 참여하는 건축가들이 이번 서울비엔날레에서 보여주는 작업들은 현대적인 도시화 과정, 새로운 영토를 확장하고 정의하는 생태·사회기반적 시스템 및 생산, 물질과 생산에 관한 문제, 개발의 대안적 모델과 유형적 혁신, 새로운 유형의 거주권과 토지 소유권 그리고 행동주의나, 시위, 중재, 참조로서의 건축을 재고하는 등 매우 다양하다.

주제전의 전시배치는 연속적인 공간에 연구 결과물과 명제, 추측을 한데 모아 연결과 중첩을 가능하게 한다. 관객들에게 제시하는 특별한 관람동선이나 관람방법은 없고 오늘날 전세계에서 이뤄지고 있는 도시 건축 행위들의 규모와 형태에 몰입하도록 돕는다. 관람객은 오늘날 세계가 마주하고 있는 과제와 우리의 거주 방식 변화의 시급성, 그리고 이러한 상황에서 건축과 형태의 역할에 대해 자신만의 방식으로 이해하고 탐험해 볼 수 있다.

작가들의 전시와 함께 주제전 큐레이팅 팀이 준비한 영상 시리즈가 상영된다. 이 영상물은 도시를 형성하고 도시에 의해 만들어지는 다양한 집합체의 삶과 행동양식을 보여준다. 기록영상과 다큐멘터리, 예술 영화와 연구프로젝트를 통해 세계 현대 도시들의 상황과 집합성을 선보이고, 시민과 그들의 삶을 담으며, 도시의

"Collective City" seeks to question how modes of collective practice and action can challenge the current paradigms of city development and offer resistance to the dominant systems of spatial production. The Seoul Biennale reflects on new models of co-existence, social practice, governance, research and speculation, to suggest alternative concepts of architecture, the city and the environment. The Thematic Exhibition is an invitation to radically reimagine the structure of our cities, a provocation to fundamentally reprioritize, shifting focus from the success of the individual and capital through real-estate speculation and the commodification of land, to foreground collective rights and to claim the city as a shared investment.

Participating architects have contributed from all around the world. Together they present varied critiques on the contemporary processes of urbanization, exploration of ecological and infrastructural systems expanding and defining new territories, questions of material and production, alternative models of development and typological innovation, new forms of tenure and landownership to architecture as a form of activism, protest, mediation and consultation.

The layout of the exhibition positions research, proposition and speculation in close proximity to one another in a continuous space, allowing for connections and overlap. There is no defined sequence of experience or orientation, rather the exhibition is intended as an immersion within the many scales and forms of action currently active in the global practice of architecture and urbanism. In this saturated space, the viewer can navigate their own encounter to best understand the challenges facing our world today, the urgent need for transformation of our occupation of the planet, and the potential of architecture and form to engage meaningfully within that context.

Accompanying the installations from invited architects, is a body of film-work exhibited by the curatorial team of the Thematic Exhibition, that seeks to use moving image to capture the

복잡성과 인간과 사회 구조를 소개한다.

lives and rituals of the communities that shape and are shaped-by our cities. A selection of archive footage, documentaries, artist films and research projects reflect upon the contemporary urban condition and collectives around the world, exposing the complexity of cities and human and societal structures, recording the lives lived within them.

인구 감소 시대와
고령사회에 접어든
지금, 소도시와 농어촌은
공동화로 소멸 위기에 처해
있다. 지역으로의 이주는
공동체와 함께 집합적
거주 계획으로 지역에
정착하고 활력을 만들어가는
방식이 대안이 될 수 있다.
오시리가름 협동조합주택,
눈뫼가름 협동조합주택,
의성고운마을 프로젝트들은
주거기반공동체와
지역공동체 형성을
위한 대안적 모델로서
지역활성화의 가능성을
발견하고자 한다.

Faced with a declining
and aging population,
small cities are losing
their inhabitants and
are under threat of
disappearing. In order
to bring vitality to
shrinking rural towns, a
population move can be
considered an alternative
solution, organized under
a collective residential
plan that can encourage
communities to settle
and build a life together.
By three projects, Osiri
gareum Cooperative
Housing, Nunmoe
gareum Cooperative
Housing and Uiseong
Gowoonmaeul, a
regional revitalization
plan could discover
its potential through
a certain process: an
alternative model comes
to form a residential-
based community and
takes root in the regional
community.

더불어 사는 일상
이엠에이건축사무소(주)

**COLLECTIVE FORM OF
THE EVERYDAY**
EMA architects &
associates

현대도시의 집합성을 물리적인 상황을 넘어서 SNS에서 교환되는 밀도에 주목하여 살펴보고자 한다. SNS에서 드러나는 건축물(어쩌다가게, 어쩌다집, 이안북스사옥)의 이미지와 밀도를 전시하고 디지털시대의 건축의 의미를 다시 생각해 보고자 한다.

This project looks beyond physical notions of the collective, focusing instead on how the city is consumed through Social Network Services (SNS). SAAI presents the way four buildings, Uhjjuhdah Shops @Mangwon, Uhjjuhdah Shops @Seogyo, Uhjjuhdah House and IANN Books are represented on SNS, to reconsider the meaning of architecture in the digital age.

SNS 집합도시
(주)건축사사무소SAAI

SNS COLLECTIVE CITY
Architects Office SAAI

두레주택
(주)조진만 건축사사무소

DU-RE HOUSE
Jo Jinman Architects
(JJA)

산새마을은 주민참여가
활발한 공동체로 세간의
주목을 받기도 하였다.
두레주택은 서울시에서
운영하는 임대용
공유주택으로 청년층과
독거노인 등의 주거 문제를
해결할 목적으로 세워졌다.
주택 내 세부 공간의 역할을
분석하고 관계맺음의
중요성을 감안해 마을의
개념을 주택에 접목하고,
주택을 다시 마을 단위로
확장시키고자 하였다.

Sited in Sansae Village,
known for its active local
community, DURE House
is a share house which
seeks to accommodate
both the young and the
elderly living alone.
Through analyzing the
ways existing local
residents occupy space
and socialise, the design
of the building seeks to
bring the village into the
house and extend the
house to the village.

"건축의 핵심은 현장성이라고 할 수 있다. 도시든 건축이든, 현장의 역사, 지형, 사람, 현장성이다. 가상적으로 혼자 꿈꾸는 것이 아닌 집단적 현실"이라고 덧붙이지 않았을까.

"the essence of architecture may be found in that realm, whether urban or architectural, the history, geography, and people of a site form this visceral reality. It is not a virtual or an individual dream...", but a collective reality.

건축의 공적 역할
정기용

THE PUBLIC ROLE OF ARCHITECTURE
Guyon Chung

난민 헤리티지
DAAR
(탈식민건축미술레지던시) /
알레산드로 페티

REFUGEE HERITAGE
DAAR (Decolonizing
Architecture Arts
Residency) / Alessandro
Petti

이번 비엔날레의 전시는
데이셰 난민촌의 유네스코
세계문화유산 등재 신청서
내용을 담은 시청각물로
구성된다. 난민 헤리티지는
고난과 실향(失鄉)의
아픔 외에도 다양한
난민의 역사를 추적,
기록, 표현, 공개함으로써
인도주의적 관점 이상의
'난민다움(refugeeness)'을
제시하고 실천하고 있다.

Refugee Heritage
constitutes of an audio/
visual installation that
narrates the UNESCO
World Heritage Site
nomination dossier of
Dheisheh Refugee Camp.
In documenting and
representing refugee
history beyond the
narrative of suffering and
displacement, Refugee
Heritage is project that
attempts to imagine and
practice refugeeness
beyond humanitarianism.

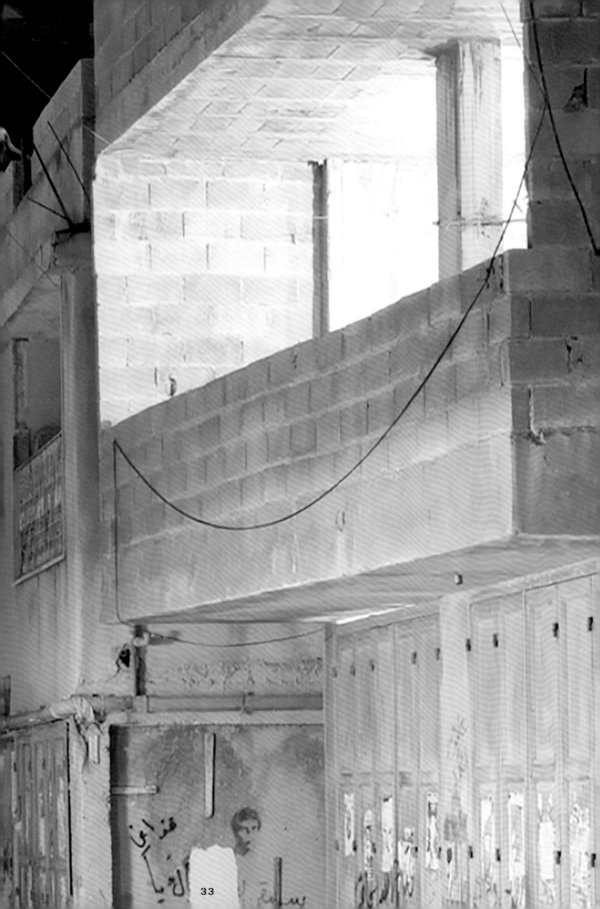

The strategy combir
infrastructure suc

두 개의 변신 전략
알레한드로 에체베리,
호르헤 페레스-하라미요

TWO TRANSFORMATION STRATEGIES
Alejandro Echeverri & Jorge Perez-Jaramillo

ge-scale transport
the gondola lift,

메데인 시의 지난 20년간의 변화과정을 두개의 프로젝트로 압축하였다. 도시, 광역시계획과 전략적 프로젝트에 기반을 둔 접근법과 중소규모 사업에 적용되는 기준이다. 우리는 본 프로젝트를 통해 비판적 시각으로 메데인 시의 변화 과정을 보다 이해하게 될 것이다. 이해를 바탕으로 오늘날까지의 교훈, 향후 주요 과제, 현재 진행 중인 모범사례 등을 살펴봄으로써 오늘날 세계 도시들이 직면한 과제 해결에 메데인 시의 사례가 도움이 되기를 희망한다.

The transformation of Medellin over the past twenty years is consolidated through two projects, one at a metropolitan, institutional scale versus one at neighbourhood level. Consolidating a critical reflection on the evolution of the city, the exhibition presents the lessons learned, prevalent challenges and good practices; contributing towards solving the global issues of the contemporary city.

건물들과 그 영역
토니 프레튼 아키텍트

BUILDINGS AND THEIR TERRITORIES
Tony Fretton Architects

도시는 정치적, 경제적, 문화적 개념들로 형성되지만, 사람들은 이 외의 각기 다양한 생각을 기준으로 도시를 점유하고 경험하고 다시 상상한다. 그러나 이 생각들이 사회·문화적 인식을 토대로 사회 전반에서 통용되는 개념이라면 소통구조(communicative architecture)의 기반이 될 수 있다. 소통구조는 도시의 건물들은 물론이고 신문, 영화, 소설 등의 대중매체가 건물을 활용, 변형, 해석하는 과정에서 건물들에 각인된 흔적 속에서 찾아볼 수 있다.

While cities are shaped by political, financial and cultural concepts, people occupy, experience and reimagine them according to very different ideas. Those ideas, when they lie in the broad social and cultural agreements of a society, can form the basis of communicative architecture. They are evident in buildings through the imprint on them of use and alteration, and through interpretation by mass media.

스틸컷, 3부: 새로운 주택,
1926-28
소스: 바우하우스 아카이브
베를린

Still, Part III: The New
House, 1926-28
Source: Bauhaus-Archiv
Berlin

기후변화대응조치 2.0 프로젝트는 기후정상회의에서 발생한 시위와 환경파괴에 대한 전례 없는 법적 소송 사례 등을 소개함으로써, 다양한 집단들의 기후변화 대응조치를 강화하기 위해 뉴욕 UN 본부와 워싱턴DC의 미국대법원에 작품을 설치할 것을 제안하였다. 아울러 이 프로젝트는 도시의 랜드마크에 전시를 제안함으로서 기후변화에 대한 대응조치를 가속화할 수 있는 '건축적 연대'를 제공한다.

Climate Actions 2.0 proposes installations at two sites: The United Nations Complex in New York and the Supreme Court in Washington D.C. They amplify diverse forms of climate actions, including targeted protests at climate summits and unprecedented litigation on environmental degradation. By proposing tactical installations at civic landmarks, the project offers architectural solidarities towards accelerated collective climate actions.

기후변화대응조치 2.0: 건축적 연대 카담바리 백시

CLIMATE ACTIONS 2.0: ARCHITECTURAL SOLIDARITIES
Kadambari Baxi

SYSTEMATIC AFFIRMATIVE
INFRINGEMENTS

본 프로젝트는 향후 멕시코 도시환경의 형태를 결정지을 공공주거정책과 금융 상황, 정치적 안건, 사회참여와 건축 간의 상관관계를 알아보고자 한다.

This project looks into the relationship of public housing policy, finance, political agenda, social engagement and architecture, as it determines the shape of Mexican urban environments for years to come.

멕시코 주거 도시화
엘 시엘로

HOUSING URBANISM
MEXICO
El Cielo

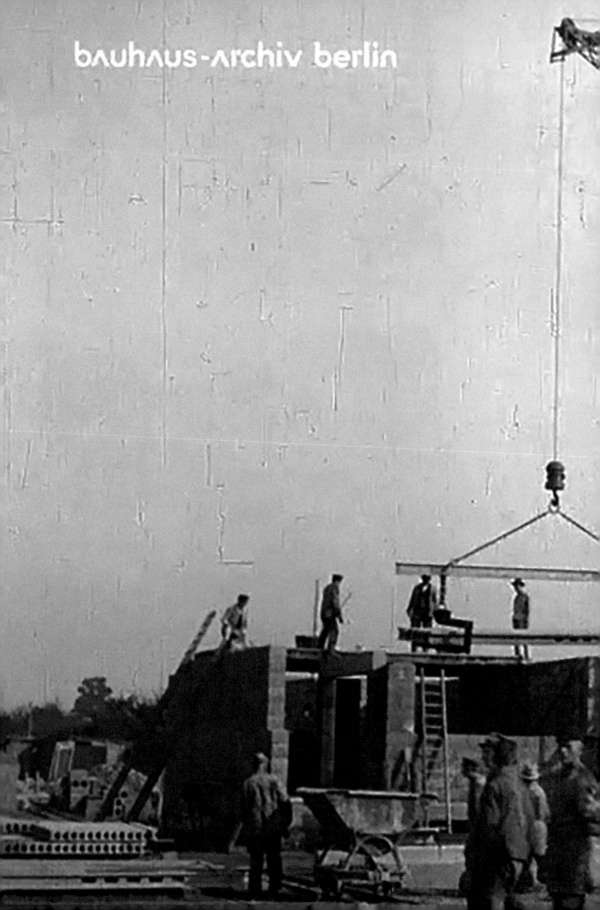

bauhaus-archiv berlin

스틸컷, 바우하우스 데사우와
건축(1부), 1926-28
소스: 바우하우스 아카이브
베를린

Still, The Bauhaus Dessau
and its construction (Part 1),
1926-28
Source: Bauhaus-Archiv
Berlin

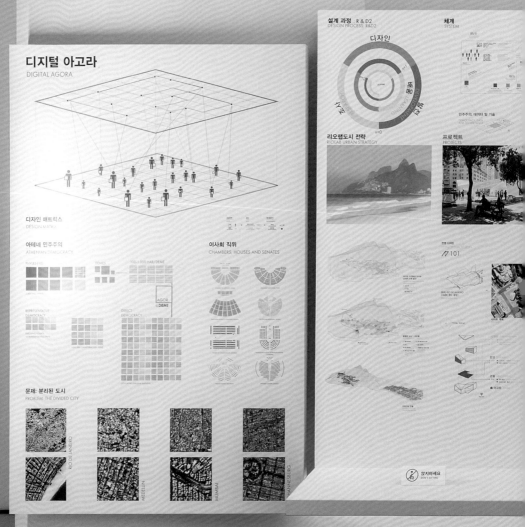

디지털 광장은 참여 민주주의를 실천하는 물리적인 가상의 공간으로, 다수의 참여를 도모하고 의견을 수렴하여 구체적인 공공정책을 통합함으로써 함께 만드는 도시를 실현하는 데 목적을 둔다.

The Digital Agora is the concept program for a physical and digital space for participatory democracy, responding to the global demand for more participation on the public decision making of cities. Integrating specific public policies with spaces for systematization and effective technologies, a new calibration between representative and direct democracy at city level can be generated.

디지털 광장
페드로 앙리크 데 크리스토 / +D

DIGITAL AGORA
Pedro Henrique de Cristo / +D

스틸컷, 아흐메드 마터, 메카
군중, 2017
작가제공

Still, Ahmed Mater, Mecca
Crowd, 2017
Courtesy of the artist

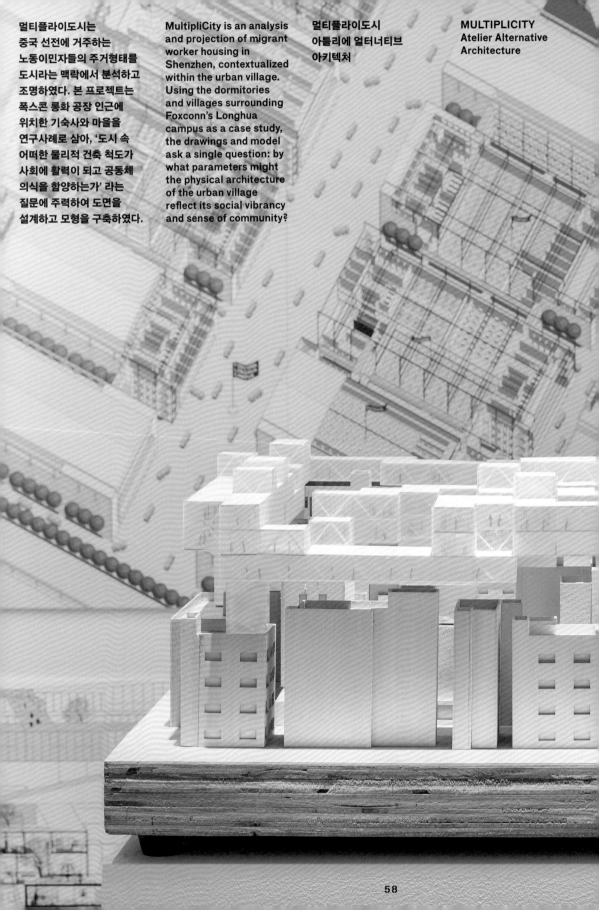

멀티플라이도시는 중국 선전에 거주하는 노동이민자들의 주거형태를 도시라는 맥락에서 분석하고 조명하였다. 본 프로젝트는 폭스콘 룽화 공장 인근에 위치한 기숙사와 마을을 연구사례로 삼아, '도시 속 어떠한 물리적 건축 척도가 사회에 활력이 되고 공동체 의식을 함양하는가' 라는 질문에 주력하여 도면을 설계하고 모형을 구축하였다.

MultipliCity is an analysis and projection of migrant worker housing in Shenzhen, contextualized within the urban village. Using the dormitories and villages surrounding Foxconn's Longhua campus as a case study, the drawings and model ask a single question: by what parameters might the physical architecture of the urban village reflect its social vibrancy and sense of community?

멀티플라이도시
아틀리에 얼터너티브
아키텍쳐

MULTIPLICITY
Atelier Alternative
Architecture

스틸컷, 폴리군 네덜란드 뉴스, 1972년 46주 방영분, 1,000명을 위한 사무실, 1972
출처: 네덜란드 소리와 영상 연구소

Still, Polygoon Hollands Nieuws, week 46, 1972, Office Building for 1000 People, 1972
Source: Netherlands Institute for Sound and Vision

스틸컷, 폴리군 네덜란드 뉴스,
1975년 23주 방영분, '데 드리
호벤', 현대식 실버타운 신축,
1975
출처: 네덜란드 소리와 영상
연구소

Still, Polygoon Hollands
Nieuws, week 23, 1975,
Opening of "De Drie Hoven,"
Modern Retirement Housing
Complex, 1975
Source: Netherlands
Institute for Sound and
Vision

본 프로젝트는 지난 수십 년 간 한국의 두가지 주거 양식에서 영감을 받아 기획한 총 6가지 시나리오로 구성되어 있다. 여기서 말하는 주거 양식은 고시원과 다양한 종류의 방을 말한다.

여러 용도의 방이 밀집된 형태의 고시원을 파헤치며, 목욕탕, PC방, 수다방 등을 갖춘 남다른 한국형 주거형태가 전 세계 여타 도시에 적용가능한지 여부를 살펴보고자 한다.

Out with a Bang presents six projective scenarios for new domestic forms inspired by architectural archetypes developed in South Korea: the goshiwon, or boarding house, and the bang, or

function-specific rentable rooms. The scenarios seek to learn from the contemporary South Korean city to formulate projects that can fit in other metropolises.

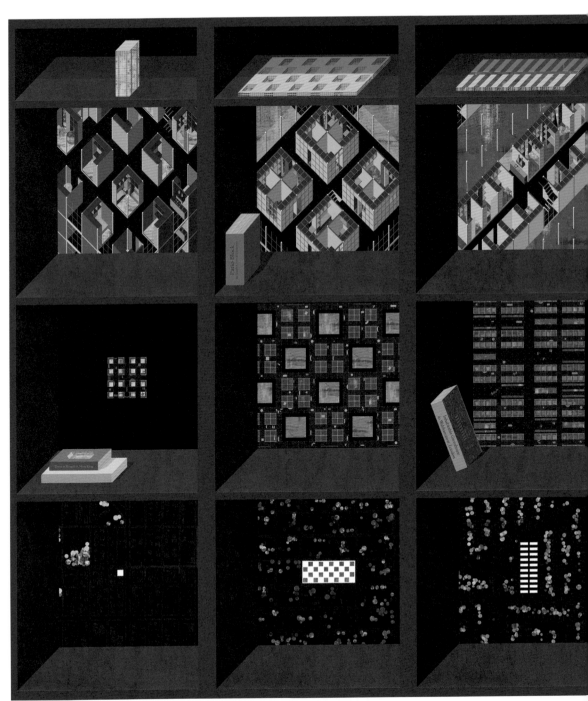

고시원 엿보기
블랙스퀘어 / 마리아
지우디치

OUT WITH A BANG
Black Square / Maria
Giudici

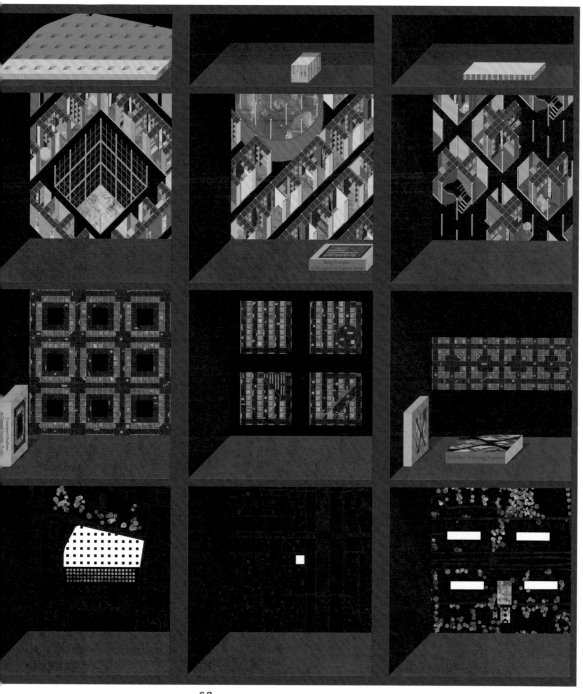

Do they want to do

집단주의 도시에서의 건축 도전

김성홍

Architectural Challenges in a City of Collectivism

Sung Hong Kim

도시는 본질적으로 집합적(collective)이다. 루이스 멈포드는 도시를 "지리적 네트워크, 경제적 조직, 제도적 프로세스이자, 사회적 행동이 펼쳐지는 극장이며, 집합적 결속의 미학적 상징"이라고 정의하였다[1]. 즉, 멈포드는 도시를 물리적, 경제적, 사회적, 미학적 요소로 구성된 집합 단위로 보았다. 하지만 오늘날의 대도시는 수많은 이질적 집합 단위를 수용하고 있으며, 이 점에서 도시가 반드시 집합적인 것은 아니다. 전 세계적으로, 현대 도시들은 공간적, 정치적으로 분열되고 있으며 경제적·사회적으로 계층화되고 있다. 이런 의미에서 제2회 2019 서울도시건축비엔날레(SBAU)의 공동감독인 임재용과 프란시스코 사닌이 제안한 주제 '집합도시'는 더 포괄적이고 통합된 도시를 향한 집합적 행동을 촉구한다. '도시를 되찾고, 건축을 재정립한다'는 부제는 건축의 역할이 이러한 도시 위기를 다루는 '정치적, 문화적 프로젝트의 대리인'이라고 제시하고 있다.

　　나는 UIA 2017 서울세계건축대회 기조 포럼인 '도시의 미래: 도시 안에서 살아가기'를 기획했다.[2] 도심에서 다른 사람들과 거주하고 일하는 것이 경제적, 사회적, 문화적 활력을 위해 점점 더 중요하다는 것을 포럼의 전제로 삼았다. 배형민과 알레한드로 자에라-폴로가 공동감독을 맡은 제1회 2017 SBAU 주제인 '공유도시(Imminent Commons)'는 '도시성'과 '도시 공동성'을 유지하고 확장하는 것이 미래 도시의 중요한 도전이라고 보았다. 또한 2020년 베니스건축비엔날레 총감독 하심 사르키스가 제안한 주제 '우리는 어떻게 같이 살아갈 것인가(How will we live together?)'는 더불어 살 수 있는 공간을 상상하고 구축하는 건축가들에게 '공간계약'을 다시 요구하고 있다.

　　최근 이와 같은 비엔날레, 포럼, 심포지엄, 전시회 등의 공통된 흐름을 어떻게 봐야 할 것인가? 도시, 공공,

Cities are collective by nature. Lewis Mumford defined the city as "a geographical plexus, an economic organization, and an institutional process, a theater of social action, and an aesthetic symbol of collective unity."[1] In short, Mumford sees the city as a collective unit consisting of physical, economic, social and aesthetic components. Today's larger cities, however, tend to house a number of discrete collective units, wherein the city as a whole is not necessarily collective. All around the world, contemporary cities are becoming more fragmented spatially, divisive politically, stratified economically and socially. 'Collective City,' the theme for the 2019 Seoul Biennale of Architecture and Urbanism (SBAU) proposed by co-curators Jae Yong Lim and Francisco Sanin, is in this sense a beacon call for collective action towards more inclusive and unified cities. The sub-title, 'reclaiming the city redefining architecture,' casts the role of architecture as "an agent of the political and cultural project" that addresses this urban crisis.

　　I moderated 'Urban Futures: Living in the Inner City,' the kick-off keynote forum of UIA 2017 Seoul.[2] The premise was that living and working with different people in the inner city is more and more important to economic, social, and cultural vitality. 'Imminent Commons,' the theme for the first SBAU in 2017, curated by Hyungmin Pai and Alejandro Zaera-polo, also maintains that the protection and extension of 'urban-ness' and 'urban commons' is the major challenge for future cities. The theme for the 2020 Venice Architecture Biennale, 'How will we live together?' proposed by Hashim Sarkis again calls for a "spatial contract" devised by architects who imagine and create spaces that we can live in together.

　　How should we look at the common trends in

1　Lewis Mumford, "What is a City?" In Richard T. LeGates and Frederic Stout, Eds., *The City Reader*, 2nd Edition, London and New York: Routledge, 2000, p. 94 (originally In *Architectural Record*, 1937).
2　"Urban Futures: Living in the Inner City – Regenerating Urban Architecture for Cohesion and Sustainability," Keynote Forum 1, Cristiane Muniz, Wilfried Wang, John Peponis, and Sung Hong Kim, September 4, UIA 2017 Seoul World Architects Congress, Program Book, 2017, p. 37.

1　Lewis Mumford, "What is a City?" In Richard T. LeGates and Frederic Stout, Eds., *The City Reader*, 2nd Edition, London and New York: Routledge, 2000, p. 94 (originally In *Architectural Record*, 1937).
2　"Urban Futures: Living in the Inner City – Regenerating Urban Architecture for Cohesion and Sustainability," Keynote Forum 1, Cristiane Muniz, Wilfried Wang, John Peponis, and Sung Hong Kim, September 4, UIA 2017 Seoul World Architects Congress, Program Book, 2017, p. 37.

공유, 공동성 등에 대한 개념들이 모든 것을 적절히 포괄할 수 있는 상위 주제로 적절하기 때문인가? 아니면, 도시 위기와 관련해 건축가들의 입지가 좁아지고 있는 상황에서 건축의 범위를 전략적으로 확장할 필요가 있기 때문인가?

도시를 가시화하고 실체화하는 것은 궁극적으로 건축가가 하는 일이다. 건축의 범위나 역할의 폭이 아무리 넓어도, 건축가가 가장 잘 할 수 있는 것은 물리적인 것을 다루는 일이다. 건축가들이 계획과 디자인을 통해 도시 불평등을 개선하리라는 기대가 있지만, 아이러니하게도 건축가들의 개입은 종종 경제적, 사회적, 문화적 평등과 상반되는 결과를 초래한다. 새로운 건물들은 도심지의 젠트리피케이션 현상을 촉발시키며, 이로 인해 중하위 계층들은 재산권의 변화, 부동산 투기 및 개발, 치솟는 임차료 등의 이유로 도시 변방으로 밀려난다. 도시 확산 때문에 많은 사람들이 도심에서 더 먼 외곽에서 출퇴근하게 되며 새로운 건축물이 주는 문화적 혜택으로부터 소외된다. 반면 대규모 개발은 중상층만을 위한 도심 속 거주지를 만들어낸다. 하이엔드 디자인은 상징권력을 이용하고자 하는 기업가와 정치인에게 경제적 이익과 정치적 의제를 가져다 주기 때문에 매력적이다. 하지만 건축가가 도시 정책을 만들고 도시의 공간적 구조를 변화시키는데 참여할 기회는 많지 않다. 지역, 국가, 도시에 따라 다르지만 건축가가 더 많은 영향력을 행사하고자 한다면, 구체적인 배경과 맥락 속에서 집합성을 방해하는 어떠한 장애물과 격차가 있는지 이해하는 것이 가장 먼저 해야 할 일이다.

서울비엔날레의 개최지인 서울은 현대 한국인들의 정신을 투영한다. 물리적으로는 이질적인 도시형태와 동질적인 건축유형이 역설적으로 결합된 도시다. 정지우의 『분노사회』에 따르면 사회학적으로는 '집단주의'와 '개인주의'가 역설적으로 결합된 도시다.[3] 집단주의는 개인보다 집단의 가치를 우선시한다. 집단과 개인의 차이에 대한 개념은 페르디난트 퇴니스와 막스베버 등 독일 사회학자들로 거슬러 올라간다. 살짝 다른 관점으로 사회학자 바실 번스타인은 사회언어학적으로 두 가지 언어 코드를 구분했다. 폐쇄적 지역사회에서는 '제한된 코드'가 발견되는 반면, 개인의 가치를 중시하는 열린 사회에서는 '정교한 코드'가 발견된다. 가장 최근에는 심리학자 기어트 홉스테드는 이전 사회학자들의 두 가지 사회, 문화적 코드와 궤도를 같이 하는 집단주의와 개인주의에 대한 '문화차원' 이론을 제시했다.

recent international biennials, forums, symposiums, and exhibitions? Is it because the idea of the city, public, sharing, and commonness are appropriate for interdisciplinary events that require big thematic umbrellas under which everything can fit nicely together? Or is it because architecture is increasingly marginalized in the urban crisis, such that the scope of architecture needs to be strategically extended?

Ultimately it is architects who make the city visible and tangible. No matter how expanded the scope and role of architecture is, what an architect can do best is deal with the physical. Architects are called on to ameliorate urban inequality *through* planning and design. Ironically, though, architects' interventions are often viewed as antithetical to economic, social, and cultural equity. Many new buildings spark gentrification of the urban center that forces the lower and middle classes to move to the peripheries because of the changes in property ownership, real estate speculation and development, and skyrocketing rents. The urban sprawl makes people commute further for work in the inner city, as those workers are alienated from the cultural benefits of the new architecture. Large-scale development creates urban enclaves only for the higher middle or upper classes. The 'high-end' design is appealing to entrepreneurs and politicians who exploit the symbolic power to entice economic activity and refine their political agenda. There appears to be little opportunity for architects to actively take part in shaping policy and changing the spatial fabric of cities. Their influence varies depending on the region, nation-state, and city. The first step for architects, if they are to exert more influence, is to understand what obstacles and discrepancies exist that impede the collective in specific settings and contexts.

Seoul, the venue of the SBAU, is the quintessential Korean city in that it portrays the contemporary Korean psyche. Physically it is characterized by a paradoxical combination of *heterogeneity* in its urban morphology and *homogeneity* in its architectural typology. Sociologically, it is characterized by a paradoxical coalescence of 'collectivism' and 'egotism', according to Jung Ji Woo's 'Angry Society' (*Bunno Sahoe*).[3] It is understood that collectivism prioritizes the group over the individual. The origin of the notion goes back to German sociologists, most notably Ferdinand Tönnies and Max Weber. In a slightly different line of thought, sociologist Basil Bernstein distinguishes the two types of

3   정지우, 『분노사회: 현대사회의 감정에 관한 철학에세이』, 부산: 이경, 2014.

3   Jung Ji Woo, Bunno Sahoe [Angry Society] (in Korean), Busan: Ikyung, 2014.

정지우는 『분노사회』에서 현대 한국 사회는 두 가지 반대되는 가치 체계인 집단주의와 개인주의의 혼합체라고 정확히 꼬집었다. 정지우는 지난 100년 동안 한국 사회의 집단주의가 지연, 학연, 혈연에 근거한 강력한 족벌주의(nepotism)와 결합하여 강화되어 왔다고 주장하고 있다.[4] 이러한 집단주의는 민간 기업뿐 아니라 정치, 관료, 학계 등 거의 모든 영역을 지배하고 있다. 이기심이 개인의 권리를 혼동하는 것처럼 집단주의 문화와 공동선을 혼동하고 있다.[5]

'퍼블릭(public)'을 한국어로 번역하면 공공(公共)이다. 이때 한자의 첫 번째 공(公)은 '개방'을, 두 번째 공(共)은 '함께' 또는 '공유'를 의미한다. 그런데 두 번째 공(共)은 공산주의(共産主義)를 의미하는 말에도 사용된다. 1960년대부터 1980년대까지 군사 독재 시기에 한국정부는 반공을 정치적 경쟁자를 탄압하기 위한 수단으로 활용하였다. 이른바 레드 콤플렉스라고 불리는 반 공산주의와 반 사회주의 이데올로기가 한국에서는 늘 존재해왔기 때문에, '공공'은 논쟁적이고 복잡한 가치관이 담긴 말이 되었다.[6]

도시계획과 건축설계 분야에서 쓰는 말 중 가장 오염된 단어 하나가 '공동체'이다. 실제 혹은 가상 공간에서 사람들이 소속감, 공유, 친밀감을 느끼는 사회적 단위를 의미하는 말이다. 하지만 공동체는 종종 개인의 차이보다는 물리적, 심리적 장벽을 만들어내는 동질성으로 묶인 집단으로 이해한다. 이는 사회학자 에밀 뒤르켐이 정의한 '기계적 연대'에 가깝다. 정지우는 한국에서 가장 강력한 공동체로 교회 관련 단체와 고급아파트 주민협의체를 꼽는다. 이런 현실과 달리 건물, 단지, 도시 블록 사이에 빈 공간을 배치하고 이곳이 사람들이 모이고 소통하는 공유공간이라고 주장하는 안들을 건축 및 도시 설계공모전에서 보는 것은 별로 놀라운 일이 아니다. 현실세계에서는 정반대로 다른 집단의 진입을 차단하는 강력하고 날카로운 경계를 만들기 위해 치밀하고 암묵적인 전략이 사용된다. 집단주의와 이기주의가 결합하여 하나가 된다.

서울의 실험과 도전
서울시는 총괄건축가와 공공 건축가 제도를 도입했고

language codes in a socio-linguistic setting: a 'restricted code' is found in a narrow community, whereas an 'elaborated code' is found in a society which values individuality.[4] Most recently, the psychologist Geert Hofstede's 'cultural dimension' theory on collectivism vs. individualism echoes the two models and codes of socio-cultural groups.[5]

What Jung Ji Woo's *Angry Society* rightly pointed out is that contemporary Korean society is an amalgam of two opposing value systems, collectivism and egotism. Jung argues that for the last 100 years collectivism has been reinforced in South Korean society combined with strong nepotism based on region, school and kinship connections. It dominates every sphere: political, bureaucratic, and academic circles as well as private enterprises. Whereas selfishness is confused with the rights of an individual, the collectivist culture is confused with communal goods.

The Korean word for 'public' is gong-gong. In ancient Chinese, the first '*gong*' (公) means 'openness' and the second '*gong* (共)' does 'togetherness' or 'sharing'. But the second-word '*gong* (共)' is also used to describe Communism (共産主義). During South Korea's era of military dictatorship from the 1960s to the 1980s, the government took advantage of anti-communist sentiment as a tactic for the suppression of political adversaries. In South Korea, anti-communist and anti-socialist sentiments (the so-called 'red complex') are ever-present, making the perception of 'public' even more controversial and value-charged.[6]

One of the misused and contaminated terms in planning and design is 'community.' It is a social unit in physical or virtual space where people feel a sense of belonging, sharing and relatedness. But the community is often understood as a group bonded by commonness rather than differences in a way that creates a strong physical and psychological enclosure. It is close to the 'mechanical solidarity' defined by Emile Durkheim. Jung Ji Woo argues that the strongest communities in South Korea are the ones affiliated with churches and upscale apartment homeowners' associations. In architectural and urban design competitions, it is no surprise to see

4   Basil Bernstein, *Class, Codes and Control, Vol. 3: Towards a Theory of Educational Transmissions (Primary Socialization, Language and Education)*, London: Routledge & Kegan Paul, 1975.
5   Geert Hofstede, *Culture's Consequences: Comparing Values, Behaviors, Institutions and Organizations Across Nations*, Beverly Hills: Sage, 1980.
6   Sung Hong Kim, "The Paradox of Public Space in the Korean Metropolis," In Limin Hee, Boontharm Davisi, and Erwin Viray (Eds.), *Future Asian Space: Projecting the Urban Space of New East Asia*, Singapore: NUS Press, 2012.

4   Basil Bernstein, *Class, Codes and Control, Vol. 3: Towards a Theory of Educational Transmissions (Primary Socialization, Language and Education)*, London: Routledge & Kegan Paul, 1975.
5   Geert Hofstede, *Culture's Consequences: Comparing Values, Behaviors, Institutions and Organizations Across Nations*, Beverly Hills: Sage, 1980.
6   Sung Hong Kim, "The Paradox of Public Space in the Korean Metropolis," In Limin Hee, Boontharm Davisi, and Erwin Viray (Eds.), *Future Asian Space: Projecting the Urban Space of New East Asia*, Singapore: NUS Press, 2012.

국제설계공모를 통해 중요한 공공 프로젝트를 추진하고 있다. 자하 하디드가 설계한 동대문 디자인 플라자(DDP)와 MVRDV가 설계한 서울로 7017이 대표적인 예라고 할 수 있다.[7] 역사적 장소에 대한 부족한 인식과 과도한 건설 비용에 대한 논란과 비판에도 불구하고, DDP는 서울에서 반드시 방문해야 할 공공 장소로 손꼽히고 있으며, 서울비엔날레의 주요 행사장이기도 하다. 서울로 7017 또는 스카이가든도 관광객이 찾는 명소로 꼽힌다.

한국판 뉴욕 하이라인으로 기획한 서울로 7017은 서울역 위를 지나는 고가도로를 재생했다. 사업 초기에는 소매업자와 노점상인, 지역 정치인들의 반대가 있었지만, 점차 도시공간에 녹아 들어 지금은 서울의 새로운 유형의 도시 공간으로 인정받고 있다.

이런 소수의 프로젝트와 달리 여러 실험적 사업은 큰 도전에 직면해 있다. DDP와 서울로 7017의 경우는 사유지와 경계가 명확히 구분된 공공부지를 활용하여 별 문제가 없었지만, 사유지에서 시도한 대규모 프로젝트는 중단되거나 교착상태에 있다. 서울의 심장부에 위치한 도시 건축 세운상가와 인접 지역은 현재 이러한 논란의 중심에 있다. 1990년대 중반 이후 역대 서울 시장들은 세운상가 지역을 활성화하는 비전을 제시했다. 2004년 최대 규모의 재개발 프로젝트 국제현상공모전에 코에터 김, 리차드 로저스, 마차도 앤드 실베티, JERDE, 렘 콜하스, 피터 아이젠만, 폰 게르칸, 테리 파렐 등 스타건축가들이 초청되었다. 이들 건축가들은 경험을 축적한 국내 대규모 건축사사무소와 협업하여 오래되고 불규칙한 도시 지역을 바꾸는 대담한 아이디어들을 제안했다. 하지만 이렇게 조명 받은 제안들은 계획 단계 이상으로 진행되지 못했다. 이 안들이 폐기된 이후 과정은 전문가들조차 이해할 수 없을 정도로 복잡하다. 2016년 세운상가 주변 일부 지역을 대상으로 또 한차례 국제현상공모가 진행되었고, 네덜란드의 KCAP의 설계안이 당선되어 다양한 과정을 거치면서 계속 수정되고 있다. 하지만 여전히 이 프로젝트가 설계안 대로 진행이 될지는 여부는 불투명하다.

유일하게 현실화된 프로젝트는 세운 몰 건물 양측에 설치된 보행 데크이다. 그 외 세운상가 인접 지역들은 별다른 계획 없이 철거되거나 방치된 채 남아있다. 최근 서울시는 오래된 가게와 식당들은 살아있는 생활 유산으로 반드시 보존되어야 한다는 논리로 민간 개발자가 시작한 재개발 프로젝트들을 성급히 중단시켰다. 재개발은 악이고 보존이 선이라는 도식은

that the premise behind allocating open space between buildings, complexes, and urban blocks is usually to create communal spaces where people gather and interact. In reality, meticulous and tacit strategies are employed to make stronger and sharper communal boundaries which shut out other groups. This consolidates collectivism and egotism together.

Seoul's Experiments and Challenges

The Seoul Metropolitan Government has adopted a system of city architects and public architects and promoted important public projects through international design competitions.[7] The DDP by Zaha Hadid and Seoullo 7017 by MVRDV are prime examples. Despite the controversy and criticism over the interpretation of the historic site and the excessive construction costs, the DDP has become a must-visit public place in Seoul. It is the main venue for the SBAU. Seoullo 7017 or Skygarden is also on the list of many visitors. Conceived as Seoul's version of New York's Highline, it renovated the former elevated overpass over Seoul Central Railway Station. Retailers, vendors and hence local politicians opposed the plan at the beginning, but it has gradually fit into the city fabric and become a new type of urban space in Seoul.

But other experiments have been confronted with greater challenges. The DDP and Seoullo 7017 are solely on public properties with clear lines of demarcation from private properties. However large-scale projects on private lands have either come to a halt or are in a deadlock. The Sewoonsangga, the earliest urban architecture in the heart of Seoul, and its adjacent areas, are at the center of controversy now. Since mid-1990, succeeding mayors of Seoul had different visions for revitalizing the Sewoonsangga areas. In 2004, an international competition for the largest redevelopment projects invited Koetter Kim, Richard Rogers, Machado and Silvetti, JERDE, Rem Koolhaas, Peter Eisenman, von Gerkan, Terry Farrell, and other renowned architects. Cooperating with seasoned local corporate firms, they proposed audacious ideas to replace the old irregular urban fabric of the city. Yet all the proposals that received the spotlight never got beyond the drawing board. The abandonment of projects, and the subsequent process and changes, were so complicated that even experts had a hard time grasping them. Another international competition for a reduced block of Sewoonsangga was implemented again in 2016, and the winning proposal by KCAP has been modified continually

7   https://project.seoul.go.kr

7   https://project.seoul.go.kr

대안을 찾는 합리적 토론과 결정을 더 복잡하고 만들고
있다.

아파트 단지와 공공주택

서울을 집합도시로 만드는데 있어 가장 어려우면서
중요한 과제는 대규모 아파트 단지의 계획 및 설계에 있다.
대표적 사례가 358,000 평방미터에 달하는 잠실5단지
아파트 주택재건축정비사업 국제설계공모와 187,000
평방미터의 중계동 본동 재개발정비정비구역 공동주택
국제지명설계공모전과 주거지보존 프로젝트다. 두
번째 지역은 일명 104마을로 불린다. 새롭고 의미 있는
디자인을 제안하는 건축가의 시도가 얼마나 쉽게 좌절
되는지를 단적으로 보여주는 사례이다. 국제설계공모에서
당선된 안의 실현은 토지소유자, 개발자, 지역 정치인,
서울시와 산하기관에 의해 이런저런 이유로 번번히
가로막혔다. 토지소유자와 주택소유자가 기존
주택시장에서 통용되는 고정관념에서 벗어난 건축가들의
배치와 평면은 경제적 이익이 줄어들 것이라고 믿는
토지와 주택소유자들에 의해 거부되었다. 건축가들이
구사하는 이미지와 말은 그들에게는 천진난만하거나
현실성 없이 비춰졌다. 명성 있는 건축가들은 소수의
상징적인 건축물들을 한정된 지역에 자유로이 구현할 수
있겠지만, 도시 전반의 질을 높이는 변화를 만드는 데에서
소외되어 있다.

집합성과 집단주의 사이의 모순과 혼란은
공공주택에서 더 심각하다. 2015년 공공주택 특별법
제정과 함께 저소득층을 위한 중장기 임대 주택을
공급하기 위한 사업을 시작했다. 서울시는 도시 전역
곳곳에 저렴한 공공주택 공급을 통해 도시 내의 경제
사회적 격차를 줄이고자 노력하고 있다. 서울시가
개발을 유도할 수 있는 최고의 도시계획적 수단은
민간의 투자 수익률을 높이는 용적률 인센티브를
제공하는 것이다. 저소득층이 모여있는 지역에서는
절반 이상의 원주민이 재개발 사업 과정에 다른 곳으로
이주하기 때문에 공공주택이 들어오는 것에 대한 반대가
상대적으로 적다는 것이 연구결과로 밝혀지고 있다.
그 결과 도시인프라가 열악한 구릉지에는 공공주택의
비율이 높은 고층고밀도의 아파트가 건설된다. 반면
대지가 평평한 부유한 지역에서 시행되는 주택재건축
사업은 토지소유자 대부분이 건설 이후에도 남기 때문에
공공주택이 들어오는 것을 결사 반대한다. 결국 양질의
도시기반시설을 내 외부에 갖춘 저밀도의 고급 아파트
단지가 지어진다. 비싼 동네의 아파트 단지는 뚜렷한
물리적 경계와 반가로적(反街路的) 저층부를 만들어
외부로부터 스스로를 물리적, 시각적으로 단절시킨다.

through various processes up until today. It is still
hard to predict whether the project will be carried
out as designed.

The only realized project is a walking deck
attached to both sides of the existing Sewoon
Mall, but the adjacent areas remain without
a plan other than demolition or negligence.
Recently, the city government hastily suspended
the redevelopment project initiated by private
developers. The rationale behind this was that
old shops and restaurants must be preserved as
a living heritage. The notion that redevelopment
is evil and preservation is good certainly makes
discussions and decisions on alternative plans
more complicated.

Apartment Complexes and Public Housing

The most difficult and important challenge
in making Seoul more of a collective is in
the planning and design of mass apartment
complexes. International design competitions for
the reconstruction of Jamsil Apartment Complex
5 that rests on 358,000 square meters, and
the redevelopment of the '104 Village' on a hilly
187,000 square-meter site are cases in point.
They demonstrate how architects' attempts to
introduce any kind of new and meaningful design
elements can be vulnerable to being rejected. The
implementation of the winning design schemes was
blocked for one reason or another by landowners,
developers, local politicians, city officials, and
planning agencies. The experimental site plans
and unit plans proposed by architects were not
accepted because landowners and homeowners
are convinced that designs deviating from the
established market stereotypes would diminish
their profits. The visual and verbal language used
by architects sounded naïve and even irrelevant to
them. While architects may be free to make a few
symbolic urban buildings here and there, they are
marginalized when it comes to making changes
that enhance the overall quality of the city itself.

The contradiction and confusion between
collectiveness against collectivism are deeper in
public housing. With the establishment of special
legislation in 2015 the public housing program
launched an effort to supply long-term rental
housing for low-income earners. The government
intended to reduce the intra-urban gap by
distributing affordable social housing throughout
the city. The best tool for boosting development
is to grant extra floor area ratios, the most direct
incentive to promote higher return-on-investment.
Research has shown that public housing received
less opposition in the less wealthy areas because
more than half of the original households move

결과적으로 단지 공동체의 기계적 연대와 지역이기주의 현상은 한층 공고해진다. 서울시 의도와는 반대로 공공주택 프로그램은 역설적으로 기반 시설이 제대로 갖춘 부유한 지역 보다는 기반시설이 열악한 도시조직이 불규칙한 구릉지에서 더 많이 시행된다.[8] 이처럼 사회적 불균형과 불평등의 징후는 도시공간과 조직의 불연속성에서 명확히 드러난다.

양극화된 스케일이 만나는 지점들

건설산업은 지난 반세기 동안 한국 경제 발전의 동력이었다. 2016년 기준으로 한국의 국민총생산(GDP) 대비 건설투자비율은 15%로 OECD평균보다 5% 가량 높다. 건설투자비율이 정점에 올랐던 1990년대 초반에는 23%에 육박했었다.[9] 민간부문의 건설투자는 경기 상황에 따라 부침이 있었지만 정부와 지방자치단체는 건설 분야에 지속적으로 투자하였다. 2008년 금융위기를 맞기 이전까지 남한 인구의 1/4이 거주하고 있는 서울은 중앙정부의 정책과 궤도를 같이하며 탑다운 방식의 도시 계획과 건설 정책을 계속해왔다. 2008년을 기점으로 대규모 개발에서 점진적 재생으로 패러다임이 전환되었다. 하지만 필지 중심의 건축설계에 집중하는 건축가들은 이러한 변화를 주도하지 못하고 언저리에서 머물러 있다.

서울의 평균 필지 면적은 250 평방미터이지만 도심과 주변 달동네에는 100평방미터보다 작은 불규칙한 필지가 많다. 반면 재개발과 재건축으로 정비한 아파트 단지는 100,000 평방미터를 넘는다. 도시조직이 소형과 대형으로 양극화된 서울에서 중간 크기의 블록에 혁신적인 건물이 들어서는 경우는 아주 드물다. 집단주의와 이기주의의 역설적 조합은 이처럼 도시형태 및 건축유형의 양극화에 뿌리 깊게 내재되어 있다.

엄격하고 복잡한 법규가 적용되는 작고 불규칙한 필지에 사는 개인들은 이웃과의 갈등을 피하기 위해 지나치게 방어적이거나 공격적인 태도를 취하게 된다. 반면 작은 필지가 합쳐져 수백 배 규모로 커진 아파트 단지의 개인들은 집단주의 공동체의 익명의 구성원으로 편입된다. 경계는 있지만 개방적이면서도 소속감을 공유하는 이웃이나 장소는 사라지고 있다. 미세한 도시입자와 동질화 된 영토 사이의 선택지는 그리 많지 않다. 이 지점에서 서울의 조밀하게 연결된 도시 조직에서 다원성과 개방성의 가능성이 있다는 한 학자의 관찰을

out after construction. The result is that the high-density public housing units are built more in areas with poor infrastructure. The opposition to public housing is stronger in affluent areas because a higher ratio of original residents stays after the construction. Less dense, higher quality housing is built, with an enhanced infrastructure inside and outside of the complex. The complex itself is separated from the outside by strong physical boundaries and anti-street frontages of buildings. The solidarity of this community is stronger, and so is the NIMBY attitude. Contrary to the city government's intentions, public housing programs go disproportionately more into the disadvantaged areas than those with better infrastructures.[8] Symptoms of social inequality are clearly manifested in the discontinuities of the spatial fabric of the city.

The Interface of Polarized Scales

The construction industry has been a driving engine of South Korea's economic growth for the past several decades. As of 2016, the ratio of the construction investment to GDP was 15 percent, which is 5 percent higher than the average for OCED countries. In the early 1990s, when it peaked, the ratio was as high as 23 percent.[9] The investment ratios of the private sectors have been up and down based on economic fluctuation, but the central and local governments continue to invest public money to keep the ratio from falling. The city of Seoul, home to a quarter of the entire population of South Korea, led the institution of top-down planning and construction policies until the global economic crisis of 2008, when the paradigm began to shift from massive development to incremental regeneration. The architects who concentrate on the design of the plot-based individual buildings have remained on the edge of these changes.

The average plot size in Seoul is 250 square meters, but there are many small and irregular plots less than 100 square meters within the downtown and surrounding hilly areas. On the other hand, the newly constructed apartment complexes are over 100,000 square meters. There are very limited number of innovative buildings on the intermediate scale of blocks. The paradoxical combination of collectivism and egotism is deeply embedded in the polarization of urban morphologies and building typologies.

Individuals living in small and irregular

8   김성홍, "고밀도의 딜레마: 서울의 아파트 개발과 도시관리계획", 『서울도시연구』제19권 제4호, 서울연구원, 2018, 1-19쪽.
9   한국은행, BOK 이슈노트, 2016-7호, 3쪽.

8   Sung Hong Kim, "High Density Dilemmas: Apartment Development vs. Urban Management Plan in Seoul," Seoul Studies, (The Seoul Institute), Vol.19, No.4. December 2018, pp. 1-19.
9   The Bank of Korea, BOK Issue Note, No 2016-7, p. 3.

주목할 필요가 있다.[10] 도시계획과 건축설계의 핵심은 점점 더 양극화되고 파편화되는 개발의 접면을 어떻게 다룰 것인가에 있다.

제약을 넘어서는 디자인

건축을 고도의 자율적인 행위와 결과물 보는 사람들은 법과 제도를 혁신을 가로막는 장애물로 생각한다. 하지만, 이들은 법령의 조(條), 항(項), 호(號) 세부 사항들이 미시적 건축 요소뿐 아니라 거시적 도시 경관에 직접적 영향을 끼칠 수 있다는 사실을 간과한다. 초고밀 도시 서울에서 수직성, 공간구성, 프로그램, 창문·발코니·입면, 아파트 평면 및 단지 계획, 가로의 배치까지도 도시 및 건축 관련 법률과 직접적으로 연관되어 있다. 고착화된 유형을 깨는 새로운 주택실험이 건축가, 도시계획가, 행정가, 개발업자, 주택시장 사이의 갈등 속에서 갇혀 있는 것은 구체적으로는 법의 문제이다. 좀 더 다양하고 포용적이며 지속 가능한 도시를 지향하는 디자인 혁신이 법의 제약을 극복하지 못한다면 불가능하다. 법과 제도는 디자인의 문턱을 설정하는 것이다. 법, 시행령, 조례, 규칙, 지침을 조금만 바꾸어도 혁신의 도화선이 될 수도 있는 디자인의 전제를 근본적으로 흔들 수 있다.

디자인은 주어진 문제를 해결하는 것을 넘어 통합하는 것이다. "디자인 해결책은 분석과 종합의 마지막 단계에서 만들어지지 않는다. 추론과 검증의 초기 단계에서 이미 생성된다."[11] 귀납과 연역의 두 가지 사고방식 사이에는 간극이 존재한다. 콰트르메르 드 껭시와 줄리오 아르간의 유형학 이론을 빌린다면 '특정한 형태의 규칙성을 비교하고 중첩'의 순간과 '형태를 결정'하는 순간은 다르다.[12] 디자인은 이 두 가지 별개의 단계를 연결하는 터널을 통과하는 과정이라 할 수 있다. 터널에 들어가기 전과 터널을 나올 때의 모습은 달라진다. 그렇지만 터널 안에서 무엇이 벌어졌는지는 논리적으로 완전히 설명할 수 없다. 이러한 퍼즐을 푸는 것이 디자인의 본질이자 힘이다.

건축에서 도시로 스케일을 전환하면, 우리는 또 다른 차원의 차이점과 마주하게 된다. 도시의 외적 힘과 건축의 내적 원리를 어떻게 조율할 것인가 하는 문제이다. 건축 설계는 내인적이고 도시계획은 외인적이다.

urban plots, where strict and complicated rules are applied, take either defensive or aggressive attitudes towards neighbors to avoid conflicts. On the other hand, when these plots are turned into complexes hundreds of times the size, individuals are incorporated as anonymous members of a collectivist community. Neighborhoods or places with distinct but open boundaries and a sense of belonging are diminishing. Between the fine urban grain and homogenized territories, there are not many choices. An urban morphologist observed the urbanism of pluralism and openness in the dense and connected urban fabric of Seoul.[10] The key issue of planning and design is how to deal with the interface of the increasingly polarized and fragmented scales of development.

Design over Constraints

Those who view architecture as an autonomous entity may consider laws and guidelines to be constraints or even obstacles to innovation. However, they overlook the fact that small details in articles, paragraphs, or subparagraphs of an act can have huge impact on the macro urban landscape as well as micro building design elements. In the hyper dense city of Seoul, the vertical, spatial and programmatic configuration, the window, balcony, and facade, the apartment unit and site plan, and even the street layout are directly related to the urban and building laws and rules. This is part of the reason why new experimental housing schemes are locked in conflicts between architects, planners, development agencies, and markets. Design innovation for a more diverse, inclusive and sustainable city will not be easy without overcoming the constraints of laws. Regulations and systems hold a threshold for design. A few promising changes in the laws, rules, and guidelines could alter the fundamental design premise which might become a 'trigger' for innovation.

Over and above problem-solving, design is essentially integration. And design solutions are "not produced at the latter stage of a process of analysis and synthesis, but right at the early stages of a process of conjecture and testing."[11] Between these two modes of thinking, there is a gap. If we use Quatremere de Quincy and Giulio Argan's typological theories, the moment of "comparison

10 John Peponis, et. al. "The City as an Interface of Scales: Gangnam Urbanism," In Sung Hong Kim et. al. (Eds.), *The FAR Game: Constraints Sparking Creativity*, Seoul: SPACE Books, 2016, pp. 102-111.
11 John Peponis, "Evaluation and Formulation in Design." In *Nordisk Arkitekturforskning* (*Nordic Journal of Architectural Research*), n.2, 1993, p. 53, 57.
12 Rafael Moneo, "On Typology", In Peter Eisenman (ed.), *Oppositions*, Summer 1978:13, Cambridge: The MIT Press, 1978, p. 36.

10 John Peponis, et. al. "The City as an Interface of Scales: Gangnam Urbanism," In Sung Hong Kim et. al. (Eds.), *The FAR Game: Constraints Sparking Creativity*, Seoul: SPACE Books, 2016, pp. 102-111.
11 John Peponis, "Evaluation and Formulation in Design." In *Nordisk Arkitekturforskning* (*Nordic Journal of Architectural Research*), n.2, 1993, p. 53, 57.

도시계획은 물리적 결과물을 만드는 것보다는 사회적 합의와 규칙에 이르는 과정이다. 개별 건물에 대한 미학적 부담에서 벗어나 현재의 제도적 결함을 인지하고 의사결정과정에서 참여의 폭을 넓혀가는 것이 건축가에게 주어진 의무이다.

서울의 개별 필지 위에 지어진 이름있는 건축가들의 작품이 방문자의 눈에는 '개인 방언의 모음'으로 읽혀지는 반면 익명의 집장사들이 지은 평범한 일상건물은 '집합적 버내큘러'으로 읽혀지기도 한다.[13] 건축을 하나하나의 떨어진 개체로 접근하는 도시는 진정한 집합도시가 될 수 없다. 건축가는 다른 전문성을 가진 분야와 협력하고 역량을 통합하는 현명한 전략과 지식을 갖고 작업을 해 나가야 한다. 집합도시에는 다른 사람들과 소통하고 경계를 넘는 태도, 방법, 언어를 구사하는 다면적이고 유연한 전문가가 많아져야 한다. 지금 서울은 건축적 브리콜뢰르(bricoleurs)를 절실히 필요로 하고 있다.

(이 글은 영어로 작성되어 한국어로 번역되었습니다.)

and overlapping of certain formal regularities" and the moment of "formal definition" are distinguished from one another.[12] Design is the process of passing through the tunnel that connects these two discrete stages. What enters the tunnel is quite different from what exits the tunnel. It cannot fully be explained rationally. Solving this puzzle is the nature and power of design expertise.

When we shift the scale from architecture to city, we face another dimension of differences. The question becomes how to reconcile the external urban forces and internal architectural principles. Architectural design is endogenous, and urban planning is exogenous. Urban planning is more about the process of reaching social consensus and rules rather than formulation. It is incumbent upon architects to get away from the aesthetic baggage of individual buildings, recognize the shortcomings of the current system of regulations, and elbow their way into the decision-making processes.

While the work of high-end architects on the individual plots in Seoul is characterized as a "collection of idiolects," the ordinary buildings of everyday life erected by anonymous local builders are seen as a "collective vernacular" in the eyes of visitors.[13] Cities cannot thrive as true collectives where there is a fundamentally individualistic approach to architecture. Architects need to operate through a systematic approach, with complex strategies and knowledge capable of integrating and fostering relationships between interdisciplinary competences. The city needs multifaceted and flexible professionals who have a different approach, tactic, and language to cross boundaries and to communicate with others. Now Seoul desperately needs to become the work of architectural bricoleurs.

12 Rafael Moneo, "On Typology", In Peter Eisenman (ed.), *Oppositions*, Summer 1978:13, Cambridge: The MIT Press, 1978, p. 36.

13 Julian Worrall, "The Nakwon Principle," In Kim et al. (eds.), *The FAR Game: Constraints Sparking Creativity*, p. 147.

13 Julian Worrall, "The Nakwon Principle," In Kim et al. (eds.), *The FAR Game: Constraints Sparking Creativity*, p. 147.

‘경기장(stadium)’이라는 단어는 계단 15단의 거리를 가리키는 고대 단어에서 유래하였다. 이후에는 경주 시합이 개최되는 장소를 의미하게 되었고, 오늘날에는 여러 행사가 가능한 다목적 장소가 되었다. 우리는 이렇게 긴 역사를 지닌 장소를 통해 스포츠 경기의 진화 과정과 스포츠 경기와 도시 간의 관계를 연구하고자 한다.

During antiquity the first meaning for 'stadium' was a measure of length of 15 steps, then it was used to designate the space where a race took place. Today it has become a field for mixed-purpose events. The project questions the evolution of sport activities and their relationship with the city by using this archetypal object.

도시의 경기장
NP2F

URBAN ENCLOSURE
NP2F

스틸컷, 코펜하겐이라 부르는
도시, 요르겐 루스, 1960
출처: 덴마크 영화연구소

Still, A City Called
Copenhagen, Jørgen Roos,
1960
Source: Danish Film
Institute, Danmarkpaafilm.
dk

스틸컷, 리버풀의 하루, 안손 다이어, 1929
© 영국영화협회 영상자료원, 브리티시 영화 연구소

Still, Day in Liverpool, Anson Dyer, 1929
© BFI National Archive, British Film Institute

인도의 열개의 건축사무소가 소개된다. 특정 단체 및 다양한 가능성과 연관되어 디자인, 예술, 법률 문서 연구, 소외된 공동체를 위한 노력, 특정 시기에 진행되는 행사, 도시 보존, 심지어는 도발적 문제 제기까지도 포함하는 다양한 활동이 한데 모여 도시 공간의 다원성을 활발하게 유지시켜 나가는 원동력이 된다.

Ten practices from India are presented here, demonstrating a variety of collective forms, each bringing with them particular agencies and projective capacities - design, art, research legal instruments, mobilization for marginalized communities, temporal events, urban conservation or even provocations - which together keep the plurality of the urban space active and engaging.

집합도시: 인도 현대 건축의 집합적 수행의 형태
사미프 파도라
건축연구소(sP+a/ sPare)
기획: 아키텍처 레드 RED, 아비지트 무쿨 키쇼어 & 로한 시브쿠마르, 반드라 콜렉티브, 버스라이드 디자인 스튜디오, St+art 인디아 재단, 아빈 디자인 스튜디오, 피케이 다스 앤 어소시에이츠, 공간적 대안을 위한 모임, 아다르카르 어소시에이츠

COLLECTIVE CITIES: Notes on Forms of Collective Practices in Contemporary Indian Architecture Ten Practices: India. Curated by Sameep Padora Architecture and Research (sP+a/sPare): Architecture RED, Avijit Mukul Kishore & Rohan Shivkumar, Bandra collective, Anthill design, The Burside design studio, St+art India Foundation, Abin Design Studio, PK Das & Associates, Collective for Spatial Alternatives (CSA), Adarkar Associates

COLLECTIVE RESISTANCE

스틸컷, 극소수부대 - 이야기의
변주, 플로리안 제이팡, 리사
슈미츠코리네트, 알렉산더 슈
무거, 2013
작가제공

Still, Microbrigades -
Variations of a Story,
Florian Zeyfang, Lisa
Schmidt-Colinet, Alexander
Schmoeger, 2013
Courtesy of the artists

스틸컷, 지상의 조직, 플로리안
제이팡, 리사 슈미츠코리네트,
알렉산더 슈 무거, 2015
작가제공

Still, Institute Above-Ground,
Florian Zeyfang, Lisa
Schmidt-Colinet, Alexander
Schmoeger, 2015
Courtesy of the artists

제너럴 아키텍처
콜라보레이티브(GAC)는
2008년부터 르완다의
마소로(Masoro) 마을의
여러 프로젝트를 진행하고
있으며 지역 공동체의
참여, 연구, 교육을 통한
협업 디자인을 추구한다.
타인과 어울리는 방법
프로젝트는 "관계성",
"사물", "과정"이라는 세
가지 결과물을 도출하였다.
"관계성"은 "프로젝트
지도"로 나타난다. 지도는
네 가지 프로젝트를 통해
개인, 조직, 기관, 정부, 관습
간의 연결성을 소개한다.
"사물"은 눈금 표기된 도표와
그림을 활용하여 연결망으로
도출해 낸 프로젝트를 통해
선보여진다. "과정"은 네
가지 프로젝트를 수행하며
동고동락한 이들의 소중한
순간과 활약상을 엮은 영상을
통해 전시된다.

Since 2008, GAC has
been working in and
around the village of
Masoro in Rwanda,
focusing on collaborative
design practices through
community engagement,
research and education.
Playing Well With
Others describes an
approach to design
through three outputs: a
map that illustrates the
networks of individuals,
organizations and
institutions engaged
with the practice; the
projects that are the
products these networks;
and a series of films that
show the involvement of
various people.

타인과 어울리는 방법
제너럴 아키텍처
콜라보레이티브
(GAC)

PLAYING WELL WITH
OTHERS
General Architecture
Collaborative (GAC)

I like the day com
I like the beaut

Different times
taking some beaut

te the day going,
hombe airport

I am from a thousand hills side
A land of history.

s come and go
keeps it standing

스틸컷, 중세시대 마을(렉스톤), J.B. 홈즈, 1935 © 영국영화협회 영상자료원, 브리티시 영화 연구소

Still, Medieaval Village (Laxton), J.B. Holmes, 1935 © BFI National Archive, British Film Institute

CBC는 농촌개발이 좀 더 깊고 섬세한 인간적인 요소들을 고려할 필요가 있다고 믿는다. 차이나 빌딩 센터(CBC)는 사회 발전, 경제, 문화, 생태계 간의 세심한 균형을 도출하고자 일련의 연구를 수행 하였으며 이를 통하여 도시와 농촌의 관계 속에서 국가가 어떠한 새로운 역할을 할 수 있는지 살펴보았다.

CBC believes that urban and rural developments need to prioritize more profound, subtle human factors. The research conducted by CBC reflects specifically on how to achieve a delicate balance between issues of social development, economy, culture and ecology, an approach that hopes to obtain a new position outside of traditional urban-rural relationships and generate new possibilities in the field.

공간적 가치의 창조
CBC (차이나 빌딩 센터)

CREATION OF SPATIAL VALUE
CBC (China Building Centre)

CREATION O

Xiamutang (夏木塘) is located in Wan'an County, Ji'an City, Jiangxi Province, in the south of China and in the middle lower reaches of the Yangtze River. When it comes to Ji'an in Jiangxi, people will think of a group of writers in the history of Chinese literature. Many renowned writers and poets like Yang Wanli (杨万里), Wen Tianxiang (文天祥), Ouyang Xiu (欧阳修) were born here. Wan'an County under its jurisdiction has a long history. The ancient city wall of Wan'an County still remains till today, which was built in Song Dynasty and rebuilt in Ming Dynasty. The country is located in the eastern foothills of the Luoxiao Mountain. It is a typical mountainous county with beautiful scenery and unique natural scenery.

The name "Xiamutang" means 'summer forest trees and water ponds'. The natural scenery and the ecological environment here are excellent. There were more than 50 households in the village. In order to promote the development of the village, the government planned to build the village as a "Chinese Folk Game Village", which has created a systematic and dynamic cultural foundation. Although Xiamutang acquired new cultural genes in the previous round of planning, there is an urgent need to balance the relationship between tourism development and cultural heritage. The local government invited **CBC (China Building Centre)** as the consultant for the rural revitalization in Xiamutang. CBC took the initiative to fully utilize the original facilities and cultural traditions, and tried to build a balance between tourism development and maintaining tradition.

XIAMUTANG 夏木塘

centralized
has accor
process i
past half
and top-
take into
human fac
research c
reflects s
balance b
culture ar
between
discoverin
traces. C
the "Crea
is to exert
whole, tak
promotion
and comm
and acupu
urban and

Among th
selected t
exhibition
problems
explore ne
of thinking
of urban a
CBC acts
different c
and city, t
strategies
the differe
villagers, a

POND

水地别场   LANDSC

MUT
水地酒吧

AN

CB
CHINA BUILDING C
CBC建筑中

CURATOR: Peng Liao
Chief Editor of Urban Environ
China Building Centre

EXHIBITION DESIGN: Wa
Partner and Principal Archit

PROJECT PARTICIPANTS:
Ryui Nemoawa, Seung H
Zhou Xinming, Wei Zhang, X
Hao Jingfang, Wang Wei, W
Qian, Wang Dan, Zhao Dai
Yang Lu, Longwen, Long J

스틸컷, 작동방식 감당하기,
조지나 힐, 2019
작가제공

Still, Coping (with)
mechanisms, Georgina Hill,
2019
Courtesy of the artist

스틸컷, 집, 몽상, 안드레아
루카 짐머만, 2015
작가제공

Still, Estate, a Reverie,
Andrea Luka Zimmerman,
2015
Courtesy of the artist

일곱가지 서적을 올린 제단은 도시에서 활용 가능한 공유지를 제시한 시범형으로, 지나친 구조적 기교와 색을 포함하는 디자인의 총력이 담긴 지붕구조를 가진 정자의 전통을 따른 것이다. 이는 포스트 자본주의 시대에 걸맞는 도시건축을 위해 '공동의 지식'을 기원하는 마음으로 개인적인 헌물을 바치는 행위라고 할 수 있다.

Baukuh presents the Altar of the Seven Books, a model prototype of a possible public space for the city, in the tradition of the Korean garden pavilions with their impressive roofing structures, where all of the design intensity is typically accumulated. Seven architecture books of the city are placed within the space, as a personal offer of collective knowledge to the construction of post-capitalistic urbanity.

도시공간이 정치적으로나 경제적으로 순기능을 하지 못하면, "상호작용"이 영향력을 발휘할 차례다. MANY는 필요에 따라 사람들의 이주를 돕는 온라인 플랫폼이다. 도시들은 MANY를 통해 필요한 인재를 유입할 수 있고, 이주를 희망하는 인재들의 필요 역시 고려함으로써 상호 이익을 추구한다.

When urban spaces fail politically or financially, they become available for interplay. MANY is an online platform to facilitate migration through an exchange of needs. It allows cities to attract a changing influx of talent matching their needs to the needs of mobile people for mutual benefit.

매니
켈러 이스터링

MANY
Keller Easterling

and global financial disaster.

Social Capital Credits turn NEEDS into currency.

But cooperative forms of land readjustment pool properties
in collective, self-financing projects.

But this protocol links densifying propertes
to reforested properties

문제는 자산이다

켈러 이스터링

PROBLEMS are ASSETS

Keller Easterling

문제는 좋은 것이다. 집합도시는 문제를 증폭시키는 경향이 있다. 문제들은 서로 상호작용하며 변화와 촉진을 거듭한다. 정치인들과 시장이 남긴 문제와 실패는 자산이 된다. 도시 공간이 정치적 또는 경제적으로 실패하면서 재정적 관리 대상에서 제외되고, 더 이상 법이나 계량 경제학과 같은 추상적인 관념에 얽매이지 않게 된다. 그리고 도시 공간은 기온과 습도가 점점 높아지는 세계에서 근접, 배열, 위험 등 공간 속에 존재하는 무겁고 물리적이며 내재된 가치에 관한 포트폴리오를 얼마든지 제공해 준다.

집합도시는 문제를 제거하기 위한 마스터플랜, 해결책, 알고리즘이 아닌, 문제들을 한 데 어우르기 위해 상호작용하는 유기체를 디자인한다. 즉, 대상이 아닌 소위 말하는 매체를 설계하는 것이다. 특정 형태나 사물이 아닌 매트릭스나 필드를 디자인한다. 이와 같은 관점의 전환을 통해 불평등과 기후변화와 관련된 문제처럼 다루기 힘든 난제들에 대한 접근 방식이 바뀔 수 있기를 희망한다.

포괄적 해결책이나 이념적 유토피아와를 꿈꾸기 보다는, 상황이 더 애매하게 얽히면서 상호작용은 더욱 실질적이며 정확하고 견고하며 신중한 방식으로 이루어진다. 이 접근방식은 특정 대상을 디자인한다기 보다 기구의 토글이나 다이얼을 조절하는 일에 가깝다고 할 수 있다. 공간의 혼합이나 재설계에 대한 예술적 호기심은 상호작용의 프로토콜을 디자인하는 데 도움이 된다. 이는 특정 대상의 개체 수를 변화시키거나, 시간이 지남에 따라 드러나는 상대적 잠재력을 구축하기 위한 플랫폼 디자인이다. 물론 상호작용이 항상 이루어진다고 볼 수는 없다. 상호작용은 항상성보다 불균형이 동반되며 늘 실현되어야 하는 것은 아니다.

가장 강력한 플레이어들은 공간에 대한 정복욕을 바탕으로 그 공간의 모습을 정반대의 방향으로 바꿔 놓기 위한 수단을 쉽게 제공해왔는가? 이들이 바꿔 놓고자 하는 공간의 매우 반복적인 배경 속에서 승수를 찾는다면 이 상호작용은 증폭될 수 있다. 이러한 승수들은 위 공식에 편승하고 이를 내장된 증폭기제로 활용하려고 하는 세균에서 발견되기도 한다.

해결책이라기 보다는 생태학으로의 상호작용은

Problems are good. The collective city likes to multiply problems. Problems leaven and catalyze each other. Problems and failures that politicians and markets leave in their wake are resources. No longer in the grips of legal or econometric abstractions, when urban spaces fail politically or financially, they drop from the financial ledgers. And they are free to offer a parallel portfolio of heavy, physical, situated values embedded in spaces—proximities, dispositions and risks in the wetter, hotter world.

The collective city designs, not master plans, solutions and algorithms to eliminate problems, but rather organs of interplay to combine problems. It works on what might be called medium design— the design of matrix or field rather than object or figure. That inverted focus hopes to alter approaches to intractable problems like those associated with inequality and climate change.

Skipping the romance with comprehensive solutions or ideological utopias, an interplay becomes more practical, precise, robust and deliberate as it becomes more indeterminate and entangled. The approach is less like designing objects and more like adjusting the toggles and dials of organization. Designing protocols of interplay benefits from an artistic curiosity about spatial mixtures or spatial rewiring. It is the design of a platform for inflecting populations of objects or setting up relative potentials within them that unfolds over time. An interplay is then something that should not always work—that relies on imbalance rather than homeostasis.

In the greed to conquer space, have the most powerful players conveniently provided the means by which to spread a countering alteration of that space? These interplays might gain scale by finding multipliers in the very repetitive landscapes of spatial products that they intend to overwrite. They can even be sited as a germ to piggyback on these formulas and use them as built-in amplification.

Interplay, as an ecology rather than a solution, also offers some additional political advantages to the activist. Spatial changes that do not always declare themselves or appear in lexical or legal registers may also slip between ideological fights

또한, 사회운동가에게 정치적인 이점을 제공한다. 스스로 드러내지 않거나 어휘 및 법률 기록에 등장하지 않는 공간 변화는 정치적 은폐를 통해 이념적 다툼 사이를 슬며시 빠져나갈 수 있다. 하지만 더 중요한 것은, 상호작용이 타불라 라사 상태, 혹은 모든 것을 아우르는 해결책을 요구하는 것이 아니라는 점이다. 뿐만 아니라 상호작용은 어떤 하나의 가치 장부에 의존하는 것이 아니다. 시급한 상황에 대체하기 위해, 모든 체제의 혁명적 변화나 정치 상황의 완전한 혁신을 기다릴 필요가 없다. 관련이 있기만 하다면, 상호작용을 통해 더 빨리 관계를 활용할 수 있다. 만일 어떠한 디자인이 조작되거나 손상되었다면, 상호작용의 시간이 확장될수록 정치적으로 더 요령 있고 기민하게 대응할 수 있다. 이는 개입에 효과가 없었을 때, 변화하는 조건이나 순간에 맞춰 디자인이 다시 조정될 수 있도록 돕는다. 상호작용은 더 큰 도시 배경의 일부를 조율하기 위한 일시적인 수단을 제공할 수는 있지만, 영구적이지는 않다는 점에서, 모호하거나 지나치게 유연하지 않으며 오히려 좀 더 직접적이고 의도적으로 불완전하다.

이제 네 가지 상호작용 프로토콜을 제시하고자 한다. 원자재, 필요, 위험, 그리고 실패. 소셜 캐피탈 크레딧은 공동체들의 '필요'를 결손이 아니라 통화로 간주하고 거래에 활용하고 있다. 참여형 농경지 정리는 공간의 재배치를 통해 부동산의 가치를 높이는 작업이다. 공제 프로토콜은 민감한 지역이나 기후변화로 위기에 처한 지역에 기존과는 완전히 상반되는 개발 방식을 제공한다. 또한 MANY는 '필요'의 교환거래를 통해 이주를 촉진시키는 온라인 플랫폼이다. 도시들은 MANY를 통해 필요한 인재를 유입하고, 이주를 희망하는 인재들의 필요 역시 MANY를 통해 고려됨으로써 상호 이익을 추구한다.

상호작용 1: 소셜 캐피탈 크레딧

배치에 내재된 가치있는 잠재력은 자원이 거의 없을 때, 심지어 작업할 자원이 전무할 경우에 가장 잘 드러날 것이다. 마치 래칫(rachet, 한쪽으로만 회전하게 되어 있는 톱니바퀴)처럼, 상호작용이 몇 차례의 점증적인 움직임으로 힘을 받을 수 있다면, 무(無)에서도 시작은 가능하다. 상호작용은 공식적인 시장에서는 교환되지 않는 어마어마한 자원을 활용할 수 있다.

빈곤 지역에서는 현금 대신 기타 자산이 거래되며, 이러한 지역을 대상으로 진행되는 많은 시범사업들에서 '소셜 캐피탈 크레딧'이라 불리는 자산이 거래에 활용되고 있다. 공동체는 '소셜 캐피탈 크레딧'을 가지고 그들의 '필요'를 함께 결정한다. 자원에 대한 결핍이나 고갈로

with an extra degree of political cover. But more importantly, interplay requires no tabula rasa or all-encompassing solution and it is not solely reliant only on any one ledger of values. It does not have to wait for the revolutionary overhaul of all systems or the perfect renovation of political conditions to respond to urgent situations. It can begin sooner to exploit a relationship for as long as it is relevant. Since any design be gamed or corrupted, the extended temporal dimension of interplay may be more politically savvy and agile. It allows design to readjust to changing conditions or to moments when the intervention is out-maneuvered. The interplay may offer temporary means to tune parts of the larger urban landscape, but by being impermanent, it is not equivocal or hopelessly flexible—just more direct and deliberately partial.

Four protocols of interplay are considered here. They start with needs, risks and failures as raw materials. Social Capital Credits allow communities to trade need as a currency rather than a deficit. Participatory Land Readjustment produces increased property value through spatial rearrangement. Subtraction protocols demonstrate how to put the development machine into reverse in sensitive landscapes and areas of climate risk. And MANY is an online platform to facilitate migration through an exchange of needs. It allows cities to attract a changing influx of talent matching their needs to the needs of mobile people for mutual benefit.

Interplay One: Social Capital Credits

The valuable potentials immanent in arrangements are perhaps best illustrated when there are few resources—even when there is very little or even less than nothing with which to work. If, like a ratchet, an interplay can gain leverage with a number of incremental moves, it may that may start from nothing. It can use the immensely fertile field of resources that are not part of formalized markets of exchange.

Many poor communities, in the absence of cash, can trade other assets, and a number of pilot projects are now using what are called Social Capital Credits for these exchanges. With Social Capital Credits, the community gets together to determine its needs, and these needs—which would otherwise be seen as a deficiency or a drain on resources—become the community's resources. The community assigns a credit value to the task or service that reflects the degree of need. There may be a need to plant a tree, clean up a waterway, paint a wall, take care of elderly, or help with children's education. Performing any of these tasks earns the designated number of credits.

비춰질 수 있는 이러한 '필요'는 해당 공동체의 자산이 된다. 공동체는 '필요'의 정도를 반영한 과제나 서비스에 크레딧 가치를 부여한다. 나무 심기나 수로 청소하기, 벽에 페인트 칠하기, 노인을 돌보기, 아이들의 교육 돕기 등에 '필요'가 발생한다. 이러한 과제를 수행하는 것은 '크레딧'을 버는 일이다. 블록체인 토큰과 마찬가지로 '크레딧'은 훔칠 수 있는 자산이 아니기 때문에 관련된 부패가 발생할 수 없다. 다만, 토큰과는 다르게 '크레딧'은 가게에서 물품을 구매하는 데에 활용될 수 없다. '크레딧'은 공동체가 필요로 하는 백신접종, 교육, 훈련, 휴대폰 통화, 주택개조와 같은 활동들로 상환될 수 있다. 즉, '필요'는 자원인 동시에 상환될 수 있는 재료이다.

'크레딧'과 관련된 활동 기록을 확인하고 관리하는 일은 복잡한 알고리즘을 필요로 하지 않는다. 전체 공동체가 변화를 목격하고 인지하며, 과제가 완성된 후 사진이나 승인 기록이 온라인 장부에 남겨지기 때문이다. 개인이 벌어들인 다섯 개의 '크레딧'은 한 개의 공동체 '크레딧'으로 전환되어 공동체 금고에 저장된다. 공동체는 공공화장실 및 학교 건설, 구획정리 등의 일에 '크레딧'을 어떻게 활용할 것인지 결정할 수 있다. 이로써 외부원조가 좀 더 효과적으로 지원될 수 있을 뿐 아니라, 거래의 수단은 합의된 변화의 재료들을 재배치한다.[1] 아시아 이니셔티브라고 불리는 NGO는, 워싱턴 D.C., 가나의 쿠마시, 케냐의 키수무, 인도의 마두라이, 암라바티, 데라둔, 뭄바이-푸네 고속도로에서 '소셜 캐피탈 크레딧'을 활용한 시범 프로젝트를 수행하고 있다.[2]

상호작용 2: 농경지 정리
도시 변두리에서 폭발적으로 성장하는 물리적 공간을 생각해보자. 많은 사람들이 도시로 몰려들지만, 그들은 밀도가 낮고 공간이 넓은 변두리 지역에 주로 거주하게 된다.[3] 연구에 따르면 2050년까지 인도 전체 크기와 맞먹는 3백십만 평방 킬로미터에 달하는 도심 주변부 지역이 개발될 것이라고 한다.[4] 이 공간은 권리가 박탈된

Like blockchain tokens, the credits foil corruption because they cannot be stolen. But unlike the tokens, they cannot be used for merchandise bought in stores. They can only be redeemed for more things that the community needs—things like vaccinations, education, training, mobile phone talk time or home improvements. A need is the resource as well as the material to be redeemed.

The business of corroborating and managing a simple ledger of these activities requires no complicated algorithm because the entire community witnesses and acknowledges the change, and there is an on-line register for images and approvals of completed tasks. For every five credits earned by individuals, one community credit is banked in the community cache. The community can decide how to spend those credits for things like public toilets, schools or street improvements. Outside aid is not only more effectively targeted, but its means of trading is already rearranging the material of an agreed upon alteration.[1] An NGO called Asia Initiatives is conducting pilot projects with Social Capital Credits in Washington, D.C.; Kumasi, Ghana; Kisumu, Kenya, and Maduari, Vinayaganpet, Amravita, Dehradun, and the Mumbai-Pune Expressway in India.[2]

Interplay Two: Land Readjustment
Consider the physical space of explosive growth in the urban periphery. More and more people live in cities, but they live in peripheral areas that are increasingly less dense and staggering in size.[3] By some predictions, by 2050, this mostly unplanned peri-urban development, now de-densifying more rapidly, will cover 3.1 million square kilometers —the size of the entire country of India.[4] This space is also now a bellwether of both inequality and climate change—the space to which the disenfranchised are relegated, the destination of some climate migrations as well as a settlement pattern that exacerbates climate trends.

1   Grassroots Economics, an NGO in Kenya, is using the Ethereum-based Bancor Protocol in Mombasa. Ben Munster, "A Mombasa Slum Looks to Digital Currency to Escape Poverty," DeCrypt Media, August 28, 2018, https://decryptmedia.com/2018/08/28/decentralizing-africa/; Asian Initiatives is currently operating pilot projects with Social Capital Credit in India, Ghana, Kenya and Costa Rica, http://asiainitiatives.org/soccs/
2   http://asiainitiatives.org/soccs/introduction-to-soccs/ (accessed 05.06.19); SoCCs overview 2018.pdf; SoCCs Manual courtesy of founder Geeta Mehta.
3   UN Habitat, "Streets as Public Spaces and Drivers of Urban Prosperity," Nairobi, United Nations Human Settlement Program, 2013, 22; Atlas of Urban Expansion (Cambridge MA: Lincoln Institute of Land Policy, 2012); http://www.atlasofurbanexpansion.org
4   http://webtv.un.org/%C2%BB/watch/joan-clos-un-habitat-on-land-use-and-urban-expansion-press-conference/5013631178001/?term=&lan=english

1   Grassroots Economics, an NGO in Kenya, is using the Ethereum-based Bancor Protocol in Mombasa. Ben Munster, "A Mombasa Slum Looks to Digital Currency to Escape Poverty," DeCrypt Media, August 28, 2018, https://decryptmedia.com/2018/08/28/decentralizing-africa/; Asian Initiatives is currently operating pilot projects with Social Capital Credit in India, Ghana, Kenya and Costa Rica, http://asiainitiatives.org/soccs/
2   http://asiainitiatives.org/soccs/introduction-to-soccs/ (accessed 05.06.19); SoCCs overview 2018.pdf; SoCCs Manual courtesy of founder Geeta Mehta.
3   UN Habitat, "Streets as Public Spaces and Drivers of Urban Prosperity," Nairobi, United Nations Human Settlement Program, 2013, 22; Atlas of Urban Expansion (Cambridge MA: Lincoln Institute of Land Policy, 2012); http://www.atlasofurbanexpansion.org
4   http://webtv.un.org/%C2%BB/watch/joan-clos-un-habitat-on-land-use-and-urban-expansion-press-conference/5013631178001/?term=&lan=english

사람들이 좌천된 장소이자 기후변화로 인한 이주 행렬의 목적지면서, 동시에 기후경향을 악화시키는 정착 패턴 등을 보이고 있다. 즉, 불평등 및 기후변화의 징후를 보여주는 장소가 되고 있다.

글로벌 거버넌스는 종종 도시 내 실체가 있는 물리적 구성보다는 금융의 추상적 개념을 다룬다. 하지만, 도시 변두리 주거지는 금융 대장의 항목으로서가 아닌 실체가 있는 대상으로서 더 중요하다. 고밀도의 비공식적 정착은 정치, 경제적 문제 뿐 아니라 형태적 문제를 야기한다. 기반시설이 갖춰지지 않은 주거지의 '도심 밀집현상'은 모든 면에서 이동을 어렵게 만든다. 이동수단이 없이 먼 길을 걸어 가야 하거나 대중교통 이용이 어려워 통근에 지나치게 많은 시간이 소요되며, 이러한 물리적 이동성의 제약은 사회적 이동성에도 영향을 끼친다. 이와 같은 형태적 문제 해결을 위해 취해지는 전형적인 방법은 주로 철거 및 박탈을 통해 거대한 도시 간선도로를 건설하는 것이다.[5]

이러한 도시 변두리 문제 해결을 위해 유엔 해비타트는 경지정리 기술 보급을 위한 프로토콜을 개발해왔다. 20세기 초반 독일은 경지정리를 시행해, 불규칙한 모양의 부지를 가진 농부들이 인접한 자투리 부지를 거래하도록 하였다. 이로 인해 농지에 기하학적 규칙성이 부여되어 농사짓기에 적합한 부지가 늘어날 수 있었다. 한편, 부동산 자산가들은 전통적인 농경지 정리 방법을 활용하여 대형 프로젝트를 위한 부동산을 일시불로 사들였다. 이 시나리오에서는 부동산 가치의 상승으로 지속적인 혜택을 얻은 부동산 자산가들이 주요 수혜자이다. 판매자는 가치에 훨씬 미치지 못하는 현금 몇 푼을 손에 쥘 뿐이었다.[6]

유엔 해비타트를 비롯한 여러 기관들은, 스스로 자금을 충당하고, 공간 형태적 가치를 거래하며, 많은 참가자들에 의해 집합적으로 시행되는 농경지 정리 상호작용의 방법들을 개발하고 있다. 이에 따르면, 공식적 또는 비공식적 토지 소유자들은 일부 부동산의 가치가 계속 높아지는 것을 대가로 자신의 땅이 새로운 사용처로 통합되고 재개발되는 것을 용인한다. 좀 더 상황이 불안정한 비공식적 토지 소유자들은 추상적인 금융개념을 위해 자신의 땅을 포기할 필요는 없다. 대신, 그들은 소규모

Global governance often manipulates financial abstractions rather than the heavy physical material of the city. But dwellings at the urban periphery may be more important as a physical objects than items in an economic ledger. Dense informal settlements often present a morphological as well as an political and economic problem. An urban clot of dwellings with no infrastructure makes mobility, in all senses of the word, extremely difficult. Long walking distances or little access to transit makes travel times to work impossibly long, and physical mobility impacts social mobility. The typical remedy to this morphological problem is often demolition and disenfranchisement to build large urban arterials.[5]

To address this peripheral fabric, UN Habitat has been developing a protocol that recuperates selected techniques of Land Readjustment. First used in Germany in the early 20th century, land readjustment allowed farmers with irregular lots to trade adjacent fragments to regularize geometry and offer more usable areas for farming. And real estate capital uses conventional land readjustment to incentivize buying out their properties for large projects with a one-time payment. But, real estate capital is the chief beneficiary in this scenario, able to continue to benefit from increased values while the seller is left with cash that will not appreciate in the same way.[6]

UN Habitat and others are developing versions of a land readjustment interplay that is self-financing, trades in spatial morphological values, and is cooperatively directed by a larger number of participants. In this version, formal or informal owners allow their land to be aggregated and redeveloped with new uses in return for a smaller portion of the property that will continue to gain in value. Informal landowners in a more precarious situation need not relinquish their property to a financial abstraction. Instead they pool their property and work with a small group of neighbors on a tangible project in which the new physical arrangement is the source of value. Calling the protocol "Participatory and Inclusive Land Readjustment," UN Habitat is experimenting with pilot projects in Columbia, India, Angola and Turkey that use this engine to convert urban land to new uses, insert streets and infrastructures into the

5   UN Habitat, "Streets as Public Spaces and Drivers of Urban Prosperity," Nairobi, United Nations Human Settlement Program, 2013, 55; Atlas of Urban Expansion (Cambridge MA: Lincoln Institute of Land Policy, 2012); http://www.atlasofurbanexpansion.org; Patrick Lamson-Hall, Shlomo Angel, Yang Liu, "The State of the Streets: New Findings from the Atlas of Urban Expansion—2016," Working Paper WP18PL1, Lincoln Institute of Land Policy 2018.
6   https://urban-regeneration.worldbank.org/Ahmedabad, accessed December 23, 2018;

5   UN Habitat, "Streets as Public Spaces and Drivers of Urban Prosperity," *Nairobi, United Nations Human Settlement Program*, 2013, 55; *Atlas of Urban Expansion* (Cambridge MA: Lincoln Institute of Land Policy, 2012); http://www.atlasofurbanexpansion.org; Patrick Lamson-Hall, Shlomo Angel, Yang Liu, "The State of the Streets: New Findings from the Atlas of Urban Expansion—2016," Working Paper WP18PL1, *Lincoln Institute of Land Policy* 2018.
6   https://urban-regeneration.worldbank.org/Ahmedabad, accessed December 23, 2018;

그룹의 이웃들과 땅을 한데 모아 구체적인 프로젝트를 함께 실행해 나갈 수 있으며, 이 프로젝트 하에서는 새로운 물리적 재배치가 가치의 원천이 된다. 이러한 프로토콜은 '참여적이고 포용적인 농경지정리'라고 불린다. 유엔 해비타트는 이러한 새로운 동력을 기반으로 콜롬비아와 인도, 앙고라, 터키에서 시범 프로젝트를 시행하고 있으며, 이를 통해 도시의 토지를 새로운 사용처로 전환하고, 비공식 정착지에 도로와 기반시설을 설치하며, 자연재해나 분쟁을 겪은 정착지를 복원하고 있다.[7]

또한 새로운 경지정리를 통해 홍수, 심각한 태풍, 산불 등과 같은 환경 변화에 조직적으로 대항하면서 위험을 줄일 수 있다. 위험한 지역을 피해 경지를 통합하거나 재배치할 수도 있다. 또한 정글이나 열대 우림 내의 땅을 소유한 사람들은 가까운 소도시와의 교류를 늘릴 수도 있다. 이는 민감한 환경을 파괴하는 농업 및 다른 사업들보다 더 큰 수입원이 될 것이다.

그러나 농경지 정리의 상호작용이 보여주는 가장 중요한 것은, 농지의 형태적 배치를 통해 경제적 전망을 바꾸고, 심지어 소득 불평등을 해결할 수도 있다는 점이다. 이와 같은 상호작용은 초기 투자에 대한 위험 없이 보통의 이웃들이 기존의 자산을 완전히 재구성할 수 있도록 돕는다. 투자를 통해 대부분의 재산을 모은 상위 1%처럼, 이 상호작용을 통해 많은 사람들이 오직 임금에만 의존해야 하는 안정성 없는 상태를 벗어날 수 있다. 즉, 모든 참가자들이 농경지 배치를 활용하여 가치를 향유하고 창출할 수 있다. 토지 소유자는 금융 기관의 다양한 규제에 의해 자산을 통제 받는 대신, 유지보수 및 관리를 통한 농경지 정리의 가치창출에 구체적으로 기여할 수 있다.

상호작용3: 홍수 및 산불에 대비한 공제 프로토콜

낮은 이율의 장기 모기지론은 20세기 무분별한 확장을 낳았고, 2007년 글로벌 금융위기의 원인이 되었다. 위기의 확산을 막는 방법은 모기지 역설계를 통해 산불에서 해수면 상승에 이르기까지 지구 온난화와 관련 있는 건물의 재배치를 촉진하는 것이며, 불평등과 고용 문제도 같은 방법으로 다룰 수 있다.

20세기 중반, 담보가치를 높여줄 디자인을 위해 수 천 개씩 미국 교외 지역 모기지가 승인됐다. 2008년 금융위기 전, 이는 다시 서브프라임 모기지로 분류됐다. 하지만 이 모기지 분류의 기준이 금융이라는 추상적 개념이 아닌, 홍수 산불 등의 총체적 기후변화 위험을 낮출 수 있는 역량 같은 중요한 환경적 가치였으면 어땠을까?

informal settlements and reconstitute settlement after a natural disaster or conflict.[7]

New arrangements can also reduce risk by organizing to cope with changing environmental conditions like floods, severe storms and wildfires. Properties can be densified and repositioned to avoid risky areas. Owners of land in jungles and rain forests might consolidate access to areas near small cities that would provide more income than farming or other enterprises that destroy sensitive landscapes.

But perhaps most importantly, a land readjustment interplay demonstrates how a morphological arrangement can alter economic prospects—even address income inequality. Without risk to the initial investment, this sort of interplay can allow a modest neighborhood to completely recast their existing property. Like the one percent who make most of their money from investment it allows many to escape the precarity of relying solely on wages. A land readjustment interplay allows any participant to participate and generate value in relationships sitting atop the land. Rather than assets being controlled by financial entities with varying degrees of regulation, an owner can also tangibly contribute to the value of that arrangement with maintenance and care.

Interplay Three: Subtraction Protocol for Flooding and Wildfire

The long-term low-interest mortgage was a multiplier of sprawl in the 20th century and germ of global financial disaster in 2007. A counter-contagion might reverse-engineer the mortgage to facilitate relocations of building related to the effects of global warming—from wildfires to rising sea levels—in ways that also address inequality and employment.

Mid-century US suburban mortgages were grouped, approved and underwritten by the thousands for designs that increased bankability. Before the financial crisis of 2008, they were grouped again in bundles of subprimes. But what if mortgages were grouped and scored, not according to financial abstractions but rather to their heavy, situated environmental values like their capacity to reduce collective climate risks related to flooding or wildfires?

Simple pairings offer one elementary way to alter the chemistry or disposition of any organization. As container shipment was taking hold, double stacking of rail cars, first designed in 1977 and more broadly used in the 1980s, led

7   *Remaking the Urban Mosaic: Participatory and Inclusive Land Readjustment*, UN Habitat 2016, 5, 17, 8.

7   *Remaking the Urban Mosaic: Participatory and Inclusive Land Readjustment*, UN Habitat 2016, 5, 17, 8.

단순 페어링은 어떤 조직의 반응이나 성향을 바꾸기 위한 기본적인 방법 중 하나이다. 컨테이너 수송이 활발해지자, 1977년에 처음 설계되고 1980년대에 더욱 널리 사용된 2단 적재 열차가 규모의 경제를 주도했고 그 결과 운송비가 급감했다.[8] 또 다른 예로, 친구나 가족에게 신장을 기증하고자 하나 적합 판정을 받지 못한 경우, 공여자 교환 프로그램에 참여함으로써 친구나 가족을 도울 수 있는 기회가 확장되었다. 이 프로그램은 신장이식을 받고자 하는 친구 및 가족들이 전체 수혜자 풀의 일부가 되고 맞는 공여자를 서로 교환함으로써 이식의 혜택을 받을 수 있다.[9]

단순히 하나의 자산을 다른 자산과 연결 짓는 것만으로도 변화의 바람을 일으킬 수 있다. 침수 위기의 집들은 물리적으로나 재정적으로 손실을 보고 있는 상태다. 그러나 시 당국이 그 집들을 매입하여 그 땅을 홍수 예방 시설을 설치하는 데 사용한다면, 이들은 집단적 위험을 예방하는 데 기여한 것이다.

이런 경우 이들의 보험비용을 줄여주면서 위와 같은 행동을 더욱 촉구할 수 있을 뿐만 아니라, 환경적 위험을 줄인 대가로 추가적인 인센티브를 제공할 수도 있다. 또한 은행들은 환경적 혜택을 제공한 모기지 그룹에 대해 융자수수료나 부동산매매 수수료를 면제해 주어야 할 수도 있다. 또는 FEMA와 같은 정부 기관이 모기지 금리를 낮추기 위해 일시불로 대신 지급할 수도 있다.

새로운 형태의 취약성은 기후변화와 불평등의 접점에서 발생한다. 기후변화의 위험에 직면한 부유층들은 재배치나 정교한 방화 및 홍수방지 시설 구축을 통해 안전을 확보할 수 있다. 하지만, 이러한 공제 상호작용에서는 홍수의 위험 때문에 활용이 불가능한 단일 자산이, 역설적으로 구매자를 유인하기 위해 그 위험을 활용한다. 만일 판매자가 고지대로 이주하고 구매자 역시 올라갈 여유가 있는 경우, 그 그룹의 점수는 올라간다. 기후변화로 인한 젠트리피케이션의 피해자는 발생하지 않으며, 판매자가 구매자의 부를 활용하여 상호 이익을 창출한다.

영향력의 다른 장점들은 이 게임이 가속화되면서 작동하기 시작한다. 만일 서너 그룹이 같은 은행을 통해 거래한다면, 그 은행은 거래 규모가 커진 것에 대한 대가로 인센티브를 제공할 수 있다. 위험을 피해 다른

to economies of scale that significantly reduced the cost of shipping.[8] Prospective kidney donors who are not a good match with a friend or family members now have a better chance of helping the recipient by entering into a paired exchange program so that all friends and relatives are part of a pool to donate to all recipients.[9]

Simply linking one property to another can create a leveraging engine of change. A group of houses at risk of flood is physically and financially underwater. But if the group sells to a city, and the city uses the land to build flood protection, that group is given a high score for its reduction of collective risk.

In addition to decreased insurance costs that already incentivize each individual move, added incentives might be awarded for a score that reduces risk. Banks might be required to waive origination points or other closing costs for mortgage groups with environmental benefits. Or a state agency like FEMA in the US might provide a one-time payment to the bank for points to reduce the interest rate.

Fresh forms of vulnerability appear at the intersection of climate change and inequality. Now in the face of climate change, wealth again has the capacity to purchase safety through relocation or elaborate fireproofing or flood proofing construction. But with this subtraction interplay, a single property that is unviable due to flood risk, paradoxically uses its risk to attract a buyer. If the seller moves to high ground and the buyer can afford to elevate, the group score goes up. Now not a victim of climate gentrification, the seller leverages the buyer's wealth to mutual advantage.

Other points of leverage might come into play as the game accelerates. If groups of three or four bring transactions to the same bank, the bank might provide any number of incentives in exchange for increased volumes of business. If a move away from risk becomes popular because of its affordability, federal money may be freed to address climate change investment rather than buy outs. Also instead of relying only on housing starts for construction jobs the deconstruction of houses offers another kind of skilled work. A relatively modest contribution to mortgage principle that reduces interest rates is significantly compounded

8   https://www.up.com/aboutup/history/chronology/index.htm, accessed December 23, 2018.
9   Alvin E. Roth, *Who Gets What—And Why?* (Houghton Mifflin Harcourt, 2015). https://www.npr.org/sections/health-shots/2015/06/11/412224854/how-an-economist-helped-patients-find-the-right-kidney-donor, accessed December 23, 2018; and https://www.pbs.org/newshour/show/the-economic-principle-that-powers-this-kidney-donor-market, accessed December 23, 2018.

8   https://www.up.com/aboutup/history/chronology/index.htm, accessed December 23, 2018.
9   Alvin E. Roth, *Who Gets What—And Why?* (Houghton Mifflin Harcourt, 2015). https://www.npr.org/sections/health-shots/2015/06/11/412224854/how-an-economist-helped-patients-find-the-right-kidney-donor, accessed December 23, 2018; and https://www.pbs.org/newshour/show/the-economic-principle-that-powers-this-kidney-donor-market, accessed December 23, 2018.

곳으로 이동하는 것이 합리적 가격으로 대중화되면, 연방 정부의 예산은 매수보다 기후변화를 해결하기 위해 더 많이 사용될 수 있다. 또한 건설업 일자리를 살리기 위해 주택 착공에만 매진하는 대신, 주택을 해체하는 작업 역시 또 다른 숙련노동을 요하는 일자리를 창출할 수 있다. 모기지 원금에 대한 상대적으로 적절한 개인분담금은 모기지 전 기간 동안 복리로 청구된다. 1천 달러가 5만 달러가 되는 식이다.

열대 우림과 같은 민감한 지역의 침해 개발 문제와 관련된 공제 프로토콜 생각해보자. 도로는 종종 발전과 접근권 확장을 위한 수단으로 간주되지만, 동시에 침습적인 개발로 이어지고, 이는 숲과 토착문화가 지닌 가치있는 자원을 파괴한다. 하지만, 모기지 프로토콜이 쉽게 자산과 연결된 것처럼, 이번 프로토콜은 나무가 울창한 숲 자산을 재조성한 조림지와 연결시킨다. 숲을 통과하는 파괴적인 도로를 만들기보다 마을의 자산을 위해 숲을 비우면, 통신, 교육, 고용 자원을 더 효과적으로 활용할 수 있게 된다. 조림지는 대학과 관광업체, 제약회사의 투자 가치를 창출하며, 수익의 원천이 된다. 조림지는 항상 소유자와 연결되고, 그들에게 배당금을 제공한다. 울창한 숲 지역은 걸어서 갈 수 있는 학교와 통신망에 연결되어 있다. 이처럼 집합 공제 프로토콜은 다시금 인류의 개발 기계에 후진 기어를 넣는다.

어떤 시나리오든 게임이 진행됨에 따라, '집'보다 '빈 공간'에 점수를 주는 바둑의 역게임 형태를 보인다. 바둑에서는 두 사람이 흰 돌과 검은 돌을 격자판에 교대로 놓으며 게임을 진행한다. 흰 돌이 집을 지으려 하면, 검은 돌이 방해하거나 검은 돌이 집을 지으려 하면 흰 돌이 방해한다. 만일 집이 아니라 빈 공간에 점수를 준다면, 이 게임은 합리적인 빈 공간을 만드는 경쟁이 될 것이다. 무게가 고정되면 제대로 작동하지 못하는 래칫처럼, 적극적인 공제 프로토콜을 통해 자산의 재조직 및 재결합을 실현할 수 있다.

MANY

세계적으로 구축된 기반시설로 수십억 개의 상품과 수천만 명의 방문객, 저임금 노동자들의 이동이 완벽할 정도로 효율적으로 이루어지고 있다. 하지만, 전 세계 7천만 명에 달하는 사람들이 집을 잃고 떠도는 시점에, 정치, 경제, 환경적 이주 문제를 해결할 방법은 그리 많지 않다. 시민권(망명)을 허가, 거부하는 국가 시스템은 제대로 작동하지 않고 있다. 또한 'NGOcracy'는, 평균 17년 구금 생활의 다른 이름에 불과한, 난민 캠프 내 체류시설을 최고의 아이디어로 내 놓았다.

MANY는 '필요'의 교환을 통해 이주를 활성화하기

over the life of the mortgage. One thousand dollars becomes fifty thousand dollars.

Or consider a subtraction protocols related to encroaching development in sensitive landscapes like rain forests. While roads are often treated as the means to deliver progress and access, they encourage this development, and that destroys valuable resources embedded in forests and indigenous cultures. But just as the mortgage protocol simply linked properties, this protocol links densifying properties to reforested properties. More than a destructive road through the forest, vacating a forest property for a village property more effectively increases access to resources for communication, education and employment. The reforested areas create value that attracts investment from universities, tourism, pharmaceutical companies so that they can become sources of revenue. Reforested areas always remain linked to original property owners and return dividends to them. Densifying areas are also linked to walkable schools and broadband. A collective subtraction protocol once again puts the development machine into reverse.

In any of these scenarios, as the game is played over time, it might seem like a reverse game of Go that values clearings rather than walls. In the Chinese game of strategy—players take turns positioning white or black "stones" on a grid. White tries to build up lines of defense against Black and vice versa. If clearings were valued over thickening fortresses, the competition is about staking out a reasonable clearing while also acquiring a spot on the surface area of the clearing. Step by step, like a ratchet that works against an immovable weight, the active forms of a subtraction protocol allow properties to be reorganized and reaggregated.

MANY

Global infrastructure space has perfectly streamlined the movements of billions of products and tens of millions of tourists and cheap laborers, but at a time when nearly 70 million people in the world are displaced, there are still so few ways to handle political, economic, or environmental migrations. The nation-state has a dumb on-off button to grant or deny citizenship/asylum. And the NGOcracy offers as its best idea storage in a refugee camp—a form of detention lasting on average 17 years.

MANY is an online platform to facilitate migration through an exchange of needs. It serves people who might defang xenophobic arguments by saying, "We don't want your citizenship or your victimhood or your segregation or your bad jobs. We don't want to stay."

위한 온라인 플랫폼이다. "우리는 당신의 시민권이나, 피해의식, 차별, 나쁜 직업을 원하지 않는다. 우리는 머무르기를 원치 않는다"라고 말하면서 외국인 혐오 주장을 무력화시킬 수 있는 사람들을 돕는다.

MANY는 국가의 방해를 피하면서, 그룹 간 연결을 공고히 하는 전략가들의 의지를 반영하고 있다. 시간과 훈련을 시장 외에서 교환하는 초단기 프로젝트 기반의 여정이 글로벌 자격인증을 위해 생성되고 통합된다.

가진 자와 못 가진 자는 없으며, 해결책도 없다. 단지 '필요'와 문제를 함께 모아놓을 뿐이다.

MANY는 공간 네트워크와 국제적 유동성 구축을 위해 존재하는 정보 시스템이다. 도시는 재능 있는 사람들과 자원의 유입을 위해 미개발된 공간을 제시하며 흥정을 하고, 이주하는 사람들의 '필요'와 도시의 '필요'를 조화시키며 상호 이익을 추구한다.

MANY는 질문한다. 또 다른 형태의 세계적 이동성이 삶의 시간 혹은 계절을 중심으로 구성되어 정치적으로 더 효과적인 여러 갈래의 선택지들을 만들어 줄 수 있을까? 이러한 교환이 세계적인 지도자 자격을 얻기 위한 수단으로 기대되고 환영받을 수 있을까?

매체 디자인

파론도의 역설은 직관에 반하는 게임이론으로 만일, 이길 확률이 낮은 게임을 한다면, 당신은 질 것이지만, 질 확률이 높은 두 가지 게임을 혼합할 경우, 이길 확률이 높아지기 시작한다고 주장한다. 이길 확률을 보여주는 그래프는 이기는 게임 수가 조금씩 점증하는 얄팍한 톱니의 모습을 보인다. 그리고 그 과정은 래칫과 유사하다. 즉, 게임의 패배가 조금씩 승률을 높이는 일종의 정지마찰력을 만들어내는 듯하다. 시급하게 해결해야 할 문제가 많은 현대 사회에서 불평등 해소를 위한 움직임을 시장이 법으로 가로막고, 기후변화를 해결할 수단이 정치적 난국에 의해 방해 받는 시점에, 매체 디자인은 기존 메커니즘들의 관계를 활용한 화학반응을 제안한다. 이를 통해 가장 무겁고 움직이기 힘들었던 것들을 시장 외 공간에서 교환할 수 있게 될 것이다.[10]

역설적으로, 이는 긴급한 문제들에 더 신속히 반응하는 수단이다. 어떤 형태의 동시대적 발전은 변화를 급속히 촉진시킨다. 상호작용의 프로토콜에서, 자본은 과연 어떻게 하는지 모르는 일을 자신도 모르게 하게 되거나, 원한지도 몰랐던 것들로 보상받을 수 있을까? 집단 도시는 자체적 문제에도 불구하고 지속해 나갈 수 있을까?

Working around national obstructions, MANY reflects the persistence of resourceful people making secure group-to-group connections. Shorter project-based journeys that trade in non-market exchanges of time and training are generated and aggregated for global credentials.

There are no haves and have nots, and no solutions—only needs and problems to put together.

MANY is a heavy information system that exists to build spatial networks and cosmopolitan mobility. Cities can bargain with their underexploited space to attract a changing influx of talent and resources—matching their needs to the needs of mobile people to generate mutual benefits.

MANY asks: Might another kind of cosmopolitan mobility organize around intervals of time or seasons of a life to form a branching set of options that is more politically agile? Might this exchange be anticipated and celebrated as the means to global leadership credentials?

Medium Design

Parrondo's Paradox is a counter-intuitive game theory positing that if you play a game with a low probably of winning, you will lose, but if you alternate between two games, each with a probability of losing you can begin to generate wins. The resulting graph of the wins resembles a shallow sawtooth of incremental increases in numbers of winning games. And the process may actually behave like a ratchet—as if the losses create a kind of traction against which to make many small gains.[10]

In the urgent present, when markets themselves legally obstruct any challenges to the inequality and when political impasses obstruct any means to address climate change, medium design suggests a chemistry of existing mechanisms that move what seem to be the heaviest and most immovable solids in non-market or extra-market exchanges. Paradoxically, this may even be a means to react more precipitously to pressing problems. And some forms of contemporary development may even provide a multiplier to catalyze the change. With a protocol of interplay, can capital be tricked into doing things it does not know how to do or rewarded with things it did not know it wanted? And can the collective city actually find sustenance even in its problems?

---

10 Gregory P. Hamer and Derek Abbott, "Game Theory: Losing Strategies Can Win by Parrondo's Paradox," *Nature*, December 1999.

10 Gregory P. Hamer and Derek Abbott, "Game Theory: Losing Strategies Can Win by Parrondo's Paradox" *Nature*, December 1999.

약속의 땅: 상품화에서 협력으로 향하는 주택단지의 변화

도그마(피에르 비토리오 아우렐리, 마르티노 타타라, 마리아파올라 미켈로토), 뉴 아카데미(레오너드 마, 투오마스 토이보넨)

아래 글은 주거공간 건축의 네 가지 중요한 요소(토지 재산, 주택 소유, 건설, 유형론)에 대해 간략히 살펴봄으로써 '저소득층을 위한 주택단지'를 분석하고자 하였다. 네 가지 요소 각각에 대해, 우리는 역사적 발달, 주요 사례 연구 및 '투기 없는' 합리적 가격의 주거 제공을 위한 대안을 간략히 서술하였다.

원죄: 전원지역에서 도시에 이르는 토지 사유화

현재의 주택 위기에 대한 논의가 있을 때, 건축가와 도시 계획가 및 대중들은 저소득층을 위한 주택 및 공공지원 주택의 부족 문제를 다루는 경우가 대부분이다. 그리고 이러한 평가 과정에서 종종 간과되는 것은, 길고 논란의 여지가 많은 토지 사유화 과정과 궤를 같이하는 문제의 근본적 원인이다. 좀 더 직설적으로 말하자면, 토지에 대한 접근 없이는 주택을 확보할 수 없다. 토지 사유화는 전혀 새로운 개념은 아니다. 토지 사유화는 국가의 적극적 지원 하에 지주들이 이행한 '합법적 절도'의 과정이었다. 이와 같은 토지의 합법적 절도는 15세기에서 18세기 사이, 민족국가(특히 영국)의 대두와 함께 시작되었다. 그 당시 영국 농촌 지역에 살던 소작농들은, 지주들이 시행한 토지 사유화 과정에서 토지를 몰수당했으며, 지주들은 국가를 이용해 소작농의 전통적 권리를 폐지하는 법률 구조를 구축했다. 이는 공동경지를 경작하는 마을에 대한 점차적인 억압을 통해 가능했으며, 이 마을에서 토지 소유권은 '셀리온(selion)', 즉 농부들이 개인적으로 또는 협력하여 경작한 산발적 경지의 형태로 구성되어 있었다.[1] 이와 같은 집단체제의 근간은 비공식적인 관습에 기인한 권리이며, 이는 통합적 토지 소유 및 사용 관습에서 유래하였다. 관습에 기인한 권리에 맞서, 지주들은 사유재산권을 근거로 들었다. 지주들은 국가가 승인한 사유재산권을 활용해 거대한 땅을 자신들의 사유지로 통합할 수 있었다. 이와 같은 사유화 과정은 개선 및 효율에 관한 담론을 통해 정당화되었다. 지주들로서는 한 명의 소유주가 안정적으로 보유하고 있는 하나의

Promised Land: The Housing Complex from Commodification to Cooperation

Dogma (Pier Vittorio Aureli, Martino Tattara with Mariapaola Michelotto) and New Academy (Leonard Ma, Tuomas Toivonen)

The following text aims to deconstruct the 'affordable housing complex' by offering a concise overview of four fundamental passages in the production of domestic space: land property, home ownership, construction and typology. For each of these passages we briefly outline their historical development, major case studies and possible alternative paths towards a 'speculation free' affordable housing.

The Original Sin: The Privatization of Land from Countryside to the City

When addressing the current housing crisis, it is common for architects, planners and the public to refer to a shortage of affordable housing and social housing. What often goes missing in this assessment is the very origin of this crisis, which coincides with the long and controversial process of the privatization of land. To put it bluntly: there is no housing affordability without access to land. Privatization of land is nothing new: it is and has always been a *legalized theft* perpetrated by landlords with the active support of the state. The legalized theft of land began with in the rise of the nation state (specifically England) between the 15th and 18th century. At this time peasants living in rural areas of England were expropriated through a process of privatization enacted by landowners who used the state to create a legal structure that revoked the peasant's traditional rights. This was possible through a gradual process of suppression of the common field village in which ownership of land was organized through the form of *selions*, scattered pieces of arable land that farmers cultivated individually and cooperatively.[1] The basis of this collective system was an informal set of rights of customs, which derived from consolidated practices concerning the occupation and use of land. Against rights of customs, landlords invoked individual property titles. Such titles, granted by the state, enabled them to unite huge swathes of land as their own *private* estates. This process of privatization was justified by discourses on

---

1   다음을 보라: Gary Fields, *Enclosure. Palestinian Landscapes in a Historical Mirror*, Los Angeles: University of California Press, 2017, p. 33.

1   See: Gary Fields, *Enclosure: Palestinian Landscapes in a Historical Mirror*, Los Angeles: University of California Press, 2017, p. 33.

통합된 땅이 관리하기 쉬울 뿐만 아니라 수확량을 늘릴 수도 있었다.[2] 토지가 사유재산으로 전환됨에 따라 나타난 중요한 결과로, 지주들은 다른 물품처럼 땅을 팔 수 있을 뿐 아니라, 금융대출 담보물로 사용할 수 있게 되었다.[3] 여전히 오늘날에도 부동산 담보대출은 신용 및 현금 마련을 위한 가장 큰 출처이다.[4] 지주들에게 있어, 재산권을 통한 토지 소유는 농업생산을 통해 더 많은 흑자를 창출하기 위한 방법일 뿐 아니라, 토지를 금융 자산으로 활용함으로써 자본을 조달하기 위한 방법이기도 했다.

19세기 중반 이후 영국 등 산업화된 국가에서는 인구 중 극히 일부(20%)만이 농업에 종사하였고 그 나머지는 제조업에 종사하며 도시에 거주하였다.[5] 이로 인해 도시 내 거주인구가 급증하였고 지주들은 주택용지에 투기할 수 있는 기회를 잡게 되었다. 이와 같은 투기에 맞서 20세기 초 영국, 스웨덴, 네덜란드 등 몇몇 유럽 국가들은 소셜 하우징을 건설하려는 지역 자치당국을 지원하기 위해 대규모로 토지를 매입하기 시작했다. 국가 소유 주택은 2차 대전 이후 활성화되기 시작했다. 당시 유럽 국가들이 재건 등을 이유로 전례 없이 많은 공공주택 건설을 위해 충분한 건축 용지를 확보할 수 있었기 때문이다. 하지만, 조쉬 리안 콜린스가 언급하였듯, 국가가 가장 많은 건설을 한 시기에 시장에서도 가장 많은 건설이 이루어졌음을 잊어서는 안 될 것이다.[6] 복지국가 전성기 시절, 유럽 내 많은 자유 민주 국가들의 암묵적 정치적 목적은 '재산소유 민주주의'를 구축하는 것이었다. 심지어 핵가족을 모델로 매우 엄격하게 거주 단위를 규정한 소셜 하우징 프로젝트마저 거주자들에게 주거공간 사유의 미덕을 전파하였다. 더욱이, 이러한 프로젝트에 활용된 토지는 공공택지화 되어 '정원 부지(garden estate)'의 형태로 접근 가능하였으나, 거주자들이 이곳을 개조할 수는 없었다. 결과적으로, 1980년대 복지국가의 종말로 소셜 하우징의 사유화가 본격화되었으며, 공익의 관리인으로써의 '자애로운' 국가는 토지 소유 역사에서는 예외임이 분명해졌다.

자산으로서의 주택: 주택 소유의 금융화

오늘날 우리는 재산권을 당연하게 받아들이지만, 재산권이 국제무대에서 영위하는 무조건적 지위는 상대적으로 최근에 발생한 변화이다. 1940년대

improvement and efficiency: for landlords only a piece of land that was securely owned by one proprietor could be easily governed while increasing production.[2] A crucial consequence of the transformation of land into private property was the possibility for the landowner not only to sell land as any other commodity, but also to use it as collateral for financial loans.[3] Still today, lending money against landed property is the largest source of credit and money creation.[4] For landlords, owning land through property titles was not just a way to extract more surplus from agricultural production, but also a way to raise capital through the use of land as a financial asset. Since the second half of the 19th century in industrialised countries like England, only a relatively small fraction of the population (20%) worked in agriculture, the rest worked in manufacturing and lived in cities.[5] This meant a dramatic increase of inhabitants in cities, which provided an opportunity for landowners to speculate on land for housing. Against this process of speculation, in the early 20th century, several European states such as England, Sweden and The Netherlands initiated massive purchases of land in order to support local municipalities to build social housing. State-owned housing gained its momentum after the 2nd World War, when the opportunity of reconstruction, allowed many European countries to secure enough public land for building an unprecedented quantity of public housing. Yet, we should not forget that as noted by Josh Ryan Collins, the years in which the state built the most were also those in which the market did the same.[6] In the heyday of the welfare state, the tacit political goal of many liberal-democratic states in Europe was to build a 'property owning democracy.' Even social housing projects with their rigidly defined housing units modelled on the nuclear family, trained dwellers in the virtues of a privately owned domestic space. Moreover the land of these projects was made public and accessible in the form of 'garden estates,' yet these gardens were not alterable by the inhabitants. Eventually, the demise of the welfare state in the 1980s pushed the privatization of social housing further, making it clear that the 'benevolent' state as custodian of public good was an exception in the history of land property.

2   Ibid. pp. 115-116.
3   다음을 보라: Josh Ryan-Collins, Toby Lloyd and Laurie Macfarlane, *Rethinking the Economics of Land and Housing*, London: ZED, 2016, p. 30.
4   Ibid.
5   Ibid., p. 25.
6   Ibid.

2   Ibid. pp. 115-116.
3   See: Josh Ryan-Collins, Toby Lloyd and Laurie Macfarlane, *Rethinking the Economics of Land and Housing*, London: ZED, 2016, p. 30.
4   Ibid.
5   Ibid., p. 25.
6   Ibid.

후반 유럽과 미국에서 해방을 둘러싼 사회적 갈등이 고조되면서, 자본가들과 국가 정책 입안자들은 국제적으로 개인 자본권을 보호하기 위한 조치를 모색하였다. 주택 소유 메커니즘은 점차 심화되는 경제 금융화를 통해 1980년대 자본의 영향력을 확장시키는 프로젝트의 중심이었다. 비록 우리가 자신의 집을 소유한다고 생각해도, 현실에서 우리가 소유하고 있는 것은 모기지이다. 이제 모기지는 빚을 떠안게 되는 매개체 이상의 의미를 넘어, 주택을 금융자본의 국제적 흐름에 포함시키는 수단이 됐다. 사스키아 사센은 이 과정을 "관련 주택 소유자들을 완전히 무시한 채 투자자 이익을 챙기는 거액융자의 새로운 회로를 구축하고자, 평범한 가정의 모기지를 금융화 하려는 시도와 마찬가지로, 체계적으로 심화되고 있는 자본적 관계의 일부"라 정의한다.[7] 오늘날 주요 서방 국가에서 경제는 더 이상 산업 생산이 아닌 FIRE(금융, 보험, 부동산) 분야의 성장에 기반하여 구축된다는 사실을 명심해야 하며, 이러한 변화는 '경제의 금융화'로 정의된다. 점차 정교해지는 금융상품들로 인해, 주택은 단순히 상품의 교환 경제를 넘어, 이반 어셔가 명명한 '담보자산 포트폴리오 사회'로 편입되고 있다.[8]

이처럼 주택을 자산으로 바꾸어 놓는 변화를 뒷받침하는 것은 무엇인가? 대처의 '공공임대구매권(Right-to-Buy)' 정책이 신 자유주의 주택 보유의 상징적 정책으로 기능했다면, 훨씬 광범위한 결과를 보인 것은 아마도 위 정책의 미국 버전인 연방주택대출모기지, 별칭 프레디 맥일 것이다. 프레디 맥은 모기지를 시장에서 거래될 수 있는 금융 자산으로 전환시켜 증권화하고자 했다. 구매자가 아무도 없을 경우 프레디 맥은 이 모기지를 만기까지 보유할 수 있었기 때문에 해당 정책은 모기지에 투자할 의향이 있는 대출기관의 풀을 크게 확장시켰다.[9] 당시 투자자들은 마음이 변했거나 상황이 바뀌었을 경우, 그들 자산을 기꺼이 구매하려는 구매자를 찾을 수 있을 것이라고 믿으며 이와 같은 모기지담보증권을 구매할 수 있었다. 모기지담보증권은 특히 엄격한 위험 요건에 부합해야 하는 연금기금 등 기관투자자들에게 상당히 매력적이었다. '모든 사람'들은 모기지를 지불하며 만일 지불할 수 없는 예상 밖의 경우에는 기초 부동산 자산에 대한 담보권이 실행되어 초기 투자금을 만회할 수 있다.

Housing as Asset: The financialization of home ownership

Though we may take the rights of property for granted today, its unconditional status on an international arena is a relatively recent development. In the late 1940s, under the pressure of emancipatory social struggle which was thriving both in Europe and United States, capitalists and state policy makers were seeking measures to protect private capital rights on a global scale. The mechanism of home ownership was central to this project of expanding capital's influence in the 1980's through the increasing financialization of the economy. Though we may think of 'owning' one's own home, the reality is more often that of owning a mortgage. The mortgage, more than an instrument through which to take on debt, is now a means through which the home becomes embedded in global flows of financial capital. A process which Saskia Sassen identifies as "part of the current systemic deepening of capitalist relations, as is the financialization of mortgages for modest-income households aimed at building a new circuit for high finance for the benefit of investors and a total disregard for the homeowners involved."[7] We have to keep in mind that today in major western countries, economies are no longer built on industrial production, but in the growth of the FIRE (Finance, Insurance, Real Estate) sectors, a transformation that is defined as the 'financialization' of the economy. Through increasingly sophisticated financial instruments, housing enters the economy beyond the mere exchange of commodities and into what Ivan Ascher terms a "portfolio society of securitized assets."[8]

What underpins this transformation of housing into assets? If Thatcher's policy of a 'Right to Buy' has served as the emblematic policy of neoliberal home ownership, it is perhaps its American counterpart, the Federal Home Loan Mortage (nicknamed Freddie Mac) that holds more far reaching consequences. Freddie Mac was meant to 'securitize' mortgages by transforming them into financial asset that could be traded on the market. This policy greatly expanded the pool of lenders willing to invest in mortgages, as Freddie Mac was able to hold these mortgages to maturity if no buyer was found.[9] Investors could then buy these mortgage-backed securities in

7   Saskia Sassen, "Expanding the Terrain for Global Capital When Local Housing Becomes an Electronic Instrument," in Manuel B. Aalbers (Eds.), *Subprime Cities: The Political Economy of Mortgage Markets*, Hoboken: Wiley-Blackwell, 2012, p. 78.
8   Ivan Ascher, *Portfolio Society: On the Capitalist Mode of Prediction*, New York: Zone Books, 2016.
9   Ascher, p. 74.

7   Saskia Sassen, "Expanding the Terrain for Global Capital When Local Housing Becomes an Electronic Instrument," in Manuel B. Aalbers (Eds.), *Subprime Cities: The Political Economy of Mortgage Markets*, Hoboken: Wiley-Blackwell, 2012, p. 78.
8   Ivan Ascher, *Portfolio Society: On the Capitalist Mode of Prediction*, New York: Zone Books, 2016.
9   Ascher, p. 74.

주택담보대출 이자에 대한 세제혜택, 주택 보조금, 국가 보증 등을 통해, 주택 소유는 규제완화 및 경제 금융화의 진전으로 실행 가능해진 신자유주의 정책의 초석을 만들어 줄 것이다. 신자유주의 조건에서 주택 소유란, 주거 사다리를 오를 수 있을 만큼 충분히 운이 좋은 사람과 주택을 연결시켜 주는 일뿐 아니라, 그 주택을 자본흐름의 글로벌 네트워크에 자산으로 포함시키는 일이기도 하다. 2017년 유엔 적정 주거 특별 보고관이 발행한 주거의 금융화에 대한 보고서에는 다음과 같이 기술되어 있다. "글로벌 투자 및 산업화된 세계 경제에서 역사적인 구조 전환이 이루어지고 있으며, 주거가 그 중심에 있다. 이는 적정 주거를 필요로 하는 사람들에게 중대한 결과를 야기한다."[10] 오늘날 우리가 직면한 심각한 주택 위기는 많은 사람들을 희생시킨 몇몇 투기적 투자자들에 의해 초래된 결과일 뿐 아니라, 부동산의 가치를 우리 경제의 기반과 결부시키는 시스템적 문제에서 기인한 것이기도 하다.

대항수단1: 주택조합의 전망

국가와 시장이 실 거주용 주택 보급에 실패한 곳에 시민사회 주도형 움직임들이 더 효과적인 것으로 밝혀졌다. 주택협동조합과 같은 움직임들은, 주택의 투기 가치에 제한을 둠으로써 사적 주택 소유에 대한 대안을 제시하고 있다. 특히, 제한형 주택협동조합(Limited Equity Housing Co-op, LEHC)의 경우, 거주자들은 주택을 직접 소유하지는 않지만, 주거단지를 소유하고 있는 기업의 주식을 보유한다. LEHC는 협회 구성원들에 의해 소유지분의 가격이 사전에 결정된다는 점에서 다른 협동조합 방식과는 다르며, 이로 인해 부동산 시장 변동에 영향을 받지 않는다. 예를 들어, 비지분형 주택협동조합(Zero Equity Housing Cooperative)에서는 거주자들이 주거 공간을 떠날 경우, 그들이 처음 지불했던 금액을 전부 돌려 받을 수 있다. LEHC는 홈 클럽(Home Club)이라고 불리던 고급 주택 개발의 형태로 1920년대에 이미 미국에서 등장했지만, 시간이 지남에 그 범위가 점차 확장되면서 정치·사회적으로 진보적이 되었다.[11] 내규의 안정성을 보장하고, 특정 거주자들의 이해관계에 맞춰 내규가 변경되는 것을 방지하기 위해, LEHC는 주로 정부기관이나 토지신탁과 협력하여 자금을 제공받고

the confidence that in the event they changed their mind, or circumstances changed, they would be able to find a willing buyer for their assets. Mortgage backed securities proved extremely attractive, especially for institutional investors like pension funds who needed to comply with stringent risk demands. 'Everyone' pays their mortgage, and in the unlikely event they did not, the underlying real estate asset could be foreclosed on, recouping the value of the initial investment.

Trough tax incentives on mortgage interest, housing subsidies and state guarantees, home ownership would form a cornerstone of neoliberal policy enabled by increasing deregulation and the financialization of the economy. Home ownership in the neoliberal condition not only linked a home to someone lucky enough to climb the housing ladder, but also embedded that home as an asset in a global network of capital flows. Writing in 2017, the UN special rapporteur on adequate housing issued a report on the financialization of housing, stating that "Housing is at the centre of an historic structural transformation in global investment and the economies of the industrialized world with profound consequences for those in need of adequate housing."[10] The acute crisis of housing that we face today, is not just the result of a few speculative investors preying on the many, but also a systemic condition that has tied the appreciation of real estate assets to the very foundation of our economies.

Countermove 1: The Landscape of Housing Cooperatives

Where states and markets have failed to deliver housing for living in, civil society initiatives have proven more effective. Initiatives such as housing co-operatives present an alternative to private home ownership by placing limitations on the speculative value of housing. Particularly in forms like the Limited Equity Housing Co-op, where residents do not own their dwellings directly, but rather own a share of the corporation that owns the housing complex they inhabit. LEHC's differ from other cooperative arrangements because the price of the ownership shares is determined beforehand by the members of the associations themselves, and is therefore not influenced by the fluctuations of the real estate market. For example, in a Zero Equity Housing Cooperative, if residents leave, they are allowed to receive back the entire sum

10 United Nations, General Assembly, *Report of the Special Rapporteur on adequate housing as a component of the right to an adequate standard of living, and on the right to non-discrimination in this context,* A/HRC/34/51 (18 January 2017), available from https://undocs.org/A/HRC/34/51
11 James DeFilippis, *Unmaking Goliath: Community Control in the Face of Global Capital,* New York: Routledge, 2004.

10 United Nations, General Assembly, *Report of the Special Rapporteur on adequate housing as a component of the right to an adequate standard of living, and on the right to non-discrimination in this context,* A/HRC/34/51 (18 January 2017), available from https://undocs.org/A/HRC/34/51

계약조건을 안정화시키려 노력하였다.

또 다른 예는 독일의 '공동주택 연합체 (Mietshäuser Syndikat)'이다. 이 단체는 1992년에 설립되었으며, 목적은 현존하는 소유권의 자본주의적 형태들을 재구성함으로써 공동체의 집합 재산을 실현하는 것이다.[12] 각 신디케이트 프로젝트의 소유 구조는 51대 49 비율의 두 개 주요 지분으로 나뉜다. 좀 더 구체적으로, 집세를 지불하는 주주들의 협동조합 구성원인 사용자/거주자들이 51%를, 신디케이트 회사가 49%를 보유하게 된다. 이러한 방식으로 거주자들은 재산의 개발 및 관리에 관한 의사결정 과정에서 자율성을 유지하는 한편, 신디케이트의 참여를 통해 착취 및 사유화의 시도를 방지할 수 있다. 따라서, 토지와 부지 사용을 사회적 통제의 네트워크 내로 제한하고, 장기간 스스로 관리되는 독립체로 확보할 수 있다.[13]

주택을 시장에서 철수시키고, 사회적 권리로 환원되도록 하는 또 다른 중요한 예시는, 공동체 토지 신탁(Community Land Trust, CLT) 모델이다. 로버트 스완에 의해 1969년 미국에서 소개된 CLT 모델은, 농지에 대한 투기성 개발을 방지하기 위한 방법으로 고안되었다. 또한 이 모델은 1980년대부터 할당(allocation), 지속(continuity), 교환(exchange)의 과정을 보호할 목적으로 도시 지역에서도 시행되고 있다.[14] CLT 모델은 토지 재산과 건물 재산의 분리를 기반으로 한다. 이 두 재산이 분리되면 토지는 영구적으로 CLT의 소유가 되기 때문에, 토지 투기를 방지하고 장기 토지 임대로 주택을 구매할 수 있게 된다. 오늘날, CLT 모델은 유럽 국가뿐 아니라, 영어권 지역 내에서 점점 더 일반적이 되고 있다. 본 모델은 민간 개발자의 개입 없이 공유지를 조직화하기 위한 몇 가지 실행 가능한 대안 중 하나로 대표된다. 벨기에를 비롯한 많은 유럽 국가에서, 본 토지 취득 모델은 여전히 초기단계로 대부분 공공보조금 지원을 받고 있다. 그렇기 때문에 주로 저소득 계층을 대상으로 한다. 하지만, 미국 버전은 훨씬 선진화 되어, 모든 이를 대상으로 주거뿐만 아니라, 비상 대피소, 특별 주거, 최저소득 계층을 위한 SRO(single room occupancy) 등 다수의 공용 시설을 제공하고 있다.[15] 스위스에서, 주택조합 운동의 성공은 한편으로는

they initially paid. LEHCs were already present in America during the 1920s as luxury developments called Home Clubs, but their scope evolved through time to become more politically and socially progressive.[11] In order to ensure the stability of bylaws, and to prevent them from being changed for the gain of some residents over others, LEHC's usually collaborate with government agencies or land trusts to receive funding and to stabilize contract conditions.

Another example is the German *Mietshäuser Syndikat* or Syndicate of Tenements; founded in 1992, its aim is that of appropriating and reprogramming existing capitalist forms of ownership, implementing a collective property of commons.[12] The ownership scheme of each syndicate project is divided into two main shares of 51 and 49 percent respectively, owned by the users/inhabitants who form a cooperative of rent paying stakeholders and the syndicate company. In this way, the residents retain autonomy over the decision-making processes regarding development and the administration of property, while participation of the syndicate prevents attempts at exploitation or privatization. Land and premises are therefore ring-fenced within a network of social control and secured long term as self-administered entities.[13]

Another important contemporary example of withdrawing housing from the market and reclaiming as a social right is the Community Land Trust model. Introduced in the United States by Robert Swann in 1969, the CLT model was conceived as a way to stem speculative development on farmlands. Since the 1980s, the CLT model has also been implemented in urban areas with the aim of protecting processes of allocation, continuity and exchange.[14] The CLT model was based on the separation of land property from that of building, after which the land was to be retained by the CLT in perpetuity, thereby preventing land speculation while allowing housing units to be purchased with long-term ground leases. Nowadays, the model of the CLT is becoming more and more common in English speaking areas, as well as in other European countries. This model represents one of the few viable alternatives for the organizing of public

12 Dogma and Realism Working Group, *Communal Villa: Production and Re-production in Artists' Housing*, Berlin: Wohnungsfrage, 2018, pp. 82-83.
13 Ibid.
14 Robert S. Swann, Shimon Gottschalk, Erick S. Hansch, Edward Webster, *The Community Land Trust – A Guide to a New Model for Land Tenure in America, Center for Community Economic Development*, Cambridge, 1972.
15 James DeFilippis, *Unmaking Goliath: Community Control in the Face of Global Capital*, New York: Routledge, 2004.

11 James DeFilippis, *Unmaking Goliath: Community Control in the Face of Global Capital*, New York: Routledge, 2004.
12 Dogma and Realism Working Group, *Communal Villa: Production and Re-production in Artists' Housing*, Berlin: Wohnungsfrage, 2018, pp. 82-83.
13 Ibid.
14 Robert S. Swann, Shimon Gottschalk, Erick S. Hansch, Edward Webster, *The Community Land Trust—A Guide to a New Model for Land Tenure in America*, Cambridge: Center for Community Economic Development, 1972.

호의적인 입법에 의해, 다른 한편으로는 금융지원 및 가용토지로 가능했다. 런던 시청은 최근 몇 년간 대안적 주택 프로젝트 실현을 위해, 영세 개발업자들과 주택 협회, 지역사회 주도형 기관 등이 장기임대계약을 통해 공공기관 소유의 작은 부지들을 활용하도록 하는 몇몇 프로그램에 착수하여 지원하고 있다.

오늘날 부동산 가격 상승과 지속적인 금융 헤게모니 확장으로 인해, 주택 관련 투기 억제 전략을 수립하고 지원하는 것이 점차 어려워지고 있다. 이러한 상황에서, 오늘날 협동 건설로의 진입은 신자유주의 금융화 환경에서 재평가가 되어야 한다. 개개인이 자신의 건물 건설 자금 조달을 위해 스스로 조직화한 독일의 바우크루펜(Baugruppen)과 핀란드의 뤼흐매라켄타미넨(Ryhmärakentaminen) 등 협동 건설 이니셔티브들은 주거 관련 특정 시장의 메커니즘을 우회하는 데 성공한 것처럼 보인다. 협동 건설을 통해 개개인이 스스로 개발자의 역할을 효과적으로 수행함으로써, 개발 과정에서 중개업자의 역할은 사라지고 시장가보다 낮은 가격의 주거 공간이 마련된다. 이러한 과정은 대안적 형태의 주거 건설을 가능하게 하고 주거 마련 시 거주자들에게 재량권을 부여하지만, 동시에 신용과 자산에 접근이 가능한 사람들에게만 한정되어 있다. 협동 건설 자산은 본질적으로 개별 벤처기업처럼 운용되기 때문에, 필연적으로 일종의 기업가적 분석의 대상이 될 수 밖에 없으며, 투자에 대한 큰 수익 혹은 임대료가 급등하는 시기의 안정적인 주거 비용을 약속하게 된다.

투기적 성향에 대응할 장치가 없다면, 협동 건설 벤처기업들은 주거의 금융화된 상태를 더 고착화시킬 뿐이다. 현대 주거 환경의 개요를 설명하면서, 우리는 주택구입능력에 대한 동시대적 접근방식에 도전하길 희망한다. 또한, 개별적인 대응 차원을 넘어 오늘날의 주거 관련 도전 과제에 적극적으로 맞서는 집산화 된 운동을 요구하는 전세계적 시스템 문제 앞에서 우리가 문제 해결을 위해 동원하고 조직화할 수 있는 방법들을 모색해 나가길 희망한다.

대항수단 2: 새로운 건설 협동 조합 모델 모색
부동산 가치 상승은 점점 더 추상화되는 금융수단을 기반으로 하고 있지만, 필연적으로 바닥, 벽, 지붕 등의 매우 구체적인 현실과도 결부되어 있다. 지난 수 년 간 건설업은 지속가능성과 건설 속도, 기술·구조적 성능 측면에서 일련의 변화를 겪어왔다. 이와 같은 변화들은 건물의 성능을 향상시키는 방향으로 추진되어 왔으나, 건물의 복잡성 또한 증가시키고 있다. 동시에,

lands without interference from private developers. While in many European countries, including Belgium, this model of land acquisition is still in its initial phase, largely supported by public subsidies and therefore mainly addressing low-income populations. However, its American version is much more advanced and able to provide—in addition to housing for everyone—a multitude of communal facilities such as emergency shelters, special needs housing, and single room occupancy (SRO) housing for individuals with very low incomes.[15] In Switzerland, the success of the housing co-operative movement has been supported and made possible on the one hand by favourable legislation and on the other, by financial assistance and land availability. In London, the municipality has initiated and supported several programs in recent years that enable small plots of land in the hand of public authorities to be made available for small developers, housing associations and community-led organizations through long-term lease contracts for the construction of alternative housing projects.

Today, as property prices rise, and the hegemony of finance continues to expand, it becomes increasingly difficult to acquire and finance a counter speculative strategy for housing. In this light, contemporary forays into cooperative building should be reassessed in the neoliberal financialized condition. Cooperative building initiatives such as Baugruppen in Germany and Ryhmärakentaminen in Finland—where individuals self-organize to finance the construction of their buildings— appear successful in circumventing certain market mechanisms when it comes to housing. By effectively operating as their own developer, cooperative building cuts out the middleman in the development process, producing housing for below market cost. While this process allows for the construction of alternative forms of housing and offers a degree of agency for residents in the arrangement of their domestic life, it is also limited to those with access to credit and capital. Because they operate essentially as individual business ventures, cooperative building assets are inevitably subject to some form of entrepreneurial calculus, promising either great returns on investment, or at least stable housing costs amidst skyrocketing rents.

Without a mechanism to counter the speculative tendency, cooperative housing ventures risk only further entrenching the financialized condition of housing. In elaborating the contours

15 James DeFilippis, *Unmaking Goliath: Community Control in the Face of Global Capital*, New York: Routledge, 2004.

기성콘크리트와 구조용집성판(CLT) 등의 발전으로 건축구성제가 현장에 도착한 후 단 몇 일 또는 몇 시간 만에 조립이 가능해졌으며, 이로 인해 건설 과정에 소요되는 시간이 크게 줄어들었다. 결과적으로 현장에 필요한 인력이 줄어 전반적인 비용이 감축되었을 뿐 아니라, 장지간(long-span) 및 줄어든 자재 두께로 경제적 타당성 또한 증가하였다.

이러한 변화를 발전의 지표 정도로 가볍게 취급해서는 안 된다. 사실, 건물의 정교함과 복잡성의 고도화는 제한된 수의 건설 사로 인한 건설 산업의 집적화 및 독점 심화와도 관련이 있다. 현장관리 및 건설, 건물조립에 대한 접근방식의 변화로 건설이 단순화 됨으로써, 자본투자가 효율적으로 이루어져 건설 인건비는 줄어들었다. 그러나 이와 같은 변화를 무비판적으로 받아들여서는 안 된다. 사실 이러한 변화는 건설 노동력과 건축가 간 미약한 관계의 오랜 잔재를 반영하고 있다. 필리포 브로넬레스키가 14세기 플로렌스에서 건설 길드의 집합적 권력를 약화시킨 이래,[16] 건축의 학문적 위상(disciplinary status)은 어떤 형태로든 단순히 무엇을 짓는 행위로부터 벗어난다는 사실에 기초하여 정립되었다. 현대 시공법이 점차 조립식의 표준화된 구성재에 의존하게 되면서, 건설회사들은 건축업자라기 보다는 노동력과 자본의 조정자 역할을 수행하고 있다.

이 같은 업계 상황에서, 닉 스르니첵이 '플랫폼 자본주의'라고 명명한 현상으로의 전환은 고려해볼 가치가 있다. 우버나 에어비엔비와 같은 수백만 달러 규모 회사들은 사실 고정자산이라고 부를 만한 것이 별로 없다. 대신 이 기업들은 두 개 이상의 그룹이 상호작용할 수 있는 '플랫폼'으로 중개자 역할을 수행하고 있다.[17] 건축업자들이 곧 개발업자가 되는 (혹은 그 반대의) 경우가 점점 더 많아지면서, 이들은 상품을 사용자와 연결시키기 위해 자체적인 마케팅 플랫폼을 구축하기 시작했다. 이와 같은 측면에서 건설 과정은 우버나 아마존 메커니컬 터크 등과 유사한 형태로, 무언가를 생산하는 산업모델이라기보다는 자본집약적 기반시설을 교환가능 노동력과 연결시켜주는 플랫폼 역할을 하고 있다. 건설 산업의 플랫폼화가 가속되는 현상은 하나의 목표로 수렴된다. 바로 건설 환경의 독점화이다. 그렇다면 우리는 그들보다 먼저 목표점에 도달해야 할 것이다.

경제의 플랫폼화는 이 분야에 협동조합 플랫폼을

of the contemporary housing condition, we hope to challenge contemporary approaches to housing affordability, and in turn, begin to consider ways to mobilize and organize to confront what has by now become a globally systemic condition that calls for more than individualized responses, but a collectivized movement that actively contests the challenges of housing today.

Countermove 2: Towards a New Model of Building Cooperative

If the appreciation of real estate value has been built on increasingly abstract layers of financial instruments, it is also tied inevitably to a very concrete reality of floors, walls and roofs. Over the years, the construction industry has undergone a series of transformations undertaken in terms of sustainability, speed of construction, and technical and structural performance. Where these transformations claim to improve performance, they have also increased the complexity of buildings. At the same time, advances in prefabricated concrete and cross laminated timber have greatly sped up the construction process, with building components arriving on site and requiring only a few days if not hours to assemble together. Such developments have reduced overall costs by limiting the amount of on-site labour, while also increasing the economic feasibility of long spans and reduced material thicknesses.

Such developments should not be taken lightly as indicators of progress. Indeed, the increasing sophistication and complexity of building has gone hand in hand with the agglomeration and increasing monopolization of the building industry by a limited number of players. Transformations in approaches to site management, construction and building assemblies have served to derive efficiencies from capital investment by de-skilling construction to reduce labour costs in building. These transformations cannot be taken uncritically, and indeed reflect a long legacy of architects' tenuous relationship with construction labour. From Filippo Brunelleschi's undermining the collective power of building guilds in fourteenth century Florence,[16] the disciplinary status of architecture has in some form always rested on its emancipation from the mere act of building. Since contemporary construction methods increasingly relies on prefabricated and standardized components, construction companies are less builders than coordinators of labour and capital.

16 Pier Vittorio Aureli, "Do You Remember Counterrevolution?: The Politics of Filippo Brunelleschi's Syntactic Architecture," *AA Files*, no. 71, 2015, pp. 147-165.
17 Nick Srnicek, *Platform Capitalism*, Cambridge: UK Polity Press 2017.

16 Pier Vittorio Aureli, "Do You Remember Counterrevolution?: The Politics of Filippo Brunelleschi's Syntactic Architecture," *AA Files*, no. 71, 2015, pp. 147-165.

만드는 것을 고려해 볼 수 있는 기회다. 고립된 협동조합 이니셔티브 이상의 협동조합 플랫폼은 건설기업들은 불가능한 수준의 노동력과 자본흐름을 조율할 수 있다. 협동조합 플랫폼은 적극적으로 참여할 동기가 있는 노동력 풀, 즉 자신의 집을 짓는 노동자들을 동원할 수 있다. 자본집약적 활동으로의 변화가 건설 산업을 크게 변화시킨 것을 고려하면, 노동집약적인 활동으로의 변화 역시 마찬가지일 것이다. 내구성과 오랜 수명, 공간 구성으로 여전히 가치를 인정 받는 건설의 대안적 형식들은, 건설 기업들이 고려하는 시장 수요와 보증 밖에서 재도입될 수 있다. 예를 들어, 조립발포제를 사용해 건물 표면에만 벽돌을 붙이고 콘크리트와 강철로 구성된 현대의 '벽돌' 건물이 보통 30년에서 50년 정도 유지되는 것과는 다르게, 견고한 벽돌담은 관리만 제대로 이루어진다면 영구적으로 유지될 수 있다.[18]

건설 협동조합 플랫폼은 다수의 방법을 통해 주택 투기 상황에 관여할 수 있다. 금융 측면에서 이 플랫폼은 주거 확장 자금을 조달하기 위해 공동 소유의 자산을 동원하고 끌어 모을 수 있으며 이는 투기 시장에서 제외될 수 있다. 또한 특정 규모에서, 건설 협동조합 플랫폼은 개별협동조합 및 신용이나 자본 이용이 어려운 사람들에게 여전히 장애물로 남아있는 신용도 구축을 위해, 모기지 담보증권의 매력을 활용할 수 있다. 정치적 측면에서, 협동조합 플랫폼은 정부를 압박하는 활동가라기보다 '투자자'로 자리매김하고 있다. 미셸 페어가 언급했듯, "소위 공유 경제를 장악할 자본주의 기업과 경쟁하기 위해, 협력적 인터페이스는 반드시 실질적 피투자자, 즉 투자가치가 있는 프로젝트로 구성되어야 한다."[19] 협동조합 플랫폼은 현재 소셜 하우징을 건설하고 자금을 조달할 능력이 없는 지자체들로부터 인센티브를 확보하기 위해 노력하고 있으며, 과거 개발업자들은 이러한 인센티브를 이미 잘 활용해 왔다.

점차 감당이 안 되는 주거 비용으로, 도시를 지탱해온 도시에서의 삶은 점점 더 고갈되고 사람이 살지 않는 투기 자산들로 가득 찬 '좀비 도시'로 바뀌고 있다.[20] 한 때 안정적 주거 확보를 위한 투쟁 운동으로 조직화된 협동 건설은, 자체적으로 소셜 하우징을 건설하고 자금을 조달할 능력이 없는 지자체들과 협상하면서, 도시와 정부의 매력적인 파트너가 되고 있다.

Against this incumbent industry, it is then worth considering the transformations towards what Nick Srnicek terms 'platform capitalism.' Multi-billion dollar companies like Uber or Airbnb actually have little in terms of fixed assets, but as 'platforms' these companies operate as intermediaries that enable two or more groups to interact.[17] As builders are increasingly developers and vice versa, they have begun to create their own marketing platforms to connect their products with users. In this sense, the construction process behaves more as a platform than as an industrial model of production, connecting capital-intensive infrastructure with an interchangeable labour force, a process similar to Uber or Amazon Mechanical Turk. The growing platformitization of the industry points towards one ambition: monopolization of the built environment. What we need then is to get there before they do.

The increasing platformitization of economy can be an opportunity to consider a cooperative platform entrant into the sector. More than isolated cooperative initiatives, a cooperative platform can coordinate labour and capital flows at a level that construction companies cannot. Cooperative platforms can mobilize a pool of engaged and motivated labour, of workers building their own homes. Where the shift to more capital-intensive activities has radically changed the building industry, so too would a shift towards more labour intensive activities. Alternative forms of construction, still valued for its durability, longevity and spatial organization can be re-introduced outside the market demands and warranties considered by construction companies. Solid brick walls for example can essentially lasts forever with proper maintenance, unlike current 'brick' buildings which consist only of a surface cladding of bricks placed over top an assembly of foam, concrete and steel that are commonly rated for only 30-50 years.[18]

A cooperative platform for construction enables multiple avenues through which to engage the speculative condition of housing. At the level of finance, it can mobilize and pool cooperatively owned assets to finance the expansion of housing that can be removed from the speculative market. At a certain scale, it can leverage the attractiveness of mortgage-backed securities to establish the creditworthiness which remains an obstacle for individual cooperatives and those without access

18 "RDH Building Science | How Long Do Buildings Last?," 2019, *RDH Building Science*, https://www.rdh.com/blog/long-buildings-last/.
19 Michel Feher, *Rated Agency: Investee Politics in a Speculative Age*, New York: Zone Books, 2018, p. 191.
20 다음을 보라: Niklas Maak, *Living Complex: From Zombie City to the New Communal*, Munich: Hirmer, 2015.

17 Nick Srnicek, *Platform Capitalism*, Cambridge: UK Polity Press 2017.
18 "RDH Building Science | How Long Do Buildings Last?," 2019, *RDH Building Science*, https://www.rdh.com/blog/long-buildings-last/.

대항수단 3: 주거공간에 대한 도전

역사적으로, 서구사회에서 가족과 집을 동일시 하는 것은 공적 공간과 사적 공간의 공간적 구별에 상응하는 것이었다.[21] 고대 시대, 공적 공간은 정치 및 사회 생활을 위한 공간이었던 반면, 사적 공간은 여성과 노예의 정서 및 가사 노동에 의존하는 생물학적 재생산의 공간이었다. 18세기와 19세기 자본주의와 산업화의 부상과 함께, 남성 생계 부양자가 행하는 일과 전업주부의 가사노동을 명확히 구별하기 위해, 이 같은 공적·사적 구분이 핵가족 내에서 부활하였다.[22] 남성들의 일에는 보수가 지불되었으나, 전업주부의 가사노동은 무보수로 이루어졌고 응당 주부라면 가족을 위해 해야 할 '사랑의 노동'으로 간주되었다. 일과 가사의 구분으로 인해 현대 가정에서는, 존 우드 더 영거의 영향력 있는 노동계급 주택 모델 등 18세기 이래 주거공간을 개조하려는 많은 시도에서 볼 수 있듯이, 작업장이나 마구간 같은 근로 활동 구역은 집 밖으로 배치되었다.

이후, 주거공간과 작업공간의 분리는 상류 및 하류층 주택에서 모두 중요해졌다. 상류층에서는, 집에서 (육체)노동을 하는 모습을 보이는 것은 수치스러운 것으로 간주되었기 때문에 이와 같은 공간의 분리는 탁월함의 표식으로 선호되었다. 또한 하류층과 관련하여, 봉급 생활자는 고용자의 부지 내에서 일해야 하는 것으로 여겨졌기 때문에, 정부는 봉급 생활자가 집에서 일을 하는 것이 부적절하다고 생각하였다.

1980년대 이래 주택 사유화 및 모기지 산업의 등장으로, 보유 주택과 가족 간 관계는 더 끈끈해졌다. 이러한 상황에 대응하여 건축가들은 지난 30년 간 생활공간을 좀 더 전문화되고 유형학적으로 정의된 구조물로 변화시켜왔다. 많은 유럽 국가들에서 이와 같은 경향이 흔히 관찰되며, 주택의 사유화 및 금융화가 주택 디자인에 대한 형식주의적인 접근 방식을 파생시켰다. 이에 따라 각 아파트 세대의 개성을 강조하기 위해 건물의 정면과 평면 등 주택 특징에 변화를 가하는 현상이 심화되고 있다.

전통적인 가족 구성을 위한 집에 대한 도전은 주택 개발의 물결로 가정집의 목가적인 이미지가 약화된 19세기 뉴욕에서 시작되었다. 뉴욕은 하층계급을 위한 악명 높은 공동 주택 외에도, 아파트와 거주용 호텔의

to credit or capital. In terms of politics, cooperative platforms position themselves less as activists making demands on the government, than as 'investors' themselves. As Michel Feher notes, "in order to compete with capitalist enterprises that dominate the so-called collaborative economy, cooperative interfaces must constitute themselves as actual investees, that is, as projects worthy of investment."[19] Cooperative platforms can position themselves to secure incentives from municipalities who are currently powerless to build and finance their own social housing, incentives that developers have already exploited so well in the past.

In the face of growing housing unaffordability, cities are increasingly depleted of the urban life that sustains them and replaced with a 'zombie city' of vacant speculative assets.[20] Cooperative building — once organized as a movement fighting for access to stable housing—becomes an attractive partner for cities and governments, negotiating with municipalities who are currently powerless to build and finance social housing of their own.

Countermove 3: Challenging Domestic Space
Historically, in the Western world the identification of the house with the family corresponded to the spatial separation between public and private space.[21] In ancient times, public space was the space of political and social life, while private space was the space of biological reproduction whose functioning depended on the affective and domestic labour of women and slaves. With the rise of capitalism and industrialization in the 18th and 19th century, this public/private separation was resurrected within the nuclear family to clearly distinguish work performed by the male breadwinner from the domestic labour performed by the housewife.[22] While work was remunerated, domestic labour was unwaged and considered a 'labour of love' expected from the housewife for the sake of the family. Because of domestic division of labour, in the modern home, working activities like workshops or stables were expelled from the house, as is visible in many attempts to reform domestic space since the 18th century such as John Wood the Younger's influential models for homes for the labouring classes.

21 공공 공간과 개인 공간의 분리와 주거 공간에 미치는 영향에 대해서는 다음을 보라:
Michael McKeon, *The Secret History of Domesticity: Public and Private Division of Knowledge*, Baltimore: John Hopkins University Press, 2005.
22 다음을 보라: Maria Mies, *Patriarchy and Accumulation on a World Scale: Women in the International Division of Labor*, London: Zed Books, 2014, pp. 74-75.

19 Feher, *Rated Agency*, p. 191
20 See: Niklas Maak, *Living Complex: From Zombie City to the New Communal*, Munich: Hirmer, 2015.
21 On the separation between public and private space and its consequences on domestic space see: Michael McKeon, *The Secret History of Domesticity: Public and Private Division of Knowledge*, Baltimore: Johns Hopkins University Press, 2005.
22 See: Maria Mies, *Patriarchy and Accumulation on a World Scale: Women in the International Division of Labor*, London: Zed Books, 2014, pp. 74-75.

도시였다. 높은 토지 비용으로 단독주택을 감당할
여력이 안되는 중산층 이민자들을 뉴욕에 수용하기 위해,
개발업자들은 역사상 처음으로 '중산층 아파트'를 지었다.
아파트라는 단어의 기원은 라틴어의 'appartare'이며,
'분리하다' 뿐 아니라 '나누어서 배분하다'라는 의미를
지니고 있다. 이와 같은 유형의 주택은 르네상스 시기에
한 명씩 사용하는 방이 연이어 있는 형태로 상류층 주거
건축에서 처음 시도되었다. 이와 같은 형태의 주거 공간이
19세기 뉴욕에서 처음 소개되었을 당시, 주거용 건물을
지칭하는 데 있어 '호텔'과 '아파트'라는 단어의 교체
사용이 가능할 정도로 아파트는 호텔 생활과 결부되었다.
하지만, 엘리자베스 콜린스 크롬리가 언급했듯, 당대의
작가들은 그러한 용어 사용에 동의하지 않았으며, 다가구
주택으로 구성된 건물을 지칭하기 위해 '프렌치 플랫'이나
'아파트 호텔' 등 다양한 단어를 사용했다.[23] 아파트가
영원히 거주할 장소로 여겨진 반면, 아파트 호텔은 보다
단기 혹은 임시 거주를 목적으로 하였다.

하지만 유형학적으로, 이 두가지 유형의 건물은
동일하지는 않아도 상당히 유사했다. 두 가지 유형 모두
로비, 식당, 응접실, 옥상 정원 등 공용 공간을 제공했을 뿐
아니라, 공동 주방 사용과 하우스키핑 등 공동 서비스도
제공하였다. 아파트와 호텔 모두 상업적으로 개발
되었지만, 전문화된 가사노동을 기반으로 하는 집합적
성격 및 라이프 스타일은, 무보수 가사노동과 가족에 대한
완전한 사생활 보장 등 전통적 가정의 가치관에 도전장을
내밀었다.

호텔 생활은 미국에서 성행하였다. 처음에는
호텔 이용이 부유층에 한정적이었으나, 19세기 말과
20세기 초반 사이에 '중간 가격 호텔,' '아파트 호텔',
'루밍 하우스'의 형태로 점차 중산층 및 하층 계급으로
확장되었다.[24] 형태학적으로, 거주용 호텔, 특히 중간
가격의 거주용 호텔은 가장 반가정적인 숙소였다. 이
유형은 공용으로 사용하는 로비와 식당, 거의 같은 크기의
독신자 용 방 등으로 구성되었다. 당시 호텔의 경우,
계층과 인종을 차별하기 위한 엄격한 규정이 있었지만, 이
유형에서는 방의 크기와 형태가 입주자의 성별과는 거의
무관했다.

저렴한 호텔 숙소와 개인 거주 아파트의 흥미로운
절충안이 '일인용 레지던스 호텔(SRO)'이었다. 이
유형은 빽빽이 들어선 작은 방들로 구성되어 있으며,
공용 공간으로 욕실과 작은 부엌 등이 설치된 경우도

23 Elizabeth Collins Cromley, *Alone Together. A History of New York's
Early Apartments*, Ithaca: Cornell University Press, 1990, pp.5-6.
24 Paul Groth, *Living Downtown, The History of Residential Hotels in
the United States*, Los Angeles: University of California Press, 1994,
p. 38.

Since then, the separation of domestic space
and workplace became crucial for the homes
of both upper and lower classes. For the upper
classes, the separation was desired as a mark of
distinction, since it was considered degrading to
be seen in one's own house in conjunction with
(manual) work. For the lower classes, governments
considered it inappropriate for salaried workers to
work at home, since they were expected to work in
their employer's premises.

The privatization of the housing sector since
the 1980s and the rise of the mortgage industry
have further reinforced the relationship between
home property and family household. Responding
to this condition, in the last thirty years, architects
have turned living space into an increasingly
specialized and typologically defined construct.
This tendency is fully visible in many European
countries in which privatization and financialization
of the home has given rise to an increasing
formalistic approach to housing design where
manipulation of housing features like facades and
floor plans is intensified in order to emphasize the
uniqueness of each apartment unit.

A challenge to the house as a family home
came from 19th century New York, where a wave
of housing developments undermined the pastoral
image of family dwelling. Apart from the infamous
tenements for lower classes, New York was also
the city of apartments and residential hotels. To
accommodate middle class immigrants to the
city who could not afford single-family houses
due the high cost of land, developers built the
first 'middle class apartments' in history. The
word apartment comes from the Latin *appartare*,
which means 'to separate' but also 'to distribute
in parts.' This typology of dwelling developed first
within upper class residential architecture during
the Renaissance as a set of rooms for the use of a
single resident. When this domestic arrangement
arrived in New York in the 19th century, it was
immediately associated with hotel life to the point
that in designating residential buildings the words
'hotel' and 'apartment' became interchangeable. As
noted by Elizabeth Collins Cromley, writers of the
period did not agree on terminology and they used
a variety of terms to address buildings that were
made of multiple households such as 'French flat'
and 'apartment hotel.'[23] While apartments were
meant to be permanent homes, apartment hotels
were more for transient living. Yet typologically,
these two types of buildings were very similar if
not the same. Both types offered collective spaces

23 Elizabeth Collins Cromley, *Alone Together. A History of New York's
Early Apartments*, Ithaca: Cornell University Press, 1990, pp.5-6.

있었다. 호텔 방과 거의 차이가 없는 SRO가 있는 반면, 침실이나 응접실 등 한 개 이상의 방이 제공되는 개인 거주 아파트와 더 흡사한 SRO도 있었다. 샌프란시스코 등 특정 도시에서는 이 유형의 주거가 광범위하게 확장되어, 필모어와 사우스 마켓 등의 지역에서는 거의 대부분의 주거형태가 SRO였을 정도였다.[25]

이와 같은 주거 개발은 처음에 상업적으로 촉발되었으나, 미국에 등장한 비 가정적 유형의 확산은 1800년대 미국에서 널리 유행한 사회주의 운동의 영향을 받았다.[26] 찰스 푸리에의 팔랑주 아이디어(성별에 상관없이 평등하게 공동 건물에서 거주하는 약 1,600명으로 구성된 사회적 단위)에서 영감을 받아, 완벽주의자 등 여러 운동 단체들은 평등과 연대에 기반한 삶의 방식을 제안했으며, 가정 영역의 사생활에 도전장을 내밀었다. 개인 주택에 대한 공격은 또한, 역사학자 돌로레스 헤이든이 '물질적 페미니스트(일반적인 가정에서 여성의 고립에 의문을 제기하는 여성)'라고 정의한 이들에 의해 계속 되었으며, 가사 노동의 사회화와 전문화를 바탕으로 한 급진적 가사 모델이 제시되었다.[27]

이러한 운동과 실용적 주거 개발이 융합된 대표적 사례가 1883년과 1890년 사이 뉴욕에서 지어진 홈 클럽(Home Club)이며, 그 중 첼시 호텔이 가장 잘 알려져 있다. '협동 조합 주택'의 창시자인 건축가 필립 휴버트에 의해 고안된 홈 클럽은 독신을 위한 부엌이 없는 아파트, 커플 및 가족을 위한 라운지, 회의실, 공동 부엌 등 여러 공유 공간을 갖춘 아파트 등 다양한 형태의 아파트를 제공하고 있다.[28] 홈 클럽은 낮은 집세로 가족 및 거주자들이 주거 비용을 충분히 감당할 수 있도록 지원하며, 거주자 간 공동으로 집안 살림을 하고 상호 지원할 수 있도록 돕는다. 홈 클럽은 공동 거주를 위한 새로운 공간 모델일 뿐 아니라, 공동 재산에 기반한 혁신적 경제 모델이기도 하다. 홈 클럽은 모든 방면으로 프로젝트에 참여한 주주들의 회사로 조직되었다.

유럽에서 협동 주거의 초기 예시는 에버니저 하워드의 협동 '사각모델'로, 에버니저 하워드는 런던 주위에 건설될 그의 새로운 '가든 도시(Garden Cities)' 개선 프로젝트의 일환으로 이를 제안하였다.[29] 이 아이디어는 그가 거주하던 래치워스 내

such as lobbies, dining rooms, parlours and roof gardens, but also collective services such as collective kitchens and housekeeping. Even though both apartments and hotels were commercial developments, their collective nature and their lifestyle based on professionalised domestic labour, challenged traditional domestic values such as unwaged domestic work and the absolute privacy of the family household.

Hotel life become widespread in US and while initially it was limited to wealthy classes, between the end of 19th century and beginning of 20th it was extended to middle and lower classes in the form of "mid-priced hotels," "apartment hotels," and "rooming houses."[24] Typologically the residential hotel—especially in its mid-priced version—was the most anti-domestic kind of lodging. It was comprised of rooms for single persons, more or less all of the same size supported by collective spaces like lobbies or restaurants. The size and type of the room was often indifferent to the gender of the occupant, even though hotels had a strict policy of class and race discrimination.

An interesting compromise between cheap hotel accommodation and the private apartment were the 'single room occupancy' houses also known as SRO's. This typology consisted of a beehive of small rooms, sometimes equipped with bathroom and a kitchenette supported by collective spaces. In some cases SRO accommodations were indistinguishable from hotel rooms, in other cases they were more similar to private apartments, especially when they included more than one room like a bedroom and a parlour. In certain cities like San Francisco, this typology spread extensively, to the point that entire neighbourhoods such as Fillmore and South Market were made almost exclusively of this typology.[25]

All these housing developments were commercially initiated, yet the spread of non-domestic typologies in the US was also influenced by communitarian movements that became very widespread in 1800s United States.[26] Inspired by Charles Fourier's idea of the Phalanges—a social unit consisting of circa 1.600 people inhabiting a common building in egalitarian manner regardless of gender—movements such as the Perfectionists proposed a way of life based on equality and solidarity which challenged the privacy of the

25 Ibid. p. 60.
26 See: Dolores Hayden, *Seven American Utopias. The Architecture of Communitarian Socialism*, Cambridge MA.: The MIT Press, 1976.
27 See: Dolores Hayden, *Grand Domestic Revolution*, Cambridge, MA.: The MIT Press, 1976.
28 Ibid. p. 65.
29 See: Norbert Schoenauer, 'Early European Collective Habitation: From Utopia to Reality' in Karen A. Franck and Sherry Ahrentzen, *New Households, New Housing*, New York: Van Nostrand Reinold, 1991, pp. 50-53.

24 Paul Groth, *Living Downtown, The History of Residential Hotels in the United States*, Los Angeles: University of California Press, 1994, p. 38.
25 Ibid. p. 60.
26 See: Dolores Hayden, *Seven American Utopias. The Architecture of Communitarian Socialism*, Cambridge, MA.: The MIT Press, 1976.

홈스가르스(Homesgarth)에서 시행되었다. '사각모델'은 해롤드 클래팸 렌더에 의해 디자인되었으며, 개인 주택의 완벽한 사생활 보호와 중앙 집중식 하우스키핑의 이점을 결합한 공동주택으로 탄생하였다. 사각모델의 도면은 기둥을 받쳐 만든 현관 지붕, 즉 포르티코 주변으로 위치한 개개의 주택을 수도원의 회랑과 유사한 방식으로 결합함으로써 주택의 개인 및 집합적 측면을 모두 반영하였다. 대규모 식당, 탁아소와 같은 공동 시설은 포르티코에서 바로 접근이 가능했다. 하지만 이 모델은, 영국의 첫 사회주택법의 유형적 근거가 되길 기대하며, 배리 파커와 레이몬드 언윈의 래치워스 가든 도시 계획에서는 더 이상 시행되지 않았다. 래치워스 가든 도시는 대신 단독 가정 오두막으로 구성되었다. 20세기 초 유럽에서 공동체 생활의 이행이 어려웠음에도 불구하고, 조합 주택과 중앙 집중형 하우스키핑을 결합한 수많은 실험이 시행됐고 구체적인 이니셔티브들이 등장했다. 대표적인 예가 독일의 부엌이 한 개 딸린 아인쿠헨하우스(Einkuchenhaus), 스웨덴의 컬렉티브후스(Kollektivus), 가장 급진적 형태의 공동주택인 소련의 돔커뮤나(Dom Kommuna) 등이었다.[30] 특히 돔커뮤나는 공동 주거와 사회화된 가사 노동의 가장 극단적인 형태로 간주되지만, 몇 몇 경우를 제외하고는, 대규모로 시행되지는 못했다.[31]

20세기 중, 많은 조합 주택 이니셔티브 및 집단 호텔 거주 유형은, '진보시대'의 미국 사회 개혁가 등 주거를 담당하는 정부 기관의 공격을 받았다. 이러한 개혁가들에게 조합 및 공동 주거는 가족 중심의 가치관과 재산에 대한 심각한 위험이었다.[32] 소셜 하우징과 합리적 가격의 주택이 처음 등장하면서, 협동조합 주택은 미미한 현상으로 치부되었다. 독일, 스위스, 덴마크 등에서는 협동조합 주택이 살아남았으나, 주택의 협동조합적 성격이 유형적 조직에 반영된 경우는 상당히 드물다.[33]

1970년대와 1980년대 사이 스웨덴에서 협동조합 및 공동주택이 대두되면서 이와 같은 트렌드에 예외가 나타났다. 사회민주주의여성 운동 및 다른 페미니스트 단체의 압력 하에, 몇몇 조합 주택 이니셔티브가 여성을 위한 가사노동 감축 이슈를 다뤘다. 앨리슨 우드워드가 언급했듯, 스웨덴 여성 운동은 1970년대 여성 건축가와

domestic realm. The attack on the private home was also pursued by what historian Dolores Hayden has defined as 'material feminists' who questioned the isolation of women with the average family home, and proposed radical models of domestic arrangements based on the socialization and professionalization of domestic labour.[27]

A clear example of the confluence of these movements and more pragmatic housing developments were the Home Clubs built in New York between 1883 and 1890 of which the Chelsea Hotel is the most well-known example. Conceived by the architect Philip G. Hubert, the inventor of 'cooperative housing,' Home Clubs offered a variety of apartments without kitchens for singles, couples and families supported by a wealth of collective spaces such as lounges, meeting rooms, and collective kitchen.[28] These clubs would allow families and groups of residents to afford housing at a lower rent while taking advantage of collective housekeeping and mutual support among the residents. Besides introducing a new spatial model for collective living, the Home Clubs were also an innovative economic model based on common property in a joint stock scheme. The Home Club was organized as a corporation of shareholders taking part in all aspects of the project.

An early example of cooperative housing in Europe was Ebenezer Howard's cooperative 'quadrangles' which he proposed as part of his reformist project for new 'Garden Cities' to be built around London.[29] Implemented at Homesgarth in Letchworth (where he also lived), the quadrangle was designed by Harold Clapham Lander as a communal house that combined the full privacy of the individual home and the benefits of centralised housekeeping. The plan of the quadrangle reflected the private and collective aspects of the house by aggregating individual houses around a portico in the manner of a monastic cloister. Communal facilities like a large dining hall and a child care centre were directly accessible from the portico. Yet this model was not further implemented in Barry Parker's and Raymond Unwin's planning for the Garden City of Letchworth which was instead made of single family cottages, anticipating the typological rationale of the first social housing act in England. Despite the difficulty of implementation of communal living in Europe at the beginning of the 20th century, there was a flourishing

30 Ibid. 55-67.
31 다음을 보라: Anatole Kopp, *Town and Revolution: Soviet Architecture and City Planning 1917-1935*, New York: George Braziller, 1970.
32 다음을 보라: Charles Hoc, Robert A. Slayton, *New Homeless and Old: Community and the Skid Row Hotel*, Philadelphia: Temple University Press, 1984, pp. 62–86.
33 다음을 보라: Kathrin M. McCamant, Charles R. Durret, "Cohousing in Denmark" in Karen A. Franck and Sherry Ahrentzen, *New Households, New Housing*, pp. 95-126.

27 See: Dolores Hayden, *Grand Domestic Revolution*, Cambridge, MA.: The MIT Press, 1976.
28 Ibid. p. 65.
29 See: Norbert Schoenauer, 'Early European Collective Habitation: From Utopia to Reality' in Karen A. Franck and Sherry Ahrentzen, *New Households, New Housing*, New York: Van Nostrand Reinold, 1991, pp. 50-53.

인테리어 디자이너, 저널리스트 등으로 이루어진 빅 컬렉티브 그룹의 활동을 통해 가장 효과적인 건축학적 표현을 발견하였으며, 세 가지 원칙(작은 규모의 지역사회, 투기 없는 주거, 연령 및 직업 등에서 다양한 세입자 구성)을 기반으로 매우 영향력 있는 공용 주택 개념을 구축하였다.[34] 흥미로운 것은 사회화된 가사노동을 기반으로 협동 체제를 추구하려는 가족 개혁의 시도가, 민간뿐만 아니라 공공 주거에서도 시행되었다는 점이다.

이러한 운동의 중요한 예시는, 스타켄(Stacken)으로 알려진 주거블록이다. 스타켄은 1969년 베르크츠존(Bergsjön) 텔레스코프가탄(Teleskopgatan)에 건설된 고층건물을 개보수하여 설립한 코 하우징(cohousing)으로, 스웨덴 밀리언 홈 프로젝트(Sweden Million Homes Project)의 일환으로 개발되었다. 고층 건물의 기준 층 평면은 중심부가 받치고 있는 다섯 개의 개별 아파트로 구성되어 있으며, 집단 시설은 존재하지 않는다. 몇 년의 사용 후, 아파트 보수를 통해 개인 방의 크기는 줄이고 일반 부엌은 작은 부엌으로 대체했으며, 공용 부엌과 식당 등 많은 공용 시설들이 추가되었다. 2002년, 사타켄 거주자들은 협동조합을 형성하여 시 당국으로부터 건물을 매입했다. 건물을 재단장하고 돌봄과 작업을 위한 더 많은 공용 공간을 마련함으로써 건물의 공용 구조를 강화하기 위해서였다.[35] 하지만, 이러한 사례에도 불구하고, 유럽이나 미국에 지어진 합리적인 가격의 많은 조합 주택들은 가족을 위한 전통적 주거 공간과 흡사하다. 몇 몇 공용 공간이 겨우 구색을 갖추고 있어 본래의 공용 아이디어를 상기시켜 줄 뿐이다.

최근 몇 년 간의 긴축 정책과 삶의 불안전성 증가로 인해 사회의 초석인 가정이 심각하게 위태로워지고 있다. 멜린다 쿠퍼가 강조했듯, 가족의 문제는 공공 투자 및 교육을 반대하는 신자유주의 논쟁의 중심이며, 민간 투자와 가계 빚에 의해 좌우되는 신경제 질서를 위해 제시하는 대안의 핵심이기도 하다.[36] 하지만, 2007~2008년 금융위기 이후 정부들이 도입한 최근의 긴축 조치로 인해, 새로운 세대 내 가족 형성이 심각하게 위협받고 있다. 이와 같은 상황에서, 유럽 도시 내 개발업자와 시 당국은 독신, 커플 또는 한 자녀 가정을 위한 거주 공간을 만들기 위해 아파트 크기를 급격히 줄일

of experiments and concrete initiatives that combined co-operative housing and centralised housekeeping. From the German *Einkuchenhaus* (one kitchen house), to the Swedish *Kollektivus*, to the most radical form of communal housing, the Soviet *Dom Kommuna*.[30] The latter example can be considered the most extreme form of communal living and socialised domestic labour, and yet apart from a few cases, this type of housing was not implemented at a large scale.[31]

During the 20th century, many co-operative housing initiatives and collective hotel living came under attack by governmental institutions responsible for housing such as the United States social reformers of the 'Progressive era.' For these reformers, co-operative and communal living was a serious threat to family values and home-property.[32] It was first the rise of social housing and that of affordable market housing that reduced cooperative housing to a marginal phenomenon. Where cooperative housing survived, like in Germany, Switzerland, and Denmark it is rare that the cooperative nature of the home was reflected in its typological organization.[33]

An exception to this trend happened in Sweden between the 1970s and 1980s when cooperative and communal housing went on the rise. Under the pressure of the Social Democratic Women's movement and other feminist groups, several co-operative housing initiatives addressed the issue of lessening housework for women. As noted by Alison Woodward the Swedish women's movement found its most effective architectural expression in the work of the BIG collective, a group of female architects, interiors designers and journalists who in the 1970s, developed a very influential concept of communal home based on three principles: small scale communities, speculation free housing and varied population of tenants in terms of age and profession.[34] What is interesting is that the attempt to reform the family household towards a more cooperative system based on socialised domestic work was implemented not only in private but also in public housing.

An important example of this movement, was

34 Alison Woodward, "Communal Housing in Sweden: A Remedy for the Stress of Everyday Life?." in Karen A. Franck and Sherry Ahrentzen, *New Households, New Housing*, p. 72.
35 다음을 보라: Karin Andersson, Caroline Glabik, Ellen Persson, Megan Prier, *Hej Stacken*, Project Portfolio, Design and Planning for Social Inclusion 2012/2013 Chalmers Architecture. https://suburbsdesign.files.wordpress.com
36 Melinda Cooper, *Family Values: Between Neoliberalism and the New Social Conservatism*, New York: Zone Books, 2017.

30 Ibid., pp. 55-67.
31 See: Anatole Kopp, *Town and Revolution: Soviet Architecture and City Planning 1917-1935*, New York: George Braziller, 1970.
32 See: Charles Hoc, Robert A. Slayton, *New Homeless and Old: Community and the Skid Row Hotel*, Philadelphia: Temple University Press, 1984, pp. 62–86.
33 See: Kathrin M. McCamant, Charles R. Durret, "Cohousing in Denmark," in Karen A. Franck and Sherry Ahrentzen, *New Households, New Housing*, pp. 95-126.
34 Alison Woodward, "Communal Housing in Sweden: A Remedy for the Stress of Everyday Life?," in Karen A. Franck and Sherry Ahrentzen, *New Households, New Housing*, p. 72.

필요성을 느끼고 있다.[37] 하지만 이러한 상황에서도, 각 주택 단위는 그 크기에 상관없이 자급자족할 수 있어야 하고 주택은 여전히 사적인 공간이어야 한다는 생각 속에 전통적인 가족상이 여전히 강력히 남아있다. 이에 따라 많은 개발업자들은 표준 크기보다 작은 '소형' 아파트를 추구하기 시작했으며, 표면적인 이유는 변화하는 인구 통계와 급증하는 주택 가격에 대응하기 위해서였다.

하지만, 전반적으로 '합리적인' 임대료를 약속한 사토(Sato)의 스튜디오 홈(StudioHome)과 같은 프로젝트에서는, 아파트 크기가 점점 작아지자 개발업자들이 평방미터 당 가격을 인상하고 있다.[38] 사토의 스튜디오 홈은 '공용 및 포괄적인 아파트 건물'로서의 공간 내 단점을, 공유 시설 설치 및 거주민 간 소통 확대를 전담하는 커뮤니티 관리자 운영으로 보강하도록 하고 있다. 웰리브(WeLive), 콜렉티브(Collective) 등의 민간 기업들도 이와 유사한 전략들을 이미 시행하고 있다. 이 기업들은 오래된 상업용 거주 호텔 유형과 유사하지만, 이와는 다르게 방 한 달 임대료가 3,000달러 정도인 숙소를, 집합 거주에 대한 과도한 미사여구를 사용해 광고하였다.[39] 이러한 회사들의 타깃은 일반적으로 젊은 전문직 종사자들과, 환영 받는 (동시에 착취당하는) '밀레니얼 세대'로, 장애인이나 노인 등 이러한 유형의 주거가 가장 절실한 사람들은 배제하였다.

위에서 짧게 살펴봤지만, 거의 예외 없이, 소유 체제와 가사 공간 유형 간 강력한 관계는 늘 존재해왔다. 테라스 하우스나 다른 집과 떨어져 존재하는 단독주택 등 특정 주거 유형은, 개인의 권리로써의 사유재산에 대한 생각을 공간적으로 조직화하기 위한 방법으로 사용되어 왔다. 테라스 하우스에서 재산의 공간화는, 하중을 견디고 구멍이 뚫리지 않는 경계벽의 원칙에 의해 실현되었다. 단독주택의 경우, 재산은 정원에 둘러싸인 주택의 분명한 고립성에 의해 구현되고 있다. 사유재산을 나타내는 또 다른 중요한 유형학적 측면은 가족들이 오직 하나의

the housing block known as Stacken, a cohousing development that was established by renovating a tower block built in 1969 in Teleskopgatan in Bergsjön as part of Sweden Million Homes Project. The typical plan of the tower was organised as a composition of five individual apartments served by a central core and with no collective facilities. After a few years of use, the apartments were retrofitted in order to decrease the size of the private rooms and replace the private kitchens with small kitchenette, while a wealth of collective facilities including a collective kitchen and a dining restaurant were added. In 2002, the inhabitants of Stacken formed a cooperative and bought the building from the municipality in order to refurbish it and reinforce the communal structure of the building by adding more communal spaces for care and work.[35] Yet in spite of these examples, a lot of cooperative affordable housing that tends to be built in Europe and North America often resembles traditional housing for families where the mere existence of some communal space is a reminder of the original communal idea.

In recent years, austerity policies and the increasing precarity of life is seriously endangering the family as social cornerstone. As emphasized by Melinda Cooper, the question of family was central to neoliberal arguments against public investment and education and key to their proposal for a new economic order powered by private investment and household debt.[36] Yet the recent wave of austerity measures introduced by governments since the 2007-08 recession has seriously endangered the formation of families among new generations. This situation has convinced developers and city authorities in many European cities to drastically reduce the size of apartments in order to accommodate, singles, couples or couples with one child.[37] Yet even within these conditions the

37 최근에는 전통적인 가족구성을 위한 주택 유형이 아닌 코리빙 주거형태가 투기를 목적으로 등장하고있다. 최근 토지와 주택 가격이 매우 높은 런던, 뉴욕, 샌프란시스코와 같은 도시에서 모기지를 기반으로 한 주택은 더 이상 중산층에 적합하지 않으며 가장 수익성이 높은 투기형 임대 주택이 돌아왔다. 이런 상황은 특히 주택 담보 대출이나 주택 자산을 감당할 수 없는 경제적으로 불안정한 젊은 세대에 영향을 미치고 있다. 이 새로운 유형의 임대 주택은 최소한으로 크기를 축소하였다. 욕실과 최소형 주방을 포함하여 1실 단위로 구성된 최소형 하우징 또는 최소형 플랫과 같은 유형이 증가하고 있다. 호텔과 같은 숙박시설의 형태와 유사하다고 볼 수 있으나 19세기 후반에 등장했던 유사형과 달리 매우 비싸다.

38 https://www.sato.fi/en/faq-rental-apartments/studiokoti

39 See: Sophie Kleeman, 'Absurd "Co-Living Space" WeLive is Jacking Up its Prices, 'Gizmodo' https://gizmodo.com/absurd-co-living-space-welive-is-jacking-up-its-prices-1789702081 (Accessed 30 August 2018).

35 See: Karin Andersson, Caroline Glabik, Ellen Persson, Megan Prier, Hej Stacken, Project Portfolio, Design and Planning for Social Inclusion 2012/2013 Chalmers Architecture, https://suburbsdesign .files.wordpress.com/2013/04/stacken-portfolio1.pdf, consulted on 18 October 2019.

36 Melinda Cooper, Family Values: Between Neoliberalism and the New Social Conservatism, New York: Zone Books, 2017.

37 More recently non-family housing typologies such as co-living are coming back, not in the form of co-operative housing, but as new speculative initiatives. In recent years, in cities like London, New York and San Francisco where land and housing prices are very high, home property based on a mortgage is no longer affordable for middle-class and rental units have returned as the most profitable means of housing speculation. This condition is affecting younger generations in particular, whose budget and increasingly precarious existence do not allow them to afford mortgages or home property. This new wave of housing for rent has dramatically reduced housing size to the bare minimum. Typologies like micro-housing or micro-flats which consist of one room units inclusive of bathroom and micro-kitchen are on the rise, and often marketed as some form of hotel living—yet, unlike their late 19th century predecessors, they are extremely expensive.

입구로 드나든다는 사실이다. 단독 주택의 경우 이 입구는 바로 거리와 연결되어 있고, 개인 거주 아파트의 경우 수직으로 서있는 중심부로 연결되어 있다. 하나의 입구 체제는 그 주택이 완전하게 개별화되어 있고, 사적/공적 영역 간 분명한 경계가 존재한다는 것을 보장한다. 방의 수와 서로 다른 크기, 위계적 배열 또한, 집을 그 자체로 완전해 보이도록 함으로써 재산 및 자급자족의 느낌을 강화시키기 위한 강력한 수단이다. 마이크로플랫의 경우, 부엌과 화장실 등 보통 플랫에서 제공되는 모든 서비스가 호텔과 유사한 방 안에 압축적으로 마련되어 있는데, 바로 이 때문에 마이크로 플랫은 개인 소유 아파트이지, 큰 공용 주택의 단위가 아님이 명확해졌다.

우리가 살펴봤듯이, 사적 소유 주택에 대한 대안들은 모든 유형적 측면을 공격했지만 전후 주택 가격의 상승으로 주택의 유형적 조직화 방식을 근본적으로 바꾸는 것은 거의 불가능하다. 예를 들어, 재산 제도뿐만 아니라, 주택이나 아파트의 평균 유형 모두 이웃끼리 방을 공유하는 것을 막고 있다. 개인 부엌에 익숙한 많은 거주자들에게, 심지어 공유의 혜택이 있더라도, 부엌을 이웃과 공유하는 일은 상상도 할 수도 없는 일이다. 공유의 문제는 개인적 보살핌에 의존해야만 하는 장애인과 노인들에게 특히 시급한 문제다. 자급자족을 추구하는 자신의 주거 공간 내에서는 이웃과 간병인을 공유하는 것이 불가능하기 때문이다. 집의 중요한 특징 중 하나는 변화에 대한 저항이다. 이러한 이유로, 보우터 베르보슈트와 힐데 하이넨 등 연구자들은 주거의 문제를 '고정성(obduracy)'이란 용어로 정의했다.[40] 주거 습관은 특히 불확실한 시대에 방향성을 제시해 주기 때문에 극단적으로 오래 지속되며 변화하기 힘들다. 또한 우리가 집에 대해 이야기 할 때 우리는 제도에 대해 이야기 한다. 그 제도 속에서 특정 공간적 조건은 사람들의 사고방식 속에 깊이 뿌리 박힌 사회 및 법리적 프레임워크와 연결되어 있다. 이러한 이유로, 가장 시급한 과제는 '새로운 유형'이나 더 스마트한 집을 생산하는 것이 아니라, 소유권의 대안적 형태를, 거주자 간 신뢰 및 연대감 강화를 목적으로 하는 구체적인 유형학적 배치와 전략적 방식으로 연결시키는 것이다.

legacy of the family household is still very strong in the idea that each housing unit, no matter how small, has to remain self-sufficient and thus still a private home. As such, many developers have begun pushing for 'mini' apartments that are below regulation size —ostensibly in response to changing demographics and rising housing prices.

However, projects such as Sato's StudioHome, which promise overall rents that are 'affordable' allow developers to increase the per square meter price of their developments since apartments are smaller.[38] Sato's StudioHome claims to make up for the shortcomings in space as a "communal and inclusive apartment building" with shared facilities and even a community 'manager' in charge of increasing social interaction between residents. Similar strategies have already been deployed by private companies like WeLive and the Collective which offer accommodation that is like the typology of the old commercial residential hotel, but, unlike the latter, pumped with an overdose of rhetoric about living collectively while rooms are rented for 3.000 USD per month if not more.[39] These companies target generally young professionals, the much celebrated (and much exploited) 'Millennials,' while they exclude people in most urgent need of these types of dwellings, such as the disabled and elderly people.

This brief overview shows that—with few notable exceptions—there has always been a strong relationship between systems of ownership and typologies of domestic space. Specific housing typologies such as the terraced house or the detached single-family house have been used as a way to spatially organize the idea of private property as individual right. In the terraced house the spatialization of property is realized by the principle of the party wall which is a load bearing wall and cannot be perforated. In the detached home, property is embodied in the apparent insularity of the home surrounded by a garden. Another important typological aspect that reflects private property is the access to the family, which is provided by one entrance. This entrance leads directly to the street in the case of the single-family home or to the vertical core in the case of the private apartment. The one-entrance system ensures that the house is perfectly individualized and there is a clear threshold between the private and the public domain. The number of rooms, their different size and their hierarchical arrangement

40 Wouter Bervoest, Hilde Heynen, 'The Obduracy of the Detached Single Family House in Flanders', in International Journal of Housing Policy, Vol. 13, No 4, 358-380.

38 https://www.sato.fi/en/faq-rental-apartments/studiokoti
39 See: Sophie Kleeman, "Absurd 'Co-Living Space' WeLive is Jacking Up its Prices," "Gizmodo" https://gizmodo.com/absurd-co-living-space-welive-is-jacking-up-its-prices-1789702081 (Accessed 30 August 2018).

is also a potent means to reinforce a sense of property and self-sufficiency by making the house appear complete in itself. In the case of the micro-flat, the fact that all the services of a normal flat including kitchen and bathroom are squeezed into a hotel-like room makes clear that the room is a privately-owned apartment and not a unit of a larger communal house.

As we have seen, alternatives to the privately-owned house attacked all these typological aspects, but with the rise of home property since the postwar period, it is nearly impossible to advance radical changes into the way houses are typologically organized. For example, both the property system and the average typology of a house or apartment prevents the sharing of rooms among neighbouring households. For many dwellers used to the private kitchen, it is unconceivable to share one even when the sharing of this facility would be beneficial. The problem of sharing becomes pressing especially with elderly and disabled people who have to rely on personal care since the self-sufficiency of their accommodation prevents the sharing of caregivers with other neighbours. An important aspect of the home is its resistance to change. This has led researchers like Wouter Bervoest and Hilde Heynen to define the problem of housing in terms of 'obduracy.'[40] Not only are domestic habits extremely enduring and hard to change since they give a sense of orientation especially within uncertain times, but when we talk about houses we talk about a system in which a specific spatial condition is linked with social and juridical frameworks that are deeply engrained in people's mentality. It is for this reason that the most urgent task is not to produce 'new typologies' or smarter homes, but to connect in a strategic manner alternative forms of ownership with specific typological arrangements that aim to reinforce a sense of trust and solidarity among dwellers.

40 Wouter Bervoest, Hilde Heynen, 'The Obduracy of the Detached Single Family House in Flanders', in International Journal of Housing Policy, Vol. 13, No 4, 358-380.

본 프로젝트는 현대사회의 주택 위기에 대응하는 새로운 저가형 주거형식을 탐구하고자한다. 이를 위해 유럽의 토지 소유, 건축, 소유권, 지형을 중심으로 주택과 관련된 사회, 경제적 문제를 살펴보았다.

This project focuses on collective habitation not as an idealist projection, but as a structural re-consideration of the entire housing production process: from land procurement to home ownership, from construction to typology. 'Promised Land' is not a solution to the current housing crisis, but an attempt to consider what it is to build affordable housing today in post-welfare Europe.

약속의 땅, 저가형 주거지와 건축에 관하여
도그마 + 뉴 아카데미

PROMISED LAND, RETHINKING TYPOLOGY AND CONSTRUCTION OF AFFORDABLE HOUSING
Dogma + New Academy

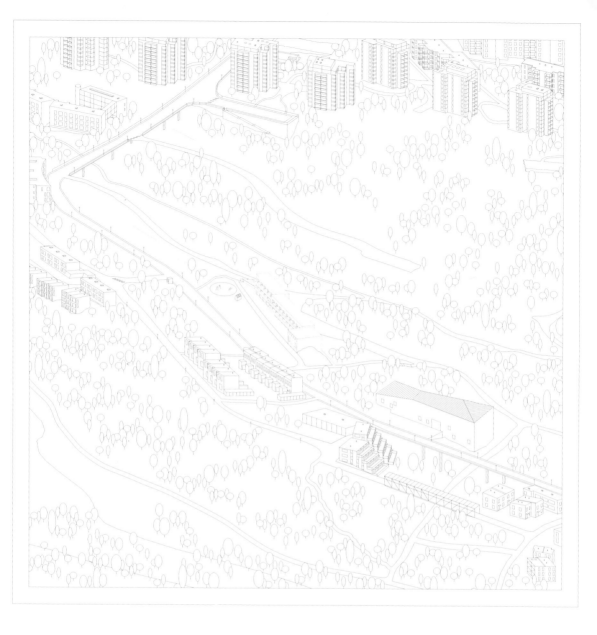

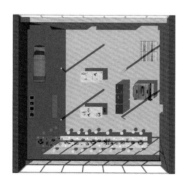

산업화와 경제발전에 따라 도시 밀도가 증가하면서 도심지역의 환경여건 개선에 대한 요구가 나날이 늘어가고 있다. Park 시리즈를 통해 인공도심공원이 어떠한 방식으로 여가생활은 물론, 활동적이고도 감성적인 생활양식을 도모하고, 공원을 주로 이용하는 시민들이 이 공간을 어떻게 소비하는지 살펴볼 수 있다.

Heightening urban density has seen an increasing call to improve and introduce nature into the urban environment. Nature-based spaces found in the urban environment have always been considered beneficial, functioning as civilization's method to reproduce nature. The Park Series is a record of artificially created urban public parks promoting leisure-based, active and emotionally affective lifestyles, and how these spaces are consumed by targeted citizens.

PARK
신경섭

PARK
Kyungsub Shin

볼스와 윌슨은 세개의
연구결과를 보여준다.
기존 유럽 도시들의 고정된
기하학적 구조와는 상반된
유동적이고 역동적인
패러다임의 도쿄 패러다임,
예전보다 밀도가 낮아진
새로운 형태의 도시를
예측하고, 정기적으로
체험하고, 도시 지도를
만드는 유로랜드샤프트
패러다임, 도시 침구학'
프로젝트의 실험실이 되어,
도쿄의 사례로부터 배우게 된
유의미한 장소들이 확산될 수
있는 코르차 마스터플랜.

Bolles and Wilson
present three bodies of
research, the Paradigm
of Tokyo as fluid and
dynamic - the anti-thesis
of the fixed geometries
of the historic European
city; the Paradigm of
the Eurolandschaft
as a networked
landscape, connected
not only by freeways
but also by digital
communication; and
the Korça Masterplan
that re-scripts the city
center with localized
interventions in a
dispersed field (learning
from Tokyo).

세 도시의 현장조사
볼스 + 윌슨

**3 URBAN FIELD
RESEARCHES
BOLLES + WILSON**

카이로도시교육환경연구소는
도시가 자아내는 격식과
거리감을 허물고 건축가,
기획가, 정책자들이 보다
자유롭게 소통할 수 있는
대안적인 체제를 추구하는
카이로 중심부와 주변부의
액션프로젝트 다수를
제시한다.

CLUSTER presents a
number of its action
projects in both
downtown Cairo and
the informal periphery
that seek to analyze and
understand the ordinary
urban landscape,
providing alternative
frameworks for
architects, planners and
policy makers to engage
informality on their own
terms.

전환기 카이로의 공공 공간
CLUSTER
(카이로도시교육환경연구소)

NEGOTIATING PUBLIC
SPACE IN CAIRO
DURING A TIME OF
TRANSITION
CLUSTER (Cairo Lab for
Urban Studies, Training
and Environmental
Research)

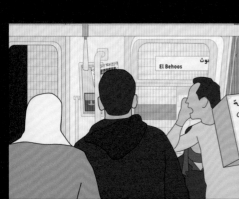

생존을 위한 집합 사업

스틸스.언리미티드(아나 죠키치, 마크 니엘렌), 예레 쿠즈마니치, 프레드라흐 밀리치

Collective Enterprises of Survival

STEALTH.unlimited (Ana Džokić, Marc Neelen) and Jere Kuzmanić, with contribution from Predrag Milić

해질녘 우리는 그 곳으로 들어섰다. 조용했고, 우리 앞에 펼쳐진 바다도 더 없이 잔잔했다. 먼지투성인 입구를 따라 벽에는 모닥불에 그을린 통발이 가지런히 매달려 있었다. 모기들은 우리를 발견하고 달려들기 시작했다. 우리 앞으로 몇 채의 오두막이 나무들 사이에 서 있었다. 오두막의 외관은 통나무에서 벗겨낸 나무 껍질이 덧대어져 울퉁불퉁 했으며, 분실물 보관소 문틀이 장착되어 있었다. 오두막은 낯익게 느껴졌지만, 덤불 속에 버려진 주거지인지, 지구 재난에 대비하는 프레퍼(preppers) 족 야영지인지, 아니면 전통 민속 재현용으로 만든 레크리에이션 오두막인지는 분명치 않아 보였다.

우리가 이 외지를 찾은 것은 2019년 7월이었다. 세모꼴 형태의 황무지였던 이곳의 이름을 번역하면 흥미롭게도 '심연'이었다. 우리는 이 곳에서 앞으로 펼쳐질 10년, 그 동안 분주히 준비해왔던 집단주택 및 공유화(communing)의 향후 실행에 대해 논의해 보고자 했다. 그 장소와 미심쩍을 만큼 거대한 목재 주방을 보았을 때, 아마 많은 사람들을 수용하기 위한 곳이라 추측되었다. 생선과 고기로 가득 찬 낡은 슈퍼마켓용 냉장고는 웅웅 소리를 내고 있었고, 냉장고 유리문 위에는 튼튼한 가죽 장갑과 원판 연삭기가 아무렇게나 놓여 있었다. 주변에는 최신 기계도 그럴듯한 장식품도 없었으며 '생태 작업장'이라고 쓰여진 표지판이 놓여 있었다. 우리는 알래스카에서 시베리아에 이르는 타이가 지대에서 발생하는 산불로 생물다양성이 심각한 수준으로 파괴되고 있다는 사실을 온라인으로 접했다. 머리 위로는 낡은 노란색의 안토노프(Antonov) 농약살포기가 뒤쪽 습지대에서 끊임없이 나타나는 마지막 모기떼에 약을 뿌리고 있었다. 작업을 하려고 앉아 있는 동안, 갑자기 전기가 나갔다. 곧 종말이 올 것 같은, 아니면 이미 종말이 온 듯한 느낌이었다. 앞으로 몇 일이 더 지나도 마음을 정하지 못할 것 같았다.

우선, 당신은 불필요하다

우리에게 주어진 10년. 과감한 조치를 취하지 않는다면 10년 이후에는 기후변화가 되돌릴 수

We enter the terrain at sunset. It is silent, and even the water in front of us does not give away any movement. Along the dusty entrance, a wall is lined with fish pots, blackened from the open fire. Mosquitos have now discovered us—and start seizing upon the fresh arrivals. Further down the road, several cabins are set between the trees. Their exteriors are rough, patches of bark peeling off the logs, a lost-and-found door frame mounted in. While the cabins appear rather familiar, it is not clear whether they are reminiscent of an abandoned bush settlement, the encampment of a doomsday preppers community, or simply a set of somewhat folkloristic recreational lodges.

It is July 2019. We have come to this remote location—at trilateral no man's land, whose name curiously translates The Abyss—to discuss the ten years ahead of us; and the future of practices of collective housing and commoning we have been busy setting up. The terrain and a suspiciously vast wooden kitchen area seemingly anticipating a large gathering. An old supermarket freezer filled with fish and meat hums away, on its glass top casually disposed of a pair of heavy-duty leather gloves and grinder disks. All around us is low-tech, without frills. A sign reads "Ecologic Workshop". Online we learn about the alarming rate of Taiga wildfires from Alaska to Siberia, and of the collapse of biodiversity. Overhead, an old yellow Antonov double-decker crop duster does multiple raids on the last mosquitos holding out in the marshland behind. While we sit to work, the electricity suddenly fails. It triggers us to wonder are we pre- or post-apocalyptic. Over the next days, we won't be able to make up our minds.

"To begin with, you are unnecessary"

Ten years—the window of time given before climate change runs off in an irreversible direction if no drastic action is made. We are in the midst of a system "runaway". Two years ago, in July 2017, Intelligencer from *New York Magazine*, featured the article "The Uninhabitable Earth", exploring worst-case scenarios of climate change. The author opens with the words: "It is, I promise, worse than you think. If your anxiety about global

없는 방향으로 악화일로를 걸을 것이다. 우리는 '고삐 풀린' 시스템의 한복판에 있다. 2년 전 2017년 7월, 『뉴욕 매거진』의 '인텔리전서'에 실린 「사람이 살 수 없는 지구」라는 제목의 기사는 기후 변화와 관련된 최악의 시나리오를 보여주었다. 필자는 "단언컨대, 상황은 생각보다 더 나쁘다. 지구온난화에 대한 걱정 중 대부분이 해수면 상승에 대한 것이라면, 지금 10대 청소년들이 살면서 어떤 끔찍한 일을 목격하게 될 지에 대해서는 실마리조차 잡지 못한 것이다."라는 문장으로 기사를 시작하였다. 10대들의 생애까지 갈 것도 없이, 우리는 최악의 시나리오 상당 부분이 우리 도시에서 점차 현실이 되고 있음을 느끼고 있다. 로테르담의 지중해성 여름, 스플리트와 베오그라드 도심부 내 열섬 지역, 1분에 40번 번개가 치는 뇌우, 사나운 날씨 영향으로 붕괴되는 기반시설, 바람 패턴의 변화 등이 그 예이다. 이는 기후변화를 환경적이고 물리적인 현상이 아닌 정치적이고 윤리적인 쟁점으로 보는 상황의 역전이자, 합의에 이르지 못한 협약들로 인해 악화된 부조리이다. 데이비스 월러스 웰스 기사가 『뉴욕 매거진』 인텔리전서에 실린 지 2년이 지났지만 기후변화는 여전히 진행 중이다. 우리는 '확실히 미래가 없는' 것들을 파악할 수 있고, 그 이면에 무엇이 있는지 지레짐작 할 수는 있지만, 그 전개 양상이 어떻게 될 지는 알 수 없다. 어찌됐든 앞으로의 변화는 급진적일 것이다.

현재 우리 사회 시스템의 다른 측면들 또한 통제불능 상태로 향하고 있다. 우리를 분열시키고 소외시킨 시스템은 노출되었고 취약하며 설명하기 힘들다. 블로거 앤 암네시아는 이러한 상황을 다음과 같이 적절히 묘사하였다. "내가 사는 곳부터, 세계는 표류하고 있다. 우리는 불안정한 것이 아니라, 불필요하다".

좋은지 싫은지 여부를 떠나, 많은 동료들이 '창의적 기업가'가 되어 자영업에 종사하는 것을 보면, 우리는 삶에서 직업 이동성과 유연성이 한층 더 높아졌음을 경험하고 있다. 이와 같은 배경에서 인간 직업의 상당 부분이 향후 몇 년 안에 자동화될 것으로 보이며, 우리를 끊임없이 추격하는 로봇에 비해 인건비가 저렴한 경우에만 일자리를 얻을 수 있을 것이다. 우리의 삶에서 주택은 궁극적인 상품이 되고 있다. 과도한 이동성과 유연성을 경험하는 삶 속 물리적 장소로서의 아파트는, 금융화되고 부동산 주도로 흘러가는 도시 경제의 중심이 되고 있다. 오늘날 우리는 에너지와 음식 등의 자원과, 의료 및 교육 같은 지원 시스템에 대한 접근이 더 이상 자명하지 않음을 경험하고 있다. 이러한 현실에 어떻게 집단적으로 대응해야 할 것인가?

우리는 과거 '현대화'를 추구하면서 도시 및 사회

warming is dominated by fears of sea-level rise, you are barely scratching the surface of what terrors are possible, even within the lifetime of a teenager today." While not exactly teenagers anymore, we feel substantial parts of those worst-case scenarios are currently becoming tangible in our very own cities. Mediterranean summers in Rotterdam, no-go heat islands in the city centres of Split and Belgrade, thunderstorms with forty lightning strikes a minute, infrastructure crumbling under the impact of violent weather events, the changes of wind patterns. Killjoy of climate justice. An absurdity propelled by the no-deal agreements. Those two years since David Wallace-Wells penned the Intelligencer article, have given us little reprieve. While we can designate what is coming as a "bold futurelessness", we can guess what lies behind—we don't know how it will play out. In any case, the future will be radical.

Other aspects of the system are also spinning out of control. A system that seems to have parted us, to have left us behind—exposed, vulnerable, and unaccounted for. The blogger Anne Amnesia has aptly described this situation: "From where I live, the world has drifted away. We aren't precarious, we're unnecessary".

In our lives, we experience the effects of advancing (job) mobility and flexibility, with many of our peers working self-employed, "creative entrepreneurs" like it or not. And that against a backdrop in which a substantial portion of jobs is about to get automated in following years: we'll be in service as long as we're cheaper than the robots haunting us. In our lives, housing has become an ultimate commodity. The apartment— as the physical site of our lives of frenzied mobility and flexibility—has become the focal point of a financialized, real-estate-driven urban economy. In our lives, we experience how access to flows of resources (energy, food) or support systems (healthcare, education) is moving beyond being self-evident. What is our collective response to this?

While the past age of "modernisation" has rolled out the vast universal infrastructures across our cities and societies that now are held responsible for doomsday to come, the current era puts us on the brink of Balkanisation: enclaves, parallel societies—the opportunities for the "do-haves", the leftovers for the rest. In front of us, cities are splintering in competing communities, each foremost catering for themselves.

"I did not know the world is so big"

While this horizon of an "unnecessariat" has started holding ground in our lives on a day

전반에 향후 다가올 재난에 책임이 있는 거대한 기반 시설을 구축했다. 현대 시대에 우리는 서로 적대적으로 여러 소규모 지역으로 분열되는 발칸화의 위험을 맞닥뜨리고 있다. 즉, 소수민족 거주지 및 평행 사회 형성으로 부유층은 기회를 차지한 반면, 나머지 사람들은 부유층이 향유하고 남은 것들만을 가질 수 있다. 도시들은 경쟁이 치열한 지역사회에서 파편화되고 있으며, 스스로 먹고 사는 데에만 급급할 뿐이다.

세상이 이토록 거대할 줄이야!

'**인**간 노동력의 불필요'의 지평이 우리의 일상을 장악하면서, 인간은 다른 전문적인 길을 탐구할 수밖에 없게 됐다. 우리가 살고 있는 로테르담, 스플리트와 베오그라드과 같은 도시는 인간에게 충분한 삶의 터전을 제공해왔다. 로테르담의 경우 한때, 수많은 노동자들이 있었기에 세계에서 가장 큰 항구가 운영될 수 있었다. (마찬가지로 그 항구 덕분에 노동자들이 생계를 유지해 나갔다.) 하지만, 최근에는 두 가지 분명한 변화가 감지된다. 항구의 자동화로 근로자가 필요 없게 되었으며, 도시 경제의 글로벌화로 컨시어지 주거공간과 같은 하이엔드 부동산 및 선진 서비스 산업이 발전하고 있다. 투기적인 도시 개발 계획에 참여할 수 없거나 시도하지 않은 사람들에게 남은 공간은 거의 없다. 지난 몇 년 간, 전통적으로 노동자들이 거주하던 도시 북쪽 부동산 가격은 세 배나 뛰었다. 도시에 계속 남아 있는 것이 점점 더 위태로워졌다.다음으로 스플리트의 경우를 보자. 지난 세기 상당 기간 동안 이 항구 도시는 해운, 조선 및 기타 산업 활동으로 성장해 왔다. 하지만 최근 몇 년간, 단기체류 방문객 수가 급증하면서 관광산업의 허브로 재탄생하고 있다. 대부분의 도심지역에 에어비엔비와 같은 숙박 플랫폼이 운영되면서 성수기에는 주민들이 머무를 곳도 찾기 힘들 지경이다. 여기서 의문점은, 어떻게 한 도시가 이와 같은 비생산적인 경제, 즉 시민들을 저숙련 서비스 노동자로 치부하는 경제에 의존하며 계속 유지될 수 있을까 하는 점이다. 이번에는 베오그라드의 경우를 보자. 베오그라드는 유럽에서 가장 소득이 낮은 도시로 대부분의 시민들이 상당히 어려운 생활을 한다. 반면, 정부는 지역 시민들을 먹여 살리는 사업과는 관련 없는 장대한 수변 개발에만 관심을 기울이고 있다. 지난 몇 년간 이곳에 지어진 새 아파트의 대부분은 거주 목적이 아닌 투기 수단으로 활용되었다. 동시에, 서비스 및 창조 '산업'에 종사하는 젊은 근로자들은 프로젝트 기반이나 정해진 노동시간 없는 임시적 계약을 기반으로 근무하고 있다. 이곳에서 젊은이들의 미래는 보장되지 않기 때문에 대부분이 다른 탈출구를 찾고 있다. 수변

to day basis, it has pushed us to explore different (professional) paths. The cities in which we live – Rotterdam, Split, Belgrade in this case—provide ample ground. Take Rotterdam, a city once characterised by a vast working-class feeding the world's largest port (and vice-versa). Nowadays shifting into two distinct operations: automation of the harbour leaving no need for workers, and globalisation of the urban economy towards high-end real-estate development (the likes of "concierge residential living") and the advanced service "industry". For those who cannot, or do not seek to participate in the latter speculative urban development schemes, there is little space left. In the past years, real-estate prices in traditionally working-class neighbourhoods in the city's North have increased threefold. Keeping our foot in the city becomes increasingly perilous. And then take Split. Over much of the last century, this port city has been developing around shipping, shipbuilding and other industrial activity. But in recent years, it has been reinvented as a tourist hub, servicing an increasing armada of short-stay visitors. Most of the inner city is put up on hospitality platforms (Airbnb and the like), with its residents challenged to find a place to live during the tourist season. In this, the bigger question is how a city can sustain on such a non-productive economy, that diminishes local population to low-skilled and servile work power. Or take Belgrade. Here, most households struggle to make ends meet on one of the lowest incomes of Europe, while the attention of the government is focused on a spectacular waterfront development obviously not catering to the local population. Over the past years, the majority of new apartments built have not been intended for living but kept as speculative vehicles. At the same time, the young pool of workers in the service and creative "industry" is working on project-basis or zero-hour contracts. Their future seems to be running out, and indeed the majority of them are seeking a way out elsewhere. In the shadow of projects like the waterfront development, a society of those dumped is developing. It is where the unnecessariat drifts together. As Anne Amnesia states: "And what's worst of all, everybody who matters seems basically pretty okay with that."

This is the context of our lives and practices. Almost ten years ago, following the escalation of the Global Financial Crisis, we set our paths following a hunch: there is an end to this broken system. The sociologist and political economist William Davies (in the article "The Big Mystique", 2017) frames the urgency like this: "The problem with viewing the future as territory to be plundered is that eventually we all have to

개발과 같은 사업의 그늘 속에서, 시민들을 외면한 사회가 발전하고 있다. 하지만 그곳은 필요 없어진 사람들이 함께 표류하는 곳이다. 앤 암네시아가 언급했듯이 "최악인 것은, 관련 있는 모든 이들이 이 상황을 큰 문제의식 없이 받아들인다는 점이다."

이는 우리의 삶과 행동에 대한 이야기다. 거의 10년 전 글로벌 금융 위기가 정점을 찍었을 당시, 우리는 이 망가진 시스템에도 끝은 있다라는 생각을 가지고 길을 모색했다. 사회학자이자 정치경제학자인 윌리엄 데이비스는 2017년 「커다란 미지」라는 기사에서 위기상황을 다음과 같이 표현했다. "미래를 우리가 앞으로 약탈할 영역이라 보는 관점의 문제는 궁극적으로 우리 모두가 그 곳, 미래에서 살아가야 한다는 사실이다. 그 미래가 도래하여 자원이 모두 고갈된 사실을 발견한다면 우리는 또 다시 미래 자원을 끌어 쓰는 일을 반복할 것이며, 결국 악순환의 고리에 빠지게 된다. 우리는 오늘날 생각지도 못한 기술, 제도, 기회들로 인해 미래 상황이 지금과는 다를 수 있다고 생각하지 못하고 있다. 모든 것을 통제하에 두어야만 마음이 편한 금융분야 종사자들은 때가 되면 찾아올 미래를 가만히 기다리는 것에 만족하지 않는다. 그들은 우리가 미래를 맞이하기 훨씬 전부터 미래의 이익을 가져와 그것을 미리 나누려 할 뿐이다."

이러한 이유로, 우리는 '해로운' 건축과 도시 개발로부터 벗어날 방법을 찾는 노력이 필요하다. 즉, 도시민들을 위험에 빠뜨리고, 경제를 더 불안정하게 하거나 투기를 조장하고, 그 부산물로 인하여 지나치게 경계가 확장되는 건축과 도시 개발을 지양해야 하는 것이다. 이는 지난 10년간 진행된 관련 노력의 근거였다. 실제로 대부분의 활동가들은 필요해 보이는 일들을 실천하며, 행동을 통해 꾸준히 학습해왔다. 로테르담의 경우, 시장이 붕괴된 직후, 모든 건물 소유주는 건물로 인해 금융 부담을 안게 되었다. 이런 상황에서, 우리는 이러한 건물 공간을 새로운 공동체, 새로운 생활과 근로 방식, 새로운 형태의 경제 재생산을 구축하기 위한 출발점으로 활용하였다. 그 결과 탄생한 '시티 인 더 메이킹'네트워크를 통해 점점 더 많은 건물, 작업장, 공용 공간과 공유활동들이 (기간이 대부분 3~10년에 불과했던) 임시적 역할을 벗어나 앞으로의 변화를 위한 '훈련 공간'으로 재탄생하고 있다.

이와 같은 활동은 장기적으로 공동체 내에 공유 공간을 더욱 많이 확보하려는 모델 실현을 위한 발판이라 할 수 있다. 여느 곳과 마찬가지로 삶의 공간이 똑같이 위협받고 있는 베오그라드에서, 우리는 '후 빌즈 더 시티'를 통해 새로운 형태, 합리적 가격의 공동소유 아파트를 마련하고자 한다. 이러한 아파트는 공동체

live there. And if, once there, finding it already plundered, we do the same thing again, we enter a vicious circle. We decline to treat the future as a time when things might be different, with yet to be imagined technologies, institutions and opportunities. The control freaks in finance aren't content to sit and wait for the future to arrive on its own terms, but intend to profit from it and parcel it out, well before the rest of us have got there."

For this reason, it is necessary to find ways out of architectures and urban developments that are in themselves "toxic": endangering the populations of the cities they act upon, making economies more volatile and speculative, and overstretching planetary boundaries with the products they generate. This was the grounding argument for our endeavours initiated in the previous decade. Most of the activities have started stubbornly, learning-by-doing: doing what seems necessary to be done. In Rotterdam, for instance, immediately following the market crash, an entire pool of buildings has become a financial burden to their owners. In this context, we have set out to use those sites as starting points to build new communities, new ways of living, working and new forms of economic reproduction. Meanwhile, the resulting network of City in the Making counts a growing number of buildings, workshops, commoned spaces and activities, intended as "training ground" beyond their temporary existence (most of the spaces are available on three to ten year terms only). They are stepping stones towards models that bring spaces in community ownership on the longer term. In Belgrade, where living space has equally come under threat, as Who Builds the City we are preparing a novel kind of affordable and cooperatively owned apartment buildings, that include community spaces and shield their residents from the perils of individual mortgages. In the next year or two, this is to lead to the first pilot building, with potential to snowball an entire ecosystem of affordable, community-owned spaces in the city.

This has forced us to delve into unexpected fields of knowledge to become: ad-hoc finance experts, autonomous energy systems pioneers, self-styled "business developers"—and more. The aim of this is to generate a more comprehensive take on the systems and infrastructure necessary to construct our futures on "ruins of the broken system". Luckily, in this endeavour we are not alone. Along the way, we have discovered daring initiatives that reinvent the world of finance and banking, for instance, the Cooperative for Ethical Finance (ZEF) from Zagreb, with whom meanwhile we have an intensive working relationship. But

공간을 포함하고 있고, 거주자들을 개인 담보대출의 위험으로부터 보호할 것이다. 향후 1~2년 후에, 첫 시범 건물이 지어질 것이며, 이로 인해 도시 내 합리적 가격의 공동소유 공간의 생태계가 빠르게 번창할 것으로 기대한다.

이러한 활동을 하면서, 우리는 앞으로 특정 지식 분야가 될 것이라 미처 예상치 못했던 분야의 지식을 파고들게 된다. 임시 금융전문가, 자율 에너지 시스템 개척자, 자칭 '사업 개발자' 등에 대해서다. 여기서의 목적은 '망가진 시스템의 잔재' 위에 미래를 건설하기 위해 필요한 시스템과 기반시설들에 대해 좀 더 포괄적으로 고려하는 것이다. 다행히, 이 같은 노력을 기울이는 것은 우리만이 아니다. 우리는 금융과 은행 분야 개혁을 위해 노력하는 과감한 이니셔티브들을 알고 있다. 그 한 예가 자그레브의 윤리 금융을 위한 조합(Cooperative for Ethical Finance, ZEF)으로, 우리와도 긴밀한 업무 관계를 맺고 있다. 하지만 이것이 끝이 아니다. 메타 수준의 협력 및 업스케일링, 공유화를 추구하는 몇몇 이니셔티브 사이에서 이미 생태계가 형성되고 있다. 우리는 세계가 이 정도로 넓다는 사실을 인지하지 못했다고 볼 수 있다.

물론, 건축이 생산되는 방식에 대해 위와는 다른 견해가 나올 수도 있다. 다양한 근무조건, 수평적 협업, 개방적인 프로토콜, 협동조합처럼 다소 옛날방식처럼 보이는 참여형태를 추구해 나가는 등, 젊은 건축가 세대 사이에서 흥미로운 유행이 다시금 관찰되고 있다. 이는 수많은 인턴들, 수직적 위계, 오래가지 못하는 의욕 등 건축업의 전형적인 특징과는 극명한 대비를 이룬다. 이러한 의미에서, 현재 진행되는 활동들은 저항적이고 정치적이며 힘든 도전이라 할 수 있다.

조합이 쓸모 없는 것은 아니다.
우리가 베오그라드에서 새로운 조합 주택 프로젝트를 추진했을 당시, (가용 자금 부족 등을 포함한) 많은 장애물들에 막혀 어떤 방향으로 나가야 할 지 몰라 혼란스러웠다. 그 때 우리는 중앙 및 남동 유럽지역 수도에서 활동하는 다른 선구자적 단체들도 비슷한 장애물들에 직면해 돌파구를 찾지 못한다는 사실을 알게 되었다. 2017년, '후 빌즈 더 시티'는 부다페스트(라코츠지 공동체)와 루블랴냐(자드루가토르), 자그레브(윤리금융조합/개방건축조합)에서 진행되는 활동들과 협력하기로 하였고, 'MOBA 하우징 네트워크'를 구성하게 되었다. 곧 우리는 집단적이며 합리적인 가격의 주택 개발을 방해하는 요인들이 모두 동일한 조직적 원인을 갖고 있음을 알게 되었다.

it does not end there. An ecosystem is already emerging between several such initiatives propelling a meta-level of collaboration, upscaling and commoning. In that sense, we did not know the world is so big.

Obviously, this also comes with a different take on how architecture gets produced. Seeking different working conditions, horizontal collaborations, open protocols and forms of participation in something seemingly archaic like co-operatives, that now make an exciting revival among generations of younger architects. It stands in stark contrast with the armies of interns, the vertical hierarchies, or the short-lived ambitions that characterise most of our profession. In that sense, these activities are both confrontational and political—and a tough challenge.

"There is nothing idle about cooperatives"
When facing a steep set of obstacles to roll out new cooperative housing projects in Belgrade (like the lack of available finance) made us wonder where to move next, we discovered that around us, in capital cities throughout Central and South-Eastern Europe, other pioneering groups are facing the same obstacles on their path to a breakthrough. In 2017, Who Builds the City together with initiatives from Budapest (Rákóczi Collective), Ljubljana (Zadrugator) and Zagreb (Cooperative for Ethical Financing / Cooperative Open Architecture) decided to join forces and form the MOBA Housing Network. Soon we recognised that the factors that hinder the development of collective affordable housing solutions locally have common, systemic causes.

The network's name, MOBA, comes from "self-build through mutual help" in South Slavic languages. What we "self-build" is an institutional framework to attract, channel and manage financial flows for a new generation of housing cooperatives. They would enable lower-income populations in the region to collectively access affordable housing. Recently MOBA has completed a portfolio to seek partners—alternative, ethical and regionally operating banks, impact investors— to finance the realisation of five pilot projects.

The initiatives coming together under MOBA have been launched by architects, sociologists, ethical finance experts and social organisers—from a generation growing up in the crisis of post-socialist transition, increasingly confronted with housing insecurity and unaffordability. The network functions on a participatory basis, addressing both conceptual issues and practical steps to be made. This requires venturing into yet-unknown fields of expertise. Thus, we have set aside time to

네트워크의 이름인 MOBA는 남슬라브어로 '상호 도움을 통해 직접 건설하기'라는 뜻이다. 우리가 '직접 건설하는' 것은, 새로운 세대의 주택조합을 위해 자금 흐름을 끌어오고 관리하는 제도적 프레임워크이다. 이는 지역 내 저소득 계층 집단이 합리적 가격의 주택을 이용할 수 있도록 해준다. 최근, MOBA는 다섯 가지 시범 프로젝트 이행 자금 마련을 위해, 대안 운영, 윤리적 경영을 지향하는 지역은행, 영향력 있는 투자자 등 파트너 모색을 위한 포트폴리오를 완성하였다.

MOBA 하에서 진행되는 이니셔티브는, 사회주의 체제전환 이후 혼란기에 성장해 주택 불안정과 비합리적 주택 가격 등에 불만을 품은 세대의 건축가, 사회학자, 윤리적 금융전문가, 각종 사회 단체들에 의해 추진되고 있다. 이 네트워크는 개념적 이슈와 앞으로 취해야 할 실질적 조치들을 다루면서 참여 기반으로 운영되고 있다. 이를 위해서 우리는 아직 잘 모르고 있는 전문분야를 개척해야 한다. 이에 따라 별도로 시간을 할애하여 사업개발 계획에 익숙해지고, 수명주기비용분석에 관한 임시 전문가가 되고, 주식과 투자펀드의 작동방식을 이해하고, 스스로 금융 중개인 역할을 수행하거나 회전투자기금을 구축하기 위해 노력하고 있다.

로테르담에서 '시티 인 더 메이킹'을 진행하면서 우리는 '브레이콥(Vrijcoop)' 일원이 되어 유사한 과정을 경험하였다. '브레이콥'은 30년 전 독일에서 시작되어 현재 각국에 수출된 획기적인 '미트쉐우서 진티카트(Mietshäuser Syndikat)'의 네덜란드 버전이다. '브레이콥'는 부동산을 시장으로부터 영구히 제외시킴으로써 회원들에게 합리적 가격의 투기성 없는 주택을 제공하였다. 또한, 주택사업에 미치는 토지가격의 영향이 점점 커지는 문제를 해결하기 위한 방법으로 '커뮤니티 랜드 트러스트 브뤼셀'과 협력하여 '토지은행'의 설립 가능성을 모색했다.

놀랍지 않게도, 이러한 노력들은 네트워크화되고 새로운 모습을 띠며 선구적인 활동가로부터 많은 에너지를 요구하는 경우가 많다. 그리고 그 중 많은 경우가 공유화 및 공동 자원의 원칙을 기반으로 진행된다. 마시모 데 엔젤리스는 그의 책 『Omnia Sunt Communia: on the commons and transformation to post-capitalism』에서 다음과 같이 서술했다. "이처럼 재생산의 공유 자산은, 결핍과 필요 또는 건강한 음식, 주택, 물, 사회보장 및 교육 등에 대한 열망을 해결하기 위해 이미 전세계 많은 서민들이 동시다발적으로 구축하고 있다. 이로 인해 복잡한 조직 속에서 새로운 공동체, 문화, 웰빙과 안보, 신뢰를 구축하는 새로운 방식의 혜택이 제공될 뿐 아니라,

get intimate with business-development plans, to become ad-hoc experts in Life Cycle Cost Analysis, to grasp the workings of shares and investment funds, to prospect becoming financial deal-brokers ourselves, or to explore setting up a revolving investment fund.

In Rotterdam, with City in the Making, we experience a similar trajectory by being part of Vrijcoop, pioneering a Dutch version of the groundbreaking Mietshäuser Syndikat (that started in Germany some thirty years ago, and now expands to other countries). It provides affordable and non-speculative housing to its members by taking real-estate indefinitely out of the market. Also, we have started exploring the possibility of a "land bank", while being in exchange with Community Land Trust Brussels about ways to tackle the rising impact of land prices on housing projects.

Not surprisingly, these efforts are often networked, emergent and demand massive amounts of energy from their pioneers. And many of them revolve around principles of commoning and the commons. In his book Omnia Sunt Communia: On the Commons and Transformation to Postcapitalism (2017) Massimo de Angelis states: "These commons of reproduction are already being set up spontaneously by many commoners around the world to address lacks and needs or aspirations for accessing healthy food, housing, water, social care and education. [...] Not only would they give us the benefit of new communities, new cultures, and new methods of establishing wellbeing, security and trust within complex organisation, they would also protect us from the whims of financial markets, and, especially, increase our security and power to refuse the exploitation of capitalist markets."

As much as the "broken system" has relied on its systemic embrace, also we—to deter it— in turn, experience the necessity for a systemic approach. Even when the latter comes with excitement and empowerment, it also comes with insecurity and instability for those willing to take a long shot into a yet uncertain future. For example, bank loans for cooperative housing are hardly available, bids to take property out of the market compete with regular profitable offers, legal frameworks are often normative but merely supportive to novel approaches, building the trust within the groups takes more energy than estimated, long term commitments are countered by the everyday challenge of providing for rent and living, etc. As a result, these efforts sometimes seem to have a limited impact. Pioneering is often an erratic process, and its outcomes depend on

금융시장의 불안정으로부터 보호해주고, 무엇보다도
자본주의 시장의 착취를 거부할 수 있을 정도의 안정성과
힘을 길러준다.”

　　‘붕괴된 시스템’이 체계적 수용에 의존하는 것만큼,
우리는 이를 억제하기 위해 체계적 접근이 필요하다고
느낀다. 체계적 접근은 기대감을 증폭시키고 권한을
강화해주는 동시에, 아직 불확실한 미래에 기꺼이
승부를 건 사람들에게는 불안감을 조성한다. 예를 들어,
조합주택을 위해 은행 융자를 받는 것은 거의 불가능하고,
소유물을 시장과 분리시키고자 하는 노력은 정기적
수익을 보장하는 제안과 경쟁해야 하며, 법체계는
규범적이지만 새로운 시도에 대해서는 단지 지지를 표할
뿐이다. 또한, 집단 내 신뢰를 구축하는 것은 생각보다
많은 노력이 요구되고, 장기적인 책임은 집세 및 생활비 등
일상적인 과제들에 의해 중요성이 퇴색된다.

　　그 결과, 이러한 노력들의 영향력에는 한계가 있어
보인다. 새로운 분야를 개척하는 것은 종종 불안정한
과정이고, 그 결과는 많은 시간을 요하는 집단 및 개인의
노력에 달려있다. 그러나 이와 같은 노력에도 불구하고
도시 및 더 큰 공동체로 수준을 향상시키는 방법을 찾지
못할 때가 많다. 이와 같은 그늘에서 벗어나 효과적인
대안을 제공하기 위해 우리는 ‘부적응자들의 섬’에서
벗어나야 한다. 갈 길은 멀고, 기후변화로 인해 우리가
원하는 목표를 제 시간에 달성할 수 있을지 확신은 없다.

　　아담 그린필드는 『Radical Technologies:
the design of everyday life』(2017)이라는 책에서,
생산력(재생산력)은 대부분 협동조합을 중심으로
조직된다는 다소 도발적인 시나리오를 제시하였다. 그는
전망과 함께 다음 경고를 덧붙였다. “20세기 청구서
지급일이 맹렬히 다가오고 있다. 경제는 때로 문제를
바로잡는 것처럼 보인다. 하지만 글로벌 환경을 보존하기
위한 대규모의 활동은 진행되지 않고 있으며, 석탄을 손에
넣은 사람들이 마음대로 석탄을 태우는 것을 막을 방도도
없다. 만일 인류가 간신히 곤경에서 벗어난다 하더라도,
‘그날’은 생각보다 훨씬 빨리 올 것이다.”

　　나는 10년 후에도 같은 일을 하고 있을까 두렵다.

에 비스에서 우리는, 10년 후에 우리가 어떤 일을 하고
있을 것인가에 대해 운명론과 낙관론을 오가며
대화하였다. 이런 생각을 하는 사람이 혼자가 아니라는
사실에 위안을 얻었지만 동시에 우리가 같은 악몽을
생각하고 있다는 사실에 두려움을 느꼈다. 10년 뒤에도
지금과는 크게 다르지 않을 것이다. 우리를 영구적으로
위기에 몰아넣는 ‘굴레’는 계속해서 작동할 것이며, 도시
속 우리를 위한 공간은 계속 줄어들어 결국 미래세대에

time-demanding collective and personal efforts
that often do not find ways in upscaling to a city or
broader community level. To get out of this shadow
and provide operating alternatives, we have to
go beyond "islands of misfits". A long road, while
climate collapse offers little assurance that we will
get there in time.

　　In the book *Radical Technologies: The Design
of Everyday Life* (2017), Adam Greenfield explores
a provocative scenario in which productive (and
reproductive) capacity, for the most part, is
organised around cooperatives. His outlook comes
with ample warning: "The bill for the twentieth
century has come due with a vengeance. The
entire economy [...] sometimes seems geared to
repairing the damage. [...] But there is no large-
scale coordinated action to preserve the global
environment, no way to prevent anyone from
burning whatever quantity of coal they can get their
hands on, and if humanity does manage to squeak
by, it's clearly going to be a closer thing than
anybody would prefer."

　　"I am afraid I will be doing the same thing ten
　　years from now"

B ack in The Abyss, the conversation on where
we will find ourselves ten years from now
perpetuates from fatalism to optimism and back.
It is assuring to know that one is not alone in these
thoughts, and at the same time alarming to realise
that we encounter the same nightmare: that in
ten years, not enough will have changed. That the
same wheels that perpetually drive us into a crisis
will keep on spinning. That the space for us, in our
cities, will continue to diminish, with little left for
generations to come. A nightmare in which we are
one step older but still far from a secured space to
live, a living income, an environment that provides
for mutual care—a future not plundered.

　　If this is the nightmare that we will wake up
to—the same "old" broken system—any attempt to
speak of a collective city will remain futile. Instead,
we expect that islands of "collectivity" will arise
out of joint resistance to the imminent collapses.
Could these resistances be seen as enterprises of
survival? Are we well accompanied for what is to
come?

I f we are to wake to a different place
ten years from now, we will have to
acknowledge the following: at the core of this
broken system is the premise that our dominant
economy (only) functions in pursuit of perpetual
growth. And as long as we are ignorant to the
impossibility of limitless growth on a finite planet,
nothing substantial will change. The mantra of
sustainability—reduce, reuse and recycle—will

별로 남겨줄 것도 없게 될 것이다. 세월이 흘러도 안전한 주거공간, 안정적인 소득, 서로를 돌보는 환경, 즉 약탈당하지 않을 미래를 확실히 보장받을 수 없다는 사실이 바로 악몽이었다.

우리가 마주하게 될 미래가 망가진 시스템이 '낡은' 채 그대로 남아있는 악몽과 같은 상황이라면, 집단도시를 주창하는 어떠한 시도도 소용없게 될 것이다. 하지만 우리는 붕괴가 임박한 순간에 '집단성(collectivity)의 섬'들이 공동으로 저항하여 떠오르기를 기대한다. 이러한 저항이 생존을 위한 사업의 형태로 실현될 수 있을까? 우리는 다가올 미래를 위해 협력하고 있는가?

만일 10년 후 상황이 지금과 다르길 원한다면, 다음의 사실을 인지해야만 할 것이다. 붕괴된 시스템의 핵심은 지배적인 경제가 끊임없는 성장을 추구하는 경우에(만) 작동한다는 전제이다. 유한한 행성에서 살면서 무한한 성장이 불가능함을 간과하는 한, 중요한 어떤 것도 변화하지 않을 것이다. 지속가능성의 만트라인 '줄이고, 재사용하고, 재활용하자'는 근본적인 문제를 해결하기에 충분치 않다. 정치인들이 환경을 보호하는 활동을 하도록 만들기 위해서, 근본적으로 생태와 관련된 이슈들이 정치적 논쟁거리가 되도록 해야 한다.

우리가 구축하고 있는 '집단성의 섬'들은, 이와 같은 유해한 환경 앞에서 건축의 도구를 재구성하는데 기여하고 있다. 재사용 대신에, 우리는 '난도질'한다. 우리는 부동산 시장의 사각지대를 난도질한다. 그렇게 하지 않으면, 우리의 노력이 단지 투기 수입을 상승시키는 데에만 도움이 되기 때문이다. 우리는 금융제도의 사각지대를 난도질한다. 이는 정치적 DNA가 다른 서민 자본 공급자를 창조해내기 위함이며, 자본의 축적에서 자본의 재생산으로 관점을 전환시키기 위함이다.

재활용 대신 우리는 '해체'한다. 우리 활동의 핵심은 어떻게 하면 현존하는 공간에 새로운 가치를 부여하고, 기존의 공간들이 (지속적으로) 확장중인 경제에 미치는 영향을 축소시킬지에 대해 고민하는 것이다. 우리는 전통적인 경제 관계를 해체하고, 참여자들이 상호호혜적인 이익을 얻을 수 있는 생태계로 재조립하고자 한다. 자원의 공유를 위해 소유권을 해체하여, 내부의 경제 집단들이 생겨난다. 전통적인 건축물은 권한을 부여해주는 새로운 관계를 창출하는 하나의 자원으로서 재전용되며, 이 관계는 시장이나 개인의 성공에 의존하지 않는다. 마지막으로, 줄이는 대신 우리는 '중단'한다. 이것이야말로 다가오는 폭풍우에 대비해 건축가들이 해야 할 일이다. 끊임없는 건설을 중단하자.

not be enough to fill this missing core. In order to "ecologise" politics, we will have to politicise the ecology radically.

The islands of collectivity that we have been setting up contribute to reframing the tools of architecture in the face of this toxic environment. Instead of reuse, we hack. We hack the blind spots of the real-estate market in which otherwise our efforts would only serve a speculative revenue spiral. We hack the blind spots of financial institutions, to create grass-roots capital providers with a politically different DNA, to ensure that we can shift from accumulation of capital to the reproduction of it. Instead of recycle, we dismantle. At the core of our operation is not only the thought of how to bring new value to existing spaces but how to downsize their impacts and their contribution to an (ever) expanding economy. We dismantle traditional economic relations to reassemble them into operating ecosystems at the mutual benefit of their participants. We dismantle ownership in favour of the common sharing of resources, creating internal economic circles. Traditional structures get reappropriated as a resource to build up a new relationship of empowerment that does not depend on their credibility to either market or personal successes. And finally, instead of reduce, we cease. And that is what architects need to do in the face of the coming storms. Just cease to build, endlessly.

What collectivity shares with the climate debate are not some abstract origins in their humanistic virtue. In the end, the planet will keep spinning without us, won't it? What climate debate and collectivity have in common is that they are both rooted in economic relations. And they reproduce the (im)possibility of stepping beyond the known paths. As long as we do not aim to change our relationship to perpetual growth—the core of the problem—we will wake up in the city of fragmented and antagonising collectives, instead of a collective city.

집단성 및 기후변화 논의는 인본주의적 가치에
내재된 추상적인 성격의 것이 아니다. 결국, 지구는 우리가
존재하지 않아도 계속 자전할 것이다. 또한, 기후변화
논의와 집단성의 공통점은 두 가지 모두 경제와 깊은
관련이 있다는 점이다. 이 두 가지는 이미 알려진 길을
넘어서는 (불)가능성을 재창조해낸다. 문제의 핵심인,
끊임없는 성장을 추구하는 관계를 변화시키고자 하지
않는 한, 우리는 집단 도시 대신, 분열되고 적대적인
공동체 도시에서 살게 될 것이다.

ARE YOU WELL AC

많은 사람들과 함께하고

2029

2029

THE FRONT YARD:
SLEEPER
CELLS

THE BACKYARD:
APARTMENTS
OF MANY

"THERE IS
NOTHING
"협동조합은 항상
바쁘게 움직이죠"
**IDLE ABOUT
COOPERATIVES.**"

A group of 'misfits', with their futures under threat,
comes up with a daring plan to adopt the wreck of
an iconic green investment gone bust, and turn it
into a giant future-proof living machine.

이재가 불투명한 자질 '부적응자' 집단이 과감한
계획을 내놓는다. 과거에 실패한 대규모 친환경
투자 모델의 잔해를 거대한 미래지속형
생활시스템으로 바꿔낸다는 것이다.

"TO BEGIN WITH,
첫째 당신은 필요
없는 존재야 U ARE
UNNECESSARY."

Residents of an abandoned tourist destination
brace themselves after the last Budget Flight
leaves town. Their starting point is rebuilding the
community in its deserted hotels.

사람들이 살지 않는 관광지. 국민관광 사업 공원의
마지막 경비 비행편이 떠나서 걸리며 다진다. 지역
사회 생산은 버려진 호텔에서부터 시작된다.

HOTEL

2029년을 기점으로 과거를 다루는 일련의 포스터를 살펴보면 이 미래형 공상 소설은 자칭 "부적응자"집단의 현실을 언뜻 보여준다. 2008년 세계경제 위기의 여파로 이들은 집단을 이루어 색다른 미래를 위한 기반시설을 구축하기 시작했다. 이는 곧 새로운 삶의 방식을 탐색하고 생존에만 국한되지 않은 직업, 성장에 목매지 않은 경제, 획기적인 에너지 자원의 전환, 공동의 이익을 우선시하는 미래인 것이다.

Looking back from 2029, through a series of posters, this future fiction gives a glimpse into the reality of a self-proclaimed pool of 'misfits'. Following the Global Financial Crisis (2008), they started building infrastructures to collectively break ground for a different future – for new forms of living, for work beyond the reality of jobs, for an economy beyond growth, for a radical energy transition, for cities based in commons.

세상이 이토록 거대할 줄이야!
스틸스.언리미티드(아나 조코치, 마크 닐렌), 제르 커즈마니크, 프레드라그 밀리크

I DID NOT KNOW THE WORLD IS SO BIG STEALTH.unlimited (Ana Džokić and Marc Neelen), Jere Kuzmanić, Predrag Milić

155

만약 도시에서의 집합이, 공적영역의 토대가 되는 열린 공간과 작은 틈새들로 구성된다면 다카라는 도시는 구조적 모순을 보여준다고 할 수 있을 것이다. 곳곳에 세워진 높은 장벽이 공과 사를 구별하고 법규의 적용을 받는 곳과 그렇지 않은 곳을 나누기 때문이다. 우리는 장벽들로 구성된 다카를 사각지대의 집단성이 발현되는 공간으로 묘사하였다.

If the collective in a city is constituted by porosity and openness, Dhaka city presents a structured contrariness, constituted by the presence and intensity of walls that separate the private from the public, the legal from the undocumented. Dhaka: A Million Stories re-describes the walls of the city as a meta-site against which the unregistered collective unfolds.

다카: 백만가지 이야기
마리나 타바시움 + 벵갈 인스티튜트 포 아키텍처, 랜드스케이프 앤 세틀먼츠

DHAKA: A MILLION STORIES
Marina Tabaussum + the Bengal Institute for Architecture, Landscapes and Settlements

CBC는 농촌개발이 좀 더 깊고 섬세한 인간적인 요소들을 고려할 필요가 있다고 믿는다. 차이나 빌딩 센터(CBC)는 사회 발전, 경제, 문화, 생태계 간의 세심한 균형을 도출하고자 일련의 연구를 수행 하였으며 이를 통하여 도시와 농촌의 관계 속에서 국가가 어떠한 새로운 역할을 할 수 있는지 살펴보았다.

CBC believes that urban and rural developments need to prioritize more profound, subtle human factors. The research conducted by CBC reflects specifically on how to achieve a delicate balance between issues of social development, economy, culture and ecology; an approach that hopes to obtain a new position outside of traditional urban-rural relationships and generate new possibilities in the field.

공간적 가치의 창조
CBC (차이나 빌딩 센터)

CREATION OF SPATIAL VALUE
CBC (China Building Centre)

Sleeping Room

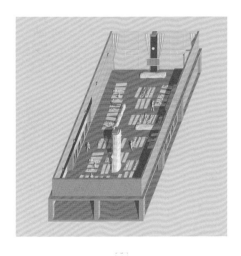

Latitude 36.3331228
Longitude 127.452525

Sauna_Salt

Latitude 37.7915662
Longitude 128.9190019

Sauna_Charcoal

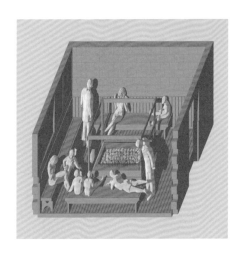

Latitude 34.8188609
Longitude 126.3834813

Sauna_Gems

Latitude 37.46838
Longitude 127.0361

"미래에는 비(非) 소유와 공동체 생활에 기반을 둔 집단 거주(collective housing) 모델이 등장할 것으로 전망된다. 전통적인 가정의 개념이 완전히 무너지고 공공성(publicness)이 이를 대체하는 거대한 공동체 주택 생활이 예상된다. 미래의 급진적인 도시형 거주의 전형은 우리가 지금 알고있는 '집'의 형태가 사라지고 공적인 공동체적 가사활동이 지배적일 것이다. "

amid.cero9 presents a model for collective housing based on dispossession and communal living, based on a deep examination of Korean Bangs, and how the presence of Jimjilbangs has drastically shifted interiors towards a non-conventional domesticity, A radical prototype for inhabiting cities in the future, the proposal celebrates the disappearance of homes as we know them and takes command of public interiors for communal domestic activities.

집 없는 문명
아미드.세로9

A CIVILIZATION WITHOUT HOMES
amid.cero9

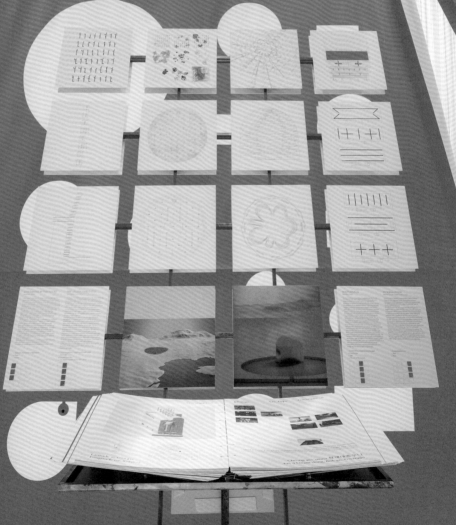

걸프 지역 최초의 알루미늄 제련소는 1968년 바레인에 건설되었다. 현재 이 제련소는 단일 제련소 중 세계에서 네 번째 규모로, 기원전 제3천년기(BC 3,000~2,001년)로 거슬러 올라가는 금속 무역의 역사를 오늘날에도 이어가고 있다. 알루미늄 생산 공정에 대한 조사를 바탕으로 한 본 설치작품은 필름, 사진, 모래 거푸집으로 만든 원형 모형(prototype) 등을 활용해 알루미늄의 생산 주기와 알루미늄을 둘러싼 각종 정치적 관계를 소개하고, 알루미늄의 새로운 활용 가능성을 연구한다.

The first aluminum smelter in the Gulf region was inaugurated in 1968 in Bahrain and is today the fourth largest single-site smelter in the world, continuing a history of metal trade rooted in the third millennium BC. An investigation into its production processes, through film, photography and a sand-casted aluminum prototype, attempts to understand the politics and production cycle of the material and to extract a different potential of its use.

생산의 장소, 알루미늄
누라 알 사예, 안네 홀트롭

PLACES OF PRODUCTION, ALUMINIUM
Noura Al-Sayeh and Anne Holtrop

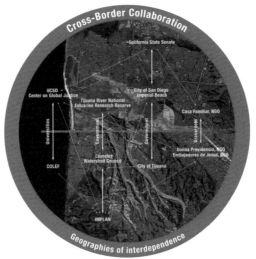

세계에서 가장 큰 국경도시이자 서반구 접경지대 기준으로 일일 교통량이 가장 많다고 알려진 샌디에고와 티후아나에 초점을 맞추고 있다. 그 결과 연구에 기반한 정치적이고 건축적 산물을 이 지역에 배출할 뿐 아니라 UCSD 국경 접경지대 공유 정류장 건설을 통해 분리된 공동체 사이의 공존을 도모할 수 있었다. 공유 정류장은 두 국가가 접하는 국경선 인근 이민자 마을에 위치하고 있다. 이곳에서 대학 연구진, 학생, 공동체 지도부가 공동으로 연구, 교육, 도시화 활동을 하고 있다.

This project focuses on the San Diego-Tijuana conurbation, with more border crossings daily than any other check point in the Western Hemisphere. Presenting the UCSD Cross-Border Community Stations as an infrastructure for co-existence between divided communities; field hubs located in immigrant neighborhoods on both sides of the border where research, teaching and urban advocacy are done collaboratively among university researchers, students and community leaders.

국경 접경지대 공유정류장
에스투디오 테디 크루즈 +
포나 포르만

CROSS-BORDER COMMUNITY STATIONS
Estudio Teddy Cruz +
Fonna Forman

두 개의 프로젝트는 전략적으로 건축과 도시계획을 책임감 있는 사회 변혁의 수단으로 활용하여 공공 공간의 인간화와 취약 부문에 대한 사회적 접근을 우선 과제로 삼아왔다. 리오네그로, 우리 역사의 색채는 콜롬비아 도시 리오네그로의 역사적 도심을 되살리기 위해 구성된 시민 참여형 민관 협력체다. 그리고 수십 년 동안 방치되고 부패와 후견주의(clientelism)로 극심한 사회적 타격을 입은 부카라망가의 공공 공간을 복원하고 활성화했다.

Two projects strategize architecture and urban planning as a tool for responsible social transformation. Rionegro the Color of our History is a public-private alliance for the revitalization of the city's historical center. In Bucaramanga, TABUÚ reactivated and recovered public spaces that had been abandoned for decades and where deep social damage has been produced by corruption and clientelism.

집합 공간의 변신: 공공 공간, 민주주의의 반영
아르키우르바노 + 타부:
존 오르티스, 이반 아세베도

TRANSFORMATION OF COLLECTIVE SPACES: PUBLIC SPACES, THE REFLECTION OF DEMOCRACY ARQUIURBANO + TABUÚ: John Ortiz and Iván Acevedo

Hong Kong protest. July 2019

저철분유리(UCG) 투쟁: 저철분유리의 환경 파괴적 도시화에 대한 개입조치

안드레스 하케 / 오피스 포 폴리티컬 이노베이션

The Ultra-Clear Glass Rebellion: An Intervention on the Toxic Urbanism of Ultra-Clear Glass

Andrés Jaque / Office for Political Innovation

432 Park Avenue magazine, 2014.

1989년, 베를린 장벽이 무너졌으며, 천안문 사태가 벌어졌고, 남아프리카 공화국에서 프레데리크 빌럼 데 클레르크가 아파르트헤이트 철폐에 착수하였다. 또한 동일한 해에 나타난 급진적인 기술·사회적 발명으로 세계를 운용하는 방법이 재편되기 시작하였으며, 이러한 발명은 금융, 환경, 문화 및 기관 간 서로 이해하는 방식을 재정립하고자 하였다. 이처럼 중대한 정치적인 변화는 동시대 권력분산 형 건축에 상당히 효과적으로 뿌리 내려 거의 인지되지 않았다. 1989년 PPG 인더스트리즈는, 최초의 산업용 저철분 유리 스타파이어로 특허를 받았다.[1] 이 놀라울 정도로 투명한 유리는, 곧 가장 선호되는 건축 자재 및 글로벌 재력의 지표가 되었다. 2008년까지 네 개의 다국적 기업 세인트 고바인, 가디언 글래스, 필킹톤, AGC는 저철분유리(UCG) 생산에 있어 연 6%의 안정적인 성장률을 보였다.[2] UCG 산업은 연간 227만 1,000톤의 생산량을 기록하며, 2021년까지 51억 달러 규모의 매출을 달성할 것으로 보인다.[3] UCG는

1989 was the year the Berlin Wall fell. The year of the protests in Tiananmen Square. The year when Frederik Willem de Klerk initiated the end of apartheid in South Africa. It was also the year that a radical techno-social invention started to reshape the way the planet operates, an invention intended to redefine the way finances, environments, cultures, and bodies relate to each other. This momentous political turnover was embedded so effectively in the contemporary architecture of power distribution that it could hardly be perceived. In 1989, PPG Industries patented Starphire; the first industrially produced low-iron floated glass.[1] A surprisingly clear glass that rapidly became a most-desired architectural component and an indicator of global financial power. By 2008, four transnational corporate giants, Saint Gobain, Guardian Glass, Pilkington, and AGC, had stabilized a steady 6% annual growth in the production of Ultra-Clear Glass (UCG).[2] With an annual production of 2,271,000 tons, the UCG industry will reach $5.1 billion in revenue by 2021.[3] It is commercialized under brands such as Starphire, Diamant, Clearvision, Optiwhite or AGC Clear.

UCG is composed of 73% SiO_2, 14% NaO_2, 10% MgO, and 1% Al_2O_3.[4] With an iron content below 0.01% (ten times less than ordinary glass) UCG reaches up to 97% of light transmittance,[5] filtering out the greenish tone that characterizes ordinary glass. The colorless quality of UCG multiplies its compatibility with an exceptionally broad range of radiation-filtering coatings, allowing a very selective management of the amount of energy contained in the incoming and outgoing radiation of a building. This capacity makes UCG the best performing clear panel used to engineer the transmittance of incoming and outgoing

1   "Sustainable in Every Light," Vitro Architectural Glass Brochure. http://www.vitroglazings.com
2   "Global Ultra Clear Glass Market 2019 to 2023", ReportsWeb. https://www.reportsweb.com
3   Idem

1   "Sustainable in Every Light," Vitro Architectural Glass Brochure. http://www.vitroglazings.com
2   "Global Ultra Clear Glass Market 2019 to 2023", ReportsWeb. https://www.reportsweb.com
3   Ibid.
4   "Starphire Technical product data sheet," Vitro Architectural Glass Brochure.
5   "Guardian Clarity," Guardian Glass Industries brochure. https://www.guardianglass.com

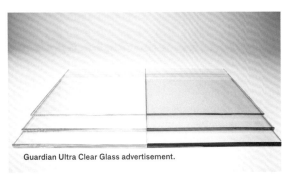

Guardian Ultra Clear Glass advertisement.

Ultra Clear Glass production in Guardian advertisement.

스타파이어, 다이아먼트, 클리어비젼, 옵티화이트 또는 AGC 클리어 등의 브랜드로 상업화되었다.

UCG는 73%의 이산화규소(SiO_2), 14%의 이산화나트륨(NaO_2), 10%의 산화마그네슘(MgO), 1%의 산화알루미늄(Al_2O_3)으로 구성되어 있다.[4] UCG는 철의 함량이 0.01% 이하(일반 유리 대비 10배 적은 수치)로 광선 투과율[5]은 97%에 달하며, 일반 유리의 특징인 녹색 톤을 걸러낸다. 무색의 UCG는 매우 광범위한 방사선 여과 코팅과의 호환성을 증대시켜, 건물에 출입하는 방사선에 포함된 에너지 양을 선별적으로 관리할 수 있도록 돕는다. 이와 같은 성능 덕분에, UCG는 유리외관의 건물에 출입하는 방사선 투과율을 조절하는 데 사용되는 최고 성능의 투명 패널이 되었다.[6]

UCG는 사물이 더 파랗게 보이게 한다. UCG에서 광선투과율은 광선 스펙트럼의 저온 부분에서 더 높게 나타나며, 이 푸른 영역의 나노미터 대역 폭은 450에서 550 사이이다.[7] 스펙트럼의 저온 부분 투과율은 90%에 달한다.[8] 스펙트럼의 고온 부분의 나노미터 대역 폭은 700이상이며, 광선 투과율은 80%로 떨어진다.[9] 결론적으로 UCG의 경우, 파란 빛 투과율이 더 좋기 때문에 하늘을 포함한 사물들이 더 파랗게 보이는 것이다.

UCG는 애플, 아마존, 알코아, 구글, 굳이어, 컴캐스트, 언스트 앤드 영, 시노팩의 본사 건물뿐 아니라[10], 센트럴 파크 사우스 220번지에 위치한 2억 3,800만 달러 팬트 하우스[11], 파크 애비뉴 432번지의 9,100만 달러 콘도미니엄, 서울 하얏트 호텔, 코넬 테크 교육 허브,

radiation in buildings with glass envelopes.[6]

Ultra-Clear Glass makes things look bluer. In UCG, light transmittance is higher in the cold fraction of the light spectrum—the blue zone that runs between the 450-550 nanometers wavelength.[7] In that part of the spectrum the transmittance reaches 90%.[8] At the hot side of the spectrum that runs above the 700-nanometer wavelength, the light transmittance drops to 80%.[9] As a result of this, UCG favors the transmittance of blue light, which makes objects appear bluer, including the sky.

UCG can be found in the headquarters of Apple, Amazon, Alcoa, Google, Goodyear, Comcast, Ernst and Young, and Sinopec.[10] In the $238 million penthouse at 220 Central Park South.[11] In the $91 million condominium of 432 Park Avenue, the Seoul Hyatt Hotel, the Cornell Tech's educational hub, Terminal 5 of Chicago Airport, Langone Medical Center in New York.[12] In Bentley, Audi, and Mercedes Benz cars.

Fifty-nine years after Buckminster Fuller and Shoji Sadao imagined a three-kilometer glass dome covering Manhattan's midtown corporate power, countless pieces of UCG, distributed worldwide, have encapsulated the daily life of the global wealth of the planet, enclosed in a consistent, blue-skied environment of apartments, offices, engineered landscapes, and transportation devices. So pervasive that it becomes invisible to those who inhabit it. Literally invisible. Ultra-clear invisible.

Sold as the top premium product of the glass industry, the price of UCG is at least 30% higher than any standard clear glass.[13] Exceptional clarity.

4   "Starphire Technical product data sheet," Vitro Architectural Glass Brochure.
5   "Guardian Clarity," Guardian Glass Industries brochure. https://www.guardianglass.com
6   "The original environmental performer," Vitro Architectural Glass webpage.
7   "Sustainable in Every Light," Vitro Architectural Glass Brochure.
8   Idem
9   Idem
10   "Vitro Architectural Glass project browser," Vitro Architectural Glass webpage.
11   Idem

6   "The Original Environmental Performer," Vitro Architectural Glass webpage.
7   "Sustainable in Every Light," Vitro Architectural Glass Brochure.
8   Ibid.
9   Ibid.
10   "Vitro Architectural Glass project browser," Vitro Architectural Glass webpage.
11   Ibid.
12   Ibid.
13   "Glass Price Calculator," Crystal Glass and Mirror Corporation. https://www.crystalglassny.com

시카고 공항 5번 터미널, 뉴욕 랭곤 메디컬 센터에서도 사용되었다[12]. 그 뿐 아니라 벤틀리, 아우디, 메르세데스 벤츠 자동차에서도 발견할 수 있다.

버크민스터 풀러와 쇼지 사다오가 맨해튼 미드타운 기업들을 덮을 수 있는 3킬로미터에 달하는 유리돔을 상상했다. 그 후 지난 59년 동안 전 세계적으로 셀 수 없이 많이 공급된 UCG는 아파트, 사무실, 조경환경, 이동수단 등 어디에서나 파란 하늘을 볼 수 있는 부유한 지역의 일상을 둘러싸고 있다. UCG가 일상 곳곳에 침투해 있는 나머지 오히려 거주자들은 잘 볼 수 없다. 문자 그대로 눈에 보이지 않는 초 투명(ultra clear) 유리인 셈이다.

유리 업계 프리미엄 제품으로 판매되기 때문에, UCG의 가격은 일반 투명 유리보다 최소 30% 높다.[13] 뛰어난 선명도, 최적의 투명도, 여과되지 않는 자연광, 제한 없는 전망, 무색 외관, 색의 순도. 이러한 문구들이 UCG를 전세계적으로 상업화하기 위해 흔히 사용된다. 건축학적 컴퓨터 생성 이미지(CGI)는 현재 다양한 이미지의 알고리즘 융합을 활용하기 때문에, 다양한 빛 조건들과 색 스펙트럼이 동시에 나타날 수 있다.[14]

HDR라고 불리는 이 기술은 하이엔드 실내를 구현하기 위해 모든 주요 건축 렌더링 회사에서 사용되고 있다.[15] 렌더링을 통해, 외부를 볼 때와 마찬가지로 선명하고 다채로운 실내를 경험할 수 있다. 여기서 헬리콥터 전망 만큼 중요한 것은, 색채와 빛의 주요 무대인 야외와 연속적으로 어우러지는 실내의 색과 빛을 구성할 수 있는 지 여부이다.

CGI 전문가인 필립 크루피에 따르면, 디지털로 만들어낸 선명도는 "실내와 실외에 대한 시각적 소유권의 감정을 (시청자에게) 동시에 발생"[16]시킬 수 있어야 한다. 아파트의 가치를 결정하는 주요 자산은 정확히 무엇인가? 선명도 정도에 따라 가치가 결정되는 것이 현실이다. 이에 따라 CGI는 선명도를 구현하는 능력으로 사회 구조를 형성하는 과정에서 핵심적인 요인으로 부상하였다.

부동산 고객을 대상으로 포괄적 서비스를 제공하며 브랜딩과 시각 스토리텔링을 전문으로 하는 크리에이티브 에이전시 네오스케이프의 아트 디렉터 라이언 콘에 따르면, "고객은 항상 하늘이 파란색이기를 바란다."[17] 고객들은 어떤 종류의 유리를 결국 사용하게 될 것인지에 관계없이 그들의 건물 렌더링 속 유리가 UCG이기를

Optimum transparency. Unfiltered natural light. Unrestricted views. Colorless appearance. Purity of color. These are the commonplace phrases used to commercialize UCG worldwide. Architectural computer generated images (CGI) now use an algorithmic blending of a wide range of images so different light conditions and color spectrums can appear as simultaneous.[14]

This technique, called high dynamic range (HDR), is used by all major architectural rendering companies to characterize high-end interiors.[15] These interiors are experienced in the renderings as being equally as luminous and colorful as the exteriors they look out at. As important as gaining access to helicopter views is the possibility for interiors to be constructed in chromatic and light continuity with those views they have a privileged domain on.

In the words of Phillip Crupi, a CGI expert, this form of digitally manufactured clarity is meant to "produce [in the viewer] the feeling of a simultaneous visual ownership of both the interior and the exterior,"[16] which is exactly the main asset the value of the apartments depends on. Clarity is the reality by which power stabilizes value. The capacity to enact clarity makes CGI a fundamental actor in the making of social structuring.

According to Ryan Cohn, art director of Neoscape Inc., a full-service creative agency specialized in branding and visual storytelling for real-estate clients, "Clients always want it blue."[17] They want the glass in the renderings of their buildings to be UCG regardless of what kind of glass they will end up using.[18]

They want to see a blue version of the city sky, both through and reflected in UCG.[19]

CGI's ultimate goal consists of assembling territorial ownership with seemingly ecological moral authority. Ownership by providing visual access to unique helicopter views and HDR light continuity between rooms and landscapes. Seemingly ecological moral authority by encapsulating high-end life into a continuous medium of skies perceived as blue and apparent energy efficiency.

Computer-generated blue skies and light interiors are both military inventions. Following the 1958 USSR launching of the Sputnik, US president Dwight Eisenhower founded the Advanced Research Projects Agency (ARPA), which later

12 Ide.
13 "Glass Price Calculator," Crystal Glass and Mirror Corporation. https://www.crystalglassny.com
14 Crupi, P. 2019, June. Personal Interview.
15 Crupi, P. 2019, June. Personal Interview.
16 Crupi, P. 2019, June. Personal Interview.
17 Cohn, R. 2019, June. Personal Interview.

14 Phillip Crupi, Personal Interview, June 24, 2019.
15 Ibid.
16 Ibid.
17 Ryan Cohn, Personal Interview, June 21, 2019.
18 Ibid.
19 Ibid.

원한다.[18]

고객들은 UCG를 통해 보는, 그리고 UCG에 반사된 도심 하늘이 파란색이기를 바란다.[19]

CGI의 궁극적인 목적 중 하나는 토지 소유권을 소위 생태학적인 도덕적 권위와 통합시키는 것이다. 여기서 소유권은, 특유의 헬리콥터 전망과 함께 방과 풍경 사이의 HDR 빛 연속성에 대한 시각적 접근을 제공함으로써 보장된다. 또한 소위 생태학적인 도덕적 권위는, 부유층의 삶을 파란 하늘과 에너지 효율성을 나타내는 연속적인 매체로 요약함으로써 만들어진다.

컴퓨터로 생성된 파란 하늘과 빛 인테리어는 모두 군사적 발명품이다. 1958년 소련이 스푸트니크를 발사한 이후, 당시 미국 대통령 드와이트 아이젠하워는 고등연구계획국(Advanced Research Projects Agency, ARPA)을 설립하였으며, 이 기관은 이후 방위고등연구계획국(Defense Advanced Research Projects Agency, DARPA)과 유타 대학교의 ARPA기금 에번스 그래픽 랩으로 분리된다.[20] 이 기관들 모두 여전히 활발히 활동하고 있다. 이 곳에서 최초의 CGI가 탄생하였으며, 현재 전세계적으로 사용되는 대부분의 디자인 툴도 이 곳에서 발명되었다.[21] 1970년대, ARPA 출신 연구자들은 첫 번째 CGI 영리 회사를 출범시켰다. 1982년, 블루 스카이 스튜디오는, 첫 CGI 상업영화인 〈트론〉을 출시하였다. 블루 스카이 스튜디오는 에번스 랩 출신 연구자들이 설립한 회사이다. 1995년 〈토이 스토리〉를 디자인할 당시, 픽사는 유리 및 유리와 빛의 상호작용 방식을 표현하면서 '사실성' 확보에 어려움을 겪었다.[22] CGI 전문가인 조셉 브레넌에 따르면, 유리는 "유일하게 남은 예술의 한계"이다.[23] 따라서 UCG를 거의 보이지 않게 만드는 광 투과율 수준으로 UCG를 묘사하는 것은 여전히 CGI 프로듀서들이 어려움을 느끼는 과제이다.[24] 마치 클래딩이 부족한 중간 문설주가 공중에 떠 있는 것처럼, CGI에서 UCG가 보이는 경우는 오직 다음의 세 가지 특징이 있을 때이다.[25] 첫째, 잔여 푸른 톤. 둘째, 무지개 같은 모서리 및 반사. 셋째, 보이지 않는 유리 표면 위 약간의 반사 및 가짜 결함.

became the Defense Advanced Research Projects Agency (DARPA) and the ARPA-funded Evans Graphics Lab at the University of Utah.[20] They are all still fully operational. It was there where the first CGIs were produced. And it was there where most design tools currently used around the world were first developed.[21] In the 1970s, former ARPA researchers launched the first CGI commercial firms. By 1982, Blue Sky Studios had launched *TRON*, the first CGI commercial movie. Blue Sky Studios (note the name) was founded by former Evans Lab researchers. When designing *Toy Story* (1995), Pixar had trouble achieving convincing "realness" in rendering glass and the way it interacted with light.[22] According to CGI expert Joseph A. Brennan, glass is the "only remaining artistic frontier."[23] The depiction of UCG with levels of light transmittance that make it almost invisible still challenges the capacities of CGI producers.[24] Depicted as a literal lack of cladding, as MULLIONS FLOATING IN THE AIR, Ultra-Clear Glass is only visible in CGI through three features.[25] First: residual blueish tones. Second: rainbowy edges and reflections. Third: slight reflections and fake imperfections on the invisible glass surface.

The rendering of UCG is by far the most time-consuming and expertise-requiring feature in high-end architectural representation. The technology that concentrates power needs to be invisible in order to be effective. Helicopters and drones, and auxiliary non-digital settings with human models, are now used widely in the production of architectural images.

VRAY or 3dsMAX, both software capable of generating realistic digital environments, are no longer used alone. Architectural CGI is no longer pure CGI, but rather composites of digital and non-digital images.[26]

No longer Photoshop, VRAY, or 3dsMax, it is Fusion that is the number one tool used to compose architectural renderings. In its commercial brochure, Fusion is said to "easily create sophisticated effects by processing images from different types and sources."[27]

18 Cohn, R. 2019, June. Personal Interview.
19 Cohn, R. 2019, June. Personal Interview.
20 Glenn Fong, "ARPA Does Windows: The Defense Underpinning of the PC Revolution," *Business and Politics*, 3(3), Cambridge: Cambridge University Press, 2017, pp. 213-237.
21 Robert Rivlin, *The Algorithmic Image: Graphic Visions of the Computer Age*, New York: Harper & Row, 1986.
22 Pat Hanrahan, "A Conversation with Ed Catmull; The head of Pixar Animation Studios talks tech with Stanford professor Pat Hanrahan," *ACMQUEUE Magazine*, Volume 8, Issue 11, November, 2010.
23 Szot, J. 2019, June. Personal Interview.
24 Cseh, J and Chappell, S, June 2019. Personal Interview.
25 Cseh, J and Chappell, S, June 2019. Personal Interview.

20 Glenn Fong, "ARPA Does Windows: The Defense Underpinning of the PC Revolution," *Business and Politics*, 3(3), Cambridge: Cambridge University Press, 2017, pp. 213-237.
21 Robert Rivlin, *The Algorithmic Image: Graphic Visions of the Computer Age*, New York: Harper & Row, 1986.
22 Pat Hanrahan, "A Conversation with Ed Catmull; The head of Pixar Animation Studios talks tech with Stanford professor Pat Hanrahan," *ACMQUEUE Magazine*, Volume 8, Issue 11, November, 2010.
23 J. Szot, Personal Interview, June 17, 2019.
24 J. Cseh and S. Chappell, Personal Interview, June 19, 2019.
25 Ibid.
26 Ibid.
27 "Fusion 16," Black Magic Design webpage. https://www.blackmagicdesign.com/

UCG의 렌더링은 하이엔드 건축 표현이나 묘사 시 확실히 가장 시간 소모가 크고, 전문성이 요구되는 일이다. 모든 권력을 결집시키는 기술은 눈에 띄지 않을 때 더 효과적이다. 헬리콥터와 드론, 사람이 등장하는 보조적 비(非) 디지털 세팅은 건축 이미지 생성에 있어 널리 활용되고 있다.

현실적인 디지털 환경을 구현할 수 있는 V-레이(VRAY)와 3ds 맥스, 이 두 소프트웨어는 이제 더 이상 단독으로 사용되지 않는다. 건축 CGI는 더 이상 순수한 CGI가 아니라, 디지털과 비-디지털 이미지의 합성이라고 할 수 있다.[26]

이제 건축 렌더링을 구성하는데 사용되는 최우선 툴은 포토샵, V-레이, 3ds 맥스가 아니라 퓨전(Fusion)이다. 광고 책자는 퓨전이 "다양한 유형과 출처에서 선택한 이미지를 가공함으로써 쉽게 정교한 효과를 만들어 낼 수 있다"라고 소개하고 있다.[27]

CGI가 사회구성기술이 되면서, 퓨전은 권력을 양성하는 기술·사회적 매개체가 되고 있다. 산화 철은 오븐의 용융 속도를 높여준다. 이 속에서 이산화규소 모래는 백열성의 덩어리가 되며 여기서 다시 투명한 유리가 만들어진다. 산화 철은 또한 주석조에서 유리의 냉각 과정을 단축시킨다. 그러나 UCG의 생산을 위해서는 이산화규소 모래에서 산화 철을 제거할 필요가 있다. 이로 인해 혼합물을 녹이는데 필요한 에너지는 30% 증가하며, 냉각 과정을 늦추기 위해 필요한 에너지는 100% 증가한다. 즉 공중에 떠 있는 중간 문설주를 만드는데 필요한 에너지의 두 배 이상이 필요하다. 이와 같은 UCG생산에 필요한 에너지의 73%가 천연가스로 사용된다.[28] 미국 내 건축에서 사용되는 UCG로 인한 에너지 소비 증가는 수압 파쇄로 추출된 천연가스의 사용으로 보충된다.

전 세계 친환경을 주창하는 글로벌 도시의 하이엔드 건물에는 코팅된 UCG가 설치되어 있다. 이 UCG의 에너지 효율 성능 영향으로, 톨레도, 오하이오, 웨드로, 일리노이와 같은 지방 도시들은 대기 중 이산화탄소(CO_2)와 질소 산화물(NO_x)의 극심한 증가를 경험하고 있다. 저 철분 유리를 생산하기 위해서는 원재료로 이산화규소 모래와 극히 소량의 산화 철이 필요하다. 페어마운트 센트롤, NYSE US 실리카 등 고품질의 저철분 이산화규소 모래를 전세계적으로

As CGI becomes a societal composing technique, Fusion became the techno-social medium where power is nurtured. The presence of iron oxide raises the melting rate in the ovens where silica sand becomes the incandescent mass out of which clear glass is produced. Iron oxide helps shorten the process of cooling the glass at the tin bath. The production of UCG requires the removal of iron oxide in the mix of silica sand. This increases by 30% the energy required to melt the mix, which together with a 100% increase in the energy required to slow down the process of cooling, more than doubles the energy required to make MULLIONS FLOAT IN THE AIR. 73% of the energy fueling the production of UCG comes from natural gas.[28] In the US the increase of energy consumption UCG brings to the built environment is fueled by the natural gas extracted by hydraulic fracturing.

As an associated effect of the energy-efficient performance of coated UCG installed in high-end buildings in global cities branded as green all around the globe, rural locations like Toledo, Ohio or Wedron, Illinois are experiencing a radical increase in the atmospheric concentration of CO2 and NOx. The production of low-iron glass requires the use of silica sand with an almost non-existing content of iron oxide as a raw material. Companies providing high-quality, low-iron silica sand worldwide, such as Fairmount Santrol or NYSE US Silica, concentrate their production in the US in the states of Illinois and Wisconsin. The product is known as St. Peter Sandstone. The growth of surface mining this sand has been boosted by an increase in the demand from both the UCG and fracking industries.[29]

The volatility of the sand led to the pollution of farmland and water bodies in the region and a non-scrutinized exponential growth of lung-affecting diseases such as silicosis and carcinoma.[30] Susquehanna Valley is the rural area in Pennsylvania where the natural gas that fuels UCG production is extracted through fracking from the Marcellus Shale.

As a result, 36,000 stream miles in Susquehanna Valley jeopardize human and animal life due to the high concentration of pollutants associated with the fracking industry. LEED (Leadership in Energy and Environmental Design) is the most widely used green building rating

26 Cseh, J and Chappell, S, June 2019. Personal Interview.
27 "Fusion 16," Black Magic Design webpage. https://www.blackmagicdesign.com/
28 "Glass Manufacturing is an Energy-Intensive Industry Mainly Fueled by Natural Gas," U.S. Energy Information Administration. https://www.eia.gov

28 "Glass Manufacturing is an Energy-Intensive Industry Mainly Fueled by Natural Gas," U.S. Energy Information Administration, https://www.eia.gov
29 "Select Sands Begins Shipping Industrial Sand; Resource Estimate Increases 91%," Seeking Alpha. https://seekingalpha.com
30 S. Younger, "Sand Rush: Fracking Boom Spurs Rush on Wisconsin Silica," National Geographic, July 4, 2013.

Landscape arrangement composed by Cabot Oil & Gas company in the Susquehanna Valley

공급하는 회사들은, 대부분의 생산을 미국 일리노이와 위스콘신 주에서 하고 있다. 이렇게 생산된 모래 제품이 세인트 피터 샌드스톤이다. 또한 UCG 및 수압 파쇄 산업 내 수요 증가로 이 모래의 노천 채광 역시 늘고 있다.[29]

모래의 휘발성으로 인해 지역의 농지와 수질이 오염되었고, 규폐증과 상피성 암 같은 원인 모를 폐 질환이 급격히 증가하였다.[30] 한편UCG 생산에 쓰이는 천연가스가 펜실베니아 교외의 서스크해너 계곡 마르셀러스 셰일층에서 수압 파쇄를 통해 추출되고 있다.

그 결과, 수압 파쇄 사업으로 배출된 높은 농도의 오염물질로 인해 수스케한나 계곡 내 3만 6천 스트림 마일에 해당하는 지역에서 인간과 동물의 삶이 위태로워졌다. 에너지 및 환경 디자인 리더십(LEED)은 세계적으로 가장 널리 활용되는 친환경 건축물 인증제도이다. 스타파이어 유리 1톤 생산 시 포틀랜드 시멘트 1톤의 생산보다 40% 더 많은 이산화탄소와 400% 더 많은 이산화황, 400% 더 많은 오존을 배출하지만, 특정 건물 평가 시 UCG가 설치되었을 경우, LEED인증 환경 제품 선언 검증(LEED-verified Environmental Product Declaration)을 받기 때문에 자동으로 2점의 가산점이 부여된다.[31]

UCG는 하나의 건설 재료일 뿐 아니라, 다양한 수준의 오염 노출 지역으로부터 인간과 동물을 격리시키고, 파란 하늘에 대한 환상을 지속적으로 품고

system in the world. Even though the production of a ton of Starphire glass implies 40% more CO2, 400% more SO2, and 400% more O3 emissions than a ton of Portland cement, when rating a specific building, two points are automatically earned when UCG is installed as these products have a LEED-verified Environmental Product Declaration.[31]

UCG is not just a construction solution, but a socio-territorial apparatus intended not only to segregate humans and non-humans in zones of diverse levels of pollution exposure, but also and mainly to provide ecological moral superiority to an elite encapsulated in a continuous bubble of blue-skied fantasy.

A blue-skied society that has offset the environmental cost of their existence to locations not visible from their UCG interiors. Neoliberal societies became addicted to BLUE. Since 2008, the consumption of UCG in the world has doubled.[32] UCG has become the material pervasively present in office, apartment, and commercial buildings in wealthy, global, urban settings.

UCG eliminates the use of iron in the composition of floated glass.[33] The glass loses its green color to become clear, with a capacity to block the transmittance of the ultraviolet, infrared, and orange spectrum of natural light when combined with coatings.[34] As a result, the

29 "Select Sands Begins Shipping Industrial Sand; Resource Estimate Increases 91%," Seeking Alpha. https://seekingalpha.com
30 Younger, S, "Sand Rush: Fracking Boom Spurs Rush n Wisconsin Silica," National Geographic, July 4, 2013.
31 "EPD Vitro Architectural Glass Flat Glass Products," Vitro Architectural Glass brochure.

31 "EPD Vitro Architectural Glass Flat Glass Products," Vitro Architectural Glass brochure.
32 "Global Ultra Clear Glass Market 2019 to 2023", ReportsWeb.
33 "The Choice is Clear," Guardian Ultra Clear Glass webpage.
34 "Sustainable in Every Light," Vitro Architectural Glass Brochure.

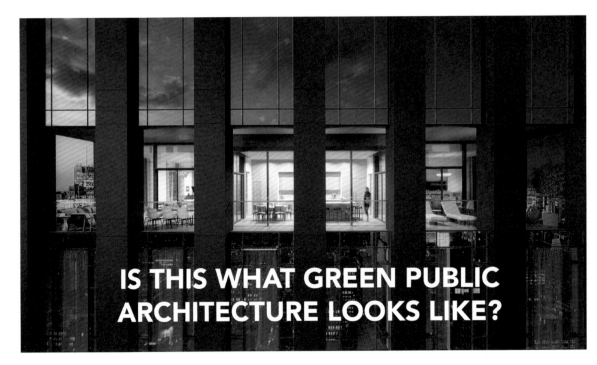

IS THIS WHAT GREEN PUBLIC ARCHITECTURE LOOKS LIKE?

있는 부유층에게 생태학적 도덕적 우월성을 제공해 주기 위해 마련된 사회적-영토적 장치이다.

그 존재의 환경적 비용을 상쇄하려는 파란 하늘 사회의 실체는 UCG 실내에서는 보이지 않는다. 신자유주의 사회는 파란 색에 중독되었다. 2008년 이래로, UCG 소비는 전 세계적으로 두 배 증가하였다.[32] UCG는 이제 부유한 글로벌 도심 속 사무실, 아파트, 상업용 건물에 흔히 사용되는 재료가 되었다.

UCG로 유동 유리 구성 요소인 철의 사용은 급감하였다.[33] 유리는 초록 빛깔을 잃고 투명해졌으며, 여기에 코팅까지 더해지면서 자연광선 중 자외선과 적외선, 주황색 가시광선의 투과를 막을 수 있게 되었다.[34] 결과적으로, 고급스러운 건물 실내에서 보는 바깥 하늘은 원래보다 더 파랗게 보인다. 2015년, UCG 소비의 증가로 유리 업계가 장기간 유지해 온 배출 감축 추세에 변화가 일기 시작했다.[35] 비록 UCG가 창문으로 인한 실내 온도 증가를 줄일 수 있다 하더라도, 이 유리를 생산하는 지역의 대기 중 이산화탄소와 질소산화물 농도가 전례없이 증가하였으며, 전 세계적으로 수압 파쇄 가스 추출을 촉진시켰다. 질소산화물이 가득한 노란색 하늘의 영역은, 파란 하늘을 추구하고 숭배하는 블룸버그 자산재단의 C40 도시나 로커펠러 재단의 세계 100대

outer sky looks bluer when seen from wealthy, global interiors. In 2015 the increase of UCG consumption reversed the long-sustained tendency in the glass industry to reduce its emissions.[35] Though UCG can reduce window-caused heat gain in building interiors, in sites of glass production it brought an unprecedented increase in CO_2 and NO_x in the air and promoted the practice of hydraulic fracturing gas extraction throughout the world. A realm of yellowish skies loaded with NOx segregated from a continuous urbanity networked by intercity initiatives—such as Bloomberg Philanthropic's C40 Cities or Rockefeller Foundation's 100 Resilient Cities—where BLUE hegemony is pursued and worshipped.

The UNBLUEING REBELLION promotes an intervention on the global urban addiction to ultra-clear BLUE, a combined action to emancipate techno-societies from segregation and the offsetting of air yellows.

Am I capable of wanting yellow, purple, green, pink, grey skies while engaged in reducing emissions?

Can others and I reengage environmentally by coexisting with the polluting effects of our daily functioning?

Digital means can be gained to sense the failure of bubbles.

This is a rebellion against the superiority of LEED-certificated, blued helicopter views. Others

32 "Global Ultra Clear Glass Market 2019 to 2023", ReportsWeb.
33 "The Choice is Clear," Guardian Ultra Clear Glass webpage.
34 "Sustainable in Every Light," Vitro Architectural Glass Brochure.
35 "Greenhouse Gases," AGC Glass webpage.
http://www.agc-glass.eu

35 "Greenhouse Gases," AGC Glass webpage.
http://www.agc-glass.eu

회복 탄력적 도시 등, 도시 간 이니셔티브로 연계된 지속적인 도시화와는 분리되었다.

'파란색 반대(UNBLUEING REBELLION)'는 전 세계 도시들이 투명한 파란색 중독 현상에 개입할 것을 촉구한다. 전 세계 도시의 공조를 통해 특정 도시들의 소외 현상을 해소하고 노란색 하늘을 상쇄하려는 압박에서 해방된 기술사회를 구현하자는 것이다.

과연 나는 배출 감축을 위해 노력하면서 노란색, 보라색, 초록색, 분홍색, 회색 하늘을 원할 수 있을까?

우리는 일상생활에서 발생한 오염의 영향과 공존하면서 환경에 다시 관여할 수 있을까?

디지털 수단이 거품의 실패를 인지하기 위해 사용될 수 있다.

이는 LEED 승인을 얻은 파란 헬리콥터 전망이 제공하는 우월함에 대한 반항이다. 우리에겐 전 지구적 도전과제들을 함께 대처해 나갈 공통 기반이 될 열등함과 함께, 헬리콥터 전망이 아닌 지면 위의 전망이 필요하다.

연구팀: 안드레스 하케, 마르코스 가르시아, 제시 매코믹, 에노 첸

and I claim inferiority and earthed vision as the common ground where a collective response to planetary challenges becomes possible.

Research Team: Andrés Jaque, Marcos García Mouronte, Jesse McCormick, Eno Chen

ARQUIURBANO + TABUÚ

RIONEGRO — 리오네그로

BUCARAMANGA — 부카라망가

TABUÚ

We are two Colombian teams that work every day for the social transformation of our cities

푸른 혁명:
저철분유리(UCG)의
환경파괴적 도시화에 대한
개입조치
안드레스 하케 / 오피스
포 폴리티컬 이노베이션
(OFFPOLINN)

BLUE REBELLION: AN INTERVENTION ON THE TOXIC URBANISM OF ULTRA CLEAR GLASS
Andrés Jaque / Office for Political Innovation (OFFPOLINN)

BLUE REBELLION An Intervention on the Toxic Urbanism of Ultra Clear Glass

Andres Jaque / Office for Political Innovation

저철분유리(UCG)의 세계 소비량은 2008년 이후 두 배 증가하였다. 판유리 생산 과정에서 철분을 제거한 것이 저철분유리다. 결과적으로 실내에서 보는 바깥 하늘을 원래보다 더 파랗게 보이게 한다. 그러나 이 유리를 생산하는 과정에서 대기 중 이산화탄소와 질소산화물 농도가 전에 없이 상승한 결과 저철분유리 생산 지역의 황색 하늘을 야기하였다.

Since 2008, the global consumption of ultra-clear glass (UCG) has doubled. The production of UCG eliminates the use of iron, changing the color of glass from green to clear, allowing the sky to look BLUE when seen from the interior. Yet this process emits vast quantities of CO2 and NOx, causing yellowish skies at sites of production.

They want the glass in the renderings of their buildings to be Ultra Clear Glass regardless what kind of glass they will use in the building.

'생각할(먹) 거리' 프로젝트는 상파울로와 도심 외곽 농경지대를 연구대상으로 한다. 환경보존, 식품, 영양보존, 지역사회경제개발 등을 통합적으로 살펴보며 이를 통해 지역 농업을 강화하고 사회의 환경지속가능성을 꾀하고자 하였다. 아울러 정책설계에 따라 도시의 위기를 완화하고 복합적인 문제 역시 해결할 수 있음을 제언하고자 하였다.

Food for Thought selects urban and peri-urban agriculture in São Paulo as an object of study in its integrative dimensions of aspects of environmental preservation, food and nutritional security and local socioeconomic development. It shows the extent to which design for policy can offer answers to complex problems in order to mitigate the crises of the metropolis.

생각할(먹) 거리
페르난도 드 멜로 프랑코

FOOD FOR THOUGHT
Fernando de Mello Franco

Organic production

of 26% per year

ARQUIUF

존 오타비오 오르티즈 로페라(Jo
아이반 다리오 에스베도 고메

RIONEGRO - 리오네

BUCARAMANGA -

We are two Colom

기록보관실은 중립적이지 않고 다양한 사회들을 반영한다. 오늘날에는 다수 이해관계자의 의견을 수렴해야만 기록물을 재배치하거나 정보를 선별 혹은 분류할 수 있다.

The archive is not neutral but reflects how different societies structured their world and acted within it. Reassembling the Archive asks how the selection and ordering of information today can be inclusive to range of voices and their corresponding forms of information.

기록의 재정립
오픈워크숍

RE-ASSEMBLING THE ARCHIVE
The Open Workshop

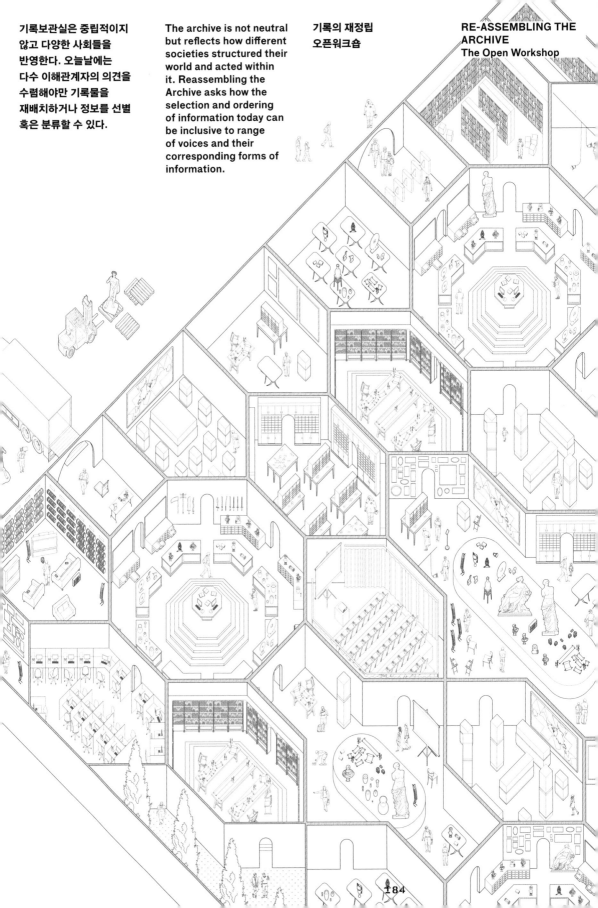

보레알 숲은 오슬로 공공지대의 정수를 보여준다. 삼림지대는 한때 나무를 기계로 베어 목재를 생산하는 곳이었다. 하지만 이러한 용도의 공간은 수십 년에 걸쳐 수많은 담론과 정책을 양산해 냈다. 그 결과는 베어져 나간 목재와 공터를 눈에 띄게 하지 않기 위한 일련의 규칙들이다. 본 프로젝트는 노르웨이 삼림 관리법의 허점을 짚어보고 이에 대한 대응책을 제시하고자 하였다.

The boreal forest is the quintessential public space in Oslo. This space is a machine to produce timber, however its use as a backdrop for all to enjoy led to decades of discussion and policymaking. The outcome is a sophisticated set of rules that hide timber extraction from sight. The project plays with the loopholes of these Norwegian forestry rules, interpreting the Norwegian forest as a large wood carpet.

숲의 모양들
LCLA 오피스, 데일 위브

SHAPES OF THE FOREST
LCLA Office with Dale Wiebe

'밤섬 당인리 라이브'는 도시의 자연적, 문화적 생태계에 관해 진행중인 매스스터디스의 두 프로젝트에 대한 수행적 현장 프레젠테이션이며 시민 참여를 수반해, 다양한 관점을 통해 '집합'의 개념을 풍부하게 하는 장이 될 것이다.

Bamseom Danginri Live is an on-site performative presentation of two in-progress projects that each deal with the natural and cultural ecology of the city, respectively. It will involve public participation, investigation, and dialogue, enriching the notion of the 'collective' through multiple views.

밤섬 당인리 라이브
매스스터디스 / 조민석

BAMSEOM DANGINRI LIVE
Mass Studies / Minsuk Cho

본 프로젝트는 리오 세코
강의 복구를 배수로에서
루안다 도시재생방안으로
제안한다.

The project proposes the
recovery of the Rio Seco
canal, from a drainage
ditch to being seen as
a solution to Luanda's
urban challenges.

리오 세코: 그린 네트워크
빌딩 소사이어티 포 아키텍쳐
/ 베라미노 산토스 "키우라"

RIO SECO: GREEN
NETWORK
Building Society for
Architecture / Belarmino
Santos "Kiwla"

RIO SECO

Building Society of Architecture

항공 및 지상 촬영된 사진을 합쳐 포토그래메터리(사진 측량) 과정을 통해 3D 모델을 구축했다.

Temple: Mam Rasha
Photogrammetric Re
Date documented: 0

이 3D 데이터를 위성사진, 역사적 기록과 교차 참조하여 해당 장소에 대한 정확한 기록과 분석을 할 수 있었다.
시간이 흐르면서 피해지가 더더욱 침식되고 중요한 증거가 사라짐에 따라 이
데이터는 복원에 필수적인 고고학적 사진 증거물로 인식된다.

Temple: Sheikh M
Photogrammetric
Date documented

2014년 습격 이후 미국에 설립된 야지디족 국제 비영리 구호단체 야즈다(Yazda)는 ISIL 대원을 상대로 제소된 국제법적 사건을 지원하기 위해 2018년 포렌식 아키텍처와 함께 당시 범죄 현장의 증거를 모아 공식적으로 문서화하기 시작했다.포렌식 아키텍처는 자체적으로 고안한 교육 프로그램을 통해 야지디족이 직접 설문 문항을 만들거나 사진을 활용한 설문조사를 진행할 수 있도록 지원해 왔다.

Forensic Architecture and Yazda partnered on an initiative to further document and analyze the destruction of Yazidi shrines and mausoleums by ISIL in 2014. A training program devised by Forensic Architecture allowed Yazda's Dohuk documentation team researchers to carry out a photographic and modelling survey themselves, generating evidence to support legal cases brought in international forums against members of ISIL.

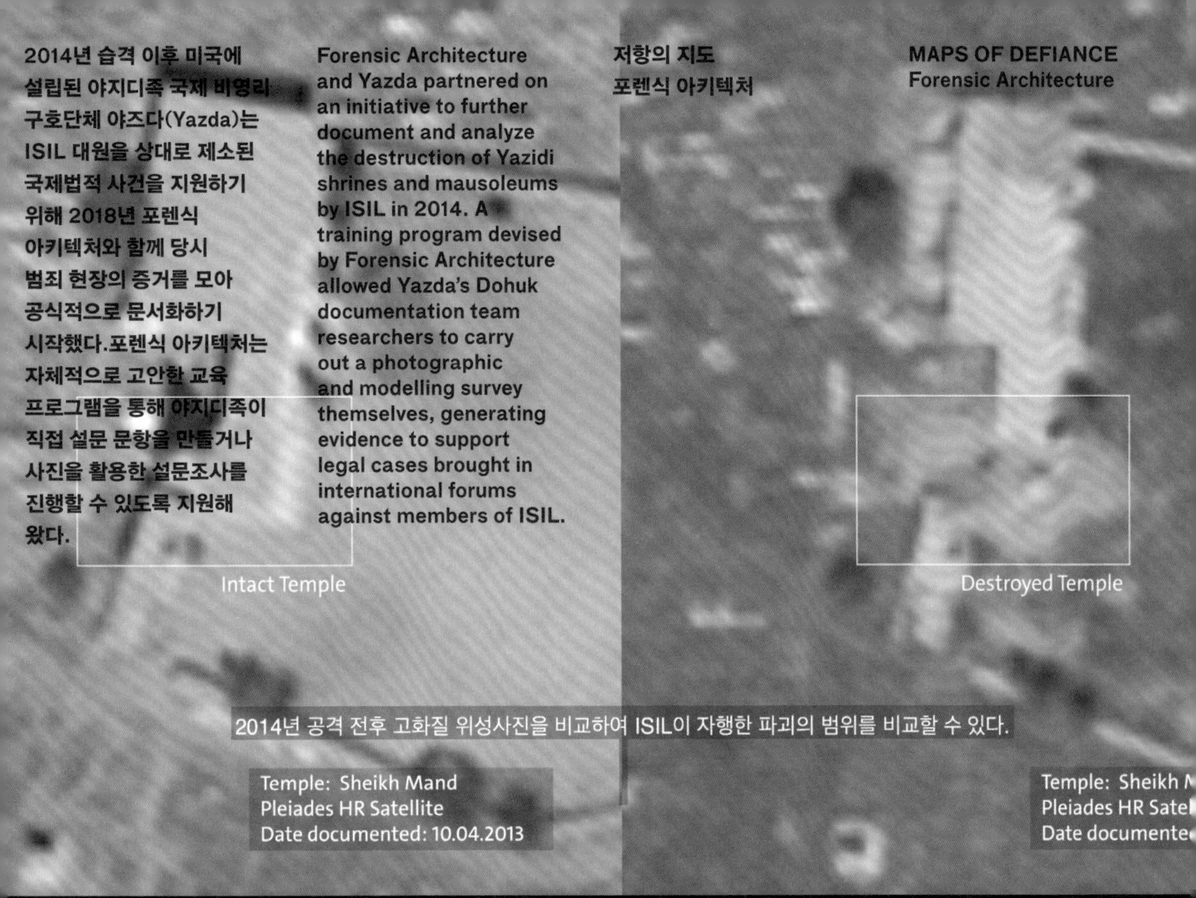

저항의 지도
포렌식 아키텍처

MAPS OF DEFIANCE
Forensic Architecture

Intact Temple

Destroyed Temple

2014년 공격 전후 고화질 위성사진을 비교하여 ISIL이 자행한 파괴의 범위를 비교할 수 있다.

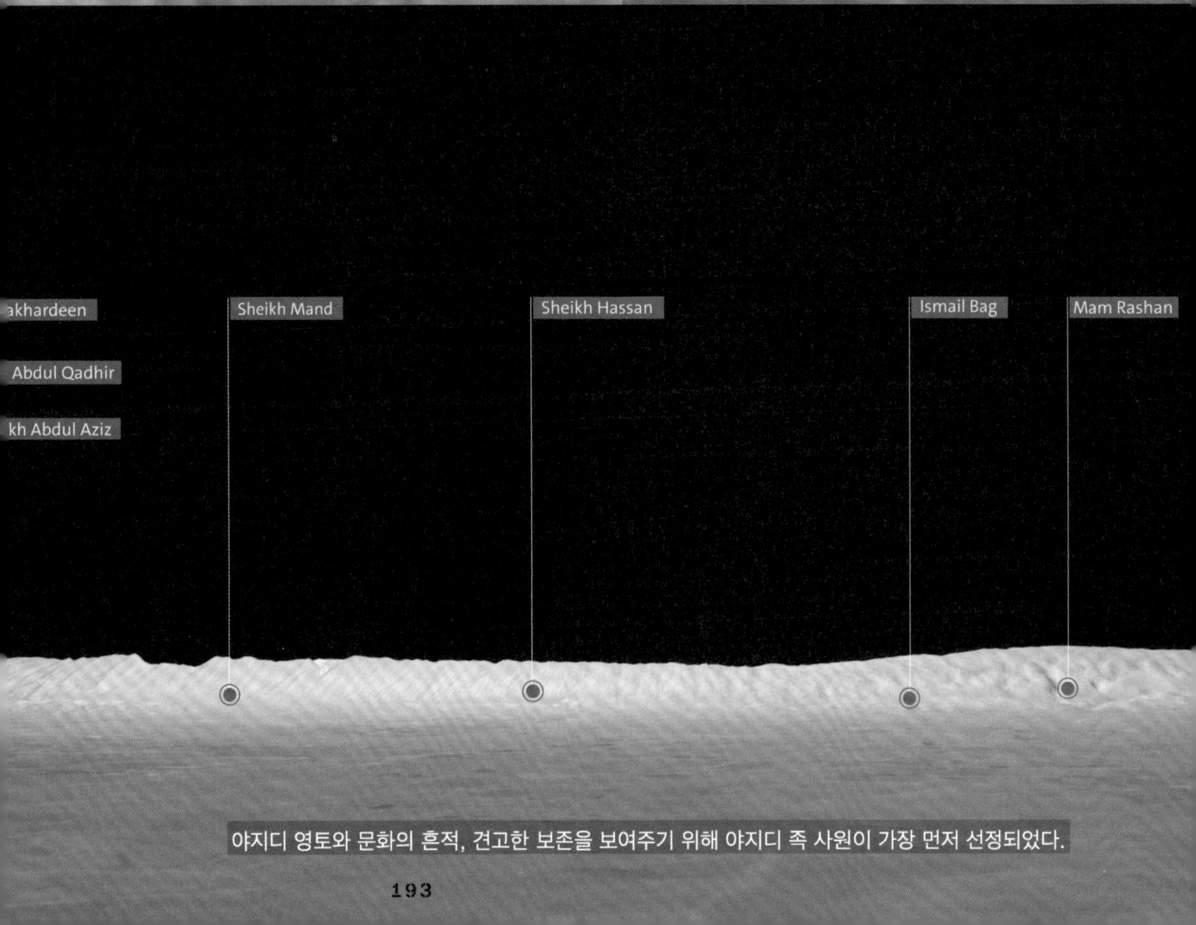

Temple: Sheikh Mand
Pleiades HR Satellite
Date documented: 10.04.2013

Temple: Sheikh M
Pleiades HR Satel
Date documente

akhardeen

Sheikh Mand

Sheikh Hassan

Ismail Bag

Mam Rashan

Abdul Qadhir

kh Abdul Aziz

야지디 영토와 문화의 흔적, 견고한 보존을 보여주기 위해 야지디 족 사원이 가장 먼저 선정되었다.

프로젝트 발자취
사미프 파도라 건축연구소
（sP+a/sPare）

PROJECTIVE HISTORIES
Sameep Padora Architecture and Research (sP+a/sPare)

프로젝트 발자취
사미프 파도라 건축연구소
（sP+a/sPare）

The practice of sP+a has evolved over the years into four distinct formats - practice, research, collaboration and collectives, focusing on projects that range from housing to the public sector. These are not siloed but are shaped to constantly feed into each other, a form of practice that can address vast multitudes of context and variations.

스틸컷, 에콰도르, 로라
우에르타스 밀란, 2012
© 르 프레누아, 프랑스
국립현대예술스튜디오 - 로라
우에르타스 밀란

Still, Aequador, Laura
Huertas Millán, 2012
© Le Fresnoy, Studio
national des arts
contemporains - Laura
Huertas Millán

스틸컷, 보이지 않는 도시의
흔적: 홍콩에 관한 세개의 노트,
판류, 보 왕, 2016
© 복 피처스

Still, Traces of an Invisible
City: Three Notes on Hong
Kong, Pan Lu and Bo Wang,
2016
© BOC Features

이미지와 건축 #11:
팔만대장경
바스 프린센

IMAGE AND
ARCHITECTURE #11:
TRIPITAKA KOREANA
Bas Princen

복제물, 사진, 예술품 등은 건축물이나 표현하고자 하는 공간을 얼마나 근접하게 묘사할 수 있는가? 이미지와 건축 시리즈에서는 건축에 활용된 재료와 재료의 특성을 이미지화하고 예술적 건축물을 놀라운 섬세함과 깊이로 재현하였다. Work #11는 1237년에 81,258의 목판으로 만들어진 세계적으로 위대한 대장경판인 팔만대장경을 다룬다.

How close can a copy, an artistic depiction, or a photographic print come to the object or space it depicts? In the series Image and Architecture, Princen seeks to capture and transfer the 'material' and 'materiality' of architectural spaces into one-to-one experiences. Work #11 in the series captures The Tripitaka Koreana, a collection of Buddhist scriptures dating back to 1237.

이러한 상황에서 출범한
'도시와 농촌 교류'
프로그램은 다양한 배경의
사람들이 농업에 도전할
기회를 제공하여 기존과는
다른 새로운 농업인을
배출하고자 한다.

Urban Rural Exchange
is a challenge to open
the farming process
and allow people with
various backgrounds to
participate.

도시와 농촌의 교류
아틀리에 바우와우

**URBAN RURAL
EXCHANGE**
Atelier Bow-Wow

스틸컷, 나이트 타임 고, 카라빙
영화 콜렉티브, 2017
작가 제공

Still, Night Time Go,
Karrabing Film Collective,
2017
Courtesy of the artists

현대 도시의 잠재력을 발견하는 능력을 키우는 것은 극단적인 도시화의 세계에서 무엇보다 중요하다. 따라서 비져너리 시티스 프로젝트는 공간적, 조직적, 도시의 힘과 압력에 의해 발생한 물질적 독창성을 탐구한다. 따라서, 도시적 독창성을 지닌 집합도시의 밑그림은 개별적이고 혼성화된 도시 지능으로 구성된 도시화를 투영하기 위해 이러한 마이크로 어바니즘(미시적 도시주의)을 활용한다.

Cultivating an ability to detect potentials in the contemporary city is paramount in our world of extreme urbanization. The Visionary Cities Project explores spatial, organizational, and material ingenuities born out of the forces and pressures of the city. The drawing of a Collective City of Urban Inventions projects an urbanism entirely composed of these individual and hybridized urban intelligences.

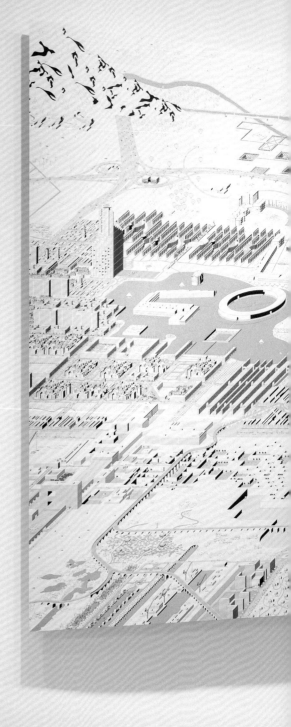

독창성의 도시
알렉산더 아이젠슈미트 /
비져너리 시티스 프로젝트

CITY OF URBAN
INVENTIONS
Alexander Eisenschmidt /
Visionary Cities Project

고대 그리스, 중세 이탈리아, 오늘날 아메리카 대륙의 도시 집합의 설계와 지속

앨릭 M. 맥클레인

Designing and Sustaining the Urban Collective in Classical Greece, Medieval Italy, and in the Americas Today

Alick M. McLean

도시의 집합성은 개개인과 장소 및 타인 간의 정서적 연결성을 의미하는 개념으로서 인류 최초의 도시가 생겨난 이래로 도시의 강점이자 약점이었다. 대부분의 도시에서 역사적으로 특정 시기에 정치와 경제 활동에 참여하지 못한 주민이 있었다는 사실은 집합 주거 구축이 사람 사이를 연결했을 뿐 아니라 분리했으며 사람들에게 권한을 부여한 동시에 종속시켰을 가능성을 암시한다. 집합 의식은 포용적이든 배타적이든 간에 공공정책, 건물, 공간, 공공 의례와 같은 집합성의 요소만큼 적극적이고도 의도적으로 설계된 개념이다. 고대 그리스와 중세 이탈리아에서 도출한 두 가지 사례를 통해 우리는 집합 주거를 확장한 혁신 기술, 확장을 제한한 요소, 그러한 한계가 공동체의 지속 가능성에 미친 영향을 더듬어 볼 수 있다. 현대 콜롬비아의 사례에서는 포용의 역사적 한계를 의도적으로 뒤바꾸는 전략을 발견할 수 있다. 과거 집합 주거의 긍정적인 속성을 합쳐 놓은 콜롬비아 모델은 역사적으로 소외된 집단의 기회를 극대화하고 유지하는 방법으로 도시 집합 주거를 확장하려는 미국과 기타 국가의 계획에 동시대적 관점에서 해결책을 제시할 수 있다.

도시가 주민 전부는 아니더라도 일부의 기회를 급속도로 극대화한 사례로는 후대 학자 사이에서 아티카 지역의 메트로폴리탄으로 일컬어지는 아테네 시에서 기원전 508년에 제정된 법률에 소개된 지도다(그림1). 해당 법률과 그 결과로 생겨난 도시 지형은 분명 정치적이고 공간적이었다. 클레이스테네스가 창안한 이 법률은 고대 아테네 부족의 사회적, 정치적 정체성의 집합체에 변화를 몰고 왔다. 클레이스테네스 이전에 존재한 아테네 4개 부족 체제는 혈연관계를 바탕으로 했다. 기원전 6세기 후반 아테네가 상업적, 군사적으로 성장하면서 일부 부족민에게 불공평한 기회가 제공되는 것이 비판의 대상이 되었다. 이러한 체제 하에서 귀족으로서의 혜택을 누리고 있던 클레이스테네스는 아테네 시민들에게 부족 체계를 해체하는 법률을 제정할 테니 자신을 도시 지도자인 집정관으로 선출해달라는 거래를 제안했다. 아테네 시민들은 이에 동의했고, 새로이 집정관으로 선출된 클레이스테네스는 정치 역사에서 가장

Urban collectivity, the emotional connection of individuals to place and to others, has been at once a strength and weakness of cities since the earliest urban settlements. The fact that nearly all cities have excluded some residents from full political or economic participation at some time in their histories introduces the possibility that collectivity can be forged as much to divide as to connect, to subordinate as to empower others. Whether inclusive or exclusive, the sense of collectivity is a construct, as actively and purposefully designed as the public policies, buildings, spaces, and public rituals that serve to shape it. Two historical examples, from ancient Greece and medieval Italy, track innovative techniques for expanding the collective, limits to such expansion, and the impact of such limitations on the sustainability of the communities. One modern example in Colombia provides strategies for deliberately reversing historical limits to inclusivity. Colombia's model, together with positive attributes of the earlier collectives, can inform contemporary initiatives to expand urban collectives in the United States and elsewhere while maximizing and sustaining opportunities for historically excluded groups.

One example of a city radically extending opportunity to some, though not all, residents is a map that is described in legislation in 508 BCE, in the city of Athens, and later rendered by modern scholars onto the metropolitan region of Attica (figure 1). The legislation and resulting urban topography were explicitly both political and spatial. They were crafted by Cleisthenes and involved reworking the nexus of social and political identity in ancient Athens, the tribes (*phylai*). Before Cleisthenes, the four tribes of Athens were based on extended family ties. By the late sixth century BCE, with the growth of Athens both commercially and militarily, unequal opportunities provided within tribes came under criticism. One of the aristocratic beneficiaries of this system, Cleisthenes, offered the citizens of Athens a deal: he would pass legislation to dismantle the historic tribal system, in exchange for his election as archon, the city leader. The Athenian citizens agreed, and the new archon Cleisthenes delivered

혁명적인 변화를 일으켰다.

클레이스테네스 치하에서 아테네는 기존 4개 부족에서 10개 부족 체제로 변화했다. 시민들은 가족 중심이 아니라 인구적, 지리적 요소를 중심으로 뭉치게 되었다. 부족들은 동일한 규모일 뿐 아니라 경제적 대표성의 다양성을 극대화하되 강력한 가족 지도자들의 영향을 최소화하는 방향으로 설계되었다. 아테네의 139개 지역과 마을은 10개 도시 공동체, 10개 내륙 공동체, 10개 해안 공동체로 분산되었는데, 이러한 공동체는 '트리튀스(trittys)'로 불렸다. 10개 부족 각각은 도시, 내륙, 해안의 트리튀스로 구성되었다.[1]

혈연관계가 유명무실해지고 지리적, 경제적 다양성이 법적으로 의무화됨에 따라 클레이스테네스와 이후 아테네 지도자들이 부족을 결집하기 위한 방안을 마련해야 했다. 그 가운데 하나가 징병제이다. 군부대는 같은 부족의 젊은이들로 구성되었다. 병역을 통해 경험을 공유하게 된 부족 구성원들은 서로 유대를 다지게 되었다. 음악극을 비롯한 연극이나 디티람보스(dithyramb, 고대 그리스의 디오니소스 찬미가)도 결속력 형성에 이용되었다.[2] 아테네 시의 관리들이 각 부족에서 가장 부유한 집안을 코레고스(choregos, 공연 후원자)로 임명하면 코레고스는 소년 50명과 성인 남자 50명을 선발해 개별적인 디티람보스를 개최했고 경연대회를 위해 전문적인 더블 리드 플루트 연주자도 고용했다.

합창단의 구성은 5세기 아테네 민주주의의 대표성 수준을 반영했다. 부족 간의 치열한 경쟁에도 불구하고 그들은 더욱 성대한 종교적 행사 무대를 마련하기 위해 서로 협력했다. 개별 부족으로 이루어진 군부대가 아테네의 군대 전체를 구성했던 것과 마찬가지였다. 아리스토텔레스의 말대로 전체는 부분의 합보다 컸다. 해안의 선원, 내륙의 농부, 도시의 상인, 제조업자, 관리, 성직자들이 결합됨에 따라 부족 내부와 아테네의 전반적인 정치계에서 모든 구성원의 발언권이 커졌다. 시민의 문해율이 높아지고 마을과 도시 연극에 시민이 참여하면서 자신을 표현하는 개개인의 역량이 강화되고 연마되었다. 극을 비평하고 분석하는 기술이 향상함에 따라 자신의 경제적 이해관계를 지키기 위한 웅변 기술

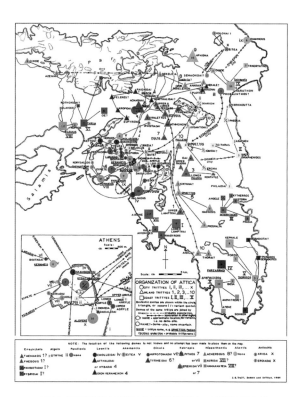

그림 1. 아고라 이미지: 2008.19.0013: "데메(deme)와 부족 지도. 각 색상은 10부족을 나타내고 각 표시는 개별 도시이다. 선은 같은 부족에 속한 도시를 잇는다. 각 표시에 붙은 숫자는 매년 도시에서 의회(Boule)로 보낸 대표자의 숫자를 나타낸다."
Figure 1: Agora Image: 2008.19.0013: "Map of demes and tribes. Each color stands for one of the 10 tribes, each mark for an individual deme. The lines link groups of demes that belong to the same tribe. The number within each mark records the number of representatives sent each year to the Council (Boule) from that deme." http://agora.ascsa.net/id/agora/image/2008.19.0013. Based on John S. Traill, The political organization of *Attica: a study of the demes, trittyes, and phylai, and their representation in the Athenian Council* [Princeton: American School of Classical Studies at Athens (ASCSA), 1975].

one of the most revolutionary political remappings in history.

The remapping of Athens under Cleisthenes consisted in replacing the four historical tribes with ten new tribes. Citizens no longer participated based on family, but according to a mix of demographic and geographical factors. Tribes were designed to be of equal size while maximizing the diversity of economic representation and minimizing the influence of powerful family leaders. The 139 neighborhoods and villages of Athens were distributed into ten urban, ten inland, and ten coastal sets of communities, known as *trittyes*. Each of the ten tribes was made up of one *trittye* from each.[1]

1   아테네 부족들의 재조직에 대한 전면적인 논의에 대해서는 다음을 보라: John S. Traill, "The political organization of Attica: a study of the demes, trittyes, and phylai, and their representation in the Athenian Council," Princeton: American School of Classical Studies at Athens (ASCSA), 1975.
2   Andronike Makres, "Dionysiac Festivals in Athens and the Financing of Comic Performances," in Michael Fontaine and Adele C. Scafuro (Eds.), *The Oxford Handbook of Greek and Roman Comedy*, Oxford: Oxford University Press, 2014, pp. 70-112.

1   For a full discussion of the reorganization of the Athenian tribes, see John S. Traill, "The political organization of Attica: a study of the demes, trittyes, and phylai, and their representation in the Athenian Council," Princeton: American School of Classical Studies at Athens (ASCSA), 1975.

또한 발전했다.[3] 디오니소스 극장, 아고라, 프닉스, 두 개의 아크로폴리스는 그러한 기술을 뽐내는 공간이었고 언어에 의한 표현이 한층 더 개발되고 교환되는 곳이었다. 이 같은 공간은 아티카 정치 체제가 혈연에 근간한 이전 4부족 체제 때보다 훨씬 더 많은 사람들의 기회를 극대화한 토대가 되었다. 이 시기에 아테네가 경제적으로 가장 역동적인 시기를 맞이한 것은 우연이 아니다. 그에 따라 문화가 꽃피고 사회가 발달했으며 토론이 활발해짐에 따라 시민의 비판적 표현과 자기변호 기량이 향상했다. 이러한 상황에 더하여 공공 법률이 제정되면서 개인과 단체가 자신의 경제적 이해관계를 주장하고 달성하기가 용이해졌다.

5세기 말에 이르러 아테네의 전성기는 전쟁 때문에 꺾이고 사양길로 들어섰다. 이러한 사실은 아테네라는 집합 주거의 강점과 약점을 동시에 보여준다. 아테네 경제는 외국인들의 참여를 허용했지만 근본적으로 '아테네인 우선주의'라는 정치적 입장 때문에 아테네인에게 유리한 환경이었다. 외국인 상인과 협력자들의 권리는 제한되었고 정복지의 외국인들은 모든 권리를 박탈하고 노예로 전락하여 농업을 비롯한 여러 분야에서 공짜로 노동력을 제공했다.[4] 펠로폰네소스 전쟁이 막바지에 이를 때쯤에는 아테네 해군 군용선에서 노 젓는 사람 대부분이 노예였다.[5]

집합성에 대한 아테네인의 추종은 부족 군부대와 연극 경연 대회를 통해 강화되었으며 배타성과 당파성의 요소를 지녔다. 디티람보스 축제의 심사위원은 경쟁이 심한 부족의 시위 때문에 어쩔 수 없이 군의 장군에게 승리자를 발표하게 하기도 했다.[6] 집합성의 어두운 면을 스스로 규제하지 못한 아테네는 외교적인 면에서 매우 무능했다. 시라쿠사 해군을 물리친 장군들이 적이 회복될 가능성을 전혀 고려하지 않은 오만한 우월주의 때문에 아테네 함대는 시칠리아 군사작전 도중에 비극으로 치달았다.[7] 아테네 사람들은 스파르타, 테베, 마케도니아, 훗날의 로마 사람들처럼 군의 당파성을 억제하지 못한 탓에 몰락했다. 인구의 절반(남편과 아들을 전쟁에 내보내고 슬퍼하는 여성들)이 전쟁을 석연치 않게 생각했음에도 아테네 사람들은 전쟁을 지속하면서

The disappearance of blood ties and the mandated geographic and economic diversity challenged Cleisthenes and later Athenian leaders to develop new ways to create tribe cohesion. One was required military service. Regiments were composed by the young men from a single tribe. Service provided shared experiences to bond members. The other way was through theater, particularly musical theater, or dithyramb.[2] City officials would appoint the wealthiest citizen in each tribe as *choregos*, and he in turn would select fifty boys and fifty men for the respective youth and adult dithyramb chorus competitions, as well as hire a professional double-reed flute player.

The composition of the chorus paralleled the degree of representation in Athenian democracy in the fifth century BCE. Despite intense competition between tribes, they worked together to create the larger religious spectacle of public theatre, just as the individual tribe regiments composed the entire Athenian army. The whole was greater than the sum of its parts. The mix of coastal seafarers, inland farmers, and urban merchants, manufacturers, administrators and priests afforded all parties greater say within tribes and in the overall politics of Athens. Each individual's capacity to express himself was bolstered by a high literacy rate among citizens, supported by universal attendance at village and urban theaters. Skills for critical analysis of dramas led to skills in self-advocacy for economic interests.[3] The Theater of Dionysus, Agora, Pynx, and Acropolis all provided spaces for such skills and vocalized interests further to be developed and exchanged, creating a public sphere that maximized opportunity for a far broader percentage of the Attic polity than the previous four blood tribes had afforded. It is no coincidence that the city entered its most dynamic period of economic growth, with parallel cultural achievements, the public development and debate of which only increased critical expression and self-advocacy within the citizenry, bolstering the capacity of individuals and groups to express and achieve economic interests through public legislation.

The fact that this Golden Age of Athens unravelled towards the end of the fifth century through war suggests at once the strength and weaknesses of the Athenian collective. Foreigners were invited to participate in Athens' economy, but

3 See Juergen Habermas, Thomas Burger (Trans.), *The Structural Transformation of the Public Sphere*, Cambridge, MA.: The MIT Press, 1989, pp. 4-5, also, pp. 24-26.
4 Sarah P. Morris and John K. Papadopoulos, "Greek Towers and Slaves: An Archaeology of Exploitation," *American Journal of Archaeology*, Vol. 109, No. 2, April, 2005, pp. 155-225.
5 A. J. Graham, "Thucydides 7.13.2 and the Crews of Athenian Triremes," *Transactions of the American Philological Association* (1974-) 122 (1992), pp. 257-270.
6 Andronike Makris, personal communication, August, 2017.
7 투키디데스, 『펠로폰네소스 전쟁사』, 6권과 7권.

2 Andronike Makres, "Dionysiac Festivals in Athens and the Financing of Comic Performances," in Michael Fontaine and Adele C. Scafuro (Eds.), *The Oxford Handbook of Greek and Roman Comedy*, Oxford: Oxford University Press, 2014, pp. 70-112.
3 See Juergen Habermas, Thomas Burger (Trans.), *The Structural Transformation of the Public Sphere*, Cambridge, MA.: The MIT Press, 1989, pp. 4-5, also, pp. 24-26.

그림 2: 보르고 알 코리노와 카스텔로 디 프라토, 1107-1157. McLean, 2008, figure 2.3.
Figure 2: Borgo al Cornio and Castello di Prato, 1107-1157. McLean, 2008, figure 2.3.

그림 3: 프라토 평면, 13세기 후반. McLean, 2008, figure 7.2.
Figure 3: Plan of Prato, late thirteenth century. McLean, 2008, figure 7.2.

그림 4: 프라토 평면, 14세기 중반. McLean, 2008, figure 7.3.
Figure 3: Plan of Prato, late thirteenth century. McLean, 2008, figure 7.2.

on Athenian terms, essentially an "Athens First" politics. Foreign traders and allies had limited rights, and other foreigners, those conquered in battle, lost all rights, becoming the slave population providing free labor to power, for instance, Athenian agro-industry,[4] and, towards the end of the Peloponnesian War, to compose a large portion of oarsmen rowing the city's naval triremes.[5]

The Athenian cult of the collective, nourished in tribal regiments and theatrical competition, had an element of exclusion and partisanship. Judges of dithyramb festivals were sometimes forced by the agitation of competing tribes to have army generals pronounce winners.[6] Athenians' inability to self-regulate the dark side of their sense of collective became catastrophic in foreign affairs. An arrogant chauvinism doomed the Athenian fleet during the Sicily campaign, when generals, having already defeated the Syracusan navy, were blind to the potential of their adversaries to recover.[7] The Athenians fell, like the Spartans, Thebans, Macedonians, and eventually Romans after them, to their inability to contain voices of militant partisanship. Athenians drained their economy and sacrificed their youth in continuous wars, despite the fact that as much as half the population questioned warfare: the future widows and grieving mothers. Enslaved foreigners were equally unenthusiastic.[8] Like the women of Athens, slaves had practically no political rights besides through their owners, the male citizenry of Athens, despite providing a large portion of the city's labor force and wealth. Were Cleisthenes' exemplary geographic and social remapping of the Athenian tribes to have emancipated and engaged the input of women and slaves, the scope of the public sphere would have sustained voices to balance military interests with concerns of gender, family, and labor.

One thousand seven hundred years later a similar remapping from noble to middle class occurred across nearly all of central and northern Italy, achieving a similar mix of positive but tragically limited results through different methods. City states from north of Rome shifted from primarily aristocratic enclaves dominated by feudal lords or high-ranking clergy to crucibles for

4 Sarah P. Morris and John K. Papadopoulos, "Greek Towers and Slaves: An Archaeology of Exploitation," *American Journal of Archaeology*, Vol. 109, No. 2, April, 2005, pp. 155-225.
5 See A. J. Graham, "Thucydides 7.13.2 and the Crews of Athenian Triremes," *Transactions of the American Philological Association* (1974-) 122 (1992), 257-70.
6 Andronike Makres, personal communication, Athens, August 28, 2017.
7 Thucydides, *The History of the Peloponnesian War*, books 6 and 7.
8 Graham, "Thucydides 7.13.2 and the Crews of Athenian Triremes," 257-70.

자신들의 경제적 부를 소비하고 젊음을 희생했다. 외국인 노예들도 전쟁에 냉담했다.[8] 노예들 역시 아테네 여성들처럼 도시의 노동력과 부에 크게 기여했음에도 자신들의 주인인 아테네 남성 시민을 통하지 않고서는 사실상 정치적인 권리가 없었다. 클레이스테네스가 시행한 아테네 부족의 지리적, 사회적 재편으로 아테네의 여성과 노예가 해방되고 사회에 참여할 수 있었다면 공론의 범위가 확대되어 군사적 이해관계와 성별, 가족, 노동에 대한 관심사를 조화시킬 수 있었을 것이다.

1,700년이 흐르고 이탈리아 중부와 북부 전역은 아테네의 사례와 마찬가지로 귀족에서 중산층 위주로 재편되었을 때 아테네와는 다른 방법을 통해 아테네와 유사한 긍정적인 성과와 비극적인 한계를 동시에 맞이했다. 로마 북쪽에 있는 도시 국가들은 주로 봉건 영주, 고위 성직자로 구성된 소수 귀족 계층이 다스리는 곳이었는데 상인과 제조업자 동업 조합의 주도 아래에 경제적 성장의 장이 되었다. 한정된 시간 동안 이탈리아는 시골 경제에서 도시 경제 체제로 변모했지만 이탈리아 도시 국가는 시골에서 도시로의 기회 재편은 물론 도시 내부에서의 기회 재편 과정에서 발생한 결함 때문에 정치적 자치 독립체로서 지속될 수 없었다.

이탈리아 중세 도시 국가 가운데서도 특징적인 곳을 들라면 토스카나에 있는 프라토가 있다(그림 2, 3, 4). 피스토이아와 피렌체 경계에 있는 이 도시는 교회와 귀족의 결합체로 시작되었다. 다른 이탈리아 도시 국가들과 마찬가지로 프라토는 두 번의 변화 과정을 겪으며 발전했다. 첫 번째 변화는 11세기와 12세기 사이에 일어났고, 두 번째는 13세기에 발생했다. 11세기에 문을 연 프라토 교구 교회는 농노의 자유를 설파하면서 시골 농노와 마을 사람들을 끌어들이기 시작했다. 새로운 주민들은 직물 산업 및 그 관련 사업, 교역, 금융업의 급성장 덕분에 일거리를 찾았다.[9] 시내 인구가 늘어나고 경제가 성장하며 다양화되면서 교구 교회인 산토 스테파노의 규모나 기념물에 대한 교구민들의 투자도 커지고 다각화되었다. 마을이 성공하면서 마을은 도시가 되고 귀족 가문이 들어왔다. 이들 중 일부는 물리적 시설이나 권한 면에서 교회 확장을 지지했고, 다른 일부는 프라토에 있는 성곽을 개발한 독일 황제를 지지하며 교회 확장을 반대했다. 이렇게 프라토의 다각적 재편이 시작되었다. 농노와 마을 주민들은 양모 공장, 도시 교역, 해외 교역, 금융업 등의 분야에서

economic growth led by constellations of merchant and manufacturing interests. For a limited time, Italy transformed itself from a rural to an urban economy, but flaws in the process of remapping opportunities from countryside to city, and within cities themselves, limited the sustainability of Italy's city-states as autonomous political entities.

One characteristic medieval city-state was Prato in Tuscany (figures 2, 3, 4). The city began as a pair of church and noble villages at the border between Pistoia and Florence. As with many Italian city-states, Prato developed through two remapping moments. The first occurred in Prato between the eleventh and twelfth centuries, and the second in the thirteenth century. Starting in the eleventh century, the parish church of Prato began to attract rural serfs and villagers, offering serfs freedom. New residents found work in the burgeoning cloth industry, in related businesses, and in trade and banking.[9] As the town population and economy grew and diversified so did parishioner investment in the scale and monumentality of the parish church, Santo Stefano. The success of the villages turned city-state attracted as well noble families. Some nobles supported the church's expansion in physical structures and authority, and others contested the church, supporting instead the German emperors, who had developed a castle stronghold in Prato. Prato therefore began with a remapping that varied according to class. Serfs and villagers retooled to perform urban labor, whether in the wool mills or urban and international trade and banking. Nobles tended less to retool, holding onto their traditional crafts of administration and warfare. Even when engaging in banking or trade, they persisted in the patriarchal leadership and highly competitive practices derived from their military background. Nobles in Prato as elsewhere in medieval Italy had multiple loyalties, to their families, to their family businesses, to their clans, which, as in the original Athenian tribes, consisted of extended families occupying entire neighborhoods. Their loyalties sometimes extended to their cities, and also to one of two larger super-clan organizations, the pro-papal Guelphs, and the pro-German imperial Ghibellines.

By the mid-twelfth century Prato developed its first quasi-democratic government composed of nobles and wealthy merchants. The main plaza projected, with arcades lining its walls and the adjoining town hall, a lay cloister and, with it, a lay life in common. At the same time the tower of the

8   Graham, "Thucydides 7.13.2 and the Crews of Athenian Triremes," 257-70.
9   이에 대해서는 다음을 보라: Alick M. McLean, *Prato: Architecture, Piety, and Political Identity in a Tuscan City-State*, New Haven: Yale University Press, 2008.

9   For this and the following, see Alick M. McLean, *Prato: Architecture, Piety, and Political Identity in a Tuscan City-State*, New Haven: Yale University Press, 2008.

도시 노동력으로 개편되었다. 귀족들은 자신들의 전통적인 관리 기술, 전쟁 기술을 고수하면서 변화에 덜 적극적인 모습을 보였다. 금융업이나 교역에 참여했지만 가부장적인 통솔력이 필요하거나 군대와 관련된 경쟁적 활동을 고집했다. 중세 이탈리아 여러 곳과 마찬가지로 프라토의 귀족들은 자기 가문과 가문의 사업은 물론 고대 아테네 부족과 마찬가지로 대가족 형태로 지역 전체를 장악한 씨족 등 여러 대상에 충성심을 보여야 했다. 때로 이들의 충성심은 도시로 확대되었을 뿐 아니라 교황을 지지하는 교황당이나 독일 황제를 지지하는 황제당 등 규모가 큰 씨족 집단으로도 확대되었다.

12세기 중반 프라토는 귀족과 부유한 상인으로 구성된 유사 민주주의 정부를 출범했다. 주요 광장은 회랑으로 둘러싸인 형태로 시청과 세속인들을 위한 수도원 가까이에 기획되었다. 동시에 교회 탑, 시청 탑, 여러 파벌의 군사 요새가 광장 안팎에 삐죽이 솟아있었다. 이런 상황에서 유추해보면 이곳의 'vita comunalis(공동생활)'이 취약했고 도전을 받았으며 사회 정치적 사실이라기보다는 건축적 허구에 가까웠음을 알 수 있다. 라틴어 vita comunalis는 현대 이탈리아어의 'comune(공중)', 영어의 'commons(평민)', 'commonwealth(연합)', 'community(공동체)' 등의 어원이 되기도 한다.

시민 광장(Platea Plebis)과 관련된 논쟁은 1260년에 황제당이 토스카나에서 추방된 친황제 가문들과 토스카나 지역의 구심점인 시에나와 군사적 동맹을 맺으면서 터무니없는 국면을 맞이했다. 이들의 동맹은 몬타페르티 전투에서 프라토와 피렌체를 비롯한 교황당 도시의 군대를 무찔렀다. 승리한 황제당 세력은 도시로 복귀하여 교황당 세력을 축출했고 교황당이 황제당에게 했던 것처럼 교황당의 저택을 파괴했다. 1266년 베네벤토 전투에서 황제당이 승리하고 프라토를 비롯한 토스카나 중심부로 돌아오면서 사태는 전환점을 맞이했다. 또다시 보복성 추방이 이루어지고 탑과 저택이 파괴되었다. 이렇게 물고 물리는 피바다 속에서 기존 시청과 산토 스테파노 주변의 건물 대부분이 파괴되어 1330년대 재건축 법안이 통과되기 전까지 제대로 발달하지 못한 채 커다란 공백으로 남았다. 60년 넘는 기간 동안 교회 공간은 훼손되고 방치된 상태였다.

귀족 당파의 몰락과 그로 인한 교회와 시민 광장 파괴는 역설적이게도 프라토의 제2차 도시 재편을 촉발했다. 주민들은 전통적인 교구 중심지에서 새로운 종교 지역으로 눈길을 돌렸다. 이러한 종교 지역은 1220년대부터 프란체스코 수도회나 도미니크 수도회 같은 탁발 수도회에 의해 개발되었으며 그 파격적인 특징

church, the tower of the town hall, and the many fortified, military towers of the *consorteria* families bristling on and behind the plaza walls, suggested that this common life, or *"vita comunalis,"* from which we get the modern Italian term of *"comune"* and the English "commons," "commonwealth," and "community," was fragile, contested, and more architectural fiction than social-political fact.

The contestation of the *Platea Plebis*, the people's plaza, came to its full absurd head in 1260, when Ghibelline families allied militarily with exiled pro-imperial clans across Tuscany and with their power center in the region, Siena. The alliance defeated Pratese, Florentine, and other Guelph city armies at the Battle of Montaperti. The victorious Ghibellines returned to banish their Guelph counterparts, and destroyed many of their houses—a practice Guelphs had earlier used against them. In 1266 the tides turned again, with the Battle of Benevento returning victorious Guelphs to Prato and other Tuscan centers, and with further reprisals of exile and tower-house destruction. At some point during this extended blood bath, the original town hall and most of the buildings around Santo Stefano were destroyed, leaving a now far larger, but unformed space until the 1330s, when laws were finally passed to rebuild the space. For over sixty years the church space remained desecrated and neglected.

The meltdown of competing noble collectives and the resulting destruction of the main church and civic plaza paradoxically initiated in Prato the community's second moment of urban remapping. Residents turned from the traditional parish center to other religious sites that had been developed in the city by mendicant orders, such as the Franciscans and Dominicans, starting in the second decade of the thirteenth century. The anomalous qualities of these sites seem to have attracted worshippers. There were no monumental church structures. Preachers dressed in simple clothes, often sack-cloths, and even presented themselves in bare feet. Officiants did not always follow traditional liturgy, and regularly led urban processions. When they preached, mendicant friars did so in a language never heard in the Roman church since its first Latin masses at the dawn of the Christian era: the vernacular. The sacred words, prayers, sermons, and songs voiced in the vulgar tongue marked the legitimization of vernacular into what was becoming the Italian language. Just as the walls of church architecture had dissolved, so had the mysteries of the Latin mass.

Due to the dense construction of the town centers, and the cursed association with structures razed by clan warfare, the spaces where

때문에 예배자들을 끌어들인 것으로 보인다. 무엇보다도 웅장한 교회 건물이 없었다. 설교자들은 마대 천으로 만든 소박한 의복을 입었고 맨발로 다녔다. 사제는 전통적인 예배 의식을 따르지 않을 때도 있었고 주기적으로 도시 행렬을 이끌었다. 탁발 수사들은 로마 교회에서는 기독교 태동 이후로 한 번도 사용하지 않은 토착어로 설교했다. 종교적인 용어, 기도문, 설교, 성가가 속된 언어로 표현됨에 따라 이탈리아어로 발전하던 토착어가 정당성을 갖추게 되었다. 교회 건축물의 벽이 와해되었듯이 라틴어 미사의 신비도 사라졌다.

시내 중심부에 밀집된 건설, 가문 간의 전투로 파괴된 구조물과의 저주받은 연관성 때문에 탁발 수도회 설교자들이 군중을 모으는 공간은 시내의 주변부나 도시 성벽 밖 시장에 자리 잡는 경향이 있었다. 모든 사람을 환영하는 탁발 수도회 설교자들은 도시 경제에 가담하는 새로운 이주자들을 설교 장소 주변으로 이끌어 그곳에서 제조나 교역에 종사하거나 관련 일자리를 찾고 거주지를 마련하도록 했다. 경제적 기회가 성장하고 이와 병행하여 인구가 폭발적으로 증가하면서 도시는 포물선 형태로 성장했다. 새로운 기회의 일부는 새로운 '교우(compagnie)'로 불리던 탁발 평수사로 말미암아 발생했다. 이미 12세기 후반부터 부유하고 유력한 가문들조차 시골에서 갓 이주한 문맹인들만큼이나 탁발 수사들에게 매력을 느끼고 있었다. 의복이나 웅장함을 따지지 않는 설교 공간에 반영된 탁발 수사들의 반계층적 조직은 일반 신도들에게까지 확대되었다. 모든 형제는 평등했고 동일한 제복을 입었으며 부활절 이전의 성 목요일 예식에서 서로의 발을 씻겨줬다. 탁발 수사들은 급진적인 소박함으로 신성함뿐만 아니라 신도들을 신에게 다가가게 하거나 신을 신도들에게 다가오게 할 것만 같은 능력을 갖춘 보였고 계층과 상관없이 누구에게나 겸손하게 행동하는 본보기가 되었다. 귀족이든 부유층이든 계층에 상관없이 누구나 탁발 수도회와 공동으로 자선 조직을 운영하고 그 과정에서 일반인들에게 일자리와 그 이외 여러 기회를 제안했다. 그에 따라 탁발 수도회는 사회적 기회의 장이 되었다.

탁발 수도회에 의해 제공된 기회는 1,300년 전 아테네의 사례와 마찬가지로 집합적인 상호작용, 담론, 직업을 유의미하게 재편했다. 외부인에 대한 프라토의 개방성은 혁신을 만들어냈다. 14세기 중반 광장의 기념행사에서 평수사로 활동하기 시작한 이주 노동자들이 14세기 말에는 새롭게 들어선 민중 정부 '세콘도 포폴로(Secondo Popolo, 제2의 민중)'의 지도자가 되었다. 이들은 귀족 이외의 주민 대다수에게까지 참정권을 확대했다. 그러나 여전히 여성은 제외되었다.

mendicant preachers drew crowds tended to be at the town's periphery, often outside city walls at marketplaces. The welcoming by mendicant preachers of all comers led many new migrants joining the city economy to participate, to find work in the expanding manufacturing and trade jobs at and around these sites, and to house themselves nearby. Urban growth became parabolic, with population explosion paralleling growing economic opportunities. New opportunities may be partly explainable by new religious "compagnie," or mendicant lay brotherhoods. As early as the late twelfth century wealthy, powerful, clan families were as attracted to the mendicant orders as new, illiterate migrants from the countryside. The anti-hierarchical organization of the mendicants, visible in their dress and their anti-monumental preaching spaces extended to lay brethren. All brothers were equal, dressed in identical vestments, and, for example, during Holy Thursday ceremonies just before Easter washed one another's feet. The radical simplicity of the mendicants rendered their sanctity, their ability potentially to draw the faithful closer to God and vice-versa, more credible, and provided models for similar acts of humility for anyone, of any class. Brotherhoods became crucibles for social opportunity, as noble or wealthy members shared with their brothers, regardless of class, experience in managing the charitable operations of the brotherhoods, and in the process offered jobs or opened other possibilities.

The opportunities afforded by the mendicants constituted as significant a remapping of collective interactions, discourse, and jobs as their parallel 1300 years earlier in Athens. Prato's openness to outsiders constituted a novelty. Immigrant workers who had started as lay brothers in mid-fourteenth century plaza celebrations became leaders of the new popular government, the *Secondo Popolo*, during the last decades of the century. They expanded suffrage to a majority of non-nobles, but still excluded women. The newly empowered governors continued to ignore Prato's destroyed central plaza, but initiated a return to monumental architecture by providing land grants and other gifts to mendicant orders, which started to interpret flexibly mendicant building moratoria. By the 1280s and 1290s mendicant plazas had churches greater in scale and monumentality than the seemingly forgotten Santo Stefano. The canons and provost of Santo Stefano responded by developing their own confraternity, around the newly rediscovered relic of the Sacred Belt of the Virgin. By 1317 they had received both private and communal funding to provide the relic a huge, monumental setting in a new transept. Shortly

새로이 권한을 부여 받은 관리자들은 계속해서 프라토의 파괴된 중심가 광장을 방치했지만 이들이 탁발 수도회에 제공한 토지 보조금과 다른 혜택 때문에 웅장한 건축물이 다시 건축되기 시작했다. 이러한 현상은 탁발 수도회에 대한 건축 유예(building moratoria)로 해석되기에 이르렀다. 1280년대에서 1290년대에 이르기까지 사람들의 뇌리에서 지워진 듯한 산토 스테파노보다 더 크고 웅장한 교회들이 탁발 수도회의 광장에 들어섰다. 산토 스테파노의 참사회 회원들과 주임 사제는 새로 발견된 신성한 유물 성모 허리띠(Sacred Belt of the Virgin)를 구심점으로 독자적인 단체를 만들어 대응했다. 1317년에 이르기까지 이들은 성모 허리띠 주위로 웅장하고 기념비적인 배경을 만들기 위해 개인과 단체 모금을 통해 트랜셉트(transept, 십자형 교회의 좌우 날개 부분)를 새로 지었다. 그 이후 얼마 지나지 않아 옥외에도 설교단을 건설했고 1332년에는 기념비적 공간을 합법화한 도시 설계 조례를 통해 새롭고 더 큰 규모로 산토 스테파노 광장을 복원하기 시작했다. 중심부 교회는 엄선된 탁발 수도회 관습과 불가사의한 힘과 기념물로의 회귀를 혼합하는 전략을 구사하여 탁발 수도회를 따르던 일반 신도들을 장엄한 그곳으로 복귀시키는 데 성공했다.

프라토에서 기념물과 신비에 바탕을 둔 신앙을 되찾는 작업은 씨족 가문 행태로의 복귀와 동시에 이루어졌다. 세콘도 포폴로는 패배한 황제당을 제거하고 관공서에 귀족뿐만 아니라 교황당 세력까지 발을 들여놓지 못하도록 하는 법을 제정했다. 다른 주변 도시도 비슷한 법을 제정했으며 오래지 않아 법을 지키지 않는 사람들이 축출되면서 씨족 가문을 배척하는 사고방식이 공동체 전체에 확산되었다. 그러면서 대내외적으로는 군사적, 경제적 경쟁의 시대가 시작되었다. 프라토의 통치 위기와 불안정성은 1351년 피렌체가 교황 세력인 앙주 가문 후원자들과 도시의 권리를 은밀히 사들이는 협상을 할 때까지 계속해서 고조되었다. 프라토는 잦아들 때도 있지만 결코 사라지지 않는 씨족주의 사고방식 때문에 일어난 당파 간의 폭력적 분쟁 문제를 해결하는 데 어려움을 겪었다. 여성이 통치에 참여할 수 없었던 데다 신규 이주자와 노동자에 대한 관용이 반전을 맞이하면서 이들의 목소리는 정치적 담론에서 배제되었다. 그렇지 않았다면 이들은 통제 불가능한 씨족 가문, 도시 간 경쟁, 침략의 문제를 해결하는 데 도움을 주었을 것이다.

한층 더 지속 가능한 도시 기회 재편 사례는 현재에서 찾을 수 있다. 콜롬비아 메데인 시가 그 사례다. 1994년 1월 1일 세르히오 파하르도가 시장에 취임한 후 그의 보좌진은 미스 메데인 선발대회 등 공공 자금으로 진행되는 미인 대회로 쏟아지던 일반인들의 관심을 재능

afterwards they added a new pulpit to the exterior, and in 1336 they began reconstruction of the new, larger Piazza Santo Stefano through urban design ordinances providing for a regularized monumental space. The central church's gambit to mix select mendicant practices with a return to magic and monumentality successfully redirected some of the pious from the mendicant periphery to their grand sacred center.

The return to a faith based on monument and magic in Prato paralleled a return to clan behavior. After banishing defeated Ghibellines, the *Secondo Popolo* passed legislation to ban all nobility, even noble Guelphs, from public office. Other nearby cities passed similar laws, and before long the mentality of clan exclusiveness, through the process of banishing its perpetrators, infected the communes themselves, initiating a period of internal and external military and economic competition. Prato's governance became increasingly contested and unstable up to 1351, when the Florentines quietly negotiated buying the rights to the city from its Guelph Angevin patrons. As with Athens, the Pratese had difficulty managing the violent partisanship engendered by a sometimes dormant, but persistent, clan mentality. The absence of representation of women in governance, and a reversal of openness to newcomers and laborers, eliminated from political discourse voices whose diverse interests may have helped manage unbridled clan and inter-city competition and aggression.

To find examples of more sustainable urban opportunity remapping it is necessary to turn to the present. The city is Medellín, Colombia. On January 1, the day he began as mayor in 2004, the first act of Sergio Fajardo and his team of advisors was to redirect public support from the Miss Medellín Beauty pageant and other publicly funded beauty contests to a new event— the Women with Talent competition, for achievements in "science, technology, entrepreneurship, culture, and arts."[10] The mayor and his team proceeded to prepare a Development Plan for Medellín 2004-7. For the plan to succeed his administration would have to get it passed within six months by the twenty-one members of the Medellín city council elected at the same time he was, only two of whom had supported him. Fajardo declined to engage in the traditional coalition building exercised by previous administrations, instead inviting councilors to

10 For this and the following see Alejandro Fajardo and Matt Andrews, "Does successful governance require heroes? The case of Sergio Fajardo and the city of Medellín: A reform case for instruction," *WIDER Working Paper* 2014/035, Helsinki: UNU-WIDER, 2014.

있는 여성들을 위한 '과학, 기술, 기업가 정신, 문화, 예술' 분야 업적 경연 대회로 돌려놓는 데 성공했다.[10] 시장과 그의 보좌진은 2004-2007 메데인 개발 계획을 마련하기 위한 과정에 돌입했다. 이 계획을 성공시키기 위해 시 행정부는 시장과 함께 선출된 메데인 시의회 의원 21명의 동의를 받아 6개월 이내에 계획안을 통과시켜야 했는데, 시의원 중 두 명만이 그를 지지했다. 파하르도는 이전 행정부와는 달리 전통적인 연립정부 구성을 거부했다. 그 대신 시의원들을 모아 자기 계획에 대해 공개적으로 논의하는 동시에 시원들에게 비판에 대한 대안을 제시하라고 압박을 가했다. 6개월이 거의 끝나갈 무렵에 의회는 만장일치로 그의 계획을 승인했다. 그의 성공은 비판자를 배제하거나 측면에서 공격하는 것이 아니라 비판자를 의사 결정 과정에 참여시켜 이전 행정부와는 달리 그 기여를 공공연하게 인정하는 방식을 통해 이루어졌다. 파하르도는 의회 내에서 스스로나 다른 사람들이 전통적인 내부 당파 집단을 구성하는 것을 거부한 가운데 자신의 행정부와 의회 전반을 연계하여 더 크고 효율적인 정치 집단을 구축했다.

　　파하르도 행정부가 대표성이 부족했던 집단을 참여시키고 당파를 초월하여 연립정부를 구성한 과정은 개발 계획 그 자체가 되었다. 개발 계획으로 기회가 도시 전체로 분배되었다. 파하르도 행정부는 교통, 주택, 일자리 접근성, 자원, 정치적 발언권의 개선과 같은 조치를 부각했을 뿐 아니라 그처럼 물질적이고 정치적인 이익을 실현하는 데 필수적인 요소를 강조함으로써 개발 계획을 통해 기회를 도시 전체로 재분배했다. 모든 주민이 존엄한 만큼 평등한 권리를 누려야 한다는 파하르도 정부의 주장이 통한 것이다. 그러한 존엄성이 지속될 수 있도록 추진된 이 계획에서는 곳곳에 창의적인 요소가 엿보인다. 우선 도시 내의 소외된 지역을 파악하고 그곳에 세계 정상급의 건축과 일류 서비스를 갖춘 도서관 등의 공공기관을 위치시킨다. 새로운 구조물, 그 주변 지역, 전체 도시 사이의 환승 연계성을 개선한 방안도 시행되었다(그림 5, 6). 시 행정부는 총체적 도시 프로젝트(Integral Urban Project)로 알려진 과정을 통해 선별된 지역과 환승선이 지나가는 지역에 고속 선로를 기반으로 한 경제적, 사회적 기회를 제공했다. 해당 프로젝트는 한층 더 확대되어 콜롬비아 알바로 우리베 대통령의 정책으로까지 연장되었는데 우리베 대통령은 2008년경까지 4,500명에 이르는 불법 무장 단체에게 무기를 버리면 관용을 베푸는 조치를 취한 바 있다. 그러한

10　이에 대해서는 다음을 보라: Alejandro Fajardo and Matt Andrews, "Does successful governance require heroes? The case of Sergio Fajardo and the city of Medellín: A reform case for instruction," WIDER Working Paper 2014/035, Helsinki: UNU-WIDER, 2014.

그림 5: 메데인의 기반 시설
Figure 5: Infrastructure at Medellín, copyright PUI (Proyectos Urbanos Integrales,Medellín). Photo downloaded from: John Drissen, "The Urban Transformation of Medellín, Colombia," Architecture In Development - News - The Urban Transformation of Medellín, Colombia, August 21, 2012.

debate openly his plan, while at the same time pushing them to offer alternatives where they had critiques. By the end of six months, councilors approved the plan unanimously. This success was achieved not by excluding or outflanking detractors, but rather by engaging their ideas in the decision process, even crediting them publicly, unlike any previous administration, for contributions. In refusing to support his or others' construction of the traditional internal, partisan collectives within the council, Fajardo joined his mayoral administration with the entire council to construct a larger and far more effective policy-based collective.

　　The Fajardo administration's process of engaging hitherto underrepresented groups and constructing coalitions above partisanship became the Development Plan itself, which remapped opportunities throughout the city by emphasizing not just improvements such as better transportation, housing, access to jobs, resources,

그림 6: PUI 지도.
Figure 6: PUI map. Copied from Elviaerosa, "Medellín's Plummeting Homicide Rates between 2002-2008: Possible Explanations," *Umusama2015*, April 21, 2015.

점에서 콜롬비아의 사례는 고대 아테네나 중세 프라토에 잠복해있던 부족주의나 씨족 우선주의와는 극명한 대조를 이룬다. 파하르도 행정부는 우리베의 무장 해제에 재통합 프로그램까지 더했다. 즉, 무장 단체 단원들에게 일대일로 정신 건강 지원을 제공하여 "그들의 능력과 열망을 파악하고 새로운 인생 계획을 설계"하도록 했다.[11] 주민의 배경에 상관없이 주민 개개인의 참여를 장려하는 정신이 파하르도를 당선시킨 풀뿌리 운동에 기반한 개발 계획의 전체 시행 과정에 스며들어 있다. 시 행정부는 시의회의 의견을 구했을 뿐 아니라 인터뷰, 정책 결정, 실행은 물론 예산 책정 과정에까지 모든 주민을 참여시켰다. 이처럼 풀뿌리 운동에서 탄생한 도시 계획의 결과 대상 지역이 탈바꿈하고 2002년만 해도 전 세계 인구 300만 명 이상의 도시 중에서 최고로 높았던 메데인의 살인율이 2006년에는 10만 명 당 32.5명으로 하락했다.

미국에서도 최근 수십 년 동안 기회 재편 계획을 통해 정부 지원이 닿지 않는 도시 내 지역으로 자원을 확산하려는 시도가 이루어지고 있다. 예를 들어 기회 재편 계획을 개발하고 있는 곳으로는 캘리포니아 대학교 버클리 캠퍼스의 하스 인스티튜트, 서던캘리포니아 대학교의 돈 인스티튜트, 오하이오 대학교의 커완 인스티튜트, 미네소타대학교 로스쿨의 메트로폴리탄

and political say, but also by emphasizing something Fajardo's administration insisted was fundamental for realizing such material and political goods: the equal right to dignity of all residents. The plan's means for sustaining such dignity had many creative components. The first was to map the sites in the city with the lowest opportunity, and to situate there public institutions, mostly libraries, with world-class architectural quality and services. Another was to improve transit connections between the new structures, their neighborhoods, and the entire city (figures 5, 6). The administration provided, on a fast-track, economic and social opportunities in the selected neighborhoods and along the connecting transit lines, in a process that became known as Integral Urban Projects (PUI). In contrast to the latent tribalism and clanism of ancient Athens and medieval Prato, the project extended Colombian President Alvaro Uribe's process of exchanging the weapons of paramilitary group members, circa 4500 by 2008, for clemency. The Fajardo administration added to Uribe's demobilization a reintegration program, providing one-to-one mental health support for former paramilitary members, "detecting their abilities and aspirations, and designing a new project of life."[11] This spirit of individual resident engagement, regardless of background, permeated the entire plan, the development and implementation of which was modeled on the grass-roots campaign that got Fajardo elected. Not only did the administration solicit the input of city councilors, but it also engaged all residents, interviewing and continually involving them in the decision-making, implementation, and even the budgeting process. The results of this grass-roots gestation of an overall urban plan was to transform targeted neighborhoods, and to drop Medellín's homicide rate from one of the highest in the planet in 2002, 185 per 100,00 residents, to 32.5 per 100,00 in 2006, for a city with a population over three million.

In recent decades in the United States a series of opportunity map initiatives have sought to promote similar diffusions of resources to underserved areas of cities. Opportunity map projects have been developed, for instance, at the HAAS Institute at UC Berkeley, the Dorn Institute at USC, the Kirwan Institute at The Ohio State University, The Institute on Metropolitan Opportunity at the University of Minnesota Law School, and the Justice Map project (figure

11 Alejandro Fajardo and Matt Andrews, "Does successful governance require heroes? The case of Sergio Fajardo and the city of Medellín: A reform case for instruction," p. 9.

11 Alejandro Fajardo and Matt Andrews, "Does successful governance require heroes? The case of Sergio Fajardo and the city of Medellín: A reform case for instruction," p. 9.

오퍼튜니티 인스티튜트, 저스티스 맵 프로젝트 등이 있다(그림 7).[12] 모두 국가 인구 조사와 지역 인구 통계 자료를 활용하여 정부 지원이 가장 부족한 지역의 지도를 만들고 주택을 골자로 한 법률을 제안한다. 최근 이들은 직업, 주택, 분리에 따른 물질적, 공간적, 사회적 격차에 관련된 법률도 제안하기 시작했다. 그러나 그 결함이 무엇이든 고대 아테네 및 중세 프라토 등의 역사적 사례나 현대 메데인의 사례와 이러한 기회 재편 계획은 실행 측면에서 다를 수밖에 없기 때문에 그 성공에 한계가 있을 수 있다. 어쨌든 미국의 기회 지도는 여러 대학 연구소에서 대규모로 연구, 개발되고 있으며 하향식 정책 체계에 영향을 줄 것이다.

아테네의 종교적인 공간과 연극 의례는 사회 공간적 현실을 개인의 차원으로 환원하여 아테네 농부, 선원, 도시의 소수 지배층에게 의례 장소, 행동, 강렬한 경험을 두루 제공했다. 이 과정에서 아테네인들은 각자의 차이를 극복하고 새로운 공간과 집합의식을 갖추게 되었다. 프라토 탁발 수사들의 단순한 토착어 설교, 행렬, 수도회는 기회를 제공한 데 그치지 않았다. 수도회의 개방성 덕분에 보복적인 씨족 우선주의와 군국주의가 단기간내에 해소되었다. 마찬가지로 메데인의 기적은 존엄성과 부흥의 교차점을 도시 전체로 확대했다. 계층과 직업을 막론한 남성뿐 아니라 여성을 참여시킨 건축 개입과 교통 환승 정책을 도시 구조와 공간 뿐 아니라 모든 이의 존엄성을 바탕으로 한 집합의식과 결합시킴으로써 그러한 일이 가능했다. 미국에서든 그 이외 나라에서든 기회 재편 지도 제작의 도전과제는 기회 재편의 수단을 가장 절실히 필요로 하는 곳을 표시하는 일이다. 인구 조사 자료나 빅데이터에만 의존해서는 안 되며 지역 조사 자료도 활용해야 한다. 정책뿐만 아니라 도시 극장, 의례, 건축, 공공장소에서 공유된 집단 경험도 참고해야 한다. 지속 가능한 기회를 정치적으로 구현하려면 위험하면서도 도시 생활에서 빼놓을 수 없는 요소인 집합의식을 과감하게 활용해야 할 것이다.

7).[12] All harness national census together with local demographic data and surveys to chart neighborhoods that are least served, and to propose legislation, particularly for housing, that has begun to address the material, spatial and social disparities of jobs, housing, and segregation. The success of these opportunity maps may be, however, limited by the difference in their implementation from those of the historical examples of ancient Athens and medieval Prato, whatever their flaws, and from the contemporary instance of remapping in Medellín. United States opportunity maps are largely researched and developed at university research centers to affect policy using a top-down model. The religious space and rites of theater of the Athenians brought the reality of social-spatial remapping to the individual level, providing ritualized sites, behavior, and intense shared experiences to allow Athenian farmers, sailors, and urban elites to see beyond their differences to create new spaces and senses of the collective. The simple vernacular sermons, processions, and brotherhoods of Prato's mendicants, displaced, for a short period, the culture of vindictive clanism and militarism, not just with opportunity, but also with an open sense of the collective. The miracle of Medellín similarly spread nodes of dignity and revival across the city, mixing architectural and transit interventions by engaging women as well as men of all classes and professions not only with the city structure and space, but also with a sense of collective based on dignity for all. The challenge for opportunity mapping in the United States or anywhere is to chart means for bringing its tools early and directly into the hands of those needing it most, relying not only on census reports and Big Data, but also on local input, not only in policies, but also in group experiences shared in urban theater, ritual, architecture, and public space. Sustainable opportunity may only be politically realizable by daring to engage that dangerous but essential component of urban life: a sense of the collective.

12 "Southern California Equity Atlas, PERE (Program for Environmental and Regional Equity) USC Dana and David Dornsife College of Letters, Arts and Sciences," PERE USC Dana and David Dornsife College of Letters, Arts and Sciences. https://dornsife.usc.edu/pere/socal-equity-atlas/

12 For a partial listing of opportunity mapping resources, see "Southern California Equity Atlas, PERE (Program for Environmental and Regional Equity) USC Dana and David Dornsife College of Letters, Arts and Sciences," PERE USC Dana and David Dornsife College of Letters, Arts and Sciences. https://dornsife.usc.edu/pere/socal-equity-atlas/

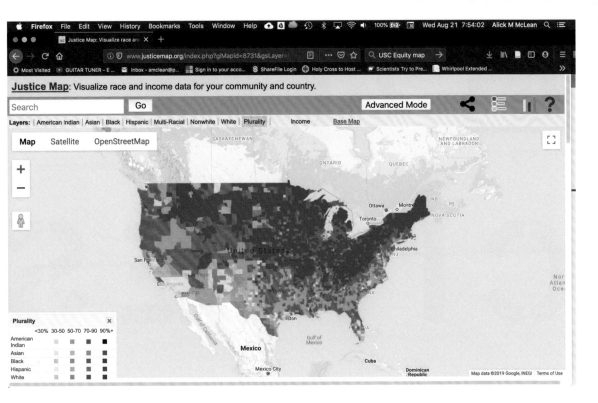

그림 7: 저스티스 지도.
Figure 7: Justice Map. Screenshot from "Justice Map - Visualize Race and Income Data for Your Community."
http://www.justicemap.org

Thanks to Francisco Sanin for his inspiration and comments, to Beth Hughes and Livia Qing Wang for their assistance, to Alice Kim for copy editing, to APH colleagues Hannah Kimberly and Bryon Williams for proofreading, to Molly Martins and Jan Healy at APH for funding and support, to Andronike Makres for her work and discussions on dithyramb, to my Syracuse and APH students, my children Sophia, Gaia, Samson, and Sofie, and especially to my wife, Michele Errie-McLean, for discussions on opportunity and identity.

도시전
CITIES
EXHIBI

TION

CITIES:

East Asia
South East Asia
Central Asia
West Asia
Middle East
North America
Central America
South America
Europe
Africa
Oceania

Beirut
Brussels
Buffalo
Cheon-an
Cheong-ju
Detroit
Madrid
Providence
Sao Paulo
Tongyeong
Ulsan
Vienna
Yeong-ju
Amman
Belgrade
Brisbane
Ganges
Addis Ababa
Casablanca
Chungjin
Dar es Salaam
Samarkand
Stockholm
Baku
Barcelona
Boston
Caracas
Eindhoven
Istanbul
Kuala Lumpur
Kuwait City
Los Angeles
Manila
Medellin
Tokyo
Belgrade
Berlin
Capetown
Christchurch
Kaohsiung
Mexico City
Paris
San Jose
Tohoku and Kumamoto
Wuhan
Jakarta
Shanghai
Toronto
Ulaanbatar
Wellington
Beijing
Haifa
Ho Chi Minh City
Lima
London
New York
Rome
Zurich
Black Rock City
Gaeseong
Hong Kong
Paris
Rotterdam
Shenzhen
Singapore
Bangkok
Barcelona
Berlin + Mumbai
Big Plans
Hong Kong
Incheon
Nairobi
Taishan
TYOLDNPAR
Amsterdam
Copenhagen
Frankfurt
Hanoi
Havana
Los Angeles
Milano
St. Petersburg
Zanzibar

INTODUCTION TO COLLECTIVE C

Industry in the 21st Century
Rivers and Waterfronts
Market Inheritance
Infrastructures
Participatory Urbanism
Density
Collective Typologies
Appropriations
Methods and Speculations
Layers and Collective Memory

As an introduction to the Cities Exhibition, this space exhibits the curatorial efforts and decisions that have been made to develop the idea of Collective Consequences as an interpretation of the main theme, Collective City. Cities are understood as an accumulation of multiple layers that result from planned as well as unplanned intentions, which we define as collective consequences. This is why even in an intensely master-planned city, the city we live in is much different from what has been drawn on paper. This is the case for the exhibition as well. Having more than 60 cities from around the globe, the Cities Exhibition intends to be curated in an open-ended fashion, instead of a fully controlled one, to create a dialogue and chemistry between individual exhibits without strict guidance or limitations of interpretation of the theme. Therefore, our intention is for the whole exhibition to be conceived as consequences that are being made by the dialogues between cities rather than as a complete singular exhibition of a city. For this, the introductory exhibition provides methods for reading a city as a platform to understand different ways for discussing contemporary topics that run our cities.

2019
SEOUL BIENNALE
OF ARCHITECTURE
AND URBANISM

COLLECTIVE
CITY

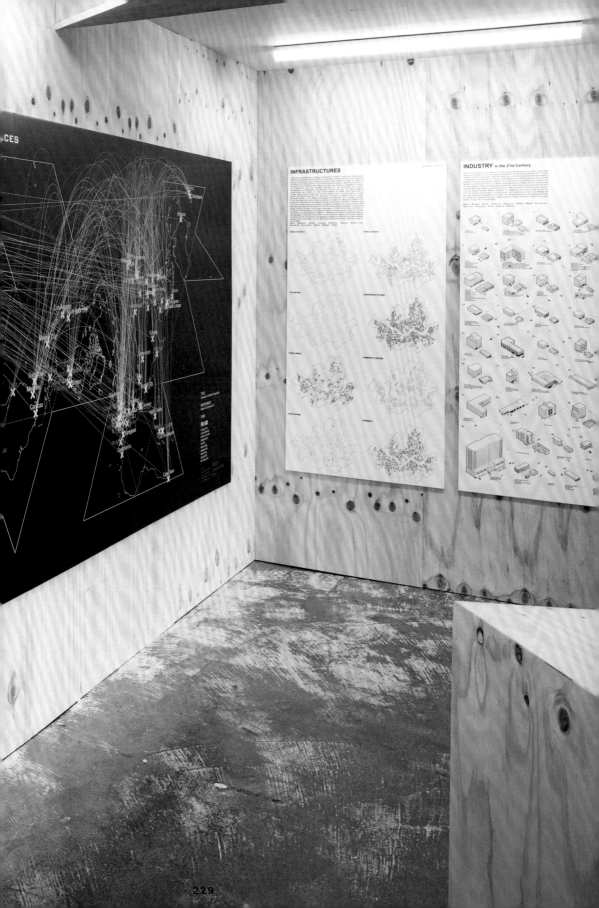

출판전시 공간은 새로운
마이크로-플랫폼 개념의
공간을 제안하며, 그곳에서
작가, 건축가, 예술가,
학생들이 도시전 출판물들을
통하여 서로의 생각을 나누고
혜안을 얻길 기대한다.

Publication pavilion
provides a new micro-
platform to share
thoughts and insights
from authors, architects,
artists and students about
the Cities Exhibition
through selected/featured
publications.

도시: 집합적 결과

임동우, 라파엘 루나, 도시전 큐레이터

Cities: A Collective Consequence

Dongwoo Yim, Rafael Luna,
Co-curators of Cities Exhibition

도시를 현대 도시 환경의 단일 징후로 논의하는 과정에서는 현재 맥락의 '도시'라는 용어의 정의 자체가 시험대에 오른다. GPS, 위성사진, 로봇 센서 등의 기술 발전으로 '도시'라는 용어는 로마 시대의 키비타스(civitas)처럼 시민들의 집합체라는 원래 의미보다는 행정적 경계의 의미에 가까워지고 있다. 도시화 프로젝트는 도시 유형의 특이성 개념을 지웠고 확장되는 기반시설 네트워크가 그 자리를 대체했다. 이러한 상황은 야간에 우주 밖에서 촬영된 인공위성 사진을 통해 관찰할 수 있다. 21세기 도시는 콘스탄티노스 독시아디스가 묘사한 에큐메노폴리스와 매우 유사하다. 이는 기술과 기반 시설 발전으로 도시 분산과 인구밀도 분산이 끊임없이 이루어지는 도시 유형이다. 이러한 시나리오에서 도시는 서로 구분되지 않고 관련성도 없으며 크기가 제각각인 파편이 연속적으로 축적된 형태다. 상파울루, 도쿄, 멕시코시티, 서울 등의 대도시권에서 퍼져나가는 광역 도시권이 그와 거의 일치한다. 이러한 현상은 전 세계적으로 '도시화'의 일반화를 불러왔다. 정형적이며 기형적인 도시 난개발이 이루어지는 현 시점에 경관의 도시화와 기반시설의 도시화는 물론 가장 최근에 이루어지고 있는 내부의 도시화는 도시 분산, 인구밀도 배분, 공공영역의 약화, 도시 효율성 등의 문제를 해결하는 방법으로서 건축의 각 전문 분야가 하는 역할을 이해하는 데만 집중하고 있다.

그러나 '도시'는 1867년 일데폰스 세르다의 『도시화 일반론』에서 계획된 도시화 프로젝트보다 훨씬 더 오랜 시간 동안 존재해왔다. 도시에는 힘, 상업, 문화, 혁신이 집중된 중심부가 있으며 심지어 비정형적인 도시 확장 영역 안에도 중심부가 있다. 우리는 아직도 정치적, 사회적, 경제적, 기술적 측면이 집중된 장소로서의 도시를 논의할 수 있다. 도시 각각은 저마다의 문제, 의사결정, 결과의 역사를 지닌다. 우리가 집합적 결과물이라는 '도시'의 새로운 정의를 받아들이려면 그 결과를 현대적으로 분석해야 한다. 이를 통해 새로운 건축적 의제가 등장할 수 있다.

20세기 말 '전형적 도시(Generic City)'는 세계화의 동의어가 되었다. 도시는 동일하게 처방된

In trying to discuss cities as a singular manifestation of the contemporary urban condition, the very definition of the term "city" is challenged in its current context. With the advancements in GPS, satellite imagery, and robotic sensing, the term "City," has become more of an administrative demarcation rather than the true collection of citizens, as referenced in Roman times as the civitas. The project of urbanization has eradicated any notion of singularity in the form of a city and replaced it with an over expanding infrastructural network. This condition is often portrayed as a world satellite image, as seen at night from outer space. Cities in the 21st century resemble more the Ecumenopolis described by Constantinos Doxiadis where the advancements of technologies and infrastructure have allowed for an endless urban dispersal and a diffusion of densities. In this scenario, cities are presented as an accumulation of scaleless fields of urban patches, indistinguishable from one another and non-contextual. One cannot differentiate the sprawling conurbation of Sao Paulo, Tokyo, Mexico City, or Seoul Metropolitan Region. This effect has spawned a series of "urbanisms" as general phenomena occurring globally. Landscape urbanism, infrastructural urbanism, and most recently interior urbanism have been preoccupied with understanding the role of architecture through its specialized fields in the current amorphous and deformed sprawling urbanization as a way of addressing the problem of urban dispersal, distribution of densities, the weakening public sphere, and urban efficiency.

Yet, "cities" have existed far longer than the project of urbanization, as planned by Ildefons Cerda's *General Theory of Urbanization* in 1867. Cities have agglomerated centers of power, commerce, culture, and innovation even within this amorphous urban expansion. We can still discuss cities as still been defined through their concentration of political, social, economic, and technological dimensions, each with their history of problems, decisions, and consequences. It is through the contemporary analysis of these consequences that we can reconcile a new

상업 개발 패턴을 따르는 것처럼 보였다. 같은 상표를 끌어들이고, 켈러 이스터링이 말한 것처럼 같은 형태의 사유화된 공공장소를 만들어냈다. 그러나 앞서 언급한 (정치적, 경제적, 사회적, 기술적) 측면의 경로 의존성 때문에 집합적 결과는 일반적 패턴과 상관없이 여전히 각 도시의 독특한 특징을 규정한다. 이 모든 측면에 필요한 기반 시설은 도시가 성숙하는 동안 수십 년, 심지어 수백 년에 걸쳐 지속된다는 특징이 있다. '집합적 결과'는 현재 환경을 형성한 다층적 결정을 수정할 것을 요구할 뿐 아니라 지속적인 성장의 새로운 패러다임을 찾아낼 것을 요구한다.

도시전은 이 요구를 충족시키기 위해 집합적 결과를 규정하는 방법론 그 자체로 기획되었다. 집합적 결과는 주제를 구체적으로 규정하거나 도시를 사전에 선정하기보다 전 세계에서 모인 건축가, 연구자, 기관, 정부 대표 간의 공통된 대화를 주재하고 이들과 개입이 필요한 도시와 그 도시의 현재 의제에 대해 상의하는 과정을 통해 발견된다. 이 과정에서 전 세계 80여개 도시에서 100명 넘는 인원이 참여했다. 참여자들의 프로젝트 제안과 의제를 교차적으로 참조함으로써 우리는 10가지 하위 주제를 결정할 수 있었다. 인프라스트럭처, 도시의 유산, 시장, 집합적 유형, 집합적인 도시기억과 레이어, 상상의 도시와 분석방법, 강과 수변, 21세기 산업도시, 도시밀도, 참여형 어바니즘과 도시공간의 점용 등이 그것이다. 이러한 방법에 따라 해당 하위 주제는 동시대 도시에 대한 상호 논의를 위한 집합적 결과들을 보여준다. 이 같은 방법을 한층 더 심층적으로 알아보기 위해 서울을 대상으로 연구를 진행했으며 전시 초입부에 서울을 사례로 한 10개 하위 주제를 전시한다. 이 사례에서 서울은 '전형적 캔버스(generic canvas)'이자 맥락적 배경 역할을 한다. 이러한 방법론으로 서울뿐만 아니라 대부분의 도시를 이해할 수 있다는 것을 보여주려는 취지에서다.

서울을 대상으로 한 하위 주제 연구는 다음과 같다.

[1]   21세기 산업도시: 주거 지역에서 떨어진 곳에 따로 조성된 산업 지구는 CIAM(근대건축국제회의)에서 요구한 근대 도시 패러다임이다. 하지만 우리가 알고 있듯이 도시가 생기와 활기를 유지하려면 산업과 주거를 분리하지 말고 통합해야 한다. 생산적 유형에 대한 연구는 이러한 생산 중심지가 서울의 일반적 건물 도시구조에서 어떻게 통합되고, 건물의 구성 요소(대형문, 들창, 창문으로 통한 통풍구, 계단, 기계용 엘리베이터 등) 변경을 통해 어떻게 생산 설비로서의 모습을 찾는지 보여준다. 이러한 건물들은 벌집과 같은 방식으로 트럭, 지게차, 오토바이, 자전거, 손수레 등을 거리로 내놓아

definition of "city" as a collective consequence where a new architectural agenda can emerge.

At the end of the 20th century, the "Generic City" became synonymous with globalization. It seemed that cities would follow the same prescribed form of commercial development, attracting the same brands, and producing the same form of a privatized public space as described by Keller Easterling. Yet, due to the path dependency of the aforementioned dimensions (political, economic, social and technical), the collective consequences still define the unique character of each city regardless of any generic patterns. The infrastructure laid out for all these dimensions has the characteristic to persist throughout the decades or even centuries that it takes for a city to mature. The "Collective Consequence" calls to action the revision of what have been the layers of decisions that form our current environment, and unearth new paradigms for the continued growth.

To achieve this, producing a Cities Exhibition has become a methodology in it on itself, for defining these collective consequences. These are found not through delegating specific definitions of themes or even pre-selecting cities, but through finding common dialogues between architects, researchers, institutions and government representatives from around the world, and having them propose cities of intervention and their current topics. This process yielded over eighty cities and over one-hundred actors from around the globe. By cross-referencing the project proposals and topics brought by the participants, we were able to devise ten subthemes: infrastructures; market inheritance; collective typologies; layers and collective memory; methods and speculations; rivers and waterfronts; industry; density; participatory urbanism; and appropriations. Under this method, these subthemes represent the collective consequence for discussing the contemporary city, interchangeably. To explore this method further using Seoul as a case study, the introductory exhibition displays the ten subthemes as studied through Seoul. Seoul, in this case, serves as a "generic canvas" as well as a contextual setting. The aim is to present these methodologies not only as a way to understand Seoul better but cities in general.

The subthemes are explored in Seoul as follows:

[1]   Industry in the 21st Century: Industrial zoning separately organized away from the residential zones was the paradigm of the modern city as mandated by the CIAM, but as we have learned, cities stay alive and busy not through the

생산 과정이 완료되도록 한다. 이것이 새로운 도시 생활이다.

[2] 강과 수변: 서울은 도시를 둘러싸고 있는 네 개의 산(북한산, 인왕산, 남산, 낙산)에서 유입되는 지천을 받으면서, 물이 형성된 지역을 중심으로 마을이 발전해왔다. 시간이 흐름에 따라 개울은 근대화에 따른 도로 확장과 지하 하수도 건설에 밀려 복개되기 시작했다. 그러나 현재의 도시 구조와 거리 배치는 과거 개울의 위치에 따라 형성되었고, 이는 청계천 복원사업을 통해 확인할 수 있다. 더 큰 규모의 예시로는 한강을 들 수 있다. 약 1km 폭에 이르는 한강은 도시의 확장을 가로막는 요소였으나, 이는 서울의 남북을 연결하는 많은 다리를 통해 한 층 더 큰 대도시로 발전하였다.

[3] 도시의 유산, 시장: 서울은 풍수지리학 전통적 논리구조에 따라 도시가 계획되었고, 공공영역보다 왕의 통치 체계에 중점을 둔 상징적인 도시 구조로 발전되었다. 당시, 공공영역은 좁은 골목길, 허가 받지 못한 시장, 개울에만 해당되었다. 서울은 이 같은 구조를 20세기에 들어서까지 유지되었고, 갑작스러운 도시 변화가 이어지면서 도시에 대한 이해와 공공 공간에 대한 의식 변화에 따라 변모하기 시작했다. 서울이 수십 년에 걸쳐 급진적인 변화를 겪는 동안 시장은 계속해서 상품 교환, 지역과 공동체 형성의 중심지 역할을 수행하였다. 이로 인해 서울은 시장의 집합체가 되었다.

[4] 인프라스트럭처: 일반적으로 기반시설은 교통, 에너지, 상하수도, 오수처리, 통신 등을 강조하지만 현대 도시는 다양한 네트워크 층을 기반으로 건설된다. 크리스토퍼 알렉산더의 『A City is Not a Tree』 라는 저서의 내용에서와 같이 다양한 공공 서비스는 서울이 다수의 세미라티스(semilattice) 구조가 가능할 수 있게 만들었다.

[5] 참여형 어바니즘: 도시가 어떻게 운영되고 발전해야 하는지에 대해 의견을 내는 과정은 시민의 가장 강력한 권한이다. 서울에서는, 도심 시위를 통해 그 권한이 표출된다. 시위 장소를 지도화하면 도시 공간이 시민들의 참여적 행위를 통해 어떻게 사용되는지 확인 할 수 있다.

[6] 도시밀도: 도시화는 주로 인구밀도를 분산하는 과정에서 가속화 된다. 특히 농촌 지역에서 도시 지역으로의 인구 이동에 초점이 맞춰진다. 이미 서울은 초인구밀도가 진행된 대도시로 성장하였다. 서울의 건폐율 연구 지도는 도시 내의 인구 분포도 비율을 볼 수 있다.

[7] 집합적 유형: 서울의 지하철은 전세계 대중교통 중에서 가장 최고라고 할 수 있다. 더불어 지하 공간에 대한 활용은 이미 일반적 교통의 기반시설을

separation of these two but their integration. The study of the productive typology reveals how these centers of production are integrated within the urban fabric of Seoul as generic buildings, only to discovered as production facilities through the adaptation of their elements: the oversize doors, overhang doors, ventilation ducts sticking through the windows, stairs, and elevators for machines. In a beehive manner, these buildings release a colony of trucks, forklifts, motorcycles, bikes, and pushcarts onto the streets to complete the production process. This is the new city life.

[2] Rivers and Waterfronts: Founded on a valley surrounded by four mountains, Bukaksan, Iwangsan, Namsan, and Naksan, Seoul became the receptor of streams and tributaries, where villages developed following the direction of the water formations. As the years passed, the streams started getting covered to make way for modernization with wider streets, and underground sewer systems. Yet, the current urban fabric and street layout found in contemporary Seoul is a reflection of the past streams that lead to the recently recovered Cheonggyecheon. At the larger scale, the Han river, being one kilometer wide, represents a barrier of expansion. Through the construction of the bridges along the Han, the north and south of Seoul have been able to stitch into the larger metropole.

[3] Market Inheritance: Originally formed as an agrarian fortress city under the urban principles of geomancy, the city of Seoul developed a symbolic urban structure focused on a framework of operation for the king with no direct relation to the public sphere. The public realm relied on the use of narrow alleys, informal market spaces and the natural stream that ran through the center of the city as the commons where the transference of goods and materials happened. There was coherence between infrastructure as public space and public space as the metabolic space of consumption. Seoul maintained this structure until the turn of the 20th century when an abrupt sequence of urban transformations radically mutated the reading of the city and its sense of public space. Throughout the several decades of abrupt changes that occur in Seoul, the markets have persisted as the metabolic centers of good transference, forming neighborhoods and communities, that define Seoul as a collection of markets.

[4] Infrastructure: Although much emphasis is usually paid to the classic infrastructures (transportation, energy, water and sewer, waste management, and telecommunications), the contemporary city is built on layers of networks.

초월하여 개발되었다. 지하상가, 행상인이 가득한 보행자용 지하도, 지하철 역 내 지하 상점, 지하 공공 편의시설 및 문화시설 등의 지하 공간은 완전한 집합 공간이 되었다. 이러한 유형의 집합 공간은 도시 곳곳에 퍼져 있으며 단순히 공학적 기반시설에 그치지 않고 계획적으로 사용되고 있다. 이 공간들은 매일 수백만 명의 사람들이 모이고 사용하는 곳이다.

[8] 도시공간의 점용: 인구밀도가 심화되고 부동산 수요가 증가되면서 임시 도시(temporal city)가 등장하고 있다. 임시 도시는 사건으로서는 한시적이고 용도로서는 영구적인 도시를 뜻한다. 인구밀도가 높아지면서 기존 도시 구조, 기반시설, 사이 공간의 용도가 이원화되는 현상이 나타났다. 공간적 의미로 사이는 두 사물의 가운데에서 일어나는 것이고, 시간적 의미로는 무엇인가가 사용되지 않을 때 일어나는 것을 의미한다. 이러한 사이 공간은 다목적으로 사용되며 전통적인 도시 표기에는 포함되지 않는다. 서울의 광화문 광장은 넓은 도로로도 사용되며, 대규모 행사의 장으로도 활용될 수 있다. 또한 건물들 사이의 공간은 대안적 상업 공간으로 이용될 수 있고, 거리는 개인 행상인 혹은 상점이 점유하여 소비자를 위한 거래와 여가 공간으로 활용된다.

[9] 상상의 도시와 분석방법: 서울은 세계적으로 뛰어난 통신 시스템을 통해 방대한 데이터를 수집하고 도시 관리에 활용하고 있다. 또한 이 데이터를 통해 현재 서울을 이해하고, 좀 더 나은 미래 서울을 만들기 위한 연구에 이용하고 있다. 우리는 인구 조사 데이터를 통해 고령화 사회의 문제를 해결하고 새로운 도시 관리 유형이 필요하다는 사실을 인식할 수 있다. 그리고 모든 교통수단에 그래픽 데이터를 더하면 대안적 공공시설이 필요하다는 사실뿐만 아니라 인구밀도와 모빌리티(mobility)의 새로운 유형이 드러난다.

[10] 집합적인 도시기억과 레이어: 서울은 현재의 세계적 대도시가 되기까지 조선시대, 대한제국시대, 일제 강점기, 전쟁, 초고속 성장 등 수 차례의 변화를 겪어왔다. 이러한 변화의 시기마다 이전 행위자가 만들어 놓은 기반시설들은 새롭게 정리되었다. 대신 남은 서로 다른 시대의 기반시설, 상징적 공간 및 건물의 다층적 구조는 고대와 현대의 것이 나란히 존재하는 역동적인 서울을 만들어냈다.

서울은 21세기형 도시를 위한 세계적 논의를 만들어낼 방법 모색의 연구 사례로서 역할을 해왔다. 하지만 이미 전세계 출판업계에서 이와 동일한 방법론을 시행하며 다양한 학술지, 정기간행물, 서적 등을 통해 도시에 대해 그 논의를 확장하고 있다. 이번 도시전 출판 전시는 인트로 전시와 함께 위의 방법론을 확인할 수 있는

Much like Christopher Alexander's *A City is not a Tree* discourse, the multiplicity of public services is what enables Seoul to work as a semilattice, with multiple connecting nodes.

[5] Participatory Urbanism: The process of voicing an opinion on how the city should work and grow is one of the most powerful rights of citizenship. In Seoul, this is manifested through demonstrations throughout the city. By mapping the locations for demonstrations, one can see how the spaces in the city are being used for a participatory event.

[6] Density: As mentioned in the beginning, the ever-increasing urbanization process deals mainly with the distribution of densities, mainly focused on the migration from rural areas to the urban areas. The city of Seoul has grown into a metropolis that condenses the whole gamut of densities within its boundaries. The display of mapping its building coverage ratios shows the distribution of population within the city.

[7] Collective Typologies: Seoul has developed an underground system considered one of the best in the world for public transportation. Yet, this underground system goes beyond the conventional transportation infrastructure. Built with a network of underground shopping districts, pedestrian underpasses filled with street vendors, and subway stations with retail shops, public amenities, and cultural facilities, these underground structures have become integral collective spaces. Spread throughout the city, the taxonomy of typologies reveals a programmatic use beyond the engineered infrastructure. These are spaces of engagement, leisure, and business where millions congregate daily.

[8] Appropriations: The increasing density and demand for real estate allows for the temporal city to emerge; temporal in terms of events, and permanence of use. The density pressure creates a duality of use in the already existing urban fabric and infrastructure, the in-between space. In-between in the sense of the space: what happens between two objects; in the sense of time: what occurs in the time when something is not used; and in the sense of governance: what is private and what is public. These in-between spaces are appropriated for alternative uses, unaccounted for by the traditional notation of a city. Seoul may use Gwanghwamun plaza as a major boulevard for traffic as much as for large temporal events. The space between buildings gets adapted as alternative commercial spaces. The public street gets occupied by private vendors and shops as their space for transaction and leisure space for their customers.

출판물 기록의 공간이 구성되었다. 이 공간은 도시전의 도입부 혹은 종지부로 자유롭게 이용될 수 있도록 하고자 했다. 관람객은 정해진 순서 없이 돈의문박물관마을 골목과 건물 안을 거닐며 10개의 하위 주제와 그 주제 속 도시에 대해 비판적 의견을 도출하거나 집합적 결과물에 대한 개개인 만의 의견과 논리를 생각해볼 수 있다.

[9]   Methods and Speculations: Being one of the most connected cities in the world, Seoul has been able to gather an immense amount of data on the city, which is being implemented for its management. Through this data, there are emerging methods for understanding Seoul, and speculative scenarios on how to intervene in the future. Census data may lead to a need for new typologies that deal with the aging society. Overlaying the graphic data for all transportation modes reveals the need for alternative public corridors and new typologies for density and mobility.

[10] Layers and Collective Memory: Since its founding, Seoul has gone through several transformations from the Joseon Dynasty, Daehan Empire, through a colonial period, devastating war, an industrial rapid growth, to its current status of a global metropolis. Each period had its system for organizing the city. Because of the path dependency of its infrastructure, each period had to deal with the foundations laid by the previous agents. This layering effect of infrastructures, symbolic spaces, and buildings from different periods is what builds the dynamism of Seoul today, where ancient juxtaposes with innovation.

Although Seoul may have served as the case study for this method for creating a global dialogue for the 21st-century city, the dialogue is expanded by implementing the same method on a collection of journals, periodicals, and books from around the world to cross-reference how the publishing industry is discussing the topic of cities. These have formed an archive paired with the introduction exhibition as a Publication's Pavilion. The aim is to have these two (Introduction and Publications) separated from the rest of the cities for them to serve interchangeably as either a beginning or an end to the whole exhibition. Visitors should meander through the alleys of Donuimun Village as well as inside the buildings in no particular order. This is in a way for visitors to be critical of the grouping of cities, reading them through a stated theme, or consume it all to ideate their logic for a collective consequence.

산업 혁명 초기 단계에서 노동의 기회가 생겨나면서 도심은 인구 과잉과 산업화에 따른 공해로 몸살을 앓았다. 산업화 초기 단계에서 이러한 과정이 이어지면서 '산업'이라는 용어에 부정적인 의미가 섞이게 되었다. 기능에 초점을 둔 근대주의 설계자들은 용도 구분 계획을 세워 도시 내의 산업과 주거 공간을 분리하고 산업 운영 공간을 별도로 마련했다. 산업 의존도가 높던 도시는 산업이 쇠락하자 파산 위기에 직면하게 되었다. 전 세계적으로 산업화가 근대화와 동일시되듯이 탈산업화 현상 역시 전 세계로 확산되었고 이로 인해 축소 도시 문제가 부각되었다. 21세기 도시는 청정 산업을 통합할 수 있는 여지가 다양하다. 청정 산업을 도시로 통합하면 도심이 활기를 되찾을 수 있을 뿐 아니라 직장인들과 거주민의 숫자가 늘어남에 따라 새로운 혼합 생산 방식의 개발이 가능할 것이다. 해당 주제로 전시되는 도시들은 시대에 뒤떨어진 포드주의 산업 모델의 잔재가 직면한 전망과 새로운 산업 동향과 더불어 도시 촉매 기법(urban catalyst)이 나아갈 방향을 보여줄 것이다.

통영,
베이루트,
빈,
디트로이트,
천안,
울산,
버팔로,
마드리드,
프로비던스,
영주,
청주,
브뤼셀,
상파울루

Industry in the 21st Century

During the initial stages of the industrial revolution, along with the opportunity of labor, many urban centers were affected by the overcrowded and polluted condition that industry brought. The term "industry" has been coined with a bad connotation ever since this process from the early stages of industrialization of the 20th century. The functional attitude of the modernist planners led to a spatial separation of industry and dwelling inside the city, producing zoning schemes where industry could operate. Cities that relied heavily on industry faced the potential of failure once an industry would become obsolete. Just as industrialization equated with modernization around the world, the post-industrial effect also became a global phenomenon, which revealed the Shrinking City condition. The 21st-century city faces different potentials for integrating clean industry back into the city. This can provide a reinvigoration to the city center, and densification of labor and dwelling that will spark new hybrid typologies of production. The cities exhibited within this theme will provide an outlook at the leftovers from an outdated industrial Fordist model and a vision for an urban catalyst through new industrial trends.

Tongyeong,
Beirut,
Vienna,
Detroit,
Cheon-an,
Ulsan,
Buffalo,
Madrid,
Providence,
Yeong-ju,
Cheong-ju,
Brussels,
São Paulo

신아 SB 조선소 도시재생 사업의 목표는 폐조선소 부지를 문화관광 허브로 되살리고, 산업 전환을 통해 통영 지역을 세계적인 관광명소로 자리매김시키는 것이다.

The purpose of the (former) Shina sb dockyard urban regeneration project is to form a cultural and tourist hub on the site of the closed dockyard, thereby establishing a global hub of tourism through industrial reorganization.

통영

Tongyeong

THE INTRODUCTION OF BROADBAND AND THE WIDE USAGE OF SOCIAL MEDIA IN LEBANON TRIGGERED CITIZENS' SELF MOBILIZATION, AND FACILITATED SOCIAL CONGREGATION IN BEIRUT'S URBAN SPACES. THIS GAVE RISE TO APOLITICAL COLLECTIVES, CIVIL SOCIETY ORGANIZATIONS, AND URBAN ACTIVISM. AN INCREASED NUMBER OF PEOPLE VOICED THEIR OPINIONS TO GOVERNANCE, PLANNING PROJECTS AND THE PROTECTION OF HERITAGE AND PUBLIC SPACE THROUGH SELF ORGANIZED RALLIES.

FIXED 32K DIAL UP INTERNET
WIRELESS BROADBAND INTERNET
FACEBOOK
TWITTER
FIXED ADSL BROADBAND

CREDITS: SAMIR JEBAJ
BEIRUT MARATHON, BEIRUT 2018

4 057 000
POPULATION IN LEBANON
3 360 000

BEIRUT ART BO
BEIRUT SPRING F

BEIRUT INTL PLATFORM OF DANCE
THE GARDEN SHOW
VINIFEST
SOUK EL TAYEB FARMER MARKET
BEIRUT MARATHON
FETE DE LA MUSIQUE
IRTIJAL FESTIVAL
NTL FILM FESTIVAL
N DU LIVRE ARABE
SALON DU LIVRE
CONCERTS

INTERNET USERS IN LEBANON
300 000   400 000   600 000

2000 2001 2002 2003 2004 2005 2006 2007 2008

ANTI-OCCUPATION
REGIONAL POLITICS
MISSING IN WAR
UNIONS
WOMEN'S RIGHTS
STUDENTS' RIGHTS
ANTI SYRIA/IRAN POLITICS
PRO-SYRIA/IRAN POLITICS
CIVIL RIGHTS
WORKERS UNION
ECONOMIC CRISIS
LANDMARKS AND PUBLIC SPACE PRESERVATION

MANIFESTATIONS PER YEAR
ISRAEL WAR ON GAZA PROTEST
ANTI US POLITICS PROTEST
ANTI IRAQ WAR PROTEST
DRIVERS UNION PROTEST
PERCENTAGE OF REGISTERED CIVIL SOCIETY ORGANISATIONS

CREATION OF MARCH 14 POLITICAL ALLIANCE
CREATION OF MARCH 8 POLITICAL ALLIANCE
2 YEARS' SIT-IN

CIVIL WAR COMMEMORATION
OPPOSITION SIT-IN, 2010, MARTYR SQUARE

ISRAEL WITHDRAWAL FROM SOUTH LEBANON

USA INVASION OF IRAQ

UN RESOLUTION 1559
SEPTEMBER 2004. CALLED ON LEBANON TO ESTABLISH ITS SOVEREIGNTY OVER ALL ITS LAND; DEMANDING SYRIA TO WITHDRAW AND TO CEASE ITS INTERVENTION IN LEBANON'S INTERNAL POLITICS. CALLED ON ALL MILITIAS IN THE COUNTRY TO DISBAND AND DECLARED SUPPORT FOR FREE ELECTIONS

PM HARIRI ASSASSINATION
CEDAR REVOLUTION
SYRIAN WITHDRAWAL
AFTER HIS SUPPORT FOR THE 1559 UN RES, PM HARIRI WAS ASSASINATED AND HIS DEATH TRIGGERED INTERNATIONAL AND LOCAL PROTESTS AND CALLS FOR THE END OF SYRIAN HEGEMONY OVER LEBANON, COUPLED WITH A POPULAR INDEPENDENCE MOVEMENT, WHERE OVER 1,000,000 PEOPLE TOOK ON THE STREETS ON MARCH 14, 2005. THE SYRIAN TROOPS WITHDREW A MONTH LATER FROM LEBANON.

ISRAEL WAR 2006-2008 PROTESTS

MAY 08 CONFLICTS
A SERIE OF ASSASSINATIONS TARGETING SYRIAN OPPOSITION POLITICAL FIGURES (11 IN TOTAL) BETWEEN 2005 AND 2008, ACCOMPANIED WITH ANTI-GOVERNMENT SIT-INS IN BEIRUT CENTRAL DISTRICT

A RIFT B
ALLIAN
GRIDLO
BEIRUT.
CLASH
AN AGR
RESULT

14 MARCH 2005 , MARTYR SQUARE.

베이루트   Beirut

INDEPENDENCE 05

Beirut can be overlaid by a network of nodes "Creative Collectives"

"창의적 집단"

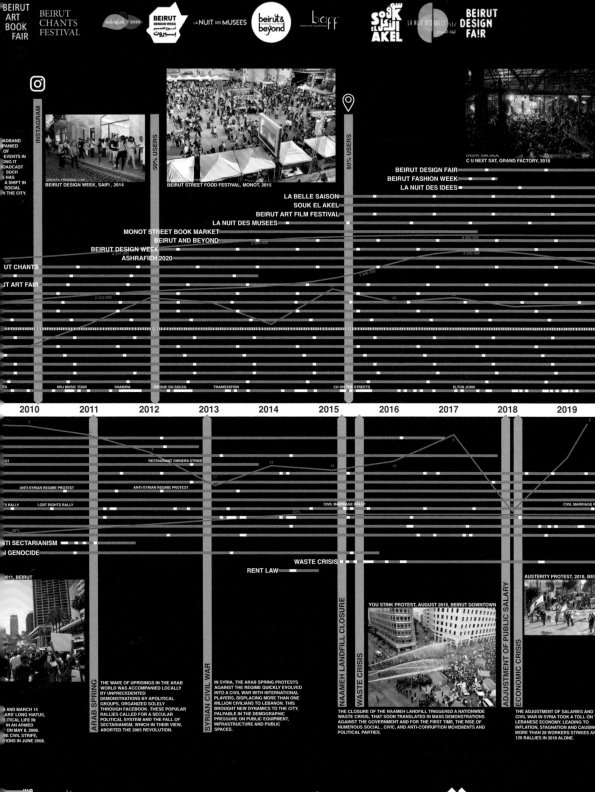

BEIRUT ART BOOK FAIR | BEIRUT CHANTS FESTIVAL | ashrafieh 2020 | BEIRUT DESIGN WEEK | LA NUIT DES MUSEES | beirut&beyond | baff | SOUK EL AKEL | LA NUIT DES IDEES | BEIRUT DESIGN FAIR

INSTAGRAM

50% USERS

80% USERS

CREDITS: CARL HALAL
C U NEXT SAT, GRAND FACTORY, 2018

CREDITS: FWMEME.COM
BEIRUT DESIGN WEEK, SAIFI , 2014

BEIRUT STREET FOOD FESTIVAL, MONOT, 2015

BEIRUT DESIGN FAIR
BEIRUT FASHION WEEK
LA NUIT DES IDEES

LA BELLE SAISON
SOUK EL AKEL
BEIRUT ART FILM FESTIVAL
LA NUIT DES MUSEES
MONOT STREET BOOK MARKET
BEIRUT AND BEYOND
BEIRUT DESIGN WEEK
ASHRAFIEH 2020

...UT CHANTS

...UT ART FAIR

NRJ MUSIC TOUR | SHAKIRA | CIRQUE DU SOLEIL | TRAINSTATION | CU ON THE STREETS | ELTON JOHN

2010 | 2011 | 2012 | 2013 | 2014 | 2015 | 2016 | 2017 | 2018 | 2019

RESTAURANT OWNERS STRIKE

ANTI-SYRIAN REGIME PROTEST

ANTI-SYRIAN REGIME PROTEST

...S' RALLY | LGBT RIGHTS RALLY

CIVIL MARRIAGE RALLY

CIVIL MARRIAGE RA...

...TI SECTARIANISM

...N GENOCIDE

WASTE CRISIS

RENT LAW

AUSTERITY PROTEST, 2018, BEIR...

YOU STINK PROTEST, AUGUST 2015, BEIRUT DOWNTOWN

NAAMEH LANDFILL CLOSURE

WASTE CRISIS

ADJUSTMENT OF PUBLIC SALARY

ECONOMIC CRISIS

...2011, BEIRUT

ARAB SPRING

THE WAVE OF UPRISINGS IN THE ARAB WORLD WAS ACCOMPANIED LOCALLY BY UNPRECEDENTED DEMONSTRATIONS BY APOLITICAL GROUPS, ORGANIZED SOLELY THROUGH FACEBOOK. THESE POPULAR RALLIES CALLED FOR A SECULAR POLITICAL SYSTEM AND THE FALL OF SECTARIANISM, WHICH IN THEIR VIEW, ABORTED THE 2005 REVOLUTION.

SYRIAN CIVIL WAR

IN SYRIA, THE ARAB SPRING PROTESTS AGAINST THE REGIME QUICKLY EVOLVED INTO A CIVIL WAR WITH INTERNATIONAL PLAYERS, DISPLACING MORE THAN ONE MILLION CIVILIANS TO LEBANON. THIS BROUGHT NEW DYNAMICS TO THE CITY, PALPABLE IN THE DEMOGRAPHIC PRESSURE ON PUBLIC EQUIPMENT, INFRASTRUCTURE AND PUBLIC SPACES.

...AND MARCH 14
...RS' LONG HIATUS,
...LITICAL LIFE IN
...IN AN ARMED
...N MAY 8, 2008.
...HE CIVIL STRIFE,
...IONS IN JUNE 2008.

THE CLOSURE OF THE NAAMEH LANDFILL TRIGGERED A NATIONWIDE WASTE CRISIS, THAT SOON TRANSLATED IN MASS DEMONSTRATIONS AGAINST THE GOVERNMENT AND FOR THE FIRST TIME, THE RISE OF NUMEROUS SOCIAL , CIVIC, AND ANTI-CORRUPTION MOVEMENTS AND POLITICAL PARTIES.

THE ADJUSTMENT OF SALARIES AND THE CIVIL WAR IN SYRIA TOOK A TOLL ON THE LEBANESE ECONOMY, LEADING TO INFLATION, STAGNATION AND CAUSING MORE THAN 28 WORKERS STRIKES AND 120 RALLIES IN 2018 ALONE.

  MARCH

...panese
...ue Pride

빈을 생산의 도시로 바꾸어 나가기 위한 특별한 '치료법'으로 근본적인 '복구 프로젝트'를 제시한다. 어떻게 하면 우리 도시가 '가치'라는 중요한 키워드를 되찾을 수 있을까? 어떻게 생산을 지속가능한 도시 발전 목표와 융합하는 도시의 핵심요소로 정착시킬수 있을까?

It stages Vienna as specific treatment of the productive city as a fundamental "repair-project": How to repair the dominant idea of "value" in our cities? How to establish the world of production as a main ingredient to be integrated in the program of sustainable city development?

빈

Vienna

산업의 쇠퇴로 고통받고
방치된 세계 여러 도시에 대한
많은 논문이 발표 되었으나
도시의 이러한 '사후공간'의
"후생"에 관한 이해는 아직
턱없이 부족하다.

Yet while much has
been written about
abandonment in
deindustrializing cities
around the world, very
little is understood about
the "after-life" of these
'leftover' spaces.

Riverbend

디트로이트

Detroit

천안 동남구청 도시재생
사업의 목표는 쇠락하는
동남구청 주변의 원도심을
새로운 지역 경제 중심지로
되살리는 것이다.

The Urban regeneration
project of Dongnam-
gu Office in Cheon-an is
aimed at creating a new
economic hub to revitalize
the declining downtown
area around the Dongnam-
gu Office, Cheon-an.

Landuse

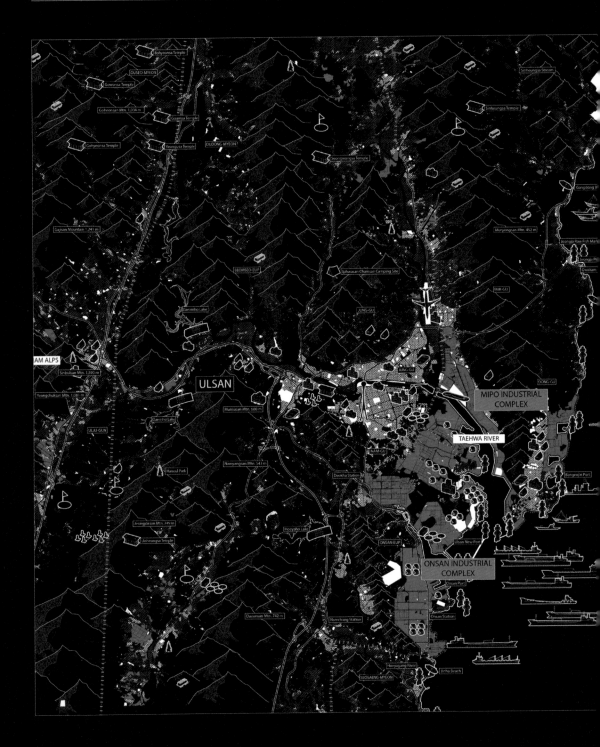

Steam Highway

Transportation

Stakeholders

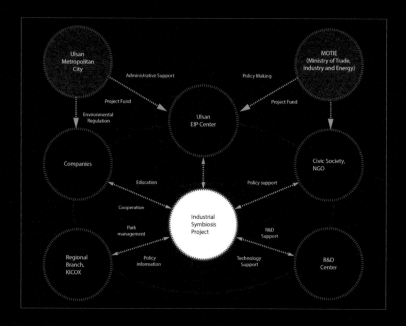

수용적인 거버넌스와
도시계획 그리고 디자인을
위한 위한 도시공업의
전반적인 접근을 다룬다

Contextualize urban
industrial symbiosis
a holistic approach to
adaptive governance,
planning, and design

울산   Ulsan

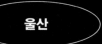

놀이 공간이 다양한
경제, 정치, 인종적 배경을
가진 사람들이 서로의 경험을
나누는 흔치 않은 무대 역할을
한다는 점과 '빈 공간을 채워
나가는 도시화'

Spaces of play are
rare moments where
people from different
economic, political and
racial backgrounds
share experiences,
and influenced by the
opportunistic infill-
urbanism

'환형도시'의 바탕에는 생산-유통-소비 사이의 새로운 관계와 (일방통행이 아닌 선형적) 순환이 있다.

The circular city is anchored to the new distances and cycles, non-linear, between production, distribution, and consumption.

21세기 산업

홀리오 델라푸엔테 마르티네스

Industry in the 21st Century

Julio de la Fuente Martínez

"획기적 지식 발전은 대양과 대륙 너머로보다 복도와 거리 너머로 더 쉽게 전파될 수 있어야 한다."
— 에드워드 글레이저

도시세(Urbanocene, 都市世)의 정점인 오늘날 도시 발전으로 도시의 성공에서 비롯된 위기가 젠트리피케이션이며 사회, 경제적 불평등과 같은 현상을 낳고 있다. 향후 수십 년 동안의 도전 과제는 제대로 작동하는 도시 경제를 구축하여 서비스 산업과 종사자 등 이제까지 중산층으로 불려온 사람들이 장거리를 통근할 필요 없이 도시에서 적당히 떨어진 곳에 살 수 있도록 하는 것이다. 오늘날 고숙련 기술자 대부분의 장거리 통근이나 도시의 제조업 일자리 부족 등 인구 압박으로 인한 문제를 감안할 때 우리는 이 같은 도시 현상을 다룰 정책을 새로이 마련해야 할 필요가 있다.

도시는 다양한 도시 계획 전략으로 생산을 은닉해왔다. 도시의 의제들은 지구 차원의 오염 문제에 대해서는 고려하지 않았으며 구획 분리에 따라 유해한 물질을 배출하는 생산 시설이 도시 외곽으로 이전되었다.

(4차 산업혁명의) 새로운 기술과 청정에너지 덕분에 도시에서 생산이 재개될 수 있었다. 소비자가 생산자가 되면서 유통 공간이 바뀌고 공간적인 측면에서 새로운 현실이 창조되고 있다. 현재의 선형적인 생산-분배-소비 체계를 회전이 빠르고 글로벌 물류 시스템과 연계된 순환 경제로 바꾸려면 도시에 가까운 공간을 발견해야 한다. 이러한 전환을 이루려면 공간의 활용과 조직에 대한 생각을 획기적으로 재고해야 할 뿐 아니라 새로운 정책과 계획 전략을 짜내어 생산까지 포함하는 진정한 집합도시를 구축해야 한다.

"Intellectual breakthroughs must cross hallways and streets more easily than oceans and continents"
— Edward Glaeser

The crisis arising from the success of cities, in the midst of the *Urbanocene* Era, produces phenomena of gentrification, social and economic inequality a.o. The challenge for the coming decades will be to find a functional urban economy where the employees of the service industry and workers whom we knew until now as middle class, can live at reasonable distances from the city, without performing endless commuting. Nowadays, the demographic pressure, the fact that most of the high-skilled people are commuters, and the scarcity of manufacturing jobs at town, make it necessary to think in new policies to integrate this phenomenon in our cities.

Cities has been hiding the production with different planning strategies. The urban agendas were not concerned with pollution at the planetary level, and thanks to zoning the harmful emissions of production were located out of the city.

New technologies and clean energy production (Industry 4.0) make possible to reintegrate production in the city. Consumers become producers, altering the distribution space and creating new realities in spatial terms. New proximities must be discovered to transform the current linear system of production-distribution-consumption into a circular one, a short- circuit economy connected to a global logistic system. That shift needs to radically rethink the use and organization of space, finding new policies and planning strategies to allow a real urban mix, including production.

An interesting approach to the "Productive City" is being exploring in Brussels and Vienna, a. o., following the idea of repositioning industrial areas, with a strategic location, into diverse and mixed-use areas, merging production, services,

'생산 도시(Productive City)'에 대한 흥미로운 접근법이 벨기에 브뤼셀과 오스트리아 빈에서 연구되고 있다. 전략적 위치에 있는 산업 지역을 생산, 서비스, 거주, 물류가 어우러진 다양한 복합 개발 지역으로 재배치해 도시 내에 생산에 적합한 공간을 만들자는 구상이 그 바탕이 된다.

이 같은 구상의 일부는 스페인 개발부가 추진한 2019 스페인 도시 의제(AUE)에도 포함되어 있다. 해당 의제는 '지속 가능한 발전을 위한 2030 의제'의 일환으로 인류세(Anthropocene)의 도전과제에 대응하려는 도시들에 유용한 지침을 제시한다. '생산 도시'는 유로판 공모전(Europan competition)의 마지막 2개 세션의 주제이기도 하다.

21세기의 집합 주거와 산업

21세기 집합 주거와 산업을 다시 연결하기 위해서는 순환과 대사의 5대 근본 원칙을 감안해야 한다.

1   정치적 측면, 왜 지금인가?
지식, 혁신, 생산을 연계할 수 있는 데다 짧은 물질 순환과 에너지 흐름뿐만 아니라 경제 균형을 고려한 도시계획 전략이 모색되어야 한다.

지금은 '기회의 창'으로 불리는 시간적 차원에 이어 순환 모델을 도시 의제에 적용해야 할 시기다(디터 라플). 리처드 플로리다 등의 전문가들은 도시 경제 성공은 3T, 즉, 기술, 인재, 관용에서 비롯된다고 주장한다. 성공뿐만 아니라 다양한 추정 성장률에 대한 개념이 다시 한 번 제시되어야 한다. 우리는 산업화 이전 도시의 장인 요소를 디지털 시대의 환경 해결책으로 삼는 역설을 수용할 수 있을까?

2   영역 분리, 어떠한 시너지 효과를 내는가?
"아무리 작더라도 모든 도시는 부자들의 영역, 가난한 자들의 영역으로 양분된다."
— 플라톤, 『국가』 중에서

도시 분리는 불평등의 원인 중 하나다. 지도 위에서 가장 많은 혜택을 받는 지역과 가장 덜 받는 지역의 위치가 나뉘는 것이 도시 분리다. 중산층이 소멸하고 교외의 빈곤이 도시보다 더 빠른 속도로 꾸준히 증가하고 있다는 사실에서 도시 분리를 확인할 수 있다. 생산 활동은 대조적인 현실을 통합하는 데 중요한 역할을 할 수 있다.

residential and logistics, making place for affordable space for production in the city.

Some of these principles are also included in the Spanish Urban Agenda 2019 (AUE) promoted by the Ministry of Development of Spain as an exemplar urban guide to address the challenges of the Anthropocene Age within the framework of the Agenda 2030. The "Productive City" is being explored in the last two sessions of the Europan competition.

The Collective and The Industry in the 21st Century

Five fundamental aspects about circularization and metabolism must be considered to relink The Collective and The Industry in the 21st Century:

1   Political scale, why now?
Planning strategies capable of linking knowledge, innovation and production must be explored, with short chains of materials and energy flows, as well as the economic balances. Now, it is time to include circular models into the urban agendas, following the temporal dimension of the so-called "Window of Opportunity" (Dieter Läpple). According to experts, Richard Florida a.o., the urban economic success is associated with the "3Ts": Technology, Talent and Tolerance. The concept of success must be raised again as well as the different growth rates that could be assumed. Could we consider the paradox of the artisan component of the pre-industrial city as the solution to improve our environments in the digital age?

2   Territorial segregation, which synergies?
"Any city, however small, is divided into two, the city of the rich, and the city of the poor," *The Republic*, Plato.

One factor of inequality is the urban segregation, the location on the map of the most privileged areas, and the least, which are currently marked by the disappearance of the middle class, and a progressive increase on the levels of impoverishment in the suburban areas, faster than in urban areas. Productive activities could be a key factor to bring together the different realities.

What synergies can be triggered so that the next economy could prosper, based on a social agenda and criteria of inclusion?

3   Urban Armatures, which frameworks?
Infrastructures come to the surface as agencies capable of articulating the frictions between masterplans and action plans, between

어떠한 시너지 효과를 통해 차세대 경제가 사회적 의제와 포용의 기준을 토대로 번영할 수 있을까?

3 도시 골조, 어떤 프레임워크를 사용할 것인가?

기본 계획과 실행 계획, 생산과 소비 사이의 불협화음이 기반시설을 통해 극명히 드러남에 따라 기반시설이 표면으로 부각하고 있다. 상위 시스템에 내재된 다양한 자연, 사회, 경제 생태계의 물질대사를 처리하려면 ('작은 것이 아름답다'는 생각의 변형인) '큰 것도 아름답다'는 생각이 필요하다.

브뤼셀 운하나 빈 지하철망 등의 사례는 4차 산업혁명 시대의 새로운 도시 이동성을 위한 공간적 프레임워크를 보여준다. 마드리드에서는 M-45 고속도로가 생산 기반시설을 위한 차세대 연구 모델을 제시한다. 새로운 이동 환경은 어떠한 방식으로 도시와 생산의 하이브리드화를 조장할 수 있을까? 또한 이러한 '새로운 근접성'은 어떠한 공간으로 전환될 수 있을까?

4 장소 만들기, 무엇을 하이브리드화할 것인가?

도시 환경의 순환성은 도시 인공물과 공공 공간의 연결에서 비롯된다. 이는 새로운 환경에서 제조 활동과 자원 관리의 가시성이 드러나고 있기 때문이다. 생활과 일을 같은 공간에서 한다는 생각은 푸대접을 받아왔지만 도시의 '불결한' 기능을 시각화하는 것이 다시 한 번 용인됨에 따라 그러한 생각에도 새로운 지평이 열리고 있다.

이처럼 공공 영역의 유기적 일부인 짧은 순환을 통해 어떠한 유형의 하이브리드가 탄생하여 주요 통치 기관들(기술 관료에 의한 통치 기관과 해방을 실행하는 통치 기관) 사이에서 중재자 역할을 할 수 있을까?

5 가상의 집합 영역은 어떠할까?

순환도시(Circular City)에서 가상의 집합 영역은 어떠한 모습을 하고 있을까? 일련의 낡은 현실과 보이지 않는 풍경이 그들을 다시 도시로 들여보내줄 새로운 환경적 감수성을 기다리고 있다.

지난 수십 년 동안 집합 주거 의제는 주택의 공동 공간을 조명하고 공동체의 개념을 강화하는 등 다양한 차원에서 모색되었다. 이제 주거 지역과 생산 지역의 공간적 시너지를 탐색할 때다. 창고가 시민 시설처럼

production and consumption. Thinking that "the great can also be beautiful" (subversion of Small is Beautiful) is necessary to manage the metabolism of the different natural, social and economic ecosystems, embedded in superior systems.

Examples, such as the Brussels Canal or the Vienna metro network, serve as spatial frameworks for a new urban mobility associated with industry 4.0. In Madrid, it could be committed to the M-45 Highway as the next study model for a productive infrastructure. How can new mobility conditions encourage hybridization between city and production, and which is the spatial translation of these "new proximities"?

4 Making place, which hybrids?

The circularization of urban environments is anchored to the contact between urban artifacts and public space, due in large part to the new conditions of visibility of the manufacturing activities and the management of resources. The visualization of the "dirty" uses in the city is once again accepted, opening new horizons to the persecuted idea of living and working in the same environment.

What hybrid typologies can emerge from these short-circuits as an organic part of the public domain to mediate between the main governance agencies: the technocratic and the emancipatory ones?

5 The collective realm, which imaginaries?

The imaginary of the collective in the Circular City is uncertain. A series of obsolete realities and invisible landscapes are waiting for a new environmental sensibility to reintroduce them into the city.

During last decades, the issue of the collective has been explored at different levels, highlighting the collective spaces in housing, reinforcing the idea of community. Now, it is an opportunity to explore the spatial synergies between the residential and the productive. Could a warehouse be as memorable as civic facilities?, or could a logistics loading-bay be the best playground during the weekend?

Oceans and Hallways

Industry in the 21st Century is about The Oceans to reduce inequalities, distances and segregation from the political scale, about world geopolitics, technocratic agencies, the 2030 Agenda, urban development and infrastructures.

깊은 인상을 남길 수 있을까? 물류 적재 구획이
주말에는 최고의 운동장으로 탈바꿈할 수 있을까?

대양과 복도

21세기 산업은 정치적 차원의 불평등, 격차, 분리를
완화하는 대양(the Oceans)뿐만 아니라 전 세계
지정학, 기술 관료 집단, 2030 의제, 도시 개발,
기반시설을 중심으로 돌아간다.

세계적 관점.

그러나 복도(the Hallways)도 21세기 산업의
핵심이다. 복도는 하이브리드 모델, 모호한 형태,
혁신적인 유형, 해방적 정부 기관, 전략적 도시화,
공동체, 망각된 풍경과 가상의 집합 영역 사이의
흐릿한 경계를 명확히 드러낸다.

지역적 관점.

산업을 다시 한 번 도시의 필수적 구성요소로
도입하고 모두가 번영할 수 있는 포용적 도시 모델을
개발하기 위해서는 대양과 복도 사이의 잃어버린
연결점들을 찾아야만 한다. 새로운 근접성의 발견을
통해 생활과 일, 자연과 인공물 사이의 균형을
개선해야 한다.

순환도시.

The global.

But it is also about The Hallways to articulate
hybrid models, ambiguous morphologies,
innovative typologies, emancipatory agencies,
tactical urbanism, communities, and the blurred
boundaries between forgotten landscapes and the
imaginary of the collective.

The local.

Only by finding the missing connections
between The Oceans and The Hallways will it be
possible to develop inclusive city models, where
industry becomes an essential urban component
again, capable of ensuring prosperity for all. New
proximities for a better balance between living and
working, between nature and artifice.

The Circular City.

경제학자와 도시계획 담당자들의 눈에 빈 공간이란 부동산 가격과 세금 정책, 다양한 인센티브와 부동산 시장 부양을 통해 없애야 하는 사회문제일 뿐이다. 하지만 이런 접근법이 간과하고 있는 사실이 있다. 도시의 빈 공간이 모두 세수창출 용도로 개발되고 채워진다면, 그로 인해 도시의 거대한 잠재력이 빛을 잃게 된다는 점이다.

Economists and planners alike see vacancy as a pejorative condition: a problem to be eradicated through pricing, taxation, incentives, and market props. What this approach neglects is the enormous potential left behind when all the empty space is efficiently filled by revenue-generating uses.

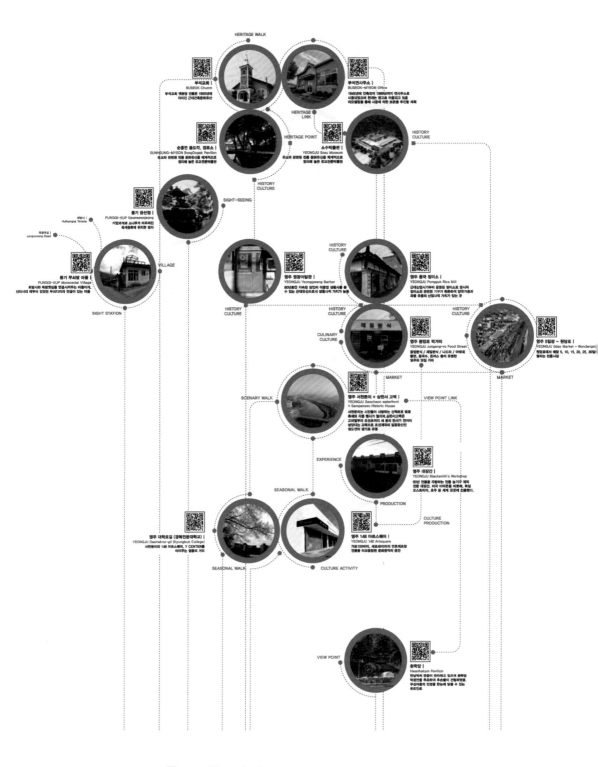

영주시의 가장 중요한 과제는 축소되는 도시에 탄력적으로 대응하는 도시공간구조로 전환하는 것이다.

The most important task that lies ahead of the city of Yeongju is its transition to a city space structure that enables the implementation of flexible response measures to the ever-decreasing size of the city.

영주

Yeongju

생산은 도시의 중요한
부분이다.

Production belongs to the
city.

브뤼셀

Brussels

산업시설이 떠나간 구역을
도심의 저렴한 복합용도
주거지구로 바꿔 나갈 방법을
제안한다.

The city can transform its
post-industrial lands into a
series of inner city mixed-
use affordable housing
districts.

상파울루

São Paulo

세계 주요 도시 대부분은 교통수단이 될 수 있는 물길을 둘러싸고 형성되었다. 그러한 물길은 천연 자원을 소비 지역에 공급하고 도시를 상업의 중심지로 만든다. 기반시설이라는 측면에서 강은 도시를 형성하는 최초의 자연 지리적 요소이자 오랜 시간 동안 형태가 변하지 않는 요소이다. 고대 이집트의 나일강부터 근대 런던의 템스강에 이르기까지 강은 도시 인구와 산업이 성장하는 데 밑받침이 되었다. 강은 자원 형태로 뿐만 아니라 지정학적 장벽을 깨뜨리는 정치 형태로도 넓은 영역을 연결한다.

그러나 탈산업화 시대에 도시의 강은 소비 가능한 도시 요소가 되고 있다. 현재 '수변'으로 이름 붙은 지역은 부동산으로 인식된 이래로 가장 높은 가치를 기록하고 있다. 따라서 전시 대상인 도시에서는 강이 도시 내 집합 공간의 필수 요소로서 뿐만 아니라 토지 가치와 도시 변화를 결정짓는 기반시설로서도 중요한 역할을 담당하고 있다.

브리스번,
암만,
갠지스

Most of the major cities around the world have formed around a body of water as they have provided means of transportation that convert them into commercial hubs and provide natural resources for consumption. Rivers become the first natural geographical element that shapes a city in an infrastructural level, and their forms persist throughout the ages. From the River Nile in Ancient Egypt to River Thames in modern London, rivers have fed the growth of cities both in population and in industries. Rivers connect expansive territories, becoming not only forms of resources but political forms that break geopolitical barriers. However, in the post-industrial era, rivers in cities have become urban elements that can be consumed. Nowadays, the term "waterfront" has become more valuable than ever as it has a connotation of real estate value. Thus, the cities view rivers not only as an integral part of collective space in a city but as an infrastructural determinant of land value and urban transformation.

Brisbane,
Amman,
Ganges

강과 수변

샌드라 카지-오그레이디, 실비아 미켈리

Rivers and Waterfronts

Sandra Kaji-O'Grady and Silvia Micheli

지난 80년 동안 진행된 세계화를 생각하지 않고는 수변 개발을 이해할 수 없다. 기술 발달로 교역이 세계화됨에 따라 쓸모 없어진 수변이 개발의 대상이 되었다. 전 세계 규모의 교역은 결과적으로 대규모 세계 관광산업의 출현으로 이어졌다.

국제 교역과 그로 인한 세계화는 복합 운송 컨테이너라는 단순한 기술 없이는 불가능했을 것이다. 1960년대 이전만 해도 바다를 따라 장거리를 가는 교역 상품은 상자, 가방, 통에 포장되어 운송되었다. 수변에서 선박에 짐을 싣고 내리는 데에는 엄청난 시간과 노동력이 들었다. 그 같은 노고와 과도한 비용은 국제 교역에 걸림돌이 되었다. 트럭이나 기차에서 운반 가능한 금속 선적 컨테이너가 발명됨에 따라 상품은 생산지 가까이에 있는 컨테이너에 실릴 수 있게 되었다. 수변 주변에 밀집된 창고나 제조업체는 무용지물이 되었다. 무거운 기계를 이용해 대형 컨테이너를 선적하는 것이 가능해지면서 노동자 수요도 줄어들었고 그 결과 수변 주변의 술집, 사창가, 싸구려 여관이 사라졌다. 1970년대에 규모의 효율을 위해 한층 더 큰 컨테이너 선박이 건조되면서 수심이 더 깊은 수변이 필요해졌다. 무엇보다도 컨테이너를 이용하는 무역에는 공간이 필요했다. 공간과 깊은 수심이 필요해지면서 전 세계적으로 수변이 도시 중심부에서 낙후된 지역으로 이동했다. 이러한 변화가 일어나지 않은 세계적 도시는 찾아볼 수 없다(뉴욕, 상하이, 런던, 시드니도 그 같은 변화를 겪었다). 역사적으로 정착지는 바다나 강의 교역 경로에 위치하기 때문이다. 여러 도시에서 불결하긴 해도 활기찼던 수변 지역 경제적 존재 이유를 상실했다.

선박 컨테이너가 이용되기 시작한 때에 비슷한 기술적 발전의 결과로 항공 수송이 저렴해졌다. 1970년대 초반 동체가 넓어 승객 수백 명을 태울 수 있는 보잉 747기가 운항을 시작했다. 첫 해에 만석을

Waterfront developments cannot be understood outside of the globalization that has taken place over the past eighty-years. They make use of ports made redundant by technological changes that themselves led to the internationalization of trade. Global trade, in turn, is a force in the advent of global mass tourism.

International commerce and the globalization it brought would have been impossible without the simple technology of the intermodal shipping container. Prior to the 1960s, goods transported by water over long distances were packaged in boxes, bags, and barrels. A great deal of time and labor was spent loading and unloading ships at portside. This effort added so significantly to costs that international trade was inhibited. The invention of a metal shipping container that could be transported from truck or train to ship meant that goods could now be loaded into containers close to where they were made. The warehouses and manufacturers that congregated around ports were no longer needed. As the large containers were loaded using heavy machinery, far fewer workers were needed and this led to the demise of portside bars, brothels and cheap lodgings. Lastly, the 1970s saw the creation of ever-larger container ships to realize efficiencies of scale and these required deeper water. More importantly, containerized trade required space. In many places, the need for space and deeper water resulted in the shifting of port operations from near city centers to less developed locations. It is hard to think of a global city where this has not occurred—New York, Shanghai, London, Sydney— since settlements have historically been located on sea or river trading routes. What had typically been a lively, albeit seedy, part of so many cities lost its economic raison d'etre.

The advent of cheap flights, the outcome of a parallel development in technology, coincided with the appearance of shipping containers. The wide-bodied Boeing 747 able to carry hundreds of passengers went into service in the early 1970s. In

기록한 747기는 항공 운임을 절반으로 줄였다. 그 즉시 사람들이 항공기를 이용하기 시작했다. 중요한 사회적 변화 시기에 보잉 747기는 항공 여행, 관광, 전 세계 사람의 운행 수단 측면 면에서 기하급수적 성장을 이끌었다. 여가 활동과 관광이 산업이 떠나간 여러 도시의 공백을 메울 수단으로 떠올랐다. 예를 들어 뉴욕에서는 수변과 제조업이 뉴저지로 옮겨간 후 대규모 재개발과 젠트리피케이션 과정이 시작되어 여전히 진행 중이다. 이와 마찬가지로 세계 여러 도시가 세계 비즈니스와 관광의 중심지로 자리매김하고자 전력을 다했다. 수변의 부활과 운영은 도시 이미지를 드높이고 세계적 열망을 충족하는 기회를 만들어냈다.

소매업을 포함한 문화 레저 산업은 수변 건축과 도시 공간의 언어와 문법에 영향을 미치며 수변에 새로운 활력을 부여하는 원동력이 되었다. 소매상점, 음식점, 도시 설치물이 대규모 관광에 다양한 활동을 제공했다. 이러한 시설을 수변 지역 건너편으로 배치해 그 구조(기술, 미학, 기능)가 쓸모없어지는 순간 새로운 구조로 신속하게 대체할 수 있도록 했다. 일부 도시 — 특히 1970년대와 1980년대에 수변을 재개발한 도시 — 는 수변에 있는 역사적 산업 구조를 모두 허물어버리고 새로 건설했다. 우리의 연구 대상인 브리즈번 강의 남쪽 기슭이 그러한 사례다. 창고를 허문 자리에서 1998년 세계 엑스포를 개최했고, 화물선은 통근 페리로 대체되었다.

현재는 역사적 도시 구조를 보존하는 것이 가치를 인정받는 추세다. 심지어 문화 관광을 활성화하기 위해 과거의 구조를 되살리기도 한다. 도시 설계와 건축은 중요한 전략이 되어 세계무대에서 존재감을 드러내는 것이 최우선 과제인 자치 정부의 소프트 파워 의제를 뒷받침하고 있다. 모든 수변 도시는 '빌바오 효과(Bilbao effect, 건축물이 지역에 미치는 영향)'를 만들어내기 위해 독자적인 구겐하임 미술관을 건설하려고 한다. 이러한 프로젝트의 목표는 쇠퇴해가는 수변을 회복하는 것이다. 각 도시는 그 목표를 이루기 위해 유명 건축가들에게 높은 수수료를 제공하고 두드러지고 미학적인 대형 프로젝트를 디자인하도록 했다. 상징적인 다리, 신기하고 멋진 정원, 대관람차, 기이한 구조물이 도시의

its first year a fully-loaded 747 cut the cost of flying a passenger by half. Flying became instantly more accessible. At a time of major societal change, the Boeing 747 drove exponential growth in air travel, tourism, and connections between people around the world. Leisure and tourism were seen by many cities as a way to fill the void left by industry. New York, for example, saw its port and manufacturing industries depart to New Jersey and embarked on a major re-development and gentrification process that is still ongoing. Likewise, cities across the world strived to position themselves as centers for global business and tourism. The reactivation and implementation of waterfronts was an opportunity to boost urban image—and fulfil global aspirations.

The cultural and leisure industry, including retail, became the main engine for the new life on waterfronts, impacting on the language and syntax of their architecture and urban spaces. The provision of retail facilities, eateries and urban installations offered a variety of activities for mass-tourism. Distributing these across waterside precincts provided the possibility of quick replacement once their structures—their technologies, aesthetics and functions—became obsolete. Some cities—particularly, those that redeveloped their waterfronts in the '70s and '80s— erased the historic industrial fabric of their ports and started again. The south bank of the Brisbane river, the focus of our research, is such a case, with the warehouses demolished to make way for a World Exposition in 1988 and the cargo boats replaced by commuter ferries.

Today, the preservation of historic fabric is more likely to be valued—even recreated to attract cultural tourism. Increasingly, too, urban design and architecture are understood as crucial strategies underpinning the soft power agenda of municipal governments that consider visibility in the global market as a priority. Every waterfront city wants its own version of the Guggenheim museum to achieve the "Bilbao effect". Such projects have are aimed at leading the revitalization of their decaying waterfronts, and this has generated numerous commissions for starchitects to design flagship projects, usually characterized by a recognizable and aesthetic. Iconic bridges, wonderland gardens, panoramic wheels and eccentric pavilions populate the rivers' banks and sea-sides of cities that have committed their waterfronts to leisure and

제방과 해안에 들어섬에 따라 수변은 세계 교역과 관광을 위한 여가 활동과 오락의 장소로 변모했다. 싱가포르의 마리나베이(Marina Bay)와 시드니의 바랑가루(Barangaroo) 재개발이 그러한 사례다.

수변은 역사적 구조물의 재사용으로 '진정성(authenticity)'을 구축하고 건축학적 상징을 통해 인기를 끌었지만 이제 수변 재개발은 — 본래의 의미가 퇴색되었을 뿐 아니라 — 갈수록 복잡해지는 도시적, 정치적 요구에 맞춰 진행되어야 한다. 시급한 과제는 재개발된 수변 공간을 전 세계 교역 종사자와 현지 주민에게 매력적인 장소로 조성하여 이익을 창출하는 방안을 마련하는 것이다. 이때 주민들의 거처를 수변의 구성요소로 반드시 포함시켜야 한다. 이렇게 하면 방문객들이 '지역 생활'을 경험할 수 있을 뿐 아니라 비수기, 성수기에 따른 관광 변동성을 조정할 수 있다. 아파트와 공원은 원만한 재개발을 추진하기 위한 정치적 수단으로서 재개발이 외국 관광객뿐 아니라 현지 주민을 위해 이루어진다는 인상을 전달한다.

기후 변화와 과학자들의 예측에 따르면 전례 없는 위기가 출현하고 있는 이 시대에 수변은 현재의 오락적 가능성을 초월하여 재고되고 재설계되는 추세다. 도시 강기슭이 예상치 못한 홍수로 피해를 입은 경우가 늘어나고 그러한 홍수 피해에 시달릴 대로 시달린 보험사가 홍수 지역에 대한 보상을 거부하면서 주요 도시의 수변 지역은 곧 쇠퇴로 접어들 수도 있다. 뉴욕은 홍수 위험이 있는 도시의 회복력과 안정성 증진 방안에 대해 세계적 논의를 주도하고 있다. 비야케 잉겔스 그룹(Bjarke Ingels Group)이 포함된 팀이 디자인한 빅유(BIG U)는 맨해튼 남부(Lower Manhattan)를 대상으로 한 가상 프로젝트이다. 해당 프로젝트는 연방 자금 지원 프로그램으로서 도시의 크고 복잡한 문제에 대한 해결을 모색하는 '디자인에 의한 재건축(Rebuild by Design)'의 후원을 받는다. 빅유는 2012년 허리케인 샌디에 의한 피해에 대응하는 과정에서 잉태된 방어적 기반시설이다. 동시에 이러한 장벽 시스템은 소셜미디어의 영향을 덜 받고 도심의 미래 생존에 초점을 둔 접근법으로서 도시 부두의 책임감 있는 디자인 모델로 떠오르고 있다.

entertainment for global business and tourism— this is the case of Marina Bay in Singapore and Barangaroo redevelopment in Sydney.

Along with the "authenticity" that might be conveyed through the re-use of historic structures or the draw gained through architectural icons, more recent waterfront redevelopments—along with ones that have lost their original sheen—have had to grapple with increasingly complex urban and political demands. The question of how to make redeveloped waterfronts attractive for both global trade and local population, and striving towards their balance, are pressing issues. It is now imperative that residential accommodation is included in the mix so as to guarantee an experience of "local life" for visitors, as well as a stable base of consumers to mediate the seasonal fluctuations of tourism. The provision of apartments and public parks is often used to smooth the political path of developments that are otherwise targeted at international transient population.

In the era of climate crisis and the unprecedented challenges that scientists predict, waterfronts are to be reconsidered—and re-designed—beyond their current entertaining potentiality. Prime urban waterfront properties may soon become decaying areas as the urban banks of the rivers are increasingly threatened by unpredictable floods – with weary insurance companies already pulling out of areas included in flooding maps. New York is leading the international discussion about how to improve the resilience and security of cities with flooding problems. Designed by a team including Bjarke Ingels Group, the Big U is a visionary project for Lower Manhattan supported by "Rebuild by Design", a federally-funded program that seeks solutions for large-scale complex problems that affect the contemporary city. The Big U is a protective infrastructure conceived in response to the damage caused by the Hurricane Sandy in 2012. At the same time, this barrier system comes across as a model of responsible design for urban waterfronts – an approach less distracted by the whispers of social media and more concerned with the future survival of our urban centers.

본 프로젝트에서는 호주에서 가장 빠르게 성장하는 도시 중 하나인 브리즈번을 가로지르는 강의 남쪽(사우스뱅크)과 북쪽지역(노스뱅크)의 도시 개발을 분석한다.

This project analyses the urban development of the south and north banks of the river that dissects one of Australia's fastest-growing cities.

물의 분배를 둘러싼 논란은 물 부족문제가 아닌 '물을 어떻게 할당하고 통제해야 하는가'에 있다.

The most controversial aspect of water distribution is not scarcity but allocation and control.

암만

Amman

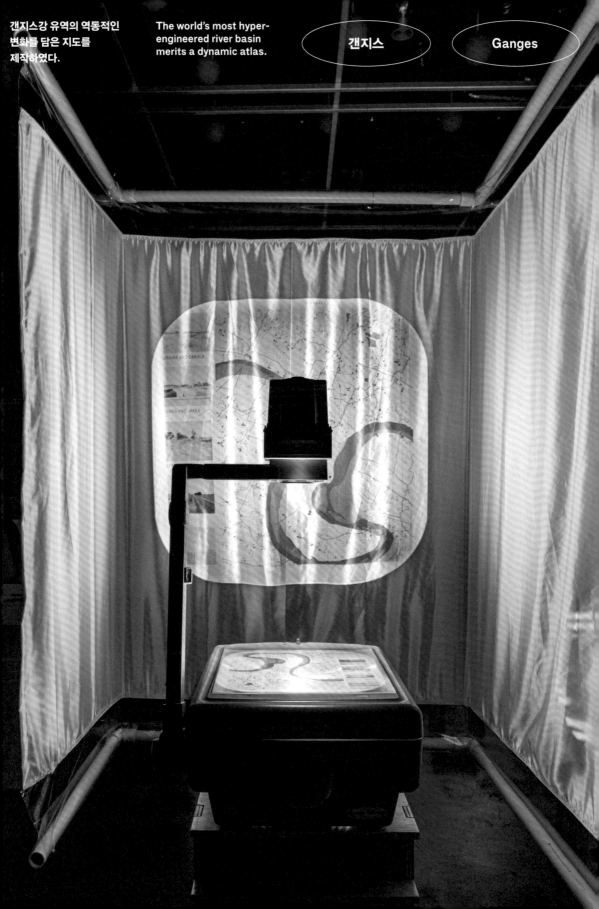

갠지스강 유역의 역동적인
변화를 담은 지도를
제작하였다.

The world's most hyper-
engineered river basin
merits a dynamic atlas.

갠지스

Ganges

도시가 경제의 중심지로 떠오르면서 상업 중심지가 나타나기 시작한다. 상업 중심지는 상업 활동이 집중된 곳으로 도시의 일부가 상품 거래의 전용 공간으로 활용된다. 그러한 측면은 지속적인 상품 교환이 이루어질 수 있는 시장이나 빽빽이 밀집된 건물을 통해 드러난다.

상업 중심지는 중심가, 번화가, 공설시장 등의 개념을 낳는다. 이러한 곳들은 단순한 소비의 공간이 아니라 사회적, 문화적 교류가 이루어지는 도시 공유지다. 도시의 특성은 이 같은 공간에 의해 형성될 수 있다.

도시(都市)라는 한자어는 상공업 발달 지역을 뜻하는 도회지와 시장이 결합된 단어다. 이 단어를 통해 경제·상업 중심지와 도시라는 개념이 불가분의 관계라는 사실을 알 수 있다. 도시가 진화하면 이러한 중심지가 도시 발전을 촉진하지만 새로운 중심지가 부상하면서 기존 중심지가 쇠락하기도 한다. 그 때문에 땅값이 상승하고 인구밀도 상승의 필요성이 생겨나 보존과 유산이 위협을 받는 것이 우리 시대의 문제다. 도시의 특정 지역이 노후화하면 그 지역을 다른 곳으로 대체해야 할까, 아니면 그 도시만의 특징을 드러내는 식으로 변화를 주어 보존해야 할까? 해당 섹션의 도시들은 역사적인 시장 지대를 재해석하고 현재의 시장을 평가하여 이 시대의 과제를 역사적 중심부로 끌어들임으로써 그러한 문제를 해결하는 중이다.

사마르칸트,
카사블랑카,
아디스아바바/
다르에스살람,
스톡홀름,
청진

As cities emerge as economic centers, commercial cores begin to appear. These are concentrations of commercial activity that have appropriated a certain part of the city to transact goods. These are manifested through markets or tightly grouped buildings that allow continuity of commercial exchanges. This gives rise to the idea of main streets, downtown, and public markets. These are not only spaces of consumption, but they represent the commons of the city where social and cultural exchanges occur. The personality of a city can be attributed to these spaces. The Chinese word [do-si], which is a combination of characters meaning city and market, reflects this inseparable condition between these centers and the concept of a city. As cities evolve, these centers catalyze development, yet sometimes they erode with time as new centers emerge. This presents the contemporary question of preservation and heritage which is contested by the increasing land value and need for higher density. Once certain areas of the city are obsolete, should they be replaced, or can they be adapted to preserve that which represents a distinct character of the city? The cities in this section tackle these questions by reinterpreting historic market sites, evaluating current markets, and pushing contemporary agendas onto historic centers.

Samarkand,
Casablanca,
Addis Ababa /
Dar es Salam,
Stockholm,
Chungjin

도시들의 도시   A city for cities

LA
MEDINA
900

The fabric of the colonial city is similar to other Moroccan towns (Rabat, Tetouan, Marrakech...) and, in many respects, has already been transformed and influenced by foreign presences starting from the end of the 19th century. The thresholds between public, common and private spaces are negotiated in a very fluid manner. There are no urban blocks and the houses are grouped and defined by streets and dead-ends. The area of the street right in front of each house is the natural continuation of the domestic space.

Site Plan 1/2000

Model Plan 1/500

Aerial View of the Medina - 1914

COLONIAL
CITY
1920

With the new city planned by Henri Prost, Casablanca becomes in a decade a new laboratory for urban and architectural experimentation. The Haussmanian urban block is introduced and adapted to this new reality and emerges is a powerful hybrid, the planning of the city is integral to its development and France is defining and new way of life, a new relationship to public space that becomes municipalized and where the common is often co-opted. This new fabric has generated some of the most fascinating issues in domestic architecture, where buildings grow to become entire cities and start developing into urban "machines".

Site Plan 1/2000

Model Plan 1/500

CITE
ISRAELITE
1940

The experimentations in Casablanca are not only urban and architectural, they are political. The French ruling started a policy of segregation amongst the different religious and social groups in order to anchor their presence. The extensive Jewish population in Casablanca played an important role in its development. This fabric is a first attempt at creating an environment dedicated to a single population group. It triggers a new moment in the urban culture, one of fragmentation and differentiation.

Site Plan 1/2000

Model Plan 1/500

카사블랑카의 다양한
이해관계자들과 거주민들이
도시의 팽창과 수축 과정에서
교집합을 형성해가는 과정을
담아내며 중립지대로서의
역할을 수행하는
카사블랑카를 소개한다.

Instances of sometimes
rapid multiplication,
contraction, dilatation,
and intersection portray
Casablanca as a territory
continuously in search of a
terrain d'entente between
its various stakeholders
and inhabitants.

CARRIE E CENTRA S
50

After the war, this notion of urban fragmentation takes a place of experimentation on a dum... intersection of a segregationist policy with... ...and tropicalize itself. The modernist... ...capes integrates housing with patios a... ...ard house. These buildings by ATBAT in... ...during the CIAM X along side the works... ...nist fabric, not streets, not urban block...

Plan 1/2000

Model Plan 1/500

AIN CHOCK
1960

After the war, this notion of urban fragmentation takes on a new dimension, it becomes a place of experimentation on a domestic scale. This "Habitat Adapte" is to adapt and tropicalize itself. The modernist bar siting in these idtereflexis landscapes integrates housing with patios as an interpretation of the Moroccan courtyard house. These buildings by ATBAT in "Carrieres Centrales" had a powerful impact during the CIAM X along side the works of the TEAM X. In this radical modernist fabric, not streets, not urban block, just buildings.

Model Plan 1/500

Ain Chock Development - 1952

DAR LAMANE
1980

Casablanca is not a layme... ...it next to the other, it is in a sense a museum of 20th ... ...tectural history. With Dar Lamane we are comfortal... ...ar fabric. It's a figure at the edge of the city, where the... ...y not taken over rural living. The planned streets becon... ...with... ground floors, communal rural structures. Since its const... ...s bee... owed by the city, yet it retains its hybrid dimension.

Site Plan 1/2000

Model Plan 1/500

Aerial View Dar Lamane - 1983

카사블랑카   Casablanca

카사블랑카: 빈 공간의 변화

도시 빈 공간의 변이와 저항
오우아라로우+초이

Casablanca: Void Transformations

On mutations and resistances of void spaces in the city
OUALALOU + CHOI

맥락의 실종. 도시를 건물의 결집체로 본다면 모든 도시가 비슷비슷하게 보일 수밖에 없다. 한 세기 넘는 기간 동안 식민지화와 글로벌화에 오염된 도시들은 똑같지는 않더라도 최소한 비슷하게는 보인다. 외관의 유사성은 맥락이라는 개념을 낳는다. 맥락은 전 세계 도시에 진입하고 개입하기 위해 그 도시를 피상적이고 포괄적으로 이해하는 데 사용되는 개념이다. 우리는 맥락을 거부한다. 우리는 두루뭉술하고 효율적이며 기반시설에 주도되는 도시의 가능성에 저항하고자 한다. 도시의 비범한 복잡성과 특수성은 도시가 몇 가지 동일한 요소(도로, 블록, 구획, 건물)로 구성되어 있는 듯 보일 때 한층 더 매력적으로 다가온다. 이러한 다양성을 이해하기 위해서는 유전적 돌연변이와 진화를 생각해보면 된다. 다원주의 식의 진화는 에토스(ethos, 관습이나 기풍)와 목적성이 있고 방향 전환이 불가능하다. 이는 동시대의 세계화된 도시로 귀결된다. 우리는 그러한 에토스를 거부한다.

빈 공간. 피상적인 진화론과 맥락에 바탕을 둔 접근법을 피하기 위해 우리는 무엇이 존재하는가, 라는 질문을 넘어서 무엇이 존재하지 않는가, 라는 질문의 답을 생각해보아야 한다. 도시는 건설되는 것이 아니라 조각되는 것이다. 도시는 다양한 특성이 있는 빈 공간에서 생겨난다. 빈 공간은 공지가 아니며 지도나 설계도에 표시되지 않는다. 빈 공간은 주인이 없거나 공공장소이거나 공유지이거나 개인 소유이거나 은밀한 곳일 수도 있지만 대부분은 무질서하게 다른 공간으로 침투한다. 우리는 이번 설치물에서 처음으로 도시의 빈 공간과 한 세기 동안의 계획을 통한 변화 모습을 보여주려고 시도했다.

명판 실험. 카사블랑카는 아름답다. 가장 20세기다운 도시이며 도시, 건축, 사회, 정치 실험을 위한 야외 실험실이다. 카사블랑카는 서로 맞닿아있지만 층을 이루지는 않는 명판의 도시이다.

벌집 모양 – 1954년과 2005년의 ATBAT AFRIQUE
Nid d'abeilles – ATBAT AFRIQUE 1954 and 2005

NO CONTEXT. When expressed as a concretion of buildings, cities become too easily comparable. After more than century of colonial and global contamination they all seem, if not alike, at the very least familiar. This apparent similarity gives rise to the notion of context, a superficial understanding and a generic way to enter and intervene globally on cities. We refuse context. We want to resist the possibility of generic, efficient, and infrastructurally driven cities. The extraordinary complexity and specificity of cities is that much more fascinating in that they seem to be made out of the same few elements (roads, blocks, parcels, buildings). This diversity is often understood in an analogy with genetic mutations and evolution. A Darwinian evolution with an Ethos, a purpose, an impossible direction to change, results in the globalized contemporary city. We refuse this ethos.

VOIDS. To avoid this superficial evolutionary

카사블랑카는 층의 형성과 식민지화와 산업화 이전 도시의 운명이었던 자기잠식을 거부했다. 역설적이게도 층이 진 도시는 조립이 불가능함에도 의심할 여지없이 아름다움의 표준이자 도시 계획의 성공적인 결과물로 간주된다. 명판의 도시는 불연속적인 파편의 도시이다. 이러한 도시에서 우리는 점진적이고 철저한 방식으로 변화를 읽어내고 측정할 수 있다.

저항과 투항. 우리는 도시 구조 내에서 빈 공간의 변화를 살펴봄으로써 무엇이 저항하고 실종되었으며 본질이 변화할 정도로 변이되었는지 파악하려고 했다. 카사블랑카는 특별한 실험의 장이었고 도시 프로젝트가 급진적이었던 만큼 프로젝트에 대한 반응도 폭력적이었다. 우리는 도시 프로젝트의 역사적 정통성보다는 사람들이 프로젝트를 바꿔놓은 방식에 주목했다. 카사블랑카에서는 균형추가 작동한다. 카사블랑카는 급진적인 실험을 환영하고 소화하는 곳이다. 도시의 빈 공간은 (전통적인 안뜰처럼) 억눌려 있다가 나중에 다시 도입되었다. 마찬가지로 (벌집모양 ATBAT에 매달려있는 파티오처럼) 조성되었다가 나중에 없어진 빈 공간도 있었다.

새로운 도시 구조의 가능성. 우리는 역사 분석 이상의 이야기를 원했고 그러한 이야기로 이 도시에서만 가능한 새로운 도시 구조를 창출한 가능성을 상상하고 싶었다. 그러한 도시 구조는 다양한 측면을 얼기설기 짜맞추기보다 20세기의 변화에 저항한 빈 공간을 중첩해놓은 구조가 될 것이다. 중첩된 부분 사이사이에 무엇인가를 건설하는 데 필요한 틀이 될 것이다.

1900년대의 명판 – 메디나. 여러 측면에서 라바트, 테투안, 마라케시 등 모로코의 다른 도시와 비슷한 식민지풍 도시 카사블랑카의 구조는 이미 19세기 말부터 이질적인 존재에 의해 변화되고 영향을 받았다. 공공장소, 공유 공간, 개인 공간 사이의 경계는 견고하지 않고 유동적이었다. 여기에는 도시의 블록이 없으며 주택들은 무리지어 있고 거리와 막다른 길에 의해 구역이 나누어진다. 각 주택의 정면 오른쪽에 있는 거리 부분은 자연스럽게 가정용 공간으로 연결된다.

1920년대의 명판 – 식민지 도시. 앙리

and contextual approach, we must look beyond what is and measure what is not. Cities are not built - they are carved. A city is made of voids of different natures. Voids are not open spaces and they cannot be described and represented in a map or a cut plan. Voids can be unclaimed, public, common, private, and intimate but mostly they all percolate with each other in an unruly fashion. In this installation, we have made a first attempt at representing the voids of a city and their transformation through a century of planning.

AN EXPERIMENT OF PLAQUES. Casablanca is beautiful. It is the ultimate 20th century city, an open-air laboratory of urban, architectural, social and political experimentations. It is a city of plaques adjacent to each other and not layered on top of each other. In that way, this city refused the layering and cannibalization that was the fate of precolonial and preindustrial cities. It is a paradoxical situation in that the layered city is impossible to fabricate, yet it is unquestionably considered the canon of beauty and the desired result of planning. A city of plaques is a city of discontinuous fragments; it allows us to read and measure transformations in a gradual and rigorous manner.

RESISTANCE AND SURRENDER. We wanted to look at the transformation of the voids in the fabric of the city and to understand what has resisted and what has disappeared and mutated to the point of changing in nature. Casablanca was an extraordinary field of experimentation and the radicality of the urban projects is only matched by the violent reactions to them. We were less attentive to the projects in their historical orthodoxy as we were to the manner in which people have transformed them. That is the Casablanca Pendulum. It is a territory that welcomed radical experimentation and metabolized them seamlessly. Urban voids were suppressed (such as the traditional courtyards) and were later reintroduced; symmetrically, voids were created (such as the suspended patios of ATBAT in Nid D'abeilles) that were later removed.

THE POSSIBILITY OF A NEW FABRIC. We wanted to go beyond this historical analysis and to use this tale to imagine the possibility of a new fabric that would only be possible in this city. It would not be a collage of dimensions, but a real superimposition of the voids that resisted this 20th century of transformation. This would become a

프로스트가 계획한 신도시 당시의 카사블랑카는 10년도 지나지 않아 도시와 건축학의 새로운 실험 공간이 되었다. 이 새로운 생태계에 후기 오스만풍의 도시 블록이 도입되고 조정된다. 이 과정에서 강력한 혼성 구조가 탄생했다. 도시 계획은 건축의 발전에 반드시 필요한 과정이며 프랑스 식민 세력은 새로운 생활 방식을 규정하고 공공장소와의 새로운 관계를 형성했다. 이제 공공장소는 시의 소유가 되고 그곳에서 공유 공간이 흡수된다. 이처럼 새로운 구조는 빌딩이 도시 블록 전체가 되고 도시의 '기계'로 발전하기 시작한 모로코 국내 건축에서 가장 매력적인 경험 일부를 만들어낸다.

1940년대의 명판 – 이스라엘인의 도시. 카사블랑카에서의 실험은 도시와 건축에 국한되지 않고 정치에서 이루어졌다. 프랑스 정부는 정부의 존재감을 확실히 드러내기 위한 방법으로 다양한 종교와 사회 집단을 대상으로 분리 정책을 시행하기 시작했다. 카사블랑카에는 수많은 유대인이 살고 있었고 그들은 도시 발전에 중요한 역할을 했다. 이 프로젝트는 단일 인구 집단의 전용 도시 구조를 만들어내려던 첫 시도다. 그에 따라 분열과 차별이 가득한 도시 문화에서 새로운 순간이 시작된다.

1950년대의 명판 – 중앙 도로. 전쟁 이후에 이곳이 모로코 국내의 실험장이 됨에 따라 도시의 파편화라는 개념이 새로운 국면을 맞이했다. '주거 공간 개조(Habitat Adapté)'는 분리주의 정책과 근대주의에 대두된 적응과 열대화의 필요성이 교차하는 지점이다. 도로가 없는 곳에 위치한 근대주의풍 술집은 안뜰이 있는 모로코식 주택 양식을 재해석한 곳이다. '중앙 도로'에 있으며 ATBAT가 설계한 건물들은 팀 텐의 작품들과 함께 근대 건축 국제회의(CIAM X) 기간 동안 강렬한 충격을 주었다. 이러한 급진적인 근대주의 구조에서는 거리와 블록은 없고 단지 건물만 있을 뿐이다.

1960년대의 명판 – 1960년대 아인초크. 세기가 바뀌고 난 후 카사블랑카의 전통적인 도시 구조는 건축가와 도시계획 전문가의 마음을 사로잡고 있으며 이러한 전통적인 도시구조를 답습하고 모방하려는 시도가 계속되고 있다. 식민지 시대 이전 도시구조의 형태가 아니라 장치와 메커니즘으로서의 필요성이

framework that would allow us to build between the creases.

PLAQUE 1900' – MEDINA. In many respects, the Fabric of the colonial city, similar to other Moroccan towns (Rabat, Tetouan, Marrakech...), has already been transformed and influenced by foreign presence since the end of the 19th century. The thresholds between public, common, and private spaces are negotiated in a very fluid manner. Here there are no urban blocks and the houses are grouped and defined by streets and dead-ends. The part of the street right in front of each house is the natural continuation of the domestic space.

PLAQUE 1920'– COLONIAL CITY. With the new city planned by Henri Prost, Casablanca becomes in less than a decade a new laboratory for urban and architectural experimentations. The post- Haussmanian urban block is introduced and adapted to this new ecology. What emerges is a powerful hybrid - the planning of the city is integral to its architectural development and the French colonial powers are defining a new way of life, a new relationship to the public space that becomes municipalized and where the common spaces are now co-opted. This new fabric has generated some of the most fascinating experiences in domestic architecture, where buildings grow to become entire urban blocks and start to develop into urban "machines."

PLAQUE 1940'– CITE ISRAELITE. The experimentation in Casablanca is not only urban and architectural, it is political. The French government has started a policy of segregation amongst the different religious and social groups as a way to anchor its presence. The Jewish population in Casablanca was extensive and they played an important role in the city's development. This project is a first attempt at creating a fabric dedicated to a single population group. It starts a new moment in the urban culture, one of fragmentation and differentiation.

PLAQUE 1950'– CARRIERES CENTRALES. After the war, this notion of urban fragmentation takes on a new dimension as it becomes a place of experimentation on a domestic scale. This "Habitat Adapté" is the intersection of a segregationist policy with the need for the modern movement to adapt and tropicalize itself. The modernist bar sitting in its streetless landscape integrates housing with patio as an interpretation of the Moroccan courtyard house. These buildings by ATBAT in "Carrières

영속적이라는 점이 흥미를 끈다. 모로코의 독립이 임박할 때쯤 새로운 전통에 대한 연구는 새로운 유형의 도시주의와 만난다. 이 새로운 유형은 기업이 직원들의 공동 주택으로 건설한 '씨떼(Cités, 주택단지)'다. 생선가시 구조의 규칙적인 패턴으로도 도시 블록이 존재하지 않는다는 사실이 감춰지지는 않는다.

1980년대의 명판 – 다르라마네. 카사블랑카는 층을 이루는 도시가 아니다. 한 곳에서 여러 시기의 도시 발전을 볼 수 있는 도시가 아니라는 뜻이다. 어떠한 의미에서 카사블랑카는 20세기 도시와 건축 역사의 박물관이라고 할 수 있다. 다르라마네에서 우리는 가장 최근에 실험된 도시 구조를 보았다. 도시 경계부의 모습에서 도시 문명이 아직 농촌 생활을 대체하지 못했다는 것을 알 수 있다. 계획된 거리는 공유 영역이 되고 1층은 공동의 농촌 건축물이 된다. 다르라마네는 지어진 이래로 카사블랑카에 잠식되고 있지만 아직 혼성 구조로서의 측면이 남아있다.

Centrales" had a powerful impact during the CIAM X alongside the works of the TEAM X. In this radical modernist fabric, there are no streets, no urban blocks - just buildings.

PLAQUE 1960'– AIN CHOCK 1960'. The traditional urban fabric has fascinated architects and urbanists alike ever since the turn of the century and there have been many attempts to emulate and imitate it. It is fascinating to see the permanence not of the figure of the precolonial fabric but of it perceived necessity, as an apparatus and a mechanism. Right around the time of Morocco's independence, this research of a new tradition meets a new kind of urbanism - the "Cités" corporate housing structures built by companies for their own employees. The regularity of the fishbone structure cannot hide the disappearance of the urban block.

PLAQUE 1980'– DAR LAMANE. Casablanca is not a layered city, each period is built next to the other. It is, in a sense, a museum of 20th Century's urban and architectural history. With Dar Lamane, we are confronted with the last experimental fabric. It is a figure at the edge of the city, where the urban civilization has not yet taken over rural living. The planned streets become shared fields and the ground floors become communal rural structures. Since it has been built, Dar Lamane has been swallowed by the city, yet it retains its hydrid dimension.

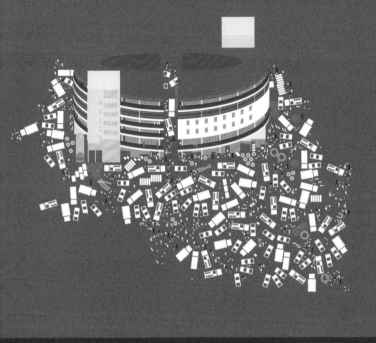

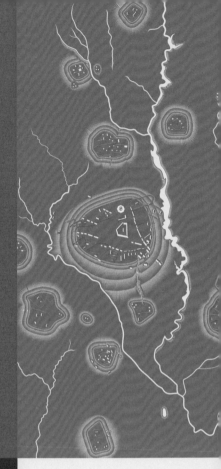

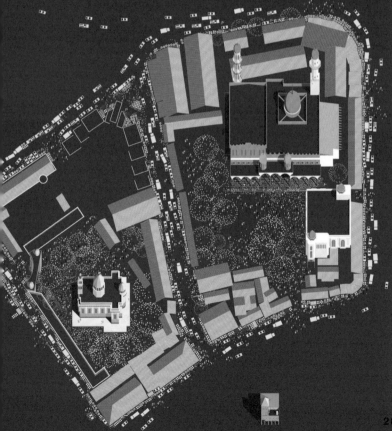

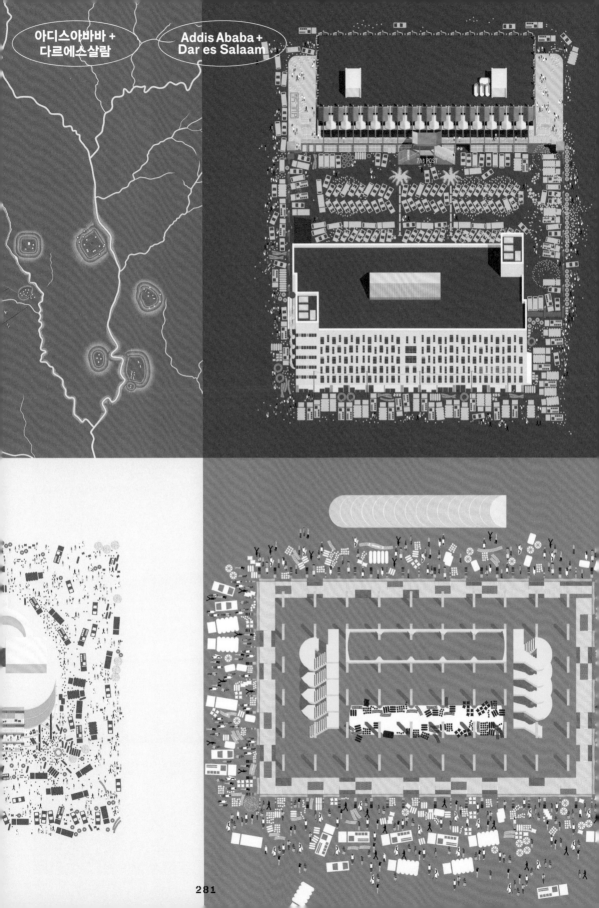

7th POST

스톡홀름을 중층적이고
다면적이며 뜻밖의 제안을
수용할 수 있는 환경으로
고찰한 건축가의 역량을 한껏
드러내며, 제시되는 환경을
통해 이 도시의 잠재력을
새로운 시각으로 바라보게
된다.

They celebrate the
architect's ability to
address the city as a
layered, multifaceted
environment capable of
absorbing unexpected
proposals and, when read
together, can help us to
see the potential of the
city in a new light.

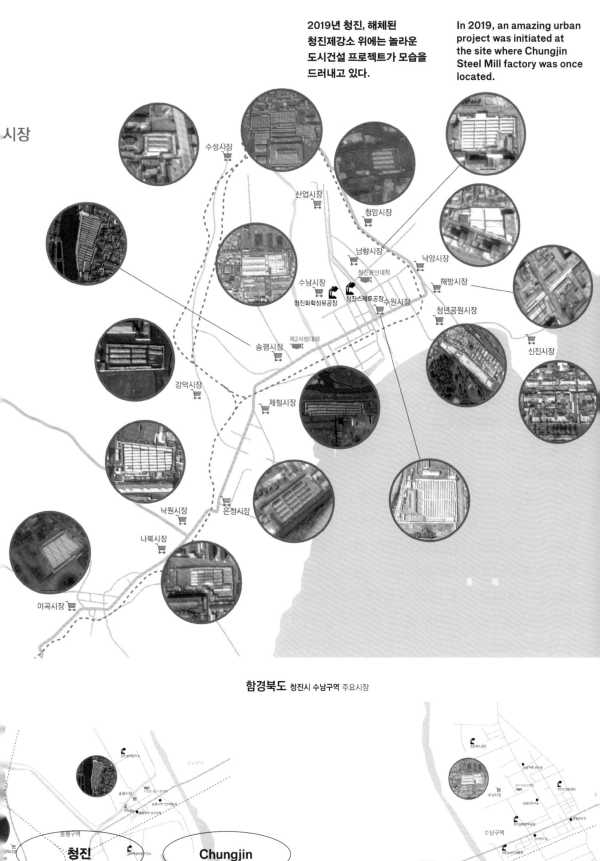

2019년 청진, 해체된 청진제강소 위에는 놀라운 도시건설 프로젝트가 모습을 드러내고 있다.

In 2019, an amazing urban project was initiated at the site where Chungjin Steel Mill factory was once located.

시장

수성시장

산업시장

청암시장

남향시장

낙양시장

청진광산대학

해방시장

수남시장

청진화학섬유공장 청진스레트공장 수원시장

청년공원시장

신진시장

제2사범대학

송평시장

강덕시장

제철시장

낙원시장

은정시장

나북시장

이곡시장

동 해

함경북도 청진시 수남구역 주요시장

청진   Chungjin

청진시

인구 667,929명
시설면적 95,379㎢
시장 수 19개

송평구역

수남구역

수남시장

수남구역

기반시설(인프라스트럭처)은 여러 측면에서 도시성을 유지하는 근본적 틀이다. 기반시설은 교통, 에너지, 통신, 쓰레기 처리, 상수도, 하수관 등 도시 운영에 필요한 서비스를 효율적으로 제공하도록 설계된 체계다. 도시 특성과 기술이 진보할수록, 지정학적 경계를 넘어서는 사회 기반시설, 데이터 기반시설 등 다양한 측면을 운영하는 기반시설도 발전한다.

기반시설 층은 건축적 층과 별개로 운영되고, 이로 인해 도시 구조 내에 빈 공간과 같은 괴리와 결핍이 발생한다. 기반시설을 점검하는 도시들은 다양한 규모로 그러한 결함을 없애는 도시 변혁에 초점을 둔다. 그 과정에서 기반시설을 공공공간으로 변경 후 재사용하여 도시 구조로 재통합하는 것, 화석 연료 기반 사회에 맞는 기반시설 구축, 전 지구적 도시계획에 요구되는 기반시설의 요건 등의 문제가 다루어진다.

반면에 주민들은 기반시설이 제공하는 도시 기능을 간과하는 일이 많다. 사용자, 제품, 물, 전기 등 일상적 서비스에 사용되는 자원의 생활주기 사이에는 단절된 부분이 있다. 주민들은 일단 자신의 영역을 떠나면 쓰레기가 어디로 가는지, 더러운 물과 하수도의 종착점이 어디인지 알 필요가 없다. 기후 변화가 임박하여 자원을 어떻게 사용하는지 인류의 존속 여부에 중요한 문제가 되고 있는 이 시기에 우리가 사용하는 자원의 생태학적 과정을 인식하는 것은 도시 환경을 위해 매우 중요하다. 우리는 지속가능한 생태계를 만들기 위해 적극적으로 집합적인 노력을 기울이면서 새로운 생활 방식과 기반시설을 고안할 뿐 아니라 물질과 구조물을 조정해서 재사용해야 한다. 그에 따라 현재 도시를 메운 일회용 소비문화를 버리기보다는, 기존에 만들어지거나 발견된 것에 새로운 이론적 접근법을 재도입하여 건축의 새로운 언어를 만들어 내야 한다.

바쿠,
보스톤,
도쿄,
로스엔젤레스,
에인트호번,
이스탄불,
카라카스,
쿠웨이트,
바르셀로나,
메데인,
마닐라,
쿠알라룸푸르

Infrastructure represents the underlying framework that sustains urbanity at its different levels. These are systems that are engineered to provide efficient services needed for urban operations such as transportation, energy, telecommunications, waste management, water, and sewer. As urbanities and technology evolve, so do infrastructures that operate at different dimensions such as social infrastructures and data infrastructures that scale beyond any geopolitical boundary.

The infrastructural layer has operated separately from the architectural layer, causing discrepancies and deficiencies in the urban fabric such as urban voids. The cities investigating infrastructures are focused on urban transformations that tackle these deficiencies at multiple scales tackling topics such as the adaptive reuse of infrastructures as public spaces that integrate back into the urban fabric, the effects of infrastructures for a fossil-fuel-based society, or the infrastructural requirements demanded by a planetary urbanism.

On the other hand, the function of the city provided by infrastructure often goes unnoticed by its inhabitants. There is a disconnect between the user, product, and the life cycle of resources used to provide everyday services such as water and electricity. Residents do not need to know where their trash goes once it has left their unit or the terminus of grey waters and sewers. In an era where climate change is imminent, and the use of resources is of critical matter for the continuation of our species, the awareness to the ecological process of materials we use is critical for our urban environments. As we progressively attempt to collectively work towards a sustainable ecology, new typologies of living and infrastructure must be considered, adapting and reusing materials and structures that may produce a new language in architecture, reintroducing a new theoretical approach to the ready-made, or found object rather than the throw away culture that plagues our current cities.

Baku,
Boston,
Tokyo,
Los Angeles,
Eindhoven,
Istanbul,
Caracas,
Kuwait City,
Barcelona,
Medellin,
Manila,
Kuala Lumpur

150년이 넘는 세월에 걸친
바쿠의 석유 산업과 도시화
사이의 밀접한 상호작용에
대한 최초의 종합적 연구를
담고 있으며, 전세계의
도시화라는 오늘날의
근본적인 변화를 이해하는 데
있어 바쿠의 역사가 얼마나
중요한지를 여실히 보여준다.

The first comprehensive
study of the close
interplay of oil industry
and urbanism in Baku
over a period of more than
150 years and reveals
the critical significance
of that history for our
understanding of the
global urbanization
processes that
are reshaping the
contemporary world.

BAKU: OIL AND URBANISM

Eve Blau and Ivan Rupnik / Photo Essay by Iwan Baan

바쿠

Baku

BIRD
HABITAT

HABITAT
ZONE

LOW LIGHT AHEAD
BOSTON UNDERSTORIES

ELECTRICITY
AVAILABLE

SOLAR PLUG-INS

PHOTOGRAPHY
ENCOURAGED

보스턴, 미국
보스턴의 하층 도시
랜딩 스튜디오 (댄 아담스 마리 로 아담스)

보스턴의 고속도로는 마치 도시를 지나는 움푹 파인 참호처럼 주변 지역보다
건설되었지만, 최근에는 그 위에 천장을 씌워 고속도로를 터널로 만들고, 그
새로운 도시가 건설되고 있다. 고속도로 고가다리 아래에는 새로운 유형의 권
고속도로망과 도시, 둘 중 어느 쪽에도 속하지 않은 이 공간에서 도시 관리와
공공 사업이 융합된 도시가 싹트고 있다. 고속도로 교각의 접합부분에 소금이
콘크리트를 뭍인하는 컨베이어 벨트가 고속도로 경사로 사이로 지나가며, 소
애호가들은 도시의 교각에서 숫자 8을 그리며 보드를 타는 흔흥의 도시이다.
고속도로 공간은 '원활한 차량 소통'이라는 유일한 목적의 신호 규제체계에
그러나 건설시장 호황과 주택시장 위기가 닥치고 토지의 가치를 최대
위한 노력이 진행되고 있는 현실에서, 본 전시작 보스턴의 하층 도시는 석유
인간의 시대라는 혁신적 변화의 순간을 포착하고 새로운 신호체계를 통해 참
보장되는 규제체계를 고속도로 고가다리 아래 공간에 적용하고자 한다. 이 공
붙어넣고, 확대하며, 그 다양성이 더욱 뚜렷하게 부각되는 신호 관리체계를 도
공간 생태계가 도시 이동링과 융합되고, 인프라 시설 정비 활동이 시민의 여가
맞물리며, 공공 사업 공간에 어린이들의 놀이터가 되고, 애완견, 강물을 유영
캠핑 애호가들이 만나 새로운 집합 공간을 형성하게 될 것이다.

BOSTON, USA
Boston Understories
Landing Studio (Dan Adams, Marie Law Adams)

Boston Understories looks at spaces and activities under highway viad
the lens of the regulatory sign. Standard markers like the ubiquitous '
have been used to limit or narrowly prescribe public access and fail to
actuality of these sub- infrastructural spaces, where the market forces
into infrastructural margins, ecological systems converge with mobilit
public works intermingle with public recreation.
Boston Understories reflects on these phenomena and introduces a ne
taxonomy of signs to encourage, amplify, and make legible actual and
collective domains of urban viaduct spaces.

도시수집:도시의 기억 상상의 도시

#더하기   #곱하기

city.nolple.com
도시수집에 참여하고 싶어하신 분은 QR코드 또는 위 링크로 접속해주세

보스턴의 하층도시는 석유-
자동차-인간의 시대라는
혁신적 변화의 순간을
포착하고 새로운 신호체계를
통해 참신하고 다양성이
보장되는 규제체계를
고속도로 고가다리 아래
공간에 적용하고자 한다.

**Boston Understories
captures a transformative
moment of the petro-auto-
human-era that envisions
viaduct landscapes with
a new, plural regulatory
framework.**

SUN

The Fukushima nuclear accident has begun to draw attention to renewable energy, especially solar energy. Sunlight and heat are important natural resources of the earth. Even in cities we can use solar energy. It is efficient and safe for each building to decentralize the generation power and not depend on a huge and risky process like nuclear power. Such autonomous energy supply is called 'off grid' and is spreading after the Great East Japan Earthquake. We can generate electricity from sunlight and power homes. We can heat water from the energy of the sun for heating and hot water supply. We realize that we live with the sun. These off-grid houses demonstrate an ecological lifestyle that matches the rhythm of the sun.

"FIRST HE TAKES AN A.V. TRANSIT SHUTTLE TO THE B.R.T. STOP."

CHAINING 3 TRANSIT VEHICLES

APPLY LIFE SUBSIDY CODE?

"THEN HE TAKES THE BUS RAPID TRANSIT THROUGH WILLBROOK AND GARDENA..."

"...AND THROUGH CARSON."

...UNTIL HE REACHES SAN PEDRO.

WOW, DAD'S COMMUTE IS LONG.

로스앤젤레스

Los Angeles

LA는 '교통 중심 도시'라는 비전에 큰 변화를 가져올 자율주행차량의 도래에 대비하고 있다.

LA is bracing for a looming arrival of automated vehicles that are bound to challenge the transit-oriented vision of the metropolis.

거리는 누구의 소유인가? 누가
거리를 공공의 혹은 개인의
소유라 규정하는가? 거리는
지난 3천 년의 역사 속에서
어떻게 변모해왔는가?

Who owns the street?
What makes it public
or private? What is
the influence of the
architecture of our streets
on its use? How have
streets developed over the
last 3000 years?

Who owns the street? What
makes it public or private?
What is the influence of the
architecture of our streets on
its use. How have streets
developed over the last 3000
years? And what does this
mean for the future of the
street?

064
New Babylon (NL)
Constant, 1959

또 하나의 상상

네이란 투란

Another Imagination

Neyran Turan

2000년대 중반의 모습을 한 번 상상해보라. 로테르담에서 한여름 늦은 밤, 참석자 대부분이 건축가인 파티가 열린다. 이러한 파티에서는 건축 이야기가 주를 이룬다. 나는 다른 사람들에게 예일 대학에 제출한 지 채 한 달도 되지 않은 나의 석사 학위 논문 주제를 신나게 이야기하고 있었다. 대화의 열기가 더해갔다. 어느 젊은 건축가 친구의 생생한 표현이 가장 기억에 남는다. "바다에 있는 유전이라. 흥미롭군. 건축과는 어떤 관련이 있지?"

그 당시에 나는 석사 논문을 쓰기 위해 1960년대 유럽 북해의 에너지 지형을 조사하고 있었다. 내 목적은 공해의 지형 차원에서 기반시설로서의 전력망과 블록의 특징을 보여주는 것이었다. 좀 더 광범위하게는 공간과 영토라는 영역을 소개하기 위한 방법으로 영토의 개념을 제시하겠다는 지적인 야심이 있었다. 1990년대 건축을 장악했고 정치를 배제한 '흐름' 논의에 대항하는 뜻이기도 했다.[1] 연구를 하는 동안 두 가지 눈에 띄는 속성이 궁금증을 불러일으켰다. 첫 번째는 수직면이 눈에 띄었다. 스타토일(Statoil,노르웨이 정유회사)의 광고 책자를 보면 해상 플랫폼(해저의 석유와 가스를 추출하기 위한 해양 구조물)이 뉴욕 록펠러 센터의 GE 빌딩과 나란히 서있다. 이 책자는 해상 플랫폼이 규모로 보나 다양한 생활, 작업 프로그램을 포함하고 있다는 점에서나 20세기 새로운 수직 구조의 상징이라고 설명한다. 두 번째로 여러 정유사와 북해 주변 국가들 사이의 북해 원유와 천연가스 분할을 보여주는 자원 배분 지도는 수평적으로 묘사된 것이 눈에 띄었다. 이 지도는 자본과 자원을 물리적 지형과 나란히 배치한 도표였다. 도시 계획이나 건물 계획과 마찬가지로

Picture a scene in mid-2000s. In Rotterdam, it is a late night summer party mostly attended by architects. Partly because of the impossibility to talk something other than architecture during most parties like this one, I was passionately talking about my masters thesis, which I had just submitted at Yale University just less than a month ago, with a small group in the party. A heated conversation followed. Eventually, a vivid expression from a young architect friend marked the highlight of the night for me: "Oil fields on the sea, interesting. What does this have to with architecture?"

At the time, as part of my master thesis, I was looking at the energy landscapes of the North Sea in Europe during the 1960s. My aim was to render the infrastructural characteristics of the grid and the block at the geographic scale of a transnational sea. The broader intellectual ambition of the work was to propose the idea of territorial form as a way to introduce the domain of the spatial and the territorial as an opposition to much pervasive and sometimes apolitical discussions of "flows" in architecture during the 1990s.[1] Two attributes were puzzling and distinct during my research. First one was a vertical. In one of the advertisement brochures of the Statoil (Norwegian oil and gas company), an offshore platform was juxtaposed with the GE Building of the Rockefeller Center of New York, with the point that offshore platforms were the new vertical icons of the twentieth century with their scale as well as with the various living and working programs they can contain. Second, a more horizontal attribute was depicted by the property allocation maps that showed the division of the North Sea oil and gas fields among many oil companies as well as their bordering countries. Here, this was an image of capital and resource juxtaposed with the physical form of geography.

1   본 연구의 요약본은 네이란 투란의 'Spatial Formats: Oil and Gas Fields of the North Sea' 참조. *Thresholds 27: Exploration* Massachusetts: MIT School of Architecture, 2004, 87-91쪽.

1   For a shorter overview of that research, see Neyran Turan, "Spatial Formats: Oil and Gas Fields of the North Sea," in *Thresholds 27: Exploration,* Massachusetts: MIT School of Architecture, 2004, pp. 87-91.

선, 곡선, 망, 블록이 해수면에 조직적 형태로 그려져 있었다. 북해의 에너지 지형도는 지리학과 기하학이, 규모와 형태가 대립되는 원시적인 형태로 그려졌다.

그 이후로 현대 건축의 논의에 많은 변화가 있었다고 할 수 있을 것이다. 예를 들어 현재의 건축 문화에서는 자원 추출 측면의 인프라스트럭처가 건축이나 도시의 주제로 타당한가라는 질문이 더 이상 제기되지 않는다. 그러나 2000년대 초반으로부터 상당한 시간이 흘렀음에도 여전히 해결해야 할 중요한 문제들이 남아 있다. 건축은 다학제간 연구를 통해 지난 20여 년 동안 광범위한 주제를 다뤘고, 세계화와 이에 동반된 사회적, 환경적 과제와 변화하는 요구에 따라 건축학적 지식의 한계를 넓혀왔다. 그 과정에서 인프라스트럭처, 지형, 영토를 정치적으로 바라보는 것이 보편화되었다. 연구와 지도 제작으로 디자인을 보여주는 다양한 '학습(learning from)' 프로젝트도 마찬가지다. 중요한 점은 이러한 주제들이 건축학에 만연해진 현재에 방법론, 표현, 기법뿐만 아니라 형평성이라는 정치 주제까지 포괄하는 한층 더 시급하고 흥미로운 질문이 전면으로 부각되었다는(예를 들어 단순히 도시 프로세스와 도시 시스템과 관련된 문제 해결을 강조하거나 정치적으로 무지한 신환경주의적 사회 개량주의와는 대조적으로) 사실이다.

게다가 자료에 접근해서 시각화할 수 있는 능력은 현대 건축학에서 매력적인 요소가 되지만 '데이터 시각화'는 그저 문제 해결을 통해 프로젝트를 정당화하기 위한 도구가 될 수도 있고, 오염된 해안, 쓸모없어진 매립지, 기후가 극심한 지역 등에 새로운 도시를 개발하는 도구가 될 수도 있다. 묘사한 것을 그려내고 공간적으로 보여주는 건축의 내재적 역량을 감안하면 건축학이 시각화 문제에 관한 담론을 통해 그러한 논의에 기여할 여지가 있다는 생각도 든다. 어쨌든 정보를 그래픽으로 전환하는 인포그래픽이나 공간적 서술을 시각화하여 기후 변화 같이 보이지 않는 것을 가시화하는 데 건축가보다 더 능숙한 사람이 있을까? 그러나 건축에서의 지구적 상상력은 지구를 최대한 정확하게 시각화하는 데 그쳐서는 안 된다.[2]

2   건축과 도시주의 맥락에서의 인포그래픽스와 데이터매핑에 대한 초기 비판은 네이란 투란의 'After Mapping: Urbanism

Much like an urban or building plan, the lines, the curves, the grids, and the blocks were drawn on the surface of the sea in order to give it an organizational form. The energy landscape of the sea portrayed a primitive form of contestation between geography and geometry, and between scale and form.

One can say that much has changed within contemporary architectural discussions since then. For one, the relevance of resource extraction infrastructure as an architectural or urban topic would not be questioned by the majority of the architectural culture now. Yet, despite having come a long way since the early 2000s there are still important questions to tackle. As interdisciplinarity has come a widespread topic expanding the limits of the architectural knowledge in relation to the changing demands of the globalizing world and its accompanying social and environmental challenges for the last two decades or so, ideas regarding politics of infrastructure, landscape and territory have become omnipresent as well as various forms of "learning from" projects that present design as research and mapping. The critical point is that now that these topics have pervaded the discipline, more imperative and interesting questions come to fore regarding questions of methodology, representation, and technique as well as political questions of equity (as opposed to a mere emphasis on problem solving in relation to urban processes and systems or politically naïve versions of neo-environmentalist do-goodism, for instance).

Moreover, while the ability to access and visualize data has embraced fascination within contemporary architectural discussions, "data visualization" sometimes becomes a tool for justifying projects through mere problem-solving, be it for a contaminated waterfront, an obsolete landfill, or a new urban development located within an extreme climate. Considering architecture's inherent capacity to illustrate narratives and provide spatial consciousness, it is tempting to think about the discipline's contribution to this discussion with a discourse on the problem of visualization. After all, who would be more adept than architects in making visible the invisibilities of climate change through the infographics of data mapping or visualizing spatial narratives? However, for architecture, the planetary imagination should not simply mean how to best visualize the

건축의 힘을 기후변화로 손상을 입은 지구의 시각화 프로젝트에 쏟기보다는 지구적 상상력의 건축적 미학을 어떻게 이야기하고 소통하며 실행하느냐는 좀 더 광범위한 문제에 돌릴 필요가 있다. 이런 식의 의문을 제기하면 우리는 낯선/지구적인(우주적인) 것과 익숙한/일상적인(건축학적인) 것 사이에서 의미 있는 상호작용을 구축할 수 있을 것이다.

우주적인 것과 건축학적인 것 사이의 얽히고설킨 관계를 보여주는 예가 현대 도시화 맥락에서 그 개념이 충분히 자리 잡지 못한 지리학(공간적/수평적) 및 지질학(용량적/부분적) 영역이다. 자원 채굴과 수송(광산 지역, 연안 시추 지역, 원유 수송관의 경로), 농업, 물류, 폐기물 처리 등 다양한 산업 활동이 일어나는 지형을 생각해보자. 이러한 '불모지' 지형은 생태학이 중점을 두는 '천연'과 황무지 지형은 물론 도시계획이 중점을 두는 도시의 바깥에 존재한다. 그러나 불모지 지형은 우리에게 익숙한 현대 구축 환경을 만들고 이해하는 데 중요한 역할을 한다.[3] 환경역사학자 윌리엄 크로논은 자신의 저서에서는 황무지라는 개념을 문제로 부각하면서 기술과 환경 (그리고 도시와 배후지) 사이의 상호작용을 제시한다. 지리학자 닐 브레너의 '지구적 도시화' 개념은 오지의 소멸과 자연의 소멸을 주장한다(산업적 자본주의의 공급망과 물류 기반시설 때문에 농촌과 자연계의 개념이 유명무실해졌다는 것이다). 스튜어트 엘든과 개빈 브릿지 등의 지리학자들은 영토 개념을 수평적 실체로 보는 데 그치지 말고 용적이나 구역으로 개념화할 것을 주장한다. 이 모든 주장에서 알 수 있듯이 지구의 불모지를 인공적이거나 자연적인

planet.[2] Instead of turning architecture's agency to a visualization project for a planet damaged by climate change, a much broader question might be at stake: How does one talk about, communicate and practice an architectural aesthetics of planetary imagination? Posing the question this way would allow us to build more meaningful interactions between what is considered to be the unfamiliar/planetary (geo-cosmic) and the familiar/quotidian (architectural).

One example of such an entanglement between the geo-cosmic and the architectural would be under-conceptualized geographic (spatial/horizontal) as well as geologic (volumetric/sectional) footprint in the context of contemporary urbanism. Think of geographies of various industrial activities such as sites of resource extraction and transportation (mining areas, offshore drilling sites, oil pipeline routes), agriculture, logistics, and waste. These "wasteland" geographies lie outside of both ecology's focus on the "natural" and wilderness landscapes and urbanism's focus on the city. However, they play a central role in the production and conception of what is known to us as our contemporary built environment.[3] From environmental historian William Cronon's work that problematizes the notion of wilderness and brings forward the interplay between technology and the environment (and thus city and the hinterland); to geographer Neil Brenner's concept of "planetary urbanization" that argues for the dissolution of the hinterland and the disappearance of nature (rendering the concepts of the rural and the natural systems obsolete due to the supply chains and logistics infrastructure of industrial capitalism); to other geographers such as Stuart Elden and Gavin Bridge's calls to conceptualize the idea of territory as volumetric or

and What Is Out There' 참조. "Surfacing Urbanisms: Recent Approaches to Metropolitan Design," ACSA West Conference Proceedings (Los Angeles: Woodbury University, 2006), 83-88에 게재됨. 보편적인 시각화 논의의 비판적 해명을 통한 인식론과 시각의 관계에 대한 과거 저작물로는 오리트 핼펀의 *Beautiful Data: A History of Vision and Reason Since 1945* (Durham, NC: Duke University Press, 2014) 참조. 그 외, 제러미 W. 크램턴의 "Maps as Social Constructions: Power, Communication and Visualization" 참조. *Progress in Human Geography* 25, no. 2 (June 2001): 235-52에 게재됨.

3 환경역사학자 윌리엄 크로논의 저작은 이러한 요소의 광범위한 개념화에 큰 영향을 끼쳤다. 윌리엄 크로논의 저서 *Nature's Metropolis: Chicago and the Great West* (New York: W. W. Norton, 1991) xvi 참조.

2 For an earlier critique of infographics and data mapping in the context of architecture and urbanism, see Neyran Turan, "After Mapping: Urbanism and 'What Is Out There,'" in Surfacing Urbanisms: Recent Approaches to Metropolitan Design, ACSA West Conference Proceedings (Los Angeles: Woodbury University, 2006), 83–88. For a historical reading of the relationship between epistemology and vision through a critical demystification of widespread visualization discussions, see Orit Halpern, Beautiful Data: A History of Vision and Reason Since 1945 (Durham, NC: Duke University Press, 2014). Also see Jeremy W. Crampton, "Maps as Social Constructions: Power, Communication and Visualization," Progress in Human Geography 25, no. 2 (June 2001): 235–52.

3 Environmental historian William Cronon's work is seminal for a wider conceptualization of this point. See William Cronon, Nature's Metropolis: Chicago and the Great West (New York: W. W. Norton, 1991), xvi.

것으로 개념화하기보다 더 광범위한 지구역사학의 일부로 개념화할 필요가 있다.[4]

어떻게 하면 자연 대 문화, 인간 대 비인간, 도시 대 황야라는 이분법에 의존하지 않고 건축과 도시계획에서 지형의 공간적 구성과 수많은 배치를 보다 분명하게 표현할 수 있을까? (건축역사학자 비토리아 디 팔마가 보여주었듯이) 불모지의 원래 개념과 아름다움, 숭고함, 자연에 대한 우리의 지식 사이에 중첩되는 부분이 있다면 어떠한 종류의 건축 미학적 상상력에 공간적, 시간적 측면으로 지구 불모지에 대한 지질학적, 지형학적 지식이 반영되어 있을까?[5] 어떻게 하면 순진한 긍정론이나 종말론적인 서사를 초월하여 건축학적 상상과 표현으로 자원 추출이나 폐기물 등의 중요한 의제에 대한 학문적 관심, 대중적 관심을 불러일으킬 수 있을까?

그 중 한 가지 가능성으로 건축과 기후 변화가 충돌하는 사례들을 제시함으로써 일상과 지구에 대한 상호 대화를 유도하여 환경에 대한 상상력을 촉발하는 방법을 들 수 있다. 이러한 환경적 상상력을 바탕으로 형태, 표현, 물질성 등 건축과 관련된 구체적 질문을 통해 도시, 환경, 지형에 대해 좀 더 광범위한 문제를 도출하는 것이다.

인간이 지질학적 행위자로 간주되는 이 시대에 건축은 세상을 평가하고 세상에 영향을 주는 잣대다. 무엇을 측정하기보다 기구를 사용해 정도를 알아내고 면밀히 조사하며 시간을 들여 내부를 깊이 들여다보며 생각하는 것이다. 내가 속한 건축가 집단은 건축의

sectional instead of merely a horizontal entity, it is critical to conceptualize our planetary wasteland as neither human nor natural, but as one quotidian aspect of a larger geohistory.[4]

Instead of resorting to dichotomies of nature vs. culture, human vs. nonhuman, urban vs. wilderness, how can we further articulate the spatial framing and myriad dispositions of these landscapes for architecture and urbanism? If the original conception of wasteland is already imbricated into our very understanding of beauty, sublimity and nature—as shown by architectural historian Vittoria Di Palma—then what kind of an architectural aesthetic imagination reflects the geologic and geographic understanding of our planetary wasteland in spatial and temporal terms?[5] Beyond naively optimistic overtones or apocalyptic narratives, how can architectural imagination and representation bring critical issues such as resource extraction and waste to disciplinary and public consciousness?

One such possibility would be to provoke such an environmental imagination that sees the categories of the quotidian and the planetary in conversation with one another by presenting a set of unconventional collisions between architecture and climate change, all of which extrapolate broader concerns of the city, the environment and geography through the lens of specific architectural questions such as form, representation and materiality.

In an era when humans are described as geological agents, architecture is a measure both to assess and to act upon the world. Measuring something means both to ascertain its degree by using an instrument, and also to scrutinize, to

4   윌리엄 크로논, 'The Trouble with Wilderness, or Getting Back to the Wrong Nature', Uncommon Ground: Rethinking the Human Place in Nature, ed. W. Cronon (New York: W. W. Norton, 1996), 69-90. 닐 브레너 및 크리스천 슈밋, 'Planetary Urbanisation', Urban Constellations, ed. M. 갠디 (Berlin: Jovis, 2011), 10-13; 스튜어트 엘든, 'Secure the Volume: Vertical Geopolitics and the Depth of Power', Political Geography 34 (2013): 35-51; 개빈 브리지, 'Territory, Now in 3D!' Political Geography 34 (2013): 55-7. 스티븐 그레이엄, The City from Satellites to Bunkers (London: Verso, 2016).
5   황무지와 자연, 아름다움, 숭고함의 개념 발전에 황무지가 끼친 중요한 영향에 대한 과거 저작으로는 비토리아 디 팔마의 Wasteland: A History (New Haven: Yale University Press, 2014) 참조. 황무지와 황야 개념을 별개로 보기보다 관련된 것으로 보는 시각에 대해서는 'Waste-Wilderness: A Conversation with Peter Galison'참조. Friends of the Pleistocene, March 31, 2011, http://fopnews.wordpress.com/2011/03/31/galison에 게재됨.

4   William Cronon, "The Trouble with Wilderness, or Getting Back to the Wrong Nature," in Uncommon Ground: Rethinking the Human Place in Nature, ed. W. Cronon (New York: W. W. Norton, 1996), 69–90. Neil Brenner and Christian Schmid, "Planetary Urbanisation," in Urban Constellations, ed. M. Gandy (Berlin: Jovis, 2011), 10–13; Stuart Elden, "Secure the Volume: Vertical Geopolitics and the Depth of Power," Political Geography 34 (2013): 35–51; Gavin Bridge, "Territory, Now in 3D!" Political Geography 34 (2013): 55–7. Also related would be Stephen Graham, The City from Satellites to Bunkers (London: Verso, 2016).
5   For a historical account of the idea of wasteland and its centrality for the development of ideas about nature, beauty and sublimity, see Vittoria Di Palma, Wasteland: A History (New Haven: Yale University Press, 2014). For a discussion of the concepts of waste and wilderness as relational instead of dichotomous terms, see "Waste-Wilderness: A Conversation with Peter Galison," Friends of the Pleistocene, March 31, 2011, http://fopnews.wordpress.com/2011/03/31/galison.

구체적 성격을 비판적이고 철저하게 재정의하는 과정을 통해 건축의 사회 참여 역량을 재규정하고자 급진적이고 실험적인 방법을 찾아내려는 열망을 품고 있다. 우리 모두는 건축학과 세계에 파장을 일으키면 건축가들이 더 이상 속도를 늦추고 내면에 초점을 맞추는 것을 거리끼지 않게 되어 외부에 매우 강력한 영향을 줄 수 있을 것이라고 믿는다.

consider with pause and inner focus. I belong to that group of architects who are eager to search for radical and experimental ways to redefine architecture's capacity to engage with the world through the critical and rigorous redefinition of its disciplinary specificities, and who all believe that shaking the discipline and the world compels us to not be afraid of slowing down and focusing inward, for the most robust outward influence.

일상적인 건축/건설 작업의
시시함과 주변 환경을 변화시키는
장대한 건설 사업의 원대한
상상력의 관계

A dialogue between
environmental imaginations
of the large scale and
the quotidian day-to-day
workings of architecture and
construction.

카라카스 Caracas

301

쿠웨이트

Kuwait City

민간 외교관 역할을 하는
주유소에 대한 이야기.

The story of the stations
that act as embassies.

METROPOLIS
BARCELONA

바르셀로나메트로폴리스
도시들의 도시

바르셀로나

Barcelona

그린 인프라는 주된 버팀목이라고 할 수 있다. 즉, 환경적 가치를 지닌 52% 이상의 도심 공간과 잠재적 녹색 지대를 연결하는 것이다.

A set of spaces of great environmental wealth that represents more than 52% of the metropolitan territory plus all those urban spaces with the potential to become green spaces.

AMB

메데인   Medellin

만약 메데인의 협곡을
집합공간으로 활용하면
어떨까?

What if Medellin uses
the ravines as collective
space?

306

도시의 '자연적' 구조를 도시
계획의 핵심 개념으로

The natural structure of
the city as a fundamental
concept for urban planning

'주어진 시점에서 구할 수
있는 자원을 최대한 활용해
계획하고 만들어 나가는
행동양식'이라고 정의할 수
있는 '상황 대응형 인프라'는
시민들이 살아 가면서 급박한
일을 겪거나 예기치 않은
기회에 직면 했을때의 적절한
대응 과정에서 탄생했다.

Defined as the coincidence
of planning and doing with
the resources available at
the moment, improvisation
emerges in response
to an urgent need or an
unforeseen opportunity.

마닐라

Manila

부동산 시장의 시 외곽 여러 직역으로 분산되는 과정과 지속적인 도로 인프라 확장이 맞물리면서 도시 곳곳에 10만 제곱미터 이상의 대규모 개발 지구가 속속 생겨나게 되었다.

The decentralization of real estate from the city core coincided with an ongoing expansion of road infrastructure and has resulted in a proliferation of large-scaled developments in the order of at least 100,000m².

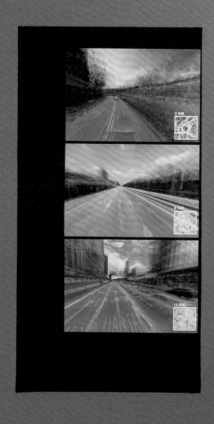

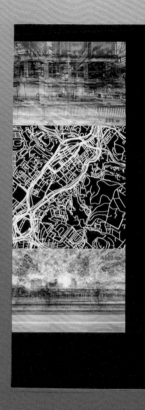

쿠알라룸푸르

Kuala Lumpur

근대주의적 도시 개발은 하향식 도시 계획 모델에 따라 이루어졌다. 이러한 경향은 기능주의의 의제인 산업 구역, 거주 구역, 오락 구역의 분리로 한층 더 심화되었다. 이 전략에는 인간의 도시 점유와 사용에 대한 기본적 이해가 결핍되어 있었다. 근대주의적 구상에 따른 마스터플랜은 일상 편의 시설이 부족한 거주 구역, 근무시간 이외에는 텅 비는 사무 구역, 도시와 단절된 생산 구역으로 대변되는 도시성을 만들어냈다.

공동체 참여는 주민 참여로 지역 형성에 관한 의견을 내는 상향식 접근법을 통해 근대주의적 도시 계획에 팽배한 유산을 파기했다. 이러한 전략은 지역에 필요한 사회적 기반시설과 편의시설을 선제적으로 마련하여 도시 효율성을 높이는 데 크게 일조한다. 주민 참여는 생활환경을 개선하기 위해 남녀노소 모두가 힘을 합치는 연합의 상징이며 때로 인종, 성별, 경제적 지위와 상관없이 사람들을 결집하는 충격적 사건을 통해 향상된다. 이 섹션에서 전시되는 도시들은 주민 참여가 어떻게 새로운 사회적 의제를 생산하고 이 시대 도시의 요구를 한층 더 제대로 반영하는 환경을 만드는지 보여주는 실제 사례들이다.

가오슝,
산호세,
케이프타운,
우한,
도호쿠 + 구마모토,
멕시코 시티,
크라이스트처치,
베를린,
베오그라드,
파리

The modernist development of cities implied a top-down approach for city planning. This was furthered plagued by the functionalist agenda that separated industrial zones, residential zones, and recreational zones. This strategy often lacked the crude understanding of human occupancy and use of the city. The idea of the modernist masterplan has produced urbanities where residential areas lack daily life amenities, office areas that are empty after work hours, and production zones that are disconnected from the city. Community involvement has reversed the prevailing heritage of the modernist planning through a bottom-up agenda that involves residents to voice their opinion on the formation of their neighborhoods. These strategies strive to produce a higher urban efficiency by proactively engaging the local needs for social infrastructure and amenities. The participation of the residents symbolizes the union across demographics for the purpose to improve living conditions. This is sometimes enhanced through traumatic events that bring people together regardless of race, gender, or economic status. The cities presented in this section showcase how the participation of residents has been able to produce new social agendas and environments that better reflect contemporary urban needs.

Kaohsiung,
San Jose,
Capetown,
Wuhan,
Tohoku +
Kumamoto,
Mexico City,
Christchurch,
Berlin,
Belgrade,
Paris

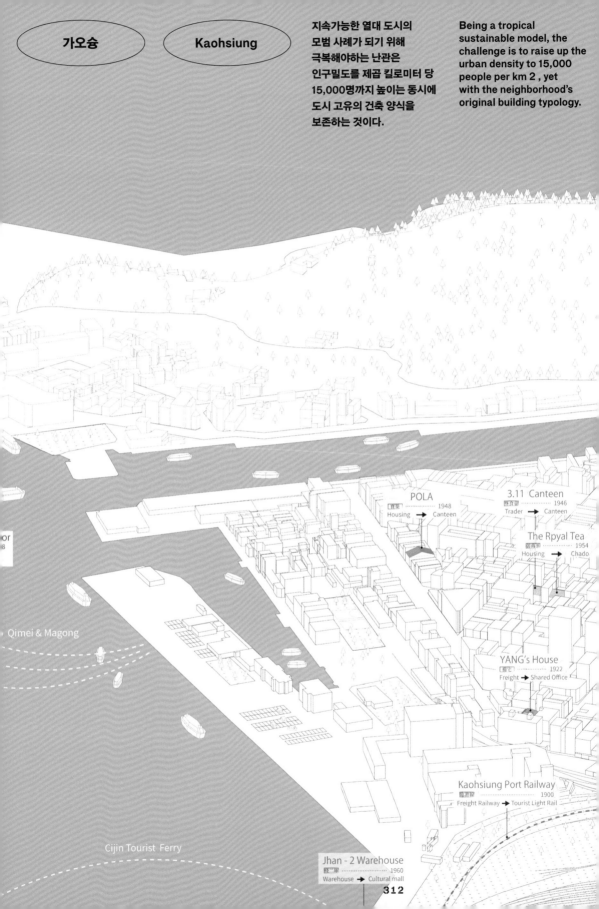

가오슝

Kaohsiung

지속가능한 열대 도시의
모범 사례가 되기 위해
극복해야하는 난관은
인구밀도를 제곱 킬로미터 당
15,000명까지 높이는 동시에
도시 고유의 건축 양식을
보존하는 것이다.

Being a tropical
sustainable model, the
challenge is to raise up the
urban density to 15,000
people per km 2 , yet
with the neighborhood's
original building typology.

POLA
寶島　　　1948
Housing → Canteen

3.11 Canteen
桃花潭　　　1946
Trader → Canteen

The Rpyal Tea
國泰茶　　　1954
Housing → Chado

YANG's House
　　　1922
Freight → Shared Office

Kaohsiung Port Railway
高雄港站　　　1900
Freight Railway → Tourist Light Rail

Qimei & Magong

or
8

Cijin Tourist Ferry

Jhan - 2 Warehouse
棧貳庫　　　1960
Warehouse → Cultural mall

312

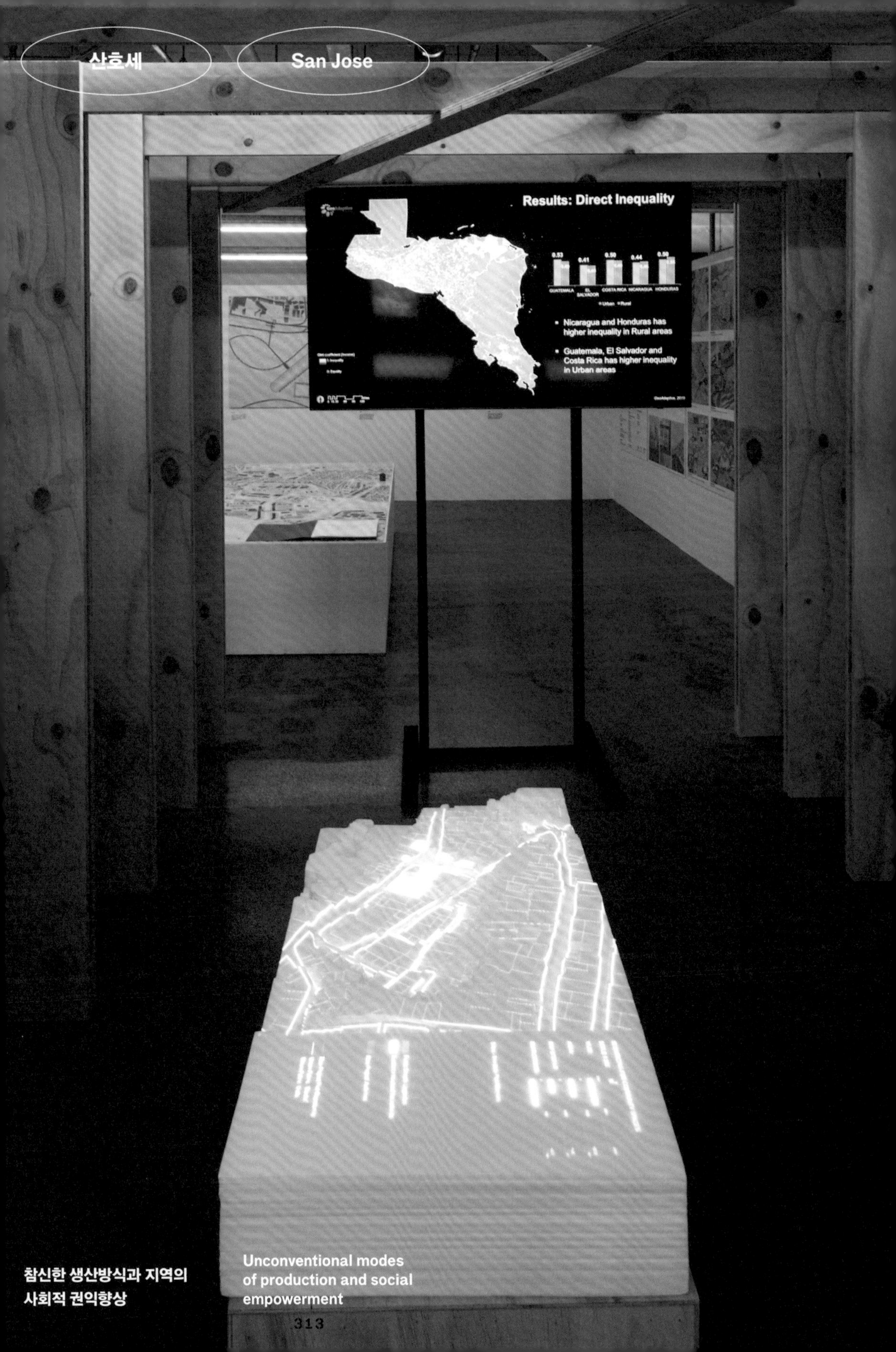

참신한 생산방식과 지역의
사회적 권익향상

Unconventional modes
of production and social
empowerment

지도가 한데 모여 주민들의 거주를 증명하는 최초의 공식 기록물이 되었으며 이는 당국으로 부터 '체류권'을 확보하는데 중요한 역할을 했다.

The drawings were then collated together to became the first official document of occupation, and a powerful tool to secure "the right to stay" upon presentation to municipal officials.

남아프리카공화국 공동체 지원

알프레도 브릴렘버그

지난 20년 동안 우리의 연구와 디자인 작업은 비교적 개별적인 개입 프로젝트를 중심으로 이루어졌다. 그 예로 수직형 체육관에서 집합 주거지에 이르는 개별 건축물, 메트로 케이블카와 같은 연결망, 남아프리카공화국의 카이얼리처(Khayelitsha) 같은 마을 프로젝트 등을 들 수 있다. 하지만 개별 프로젝트를 하면서도 우리는 줄곧 도시에 대해 생각했고 21세기에 참여적 도시주의가 어떠한 의미와 형태를 지니게 될지 논의했다.

우리는 공식과 비공식, 계획과 임의, 부와 빈곤이 엄격하게 분리될 수 있다고 생각하지 않는다. 이런 구분은 본질적으로 정치, 경제, 지형에 의해 흔들릴 수 있다. 소외는 사회, 물리적 현상이고 시민사회를 괴롭히는 일종의 질병이다. 공식과 비공식의 분리가 생겨나는 근본 이유로는 두 가지를 들 수 있다. 하나는 유기적이다. 연못에 돌멩이를 떨어뜨리면 잔물결이 퍼져나가는 것처럼 도시는 점점 더 밖으로 확장된다. 잔물결이 그러하듯이 주변부 지역은 도심에서 멀어질수록 약화되고 일관성을 잃는다. 또 다른 원인은 예기치 않은 결과의 법칙에서 찾을 수 있다. 기반시설 중에서도 교통은 자동차의 이동성은 개선하지만 가진 자와 못 가진 자 사이에 장벽을 만든다. 대중교통과 육교를 이용해 고속도로와 6차선 대로를 건너갈 수 있는 지역이라 하더라도 도로를 사이에 둔 지역들은 격리되고 다른 지역과 섞이지 못하게 된다.

그래서 우리는 이렇게 질문한다. 우리가 익히 알고 있는 형태의 도시가 존재하지 않는다면, 우리는 무엇을 만들어내고 우리가 만들어 낸 것은 어떤 식으로 참여적 존재가 될까?

불일치의 일치

백인 정착민이 처음 남아프리카공화국에

Empowering Communities in South Africa

Alfredo Brillembourg

For much of the past 20 years, our research and design work has primarily dealt with relatively discrete interventions: individual structures, from the vertical gym to collective housing; connective tissue, such as the Metro cable car; and neighborhoods, like Khayelitsha in South Africa. But all the while, we were thinking and talking about the city, about what participatory urbanism might mean and be in the 21st century.

The rigid separation of formal and informal, planned and ad hoc, wealth and poverty, makes no sense to us. Those distinctions are inherently unstable politically, economically, and geographically; marginalization is a social and physical phenomenon, a kind of illness afflicting the civic body. The disconnect between formal and informal has at least two root causes. One is organic: cities grow outward, like the ripples in a pond when one drops a pebble in the middle. Like the ripples, the encircling neighborhoods grow weaker and less coherent the farther they are from the center. The other cause follows the law of unintended consequences: infrastructure, especially transportation, creates barriers between the haves and the have-nots, in the interest of improving vehicular movement. Even where public transportation and pedestrian bridges provide access across highways and six-lane boulevards, neighborhoods are still cut off from one another, preventing mingling and sharing.

And so we ask: if the city as we know it doesn't exist, what would we invent and how would it be participatory?

Discordia Concors

It took nearly 50 years for South Africa to abandon the official principle of apartheid, though it had dominated black/white relations ever since the first white settlers arrived. In 1990, the ban on the African National Congress (anc) was repealed and, after 27 years of incarceration, Nelson Mandela was released from prison. It took another five years

도착한 이래로 흑인과 백인 관계를 지배한 것은
아파르트헤이트였고 남아프리카공화국이
이를 공식적으로 폐지하기까지는 거의 50년이
걸렸다. 1990년 아프리카 민족회의 금지법이
폐지되었고, 27년 동안 투옥되었던 넬슨 만델라가
풀려났다. 남아프리카공화국은 폭동과 소요 사태
때문에 전 국민에게 (실행 여부는 둘째치고라도
문서상으로나마) 평등한 지위를 보장하는 새 헌법을
비준하기까지 5년에 이르는 협상과 과도기적
정치체제를 거쳐야 했다.

새로운 헌법에 성문화된 권리 중 주택에 관한
권리가 있다. 그러나 법률이 공포되고 실행되기까지는
몇 년씩 걸리는 일이 다반사였고, (만델라가
대통령으로 선출된) 1994년의 환희는 현실적
변화로 이어지지 못했다. 새 정부는 첫 5년 동안
저가 주택 100만 채를 공급한다는 국가적 재건 및
개발 프로그램(Reconstruction and Development
Programme, RDP)을 시행했다. 해당 프로그램은 어느
정도나마 진전을 이룬 것이 사실이다. 변형된 형태의
RDP 모델은 남아프리카공화국의 주택난에 대한 핵심
대응책이 되었으며, 1994년 이후 약 280만 채의
주택이 건설되었다. 그러나 여전히 약 150만 가구의
750만 명이 비공식 정착지 2,700여 곳에서 살고
있다. 남아프리카공화국의 부족한 주택의 수는 250만
채가 넘는다.

2004년 이후 남아프리카공화국의 사회적
의제에는 비공식 정착지를 개선하는 작업이
포함되어왔다. 보통 읍내나 도시 변두리에 있으며
비백인 거주지로 분리된 흑인 거주 지역이 그
대상이다. 우리는 이는 모범적인 해결책을 구상하고
시행하기 위한 준비 단계로서 사회적, 정치적,
경제적 변화와 격변이 교차하는 지점의 공식적,
비공식적 주택에 대한 연구를 사명으로 삼아왔다.
우리는 남아프리카공화국에서 10여년의 경험을
바탕으로 새로운 시도를 할 때마다 연구 결과와 (때로
성공적이기도 하고 그렇지 못하기도 한) 해결책을
적용하고 있다.

비공식 주택: 블로킹아웃, 기본적·발전적 참여
2012년 우리는 흑인 거주 지역의 실태를 파악하고

of negotiations and transitional politics, disrupted
by violence and riots, for a new constitution to be
ratified, giving equal status to all South Africans—
on paper, if not always in fact.

Among the rights codified in the new
constitution is that of housing. But declaration
and implementation are often years apart, and
the euphoria of 1994—the year of Mandela's
election as President—didn't translate into action.
Among the new government's initiatives was the
establishment of the national Reconstruction and
Development Programme (rdp) for public housing,
which aimed to provide one million low-cost houses
within the first five years. There has, to be sure,
been some progress: variations of the rdp model
are the centerpiece of the country's approach to
the housing crisis, and some 2.8 million houses
have been built since 1994. However, there are
approximately 7.5 million people, distributed in
1.5 million households, who still live in 2,700
informal settlements. The housing shortage across
South Africa amounts to more than 2.5 million
homes.

Since 2004, South Africa's social agenda has
included the upgrading of informal settlements: the
townships, segregated urban areas, typically on the
periphery of towns and cities and reserved for non-
whites. We have made it our mission to research
public and informal housing at the intersection
of social, political, and economic change
and upheaval, as a prelude to conceiving and
implementing prototypical solutions. To each new
endeavor we bring the research and solutions—
successful and otherwise—of more than a decade of
experience. So, it was in South Africa.

Informal Housing: Blocking-Out, Basic and Advanced Participation
In 2012 wanted to map the townships and create
provocative images that could serve as a call to
action for anyone and everyone who might enlist in
our cause. As part of the research, we spent several
days meeting residents and community leaders
in the Philippi township, in the company of Andy
Bolnick of Ikhayalami. Andy was working on post-
disaster reconstruction following a major fire in a
small area of the township called Sheffield Road,
implementing a "blocking-out" scheme.

Blocking-out describes a design and
implementation process pioneered by the ngo
Ikhayalami and Slum Dwellers International. All

도발적인 이미지를 만들어내어 우리 운동에 참여할 만한 사람들의 행동을 촉구하고자 했다. 연구의 일환으로 우리는 며칠 동안 익하얄라미(Ikhayalami, 케이프타운의 비영리단체)의 앤디 볼닉이 동석한 자리에서 흑인 거주 지역 필리피의 주민들과 공동체 대표들을 만났다. 앤디는 셰필드 로드로 불리는 흑인 거주지의 한 구역에서 대형 화재가 발생한 이후 '블록킹아웃' 계획을 시행하며 재난 후 복원 사업에 참여하고 있었다.

블록킹아웃은 비영리단체인 익하얄라미와 슬럼드웰러인터네셔널이 설계하고 실행한 절차다. 해당 절차의 모든 단계(필요성 인지에서 설계, 실행까지)는 공동체와 도시 빈민으로 조직된 네트워크에 의해 주도되고 비영리 단체와 국가의 후원을 받는다.

이 계획은 비공식 정착지의 공간 배치를 보다 합리적인 배치로 재구성하여 경계가 있는 길이나 도로를 닦고 공공 또는 반(半)공공 장소를 만들며 응급 구조대의 접근 경로를 확보한다. 이렇게 하면 기반시설 공급이 원활해지고 사회활동이 강화되고 안전이 강화되며 기본 서비스의 이용이 용이해질 뿐 아니라 무엇보다도 불법 거주민들이 '남아있을 수 있는 권리'를 법적으로 인정받게 된다. 이러한 접근법은 정부로부터 모범 사례로 인정받았고 비공식 정착지 네트워크(Informal Settlement Network)와 연관된 판자촌 사람들 수천 명으로부터 호응을 얻었다.

블록킹아웃 절차의 첫 단계는 개별 판잣집을 물리적으로 해체하는 것이다. 그런 다음에 개선된 모델로 대체하거나 합의된 도시 계획에 맞게 새로 조립한다. 개조된 판잣집을 선택한 거주자들은 익하얄라미가 제공하는 기본형 중에서 자신의 재정 상태에 맞춰 원하는 것을 고른다. 중요한 것은 판잣집이 해체되고 대체되거나 그 자리에 새로운 집이 지어지는 전 과정이 하루 안에 이루어진다는 점이다. 건설 과정 중에 집 없이 지내야 하는 주민들에게 하루라는 짧은 기간은 크나큰 동기 부여가 된다.

자금 조달은 항상 너무나 중요하며 대개 재개발의 걸림돌로 작용한다. 블록킹아웃 계획은 보통 응집력이 있고 조직화된 공동체에 적용된다. 특히 도시 빈민 연합(Federation of the Urban Poor)과 연결된 공동체

phases of the process—from identification of need, to design and implementation—are led by the community and organized networks of the urban poor, supported by ngos and in partnership with the state.

The scheme involves the reconfiguration of the spatial layout of an informal settlement into one that is more rationalized, allowing for the creation of demarcated pathways or roads, public and semi-public spaces, and emergency services access. This would facilitate the provision of infrastructure and result in enhanced circulation, greater safety, access to basic services, and, above all, the recognized "right-to-remain" legal status for squatter communities. The approach has been acknowledged by the government as a "best practice" and supported and endorsed by thousands of shack dwellers connected to the Informal Settlement Network.

In its implementation, blocking-out consists of the physical dismantling of individual shacks, which are then either replaced with an upgraded model or reassembled according to the improved and agreed-upon urban scheme. Residents opting for an upgraded shack have a range of prototypes offered by Ikhayalami from which to choose, depending on their financial means. Significantly, shacks are dismantled and either replaced or rebuilt in situ in just one day, a tremendous incentive for residents who would otherwise be homeless during construction.

Financing is always critical and often a stumbling block for any redevelopment. The blocking-out scheme is generally employed in cohesive and organized communities—ideally those where community-based saving schemes networked to the Federation of the Urban Poor are already in operation. The program allows residents to start saving towards the 20 percent cost requirement for their shack upgrade; the rest is subsidized through donor contributions.What came to mind was to design a blocking out unit that would be double-storied and possibly involve new structural techniques and materials that would allow residents to upgrade their homes in a safe way.

Test Case: Phumezo's Shack
We asked Andy Bolnick if she knew of a place where we could try to implement the new blocking-out strategy of upgrading that could serve as an

기반의 구제 계획이 이미 운영 중인 곳이 이상적이다. 해당 프로그램을 이용하면 주민들은 판잣집 개선에 필요한 비용의 20 퍼센트를 부담하고 나머지는 기부금으로 지원 받는다. 우리의 구상은 주민들이 자신들의 집을 안전한 방식으로 개조할 수 있도록 가능하면 새로운 구조 기법과 자재를 적용한 2층집을 설계하는 것이었다.

시험 사례: 푸메조의 판잣집

우리는 앤디 볼닉에게 2층 구조의 새로운 블로킹아웃 전략을 적용할 첫 사례로 적합한 곳을 아느냐고 물었다. 앤디는 케이프타운에서 약 35 킬로미터 떨어진 코사(Xhosa) 지역에 있으며 '우리의 새로운 집'을 의미하는 정착지 카이얼리처를 알려주었다. 구체적으로 C구역(Site C)의 일부인 BT 지구(BT-Section)를 지목했다. 해당 지구는 지역 계획의 대상이기는 하지만 정부의 관심에서 벗어난 탓에 우리 계획을 실행하기에 매우 이상적인 곳이다. 거리는 활기차고 공동체 기반의 경제활동이 활발하게 이루어진다. 사람들은 과일, 채소, 중고 가구, 헌 옷 등을 팔고 아이들을 돌보며 자동차를 수리한다. 선술집과 예배를 드리는 주택(더 정확하게는 방)이 있고, 젊은이들이 가는 클럽과 노년층을 위한 클럽이 있다. 다른 한편으로 BT 지구는 매우 혼잡한 곳이다. 4,000 평방미터에 280명이 넘는 사람들이 지내고 있다. 주변에서 재료를 구해 직접 만든 판잣집은 허술할 뿐 아니라 지역 내에 무질서하게 분포되어 있어 구역 자체가 출구를 찾을 수 없는 미로와 같다. 기본적 기반시설이나 서비스가 구축되어 있지 않고 갑작스런 화재와 홍수가 반복되는 위험한 환경에서 주민들은 폭력의 위험에 노출되어 있다. 실업률은 높은데 케이프타운은 멀리 떨어져 있어 거기까지 가려면 비싼 택시를 이용해야 한다. 그래서 사람들은 판잣집 근처에서 영세 사업을 한다.

즉, 이곳에서는 판자촌 지원(Empower Shack's) 계획이 성공적인 전환점이 될 수 있었다. 그러나 그에 앞서 우리는 건물을 통해 우리의 생각을 현실화하고 증명해야 했다.

기초 공사 후 우리는 2층 판잣집의 기본형에 필요한 자재 구입을 위해 민간 보조금을 확보할 수

example of the first double-storied shack upgrade. She pointed us to Khayelitsha, meaning "our new home" in Xhosa, located about 35 kilometers from Cape Town, and specifically to an area known as the bt-Section, which is part of Site C. For our purposes, the site is a nearly ideal mix of local initiative and governmental neglect. On one hand, it has a lively street-life and community-based economy: people sell fruits and vegetables or second-hand furniture and clothing; they provide child-care; they fix cars. There are taverns, houses—more properly rooms—of worship, clubs for youth and for senior citizens. On the other hand, it is very crowded— more than 280 individuals occupy about 4,000 square meters. The shacks they have built with found materials are of poor quality and distributed helter-skelter throughout the site in a nearly impenetrable maze. There is no basic infrastructure or services, flash fires and flooding are persistent environmental threats, and residents are at risk of predatory violence. There is high unemployment and Cape Town is a long and costly taxi ride away—hence the small businesses tucked into nearly every shack. In other words, this was a place where the Empower Shack's innovations could be an effective game-changer. But first, we had to make our idea real: the proof of the concept would be in the building.

After much ground work we were able to secure a private grant to subsidize the materials for a two-story shack prototype. Phumezo, the community leader, had already expressed interest in testing the prototype. The one-story shack he had built in 1987 needed major repairs, and his wife and two daughters were pressing him for additional space. In December, 2013, Phumezo became the Empower Shack's first client. Over a period of four days, a team of four to six people, composed of regular Ikhayalami builders, UTT representatives, local activists, and of course Phumezo, assembled the latter's eye-opening new two-story shack.

Without displaying and promoting plans, slick renderings, or elegant architectural models, the in situ prototype has spurred other residents to reimagine what their homes could be. One resident even petitioned to send the money she saved for an rdp house on a double story shack. In that same mode, the build project inspired leaders in BT settlement to resume meetings to discuss a community-wide blocking out plan. Phumezo aligned local political forces and used his new shack

있었다. 그 이전에 공동체 대표인 푸메조(Phumezo)가
기본형 판잣집을 시험해보고 싶어 했다. 그가
1987년에 지은 1층짜리 판잣집은 대대적인 수리가
필요했고, 그의 아내와 두 딸은 공간이 비좁다고
불평했다. 2013년 12월, 푸메조는 판자촌 지원
계획의 첫 번째 의뢰인이 되었다.

익하얄라미의 정규 건설업자, UTT 대표, 지역
운동가, 푸메조 등 4명에서 6명으로 구성된 팀이
나흘 동안 새로운 2층 판잣집을 조립했고, 푸메조는
결과물에 눈이 휘둥그래졌다.

계획을 전시하거나 홍보하지 않았고 번드르한
렌더링이나 우아한 건축학적 모델도 없었지만 그
자리에 지어진 기본형을 본 주민들은 자신들의
집이 어떻게 바뀔지 상상할 수 있게 되었다. 심지어
한 주민은 RDP 주택을 위해 저축한 돈을 빼서 2층
판잣집을 지으려고 했다. 이 건설 프로젝트에서
자극 받은 BT 거주지 대표들은 회의를 열어 공동체
전체에 블로킹아웃 계획을 적용하는 것을 논의했다.
푸메조는 이를 지지했고 그의 새로운 판잣집을 지역
개선 프로젝트의 타당성을 보여주는 사례로 제시했다.
이러한 활동이 우리 판자촌 지원 계획의 핵심이다.
우리는 특정한 생산물이나 완벽한 '해결책'을 내놓는
방식을 취하지는 않는다. 그러나 절차 적용이 용이하며
다양한 규모로 실행과 모방이 가능한 방법론을
구축하는 데 도움을 준다.

아이디어와 장애물

다음 문제는 "어떻게 하면 우리가 설계한 기본형을
BT 남쪽의 68가구에게 제공할 수 있는가"였다.
케이프타운을 비롯한 전 세계 도시와 교외 지역은
공동공간보다 개인 공간을 우선시해서 개발에
어려움을 겪고 있다. 해결책은 명확하지 않다.
지금까지도 공유 영역에 대한 거부감은 케이프타운
지역 개발에 장애물이 되고 있다. 결국 우리는 시범
사례로 시행된 블로킹 아웃 계획을 발전시켜 공동
공간과 공유 공간을 통합하여 공유 영역을 바람직하고
사람들이 원하는 곳으로 만들고자 했다. 단기적으로
남아프리카공화국에서 우리의 목표는 남아프리카에서
현대적 건설 실무에 필요한 (가격 책정, 중량, 부피 등의
세부사항을 포함한) 지식 기반을 구축하는 것이었다.

as an example to start pushing for a neighborhood-
upgrading project. Such activity is what lies at the
heart of our Empower Shack ambitions, which are
not to deliver a product or definitive "solution," but
rather facilitate a process and build a methodology
that could be scaled and replicated.

Ideas and Obstacles

The next question was how could we make this
prototype available to the 68 families living in bt
South? Cities all over the world, including Cape
Town and its sprawl, suffer from prioritizing private
space over the collective. The solutions aren't
obvious. For now, the aversion to a shared domain
remains an obstacle to community-building in Cape
Town. Eventually, we would hope, some version of
the advanced Blocking-Out scheme, implemented
as a test case, could incorporate communal, shared
spaces in such a way as to make them desirable
and sought-after. In the shorter term, our goal
became the establishment of a knowledge base
of contemporary building practices in South
Africa—including details on pricing, weight,
and dimensions—that could inform the second
phase of the design process. By testing different
combinations of components, we wanted to identify
and promulgate an overarching modular system for
the housing unit that could leverage the availability
and economy of prefabricated industry-standard
materials while offering a variety of unit sizes to
meet spatial requirements in an efficient way.

What Does the Future Look Like?

The redesigned bt settlement would eventually be
home to a population of between 250 and 300
people, in dwellings of different sizes, costs, and
legal status (rental and owner-occupied), and with
space for mixed uses, commercial activity, and
communal functions. The neighboring school could
incorporate a vertical gymnasium and a teaching
farm. And we wanted to build a community
theater, to be called Makhukhanye Art Room, and
a health center. The strategic plan was designed
in three phases around a central triangle of public
space. Six row house sectors, five of replacement
housing and one for new rental units, bisect the
central plaza. Each house contains a courtyard in
the front as a transition from the street; a vertical
cluster of three units is interlocked on two floors
that open onto a street. We worked with okra, a
Dutch firm, on a landscaping design that would

다양한 구성 요소를 조합해 보면서 우리는 주택 조립 단위를 바탕으로 한 포괄적 모듈형 시스템을 개발하고 보급하고자 했다. 이러한 시스템은 조립식 산업 표준 자재의 유용성과 경제성을 활용하도록 하는 동시에 공간 여건에 맞는 다양한 크기의 조립 단위를 효율적으로 제공한다.

프로젝트 후의 미래는 어떠한 모습일까?

다시 설계된 BT 거주지에 들어서는 주택은 크기, 비용, 법적 상태(임대 또는 자가) 면에서 다양성을 띨 것이며 거주지에는 주택 이외에도 다목적 공간, 상업 공간, 공동체 행사를 위한 공간이 생겨 250명에서 300명에 이르는 사람들의 거처가 될 것이다. 인근 학교에는 수직형 체육관과 교육용 텃밭을 만들 수 있다. 그뿐만 아니라 우리는

마쿠카니에 아트룸(Makhukhanye Art Room)로 불리게 될 공동체 극장과 보건소를 짓고 싶었다. 공공장소의 핵심 삼각형을 둘러싼 3단계 전략적 계획이 고안되었다. 대체 주택 5열과 새로운 임대용 주택 1열로 구성된 6열 주택 구획이 광장을 양분한다. 주택마다 거리와 격리된 앞마당이 있다. 주택 3채가 거리로 통하는 1층과 2층에서 수직으로 맞물린다. 우리는 네덜란드 회사인 오크라(okra)와 함께 조경 디자인 작업을 했다. 이 조경 디자인은 흑인 거주 지역의 소극적인 물 관리를 활성화하고 소규모 텃밭을 발전시키기 위한 촉매 형태의 시범사업이었다. 나무를 심고 관리하는 통합 프로그램을 통해 주민들은 환경 품질에 기여하는 조경 관리 기술을 배웠다.

현재 네 종류의 원형 판잣집과 주택 30열이 건설되어 공공 공간과 개인 공간 사이의 관계, 사회적 범주, 문화적 정체성을 시험하고 있다. 우리가 추진한 프로젝트의 영향으로는 지역 차원에서 급속도로 생활수준을 바꿨고 주민들에게 침수 우려 없이 안전하고 각 층마다 20-30평방미터의 공간이 있는 2층짜리 집을 공급한 것을 들 수 있다. 전 세계 수백만 (지리적, 정치적, 경제적으로) 소외된 계층에게는 우리가 배우고 성취한 것이 보다 큰 의미가 있다. 블로킹아웃의 핵심 원칙은 공동체가 프로젝트의 모든 측면에 참여하는 것이다. 실제로 우리는 항상 직접적인 영향을 받는 사람들이 프로젝트에 참여해야

serve as a pilot project that we thought of as a catalyst for further developments in passive water management and micro-farming in townships. A planting and maintenance program was integrated to teach residents techniques for maintaining the landscaping as contribution to the quality of the environment.

The first four test prototype shacks and the 30 row houses built currently test the relationship between public and private space, social categories, and cultural identity. On a local scale, it changed living standards very quickly and gave the residents a two-floor house, safe above the waterline, with between 20 to 32 square meters of space on each floor. In terms of the millions of people around the world who are marginalized—geographically, politically, economically— what we learned and what we accomplished have much broader implications.

A core principle of Blocking Out is that the community is involved in all aspects of the project. Indeed, we have always insisted on participation in our projects by the people who are directly affected. This had everything to do with our belief that the residents of a favela or barrio or township know best what they need and come up with solutions and techniques singularly appropriate to their circumstances. Typically, nothing taught in schools of architecture or learned through practice touches on these issues. Equally important, no matter how much our experiences in one place form the basis for work in another, context is everything: townships and barrios are informal settlements, but they come into being as a result of location-specific issues and cultures. On the other hand, if we began to operate as a collective, creating houses that are individual, but joined together to create a collective street, the result would be a better neighborhood. In an area of 4,263 square meters, built space, containing 72 shacks, comprised 2,500 square meters; there were 1,500 square meters of open space. Reconfiguring the shacks into tightly packed row houses would yield more than 2,100 square meters of open space, while improving thermal performance, ventilation quality, public-private distinction, and passive security.

The Empower Shack solution has three goals: to increase the housing density to improve the urban environment and the community's living standards; to create incentives such that all participation is voluntary; and to consolidate

한다고 주장해왔다. 자신들에게 필요한 것을 가장 잘 알고 자신들 환경에 잘 맞는 해결책과 기술을 제시할 수 있는 사람이 빈민가 주민 자신이라고 믿기 때문이다. 건축 학교에서 배우는 것이나 실제 현장을 통해 배운 것으로는 부족하다.

그러나 그 이외에도 중요한 것은 한 곳에서의 경험이 다른 곳의 작업을 왜곡할 수 있다는 점이다. 중요한 것은 맥락이다. 흑인 거주 구역(township)과 스페인어 사용자 구역(barrio)은 비공식 정착지이지만 지역 특유의 문제와 문화로 인한 산물이다. 하지만 주택을 개별적으로 지으면서도 주택을 모아 공공 거리를 조성하고 집합적으로 운영하면 한층 더 바람직한 지역사회를 만들 수 있다. 4,263 평방미터 지역 안에 판잣집 72채를 포함한 구축 공간이 2,500 평방미터를 차지했다. 1,500 평방미터가 녹지였다. 판잣집이 밀집된 열을 이룬 주택으로 변경되면서 2,100 평방미터가 넘는 공간이 녹지가 되었고 이로 말미암아 열효율과 환기가 개선되고 공공공간과 개인 공간이 구분되며 소극적 안전성이 강화된다.

판자촌 지원 해결책은 세 가지 목표를 두고 있다. 주택 공급을 늘려 도시 환경과 공동체의 생활수준을 개선하는 것, 주민의 자발적인 참여를 위한 동기를 부여하는 것, 공동체 자원(시간, 토지, 수입)을 강화하여 성공적인 결과를 이루어 내는 것이 그 목표다. 여기에서 공동체 참여는 단순히 개인과 집단의 희망과 필요를 설명하고 협의하는 수준에서 그치지 않는다. 전체의 장기적 이익을 위해서는 투자가 필요하고 개인과 가족의 희생이 얼마간 필요하다. 이러한 과정이 어떻게 진행되느냐는 시험 지역 특유의 환경에 달려있다.

BT 인구에 대한 초기 설문 조사에서는 토지와 돈의 분배가 모두 편향되었다는 것이 드러났다. 소수의 사람들이 전체 지역의 상당 부분을 장악하고 있었고 그와 다른 소수가 보수가 양질의 일자리 대부분을 차지하며 다수의 소득은 매우 적은 상태였다. 판자촌 사람들 사이의 자원 격차를 고려하여 우리 팀은 모든 주민들이 각자의 방법으로 기여할 수 있는 계획을 제시해야 했다. 그것은 마르크스가 말한 "능력에 따른 노동, 필요에 따른 분배"의 변형이었다. 우리 사례에서 분배의 수혜자는 공동체다.

community resources—time, land, and income—for a successful outcome. Here, community involvement is not just a matter of consultation to ensure that individual and collective hopes and needs are accounted for; it requires investment and a degree of sacrifice of short-term personal or familial objectives for the long-term benefit of the whole. How, exactly, this would work is complicated by conditions in our test-area. An initial survey of the bt population showed that the distribution of land and money were both quite skewed: a small percentage of the population controlled a large part of the total occupied area; a different but equally small population held the majority of the jobs that paid well and thus the larger income pool. Given this disparity in resources among the individual shack-dwellers, our group needed to come up with a scheme that would enable all residents to contribute according to their means—admittedly a variation on Marx's "from each according to his ability, to each according to his needs," the recipient, in our case, being the community.

우한

Wuhan

정부가 공간 설계를 수단으로 활용하여 집합적 형태, 집합적 공간, 집합적 주관성, 혹은 오늘날의 맥락에서는 공동체 자체를 구성하는 현상을 탐구한다.

The study investigates the instrumentalisation of spatial design by government to shape collective forms, collective spaces, and collective subjectivities, or, in today's context, the building of communities.

미래의 공공건축과 사회에
역동적인 본보기

Aspirational models for
both tomorrow's public
architecture and for the
society to come.

멕시코 시티   Mexico City

관람자들이 인식이라는 관점에서 우리의 원초적 생존 본능을 연구하는 이 작품을 통해 인류가 만들어내는 건축물 속에서 스스로가 소외당하는 현실에 대해 다시 한 번 생각해 보는 기회가 되기를 바란다. 건축물 속에서 인간은 자연에게, 사회는 환경에게, 사람들은 서로에게 낯선 존재가 되는 것은 아닐까?

By examining our basic instinct of survival from a perceptual standpoint, we hope to bring awareness to the constructs we create that alienate us - the human as alien to the natural, the social as alien to the ecological and humans as aliens to each other.

22 February 2011
At 12.51pm on the 22nd of February, a deadly earthquake struck the city of Christchurch, population 400,000. This earthquake was the most violent of a series that hit the area of Canterbury, on the east coast of the South Island of New Zealand between September 2011 and mid 2013.
185 people died. 115 were killed when the CTY building collapsed near the middle of the city, and another 18 were killed in another concrete building nearby.
The damage caused by the quake was severe and widespread. Almost every house was damaged 3 or 4 times over many months. Over 70% of the central city was demolished. Over 10,000 houses were razed, including 7,000 along the river that runs from the central city east to the sea. The city's schools, hospitals, courts, roads, supermarkets, universities, central business district, and water supply were all damaged or broken.
The damage to the city, the inconvenience caused by the severe damage to the city, and the stress that came from experiencing an extended series of earthquakes led to acute levels of suffering and poor mental health.
The response from the people of the city was compelling. They supported their neighbours, created communities groups, lobbied the government, and took on the long process of cleaning up and repair. In the central city they kept the city alive by creating hundreds of small urban projects, many of them temporary and ephemeral. Dance floors, shopping malls, bicycle powered cinemas, community libraries, arcades, urban farms, restaurants, artworks, and memorials sprung up around the city. These created places to spend time together, to mourn, to sustain the functions of the city, and to create new experiences and public experiments.
Often these projects were created by organisations that saw specific problems and opportunities in the broken city. The projects were part of ethical strategies of activating space, resourcefulness, rethinking transport, decolonising the city, resisting gentrification, local food, public performances, play + joyfulness, and creating a gardening city. Together these projects, and the organisations that created them, created extended moments of enchantment and new ways of being.
Christchurch City
Activating Space
Resisting Gentrification
Free Theatre Christchurch
Play + Joyfulness
Greening the Rubble
Welco

크라이스트처치

Christchurch

이번 전시회를 통해 우리는 고통스러운 사건 이후 트라우마를 겪는 도시인의 경험이(흔히 생각하는 바와는 반대로) 창의적 정신과 번거로움이 있는 큰 기쁨을 자아내고, 이를 통해 도시에 영속적인 변화를 가져올 수 있다는 점을 시사하고자 한다.

This exhibition suggests that post-traumatic urbanism experiences, perhaps counter-intuitively, offer creative, joyful experiences of enchantment that can affect the permanent state of the city.

베를린   Berlin

The Project
프로젝트

The Process
과정

The Partners
파트너

HAUS DER STATISTIK

1969 SYNTHURBANISM PLAN

HISTORICAL TIME

Of forgotten ruins, failed utopias, monuments to the aspirations of socialism. You pass amid a dense groves of old trees, not sure where exactly you're headed, but you follow the voice of a friend. The din of the city dies away as the trees close in around you. Anything could transpire. You find a trunk with an inviting dimple and doze off.

베오그라드 Belgrade

변화하는 광장,

시간과 공간을 넘나드는
이들의 개입 방식은
파리를 하나의 생태계로
인식함으로써 시민의
주도권에 초점을 둔다.

By considering the city
as an ecosystem, their
ways of intervention cross
the extent of space and
time, all to offer a place
on the initiatives of the
populations.

교외(Suburbanism)는 갈수록 도시 특징의 집결지가 되어가고 있다. 도심의 높은 생활비를 감당하기 어려운 사람들은 교통 기반시설이 확장되자 도시 주변부로 옮겨갔다. 이러한 재배치와 더불어 교외에는 도시의 주요 특징이 나타났으며 이러한 경향은 계속해서 확대되고 있다. 분산되는 도시들은 도시 중심부, 교외 중심부, 그리고 그 사이 공간의 관계에서 나타나는 문제에 부딪히고 있다.

반면에 전 세계적인 도시화 확대로 농촌에서 도시로의 대규모 인구 이동이 가속화되고 있다. 공간이 제한된 도시의 성장 역량은 이러한 인구 밀도 증가를 지탱할 수 있는 건축 유형의 혁신 여부에 달려있다. 집합적 공공 공간의 가치는 미래의 건물 재고(building stock, 일정 구역 내의 건물 수)에 대한 개발 수요와 충돌한다. 밀도가 지나치게 높은 도시도 분산된 도시와 마찬가지로 이러한 문제에 직면해 있는데, 주민들이 자체적으로 중요한 해결책을 제시하기도 한다.

토론토,
웰링턴,
울란바토르,
타이후 호,
자카르타

Suburbanism has increasingly become enclaves of urban qualities. The high living expenses associated with living in the city center, paired with the expansion of transportation infrastructure, have allowed for redistribution of the population to the fringe areas of cities. This redistribution has brought urban core qualities that keep on expanding. Cities of dispersion face these issues by presenting a relationship between urban cores and suburban cores, and the space in-between.

On the other hand, the increasing urbanization around the world has spurred large population shifts from the rural to the urban. In cities with limited space, the capacity to grow relies on the innovation of architectural typologies that can support such population densities. The value of the collective public space is dependent on the potential development needs for future building stock. Hyper-dense cities, as well as dispersed cities, face these problems, and it is often the residents of these cities who present radical solutions to these problems.

Toronto,
Wellington,
Ulaanbatar,
Taihu Lake,
Jakarta

토론토　　　　　Toronto

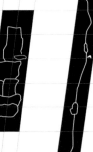

토론토의 도시화는
사방으로 확산되는 형태를
띠었지만, 사실 겉보기만큼
무질서하지는 않다.

Dispersed urbanization
in Toronto is not as
shapeless as it seems.

혼잡의 다른 문화

마이클 파이퍼

Other Cultures of Congestion

Michael Piper

유엔이 세계 인구 대다수가 도시 지역으로 이동했다고 보고한 날,『뉴욕타임즈』는 '사람들 대부분이 이제 도시에 산다'는 제목으로 기사를 냈다. 아마 이 편집상의 기술은 '도시 지역'이라는 용어가 '도시'라는 단어만큼 대중을 사로잡지 않기 때문에 생겼을 것이다. 이유와 상관없이 수많은 언론 매체, 보조금 신청서, TED 강연, 전문 출판물이 이 저명한 뉴스 출처에서 사용된 그 용어를 반복했다.

그러나 실제로 사람들은 도시라는 말에서 연상되는, 인구 밀도가 높은 중심부로 이동하는 것이 아니었다. 유엔이 밝혔듯이, 오히려 교외 주변부, 중소 마을, 빈민가처럼 보이는 거주지, 준교외 지역 등의 장소로 이동하는 것이다. 상대적으로 미묘한 차이지만 도시화 담론에서 보면 미리 생각해 놓은 특정 목표로 향하는 극단적인 왜곡이 드러난다. 이런 왜곡 중 최고는 밀도에 대한 집착이다. 이런 집착이 이미 고착화되었더라도 왜, 언제 그것이 생겨났고 어떻게 현재 우리가 교육과 일자리를 지배하는 고밀도 도시가 아닌 지역들을 바라보는 방법에 영향을 끼치는지 생각해볼 가치가 있다.

현재 많은 건축가들과 도시계획가들이 도시(urbanity)의 전제 조건으로 높은 밀도를 고려한다. 적어도 유럽, 미국, 캐나다의 건축가와 계획가들은 그렇다. 하지만 항상 그런 것은 아니었다. 이 맥락에서 근대주의 도시화는 녹지를 더 많이 확보할 수 있는 낮은 밀도 계획을 선호했다. 사실 이런 움직임은 밀도가 지나치게 높은 환경에서 생겨나는 문제에 대한 비판을 옹호하면서 생겨났다. 1800년대 중반 산업화된 도시는 공장 일자리에 동반된 인구 증가와 심각한 오염에 대비하지 못했다. 이에 대응하여 이 시기의 많은 이상적인 제안(이들은 지금도 계속해서 아방가르드적 야심에 영감을 주고 있다)은 산업적 자본주의에서 밀도가

On the same day that the UN reported that the majority of the world's population had moved to urban areas, The New York Times released an article with the headline that "most people now live in cities." Perhaps this editorial sleight of hand is due to the fact that the term "urban area" does not necessarily capture the public imagination like the word "city" does. Regardless of the reason, countless news outlets, grant applications, TED talks and scholarly publication repeated the terms of this well-regarded news source.

As it turns out, most people aren't moving to the dense centers associated with the idea of the city, but rather, as the UN notes, to a range of built areas that include among other places: the suburban periphery, small to medium towns, favela-like settlements, or exurban landscapes. It is a relatively subtle terminological substitution that, however, reveals extreme biases within the discourse on urbanization toward certain preconceived ends. Chief among these is a commitment to density, and while this commitment may be well-founded it is worth considering why and when it came about, and how it presently influences the way we look at those parts of the world that don't necessarily measure up to the kind of dense urbanity that has come to dominate scholarship and practice.

Many architects and planners today consider high density to be a precondition for urbanity or at least many of those in Europe, the United States and Canada do. But this wasn't always the case. Modernist urbanism in these contexts favored lower density schemes that afforded more open space. Indeed, the very premise of this movement centered on its critical response to problems its proponents attributed to the condition of excessive density. Industrialized cities during the mid-1800s were not equipped to deal with the surge in population and pollution that accompanied factory work. In response, many of the utopian proposals at this time—those that continue

지나치게 높은 도시와 연관된 숨 막힐 듯한 환경을 피하는 것에 기반을 두었다. 심지어 도시를 찬양하고 도시에 존재하는 밀도를 보존하는 계획을 세웠던 피에르 잔느레마저도 엉망이고 혼란 상태인 도시를 깔끔하고 조경이 훌륭한 이상적인 장소로 바꾸자고 제안했다. 근대주의적 도시설계를 개혁하려는 방법이 '실패했다'고 여겨진 1960년대와 1970년대 이후, 학문과 실제 업무의 주류는 도시의 밀도를 더 높은 수준으로 올리는 쪽으로 방향을 바꿨다.

1978년 렘 콜하스의 『정신착란증 뉴욕』이 출판되었다. 이 책은 맨해튼의 밀도를 역사적으로 해석하고 있고, 저자 주석에서 밀도의 "두 번째 도래... 혼잡의 문화(Culture of Congestion)에 대한 청사진"에 대한 논거를 보여준다. 콜하스는 자신의 사무실에서 맡은 디자인 프로젝트에 일부에서 드러나는 밀도의 사회적 측면에 대한 관심을 드러낸다. 이들 중 다수의 프로젝트는 주민들을 위한 압력솥 역할을 하기를 열망한다. 주민들을 가깝게 모으고, 만날 수 있는 기회를 제공해 사회적 연결망과 구조를 만들어 낼 수 있다는 것이다. 다른 도시 전문 건축가들, 특히 OMA에서 일을 했던 건축가들은 이런 매력을 극단적인 밀도의 산물로 우연한 사회적 사건으로 여긴다. 아마 밀도의 이미지를 만들어내려는 이런 건축가들의 노력이 그들이 진행한 프로젝트보다 더 큰 영향력이 있었을 것이다. 압축된 환경에서 인간의 상호작용에 의한 남을 의식하지 않는 산물로 여겨졌을 혼란스러움(Messiness)은 디자인 트롭이 되었다. '혼잡의 문화'에 영감을 받아 그들의 건물 설계는 밀도를 보여주었다. 1980년대와 90년대의 충돌면, 겹치는 형태, 잔뜩 쌓인 상자, 위치가 이동된 격자판은 이른바 '종이 건축'에서 나타났고, 곧 유행이 되었다. 그러나 (사회적 관계에서 방식으로) 밀도를 끌어내는 것은 대도시 중심부를 넘어서거나 초대형 빌딩 밖의 장소에서 사회적 구조를 만들어 내는 디자인 도구 발전을 미연에 방지해서 콜하스의 독창적인 문화적 프로젝트를 옆길로 새게 만들었다.

『정신착란증 뉴욕』이 출판된 시기 미국에서는 북아메리카 특유의 '무질서하게 뻗어나가는' 변두리 풍경을 빽빽한 도시 중심부의 이미지에 맞춰 종합적으로 다시 만드려는 야심찬 계획들이

to inspire avant-garde ambitions among practices today—were based on escaping the oppressive conditions associated with the hyper-dense city of industrial capitalism. Even Pierre Jeanneret (Le Corbusier), whose admiration for the city motivated schemes to preserve its existing densities, proposed to replace its messy and chaotic qualities with a sanitized and landscaped ideal. It wasn't until the 1960s and '70s, when modernist urbanism's reformist measures were deemed to have "failed," that mainstream scholarship and practice shifted its preference to higher levels of density in cities.

In 1978, Rem Koolhaas's *Delirious New York* was published. The book is part historical account of density in Manhattan and also, as the author notes, an argument for its "second coming... a blueprint for the Culture of Congestion." He suggests an interest in the social dimensions of density that is evident in some of the design projects in his office. Many of these projects aspire to perform as pressure cookers for inhabitants, bringing them into close proximity, enabling chance encounters from which it is thought that social networks and social fabric might emerge. Other urbanist architects, particularly those who had worked for OMA, share this fascination with unplanned social happenings as a product of extreme density. Perhaps even more influential than the built projects of this group of architects has been their effort to produce images of density. Messiness, which might otherwise be understood as an unselfconscious product of human interaction in compressed environments, became a design trope. Inspired by the "culture of congestion," their building designs indexed density. Colliding planes, overlapping shapes, stacked boxes, or shifted grids appeared in so-called "paper architecture" of the 1980s and 90s and made its way into popular practice soon after. However, abstracting density —from social relationships into style—has sidetracked Koolhaas' original cultural project, forestalling the development of design tools for producing social fabrics in locations beyond the metropolitan centre or outside of extra-large buildings.

Around the same time that *Delirious New York* was published, a practice of physical planning emerged in the U.S. that has taken on the ambitious, if not impossible, task of comprehensively remaking North America's "sprawling" peripheral landscapes in the image

발달했다. 그 이름에 맞게, 신도시주의는 1800년대 유럽과 북아메리카의 밀도가 높고, 결절점 같은, 자본주의 도시의 전례를 따랐다. 각 도시의 결절점의 작은 건물들로 이루어진 있는 비교적 작은 블록에는 신도시주의자들이 최근 높게 평가하는 활기찬 1층 가로 상가가 있었다. 지역적 측면에서, 신도시주의자들은 오늘날 캐나다와 미국 도시의 특징인 획일적이고 평평한 거대도시 체계에 이전 세기의 영역 체계를 기반으로 한 교통 결절점들이 생성되리라 상상한다. 계획, 개발의 화용론에 푹 빠진 신도시주의는 과거에 비해 도시 경제 규모가 양적으로 과도하게 증가함에도 불구하고 밀도에 대한 감상적인 아이디어를 주류 상업 활동에 가까스로 주입해왔다. 캐나다와 미국에서 그들의 영향력은 일반화되어, 그들의 이상에 영향을 받지 않고 도시화된 지역을 찾기가 힘들다.

　　아마『뉴욕타임즈』의 '도시'라는 단어 선택은 세계 인구가 실제로 도시 지역, 밀도 높은 도시 중심부 주변 영역까지 포함하는 그 지역에서 살고 있을 것이라는 더욱 넓은 문화적 맹점을 반영할 것이다. 문화 창조를 위한 계획이 ('도시 지역'과 대조되는 '도시'의 두 가지 특징인) 혼란스러움의 미학에 함축되었거나, 분주한 거리의 이상적인 혼잡을 통한 것이라면, 도시와 자주 연관지어지는 일종의 물리적 밀도를 유지할 수 없는 위치에서 가능한 미래를 계획할 때, 디자인 전문가들은 어떤 문화적 상상력을 사용할 수 있을까? 그래서 이러한 주변 지역이 너무 퍼져나가 시내 중심 같은 밀도에 도달하지 못하는 반면, 주변 지역의 증가하는 인구는 새로운 종류의 밀도를 보이게 된다.실제로 이 도시 이동은 20세기 초반 대도시 체계를 만들어 낸 것과는 다르다. 변화에 참여하기 위해서,그리고 밀도의 사회적이고 미학적인 프로젝트와 도시 생활을 구성하는 것에서 디자인 타당성을 새롭게 하기 위해서 그 논리를 이해하는 것이 필요하다.

of its compact city centers. New Urbanism, as it has been named, takes the cue from dense, node-like, mercantile capitalist cities in Europe and North America during the 1800s. Within each city node, relatively small blocks with comparably small buildings presented street walls with active commercial ground floors that New Urbanist today hold in high regard. At the regional scale, they imagine that the undifferentiated and flat megalopolitan order that characterizes urbanization in Canada and the U.S. today will be punctuated with dense transit-oriented nodes that are based on the territorial order of the previous century. Steeped in the pragmatics of planning and development, New Urbanism has managed to embed its sentimental ideas about density into mainstream commercial practice in spite of the fact that its economies tend towards a scale that far exceeds the quaintness of its precedent. Their reach is so ubiquitous in Canada and the U.S. that one is hard-pressed to find an urbanized area in this context whose planning policy does not bear the influence of their ideals.

Perhaps the *New York Times'* selection of the word "city" is reflective of a broader cultural blind spot towards where the world's population actually lives —in urban areas, here understood to include territories peripheral to dense city centers. If the plan for creating culture is through congestion, implicated by the aesthetics of messiness or the ideal of bustling streets—both characteristics of the city as opposed to the urban area—then what cultural imaginaries might design professionals engage when projecting possible futures in locations that will not be able to sustain the kinds of physical densities more often associated with the city? So while these peripheral sites are too expansive to reach downtown-like densities, their mounting population growth suggests new kinds of density. Indeed, this urban migration differs from the one that brought about metropolitan orders of the earlier twentieth century. Understanding its logic is necessary not only for participating in its transformation but also for renewing design's relevance in the social and aesthetic project of density and what constitutes urban life.

웰링턴　Wellington

생산적 농장에서 집합적
도시공간으로.

From productive farm
to collective urban
environment.

새로운 도시 계획 모델
원형 대도시는 선형 도시와
네트워크 도시 사이의 흔히
않은 중간단계이다. 호수를
둘러싸는 원형의 도시에서
미래 성장 모델이 구체화된다.

The circular metropolis
presents a rare
intermediary stage
between linear and
network city, a state of
liminality in which future
growth is consolidated
around urban ribbon
enveloping the lake.

342

건축 이론은 수 세기에 걸쳐 도시 조직의 중요한 요소인 건축적 　　취리히,
유형학 개념에 천착해왔다. '유형학(typology)'이라는 단어의 정의　뉴욕,
자체가 그 의미를 해석하는 파생 연구를 낳기도 했다. 유형학은 건축　런던,
요소, 건축 체계, 공공 공간과 개인공간을 구분 짓는 형식적 속성을　하이파,
분류하는 체계다. 전경-배경(figure-ground), 포셰(poche),　　　리마,
브리콜라주(bricolage), 꽉 찬 공간(solid), 빈 공간(void) 사이의　로마,
관계는 도시 형성과 도시의 공공 영역 형성을 위해 지속적으로　　호치민 시
연구되어 왔다.

로시(Rossi)는 유형학을 더 이상 축소될 수 없는 요소의 유형에
대한 연구로 본다. 주거 유형을 핵심 요소까지 파고 들면 사회,
경제학적 맥락이 드러날 것이다. 성장 중인 도시 지역의 부동산
압력은 집합성이 구축된 형태를 통해 그 모습을 드러내는
공동주택(cohabitation)이라는 새로운 건축 유형을 만들어
내고 있다. 여기에 전시되는 도시들은 공공공간과 개인 주거지를
막론하고 도시의 사회적, 경제적 압력으로 생겨난 집합적 유형을
점검한다.

The notion of architectural typologies as the prime organizers of the city has plagued architectural theory for centuries. The very definition of the word "typology" has also produced a ramification of studies to interpret its meaning. Typology has been classified as taxonomies of architectural elements, systems in architecture, and formal attributes that distinguishes public space from private space. This relationship between the figure-ground, poche, the bricolage, solid and void, has been a persistent study for the formation of cities and its public realm.

For Rossi, typology presents itself as the study of types of elements that cannot be further reduced. A residential typology, if reduced to its core elements, would reveal the socio-economic context. The current real estate pressures in the growing urban world have also produced new typologies of cohabitation, where the collective is manifested through the built form. The cities presented here look at these typologies occurring due to the social and economic pressures of the city, be it for public space, as much as the private residential type.

Zurich,
New York,
London,
Haifa,
Lima,
Rome,
Ho Chi Minh City

중심부에 남아있는 '소규모 단독 건물들'은 오늘날의 취리히 도시 구조에 스마트한 인구밀도 증가를 추진할 여지가 아직 많이 남아있다는 사실을 반증한다.

Centrally located 'odd lots' represent an unrealized opportunity for smart densification amid the existing urban fabric.

청년층은 현대 사회의
역동적이지만 불안정한
부분을 상징한다.

Youths represent a
dynamic yet precarious
section of today's
populations.

런던

London

생산의 장소에서 소통의 장소로

From a place of production to a place of human interaction.

집합적 유형

라피 시갈

Collective Typologies

Rafi Segal

이 글은 집합성의 새로운 개념(기존의 공공공간과 개인 공간에 포함되지 않는 다양한 공유 공간)에 맞춰 설계된 건물 유형과 도시 요소를 연구할 필요성을 간략하게 논한다.

공유 경제, 디지털 기술, 새로운 개념의 공동체는 우리가 일하고 생활하는 방식을 변화시키고 있다. 심지어 사유권을 사회의 근본 요소로 생각하는 미국에서도 견고했던 공과 사의 분리가 허물어지고 있다. 공공부문이 빠져나가고 남은 빈 공간은 생산자/소비자라는 이분법 도식과 결합하여 집합적 절차에 대한 수요를 만들어내고 있다. 이러한 집합적 절차는 새로운 형태의 자기조직화와 실행을 토대로 한다. 공동체를 기반으로 하는 교환과 협력 플랫폼이 소유물의 개념을 흔들고 공유 공간 개념을 널리 퍼뜨리면서 경제적, 사회적 양식도 바뀌고 있다.

회의주의자들은 공유 경제가 결핍이나 생활비 인상의 산물이라 주장하겠지만, 다른 동기도 있을 수 있다. 한층 더 지속가능한 생활 방식을 향한 바람, 소비자 중심주의에 대한 반작용, 인간관계 개선에 대한 욕구, 물리적 공간의 구분과 통제로는 더 이상 사생활을 지킬 수 없다는 인식 등이 그 동기일 가능성이 있다. 가정 영역에서는 코하우징(co-housing, 공유 집합 주택)이나 코리빙(co-living, 공유 주택)이 시장 경제에서 실행 가능한 모델이자 이용자, 설계자, 개발자 사이의 일방적인 관계를 극복할 수 있는 모델로 재등장했다. 그뿐만 아니라 이런 몇몇 모델은 세심하며 사용자 중심인 데다 공간적으로 혁신적인 형태의 주택을 만들 수 있음을 입증했다. 그러나 이 같은 모델이 도시 형태에 미칠 영향은 대부분 연구되지 않은 채로 남아있다. 건축가들은 공유라는 개념이 개별 건물 차원을 넘어 도시 유형까지도 바꿀 가능에 대해 어떻게 전망할까?

건물과 도시 각각의 차원에서 유형이란 인간

The article outlines the need to explore building types and urban elements designed in response to new notions of collectivity – various degrees of shared spaces that are not addressed through conventions of public and private space.

Sharing economies, digital technologies, and new notions of community are altering the way we live and work, diffusing the hard public-private divide even in places such as the US, where private ownership has been a foundational pillar of society. The void left by a retreating public sector, combined with the rigidity of the producer/consumer binomial, has created a demand for collective processes that reflect new forms of self-organization and action. Community-based exchanges and collaborative platforms are transforming economic and social patterns, contesting notions of property and promoting ideas of shared space.

Skeptics may claim that the sharing economy is a result of scarcity or increasing costs of living, but perhaps there are other motivations at play: a call for a more sustainable way of living, a reaction against consumerism, a desire to enhance human interaction, and growing realization that privacy is no longer achieved through the demarcation and full control of one's own physical space. In the domestic realm, co-housing and co-living are re-emerging as viable models that operate within the market economy and overcome the unilateral relationship between user, designer, and developer. And while some of these models have proven capable of producing more sensitive, user oriented, and spatially innovative forms of housing, the implications for urban form remain largely unexplored. How will architects look beyond individual buildings to envision how sharing can change urban typologies as well?

On both the building and urban scale, type refers to prototypical spatial organizations and built forms that reflect patterns and conventions of human activity or use. In relation to living, for

활동이나 쓰임새의 패턴과 관례를 반영하는 원형적 공간 조직화 및 건설된 구조를 나타낸다. 예를 들어 생활과 관련하여 교외의 단독 주택,연립 주택, 중형 아파트는 다양한 주거 건물 중에서 몇 가지 유형에 불과하다. 개별 유형은 단독으로는 생활양식을 반영하지만 이것이 반복되면 크게는 도시 형태에 영향을 미친다. 한 예로 단독 주택이 있는 지역은 아파트가 있는 지역과는 다르게 배열되며 이러한 배치가 거리 배치, 생활 편의 시설 등에 영향을 줄 수 있다. 그렇다고 유형이 고정된 것이라는 뜻은 아니다. 오히려 그와 대조적으로 그 정의를 생각해보면 유형은 변화를 허용하고 촉구한다. 유형은 조직의 핵심 구조이지만 변화에 개방적이다. 라파엘 모네오는 다음과 같이 주장한다.

> "유형이란 그 안에서 변화가 일어나는 틀로 생각할 수 있다. 변화는 역사가 요구하는 지속적인 변증법의 필수 요건이다. 이러한 시각에서 유형은 건축물을 생산하기 위한 '얼어붙은 메커니즘'이라기보다 과거를 부정하고 미래를 바라보는 방식이 된다."[1]

건축과 도시계획에서 새로운 유형은 기존 형태나 공간에 새로운 상황을 적응시키려는 시도를 하고 나서야 등장하는 경우가 많다. 사회·경제적 변화가 건축과 도시 관습의 변화보다 더 빠르게 일어나기 때문에 이미 정해진 유형에 맞춘 설계에 새로운 용도를 끼워 넣는 일이 종종 발생한다. 생활 주변에서 찾아보면 아파트 꼭대기 공간을 그 사례로 들 수 있다. 창고로 사용되었던 고미다락(loft)은 저렴하면서도 넓은 생활공간 겸 작업실을 찾는 예술가들의 소규모 창작 공간으로 사용되기 시작했다. 이것이 변화의 첫 번째 물결이었고, 그 이후 다양한 이용자를 위한 새로운 주거 유형으로 진화하였다.

또 다른 사례로 룸메이트와 함께 사는 것을 들 수 있다. 완전히 새로운 현상은 아니지만 핵가족을 위한 공간이었던 주택이나 아파트에 관계없는 사람들이 함께 사는 경우가 증가하고 있다. 단독 주택은 공동 주거라는 새로운 역동성에 맞춰 변화하고 있다. 공동

example, the single family detached suburban home, row housing, and mid-size apartment towers are but a few of many residential building types. Each type reflects an idea of living at a singular scale that by repetition impacts urban form at large. For example, a neighborhood made of single-family homes will be arranged differently than one made of towers, impacting street layout, shared amenities, etc. This is not to say that the type is a fixed entity. On the contrary, by definition, type allows and calls for adaptation. It offers a core structure of organization, yet is open to change. As Rafael Moneo argues,

> "The type can thus be thought of as the frame within which change operates, a necessary term to the continuing dialectic required by history. From this point of view, the type, rather than being a "frozen mechanism" to produce architecture, becomes a way of denying the past, as well as a way of looking at the future."[1]

In architecture and urbanism, new types often emerge after attempts to adapt an existing form or space to a new condition. Since many socio-economic changes happen more quickly than architecture and urban conventions change, we often fit new uses into designs meant for other types. In the context of living, loft apartments are a common example. Lofts began as an adaptation of warehouses and small-scale production spaces for artists looking for affordable, spacious conditions for their life and practice. While this was the first wave of transformation, lofts have since evolved into a new residential type for a variety of users.

Another example is the case of living with roommates. Although not at all a new phenomenon, unrelated groups of people increasingly find themselves living together in houses or apartments designed for nuclear families. Single-family homes are thus being adapted for a new dynamic of shared living. While this arrangement might work well for those who are interested in creating a 'family-like' environment, such arrangements do not always work well for those who prefer more privacy, variety and control in degrees of sharing. Can we not learn from these cases and develop new types of housing better designed for such conditions?

1   Rafael Moneo, "On Typology," in *Oppositions*, no. 13, Massachusetts: The M.I.T. Press, 1978.

1   Rafael Moneo, "On Typology," in *Oppositions*, no. 13, Massachusetts: The M.I.T. Press, 1978.

주거 방식은 '가족 같은' 환경을 만들고 싶어 하는 사람들에게는 잘 맞겠지만, 공동생활에서도 어느 정도의 사생활, 다양성, 통제를 중요하게 생각하는 사람에게는 맞지 않을 수 있다. 이러한 사례를 통해 얻은 교훈으로 새로운 상황에 맞게 설계된 새로운 주택 유형을 발전시킬 수는 없을까?

통근, 업무, 생활 방식이 변하면서 오늘날의 변화하는 도시 내 공간은 주택과 비슷한 위기를 맞고 있다. 새로운 유형의 집합성을 어떠한 방식으로 뒷받침할 수 있는지 이해하려면 '집합적 유형' 개념을 건물을 넘어서 길, 거리, 공원, 그 이외 도시 공간, 프로그램 등의 요소를 포함하는 도시 규모로 확장해야 한다. 또한 공공 공간, 개인 공간이라는 이분법을 일련의 '점진적인' 공간 개념으로 대체하여 변화에 대한 접근법을 마련해야 한다. 점진적 공간이란 사람, 관습, 만남의 이질성을 포용하고 생활 속의 새로운 참여 형태를 보여주는 중간 지대를 뜻한다. 여기에서 생활은 거주 공간(가정)만을 의미하는 것이 아니다. 가족이나 (가장 사적인) 개인에서 (가장 공적인) 폭넓은 사회 변화를 반영하는 도시에 이르기까지 확장된 개념을 포함한다. 이처럼 점진적인 도시 집합 공간은 공공 영역으로 확장되어 깊숙이 침투하고 우리의 사용 방식, 행동 방식, 도시에 대한 사고방식을 바꾼다. 결국 집합은 인간관계를 대표한다. '나의 것'과 '남의 것' 이라는 익숙한 개념에 도전해서 '우리 것'을 촉진하는 도시 프로젝트를 발전시킨다.

Urban spaces face similar challenges to housing in today's changing cities as commuting, working and living patterns change. To understand how new forms of collectivity can be supported, I would suggest here to extend the idea of 'collective type' beyond the building to the urban scale, including elements such as paths, streets, parks, and other urban spaces and programs. I propose to approach this transformation by replacing the binary divide of public-private with a set of 'gradient' spaces– a mediated field which speaks to a heterogeneity of people, practices, and encounters, and expresses a new form of participation in living. Here, living refers not just to the place of residence (the home) but includes and extends outward into the city, reflecting broader social changes (the most public), rather than just the familial or the self (the most private). This gradient of urban collective spaces thus extends deeper into the public realm and alters the way we use, move through, and perceive the city. Collectives, are, after all, a representation of relationships between people – they provide an opportunity to challenge prevailing notions of 'mine' and 'theirs' to advance an urban project promoting the 'ours.'

하이파   Haifa

전시 작품 '정원 도시의 계단'은 하이파를 연결하는 1,000개의 계단을 도시의 현재와 미래를 공유하는 공간이라는 관점에서 들여다보고, 계단이 도시의 유적, 시장, 영화관, 주택 지역을 가로지른다는 사실에 특별한 의미를 부여한다.

"Sit(e)lines of a Garden City" explores Haifa's system of urban stairs as present and future shared spaces, monumentalizing the city's 1000 stairs which traverse monuments, markets, cinemas, and residential homes.

리마 Lima

'공공'의 문화를 시작하기 가장 효과적인 '가능성'은 집합성을 구현하는 것에 있다.

Contributing to the construction of "the collective" is the most effective "possibility" of starting a culture of the "public".

로마

Rome

전체로서의 하나가 아닌
수많은 조각들과 부분으로
축적되어 만들어진 정물과
같다.

A still life, a city made
by the accumulation of
pieces, of parts without a
whole.

도시는 발전 속도와 상관없이 기반시설과 소구획화(parcelization)를 통해 효율성을 극대화해야 하는 도전과제에 직면해 있다. 지형, 부동산 경제, 인구밀도 유입 등의 다양한 요인 등 다양한 요소의 차이로 말미암아 조각난 자투리땅, 공간 사이의 건축 후퇴선(setback), 이용할 수 없는 부지 등의 잔여 공간이 발생하는 일이 많다. 이러한 자투리 공간들은 잔여층의 현실을 보여주거나 점용의 배경으로서의 독특한 용도로 사용된다. 도시공간의 점용이 사이 공간에서 소규모로 이루어진다고 생각하는 경우가 많지만 특정 행사와 상황은 도시 전반에 걸친 점용을 유발하기도 한다. 정치적 갈등으로 인한 대규모 군중의 이동이 난민의 집단 탈출로 이어지고 특정 지역이 임시 거주지로 전용되기도 한다. 대규모 종교 순례 때문에 유입되는 인구를 감당하기 위해 임시 기반시설이 계획될 때도 있다. 이러한 공간 점유는 영속적이지 않고 비공식인 도시 특성을 낳는다. 여기에서 전시되는 도시들은 이 모든 시나리오상의 도시공간의 점용을 경험하고 있다. 이러한 상황은 여기에서 따로 다루어지지는 않지만 장소 만들기(placemaking)의 고유한 특징이 되는 도시의 여러 층에 대한 논쟁으로 이어진다.

개성,
블랙 록 시티,
홍콩,
파리,
선전,
싱가포르,
로테르담

Appropriations

Regardless of the speed of development, cities face the challenge of optimizing their efficiency through their infrastructure and parcelization. The discrepancy that results between both due to geography, the economy of real estate, population density influx, and other parameters, often result in residual spaces like sliver parcels, setback in-between spaces, and odd lots. These spaces often result in appropriations of leftover spaces and unusual uses portraying the reality of residual layers and the setting for appropriations.

Although appropriations are often considered at the smaller scale of the in-between space, certain events and occurrences also trigger urban scale appropriations. The displacement of large crowds due to political conflicts force a mass exodus of refugees and the appropriations of terrains for temporal settlements. Large religious pilgrimages are planned with temporal infrastructures to sustain the additional population. These occupations become informal urbanities with no sense of permanence. The cities presented here, tackle appropriations through all these scenarios that generate a polemic for layers of the city that are unaccounted for but serve as the unique characteristic for placemaking.

Gaesung,
Black Rock City,
Hong Kong,
Paris,
Shenzhen,
Singapore,
Rotterdam

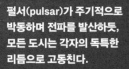

펄서(pulsar)가 주기적으로
박동하며 전파를 발산하듯,
모든 도시는 각자의 독특한
리듬으로 고동친다.

Each city is like a pulsar, a
star with a unique pattern
of pulsed appearance of
emissions.

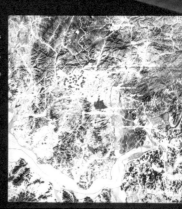

A Week at the Gaeseong Industrial District (GID)

North Korean Workers
· Leave for home (3th, 4thi group)
퇴근 (세 번째, 네 번째 조)

South Korean Workers
· Rest and Sleep at GID
일부 개성공업지구 체류
휴식 및 취침

개성

Gaesong

블랙 록 시티

Black Rock City

블랙 록 시티는 매년 미국
네바다 사막에 건설되었다
해체되는 도시이다.

Black Rock City is
constructed and
deconstructed every year
in the desert of Nevada.

누가 자기 거실을 망가뜨리겠는가

제랄딘 보리오

Who Would Destroy It's Own Living Room?

Géraldine Borio

우리가 쉽게 윤곽을 그려볼 수 있는 합법적 도시 이면에는 '사이 공간'으로 이루어진 도시가 있다. 사이 공간의 도시는 미완성이고 규정되지 않아서 변화가 계속 진행 중인 도시이자 미래나 과거를 상상할 수 있는 여지가 남아있는 도시다. 거시적인 차원에서 볼 때 형태학적인 '사이 공간'은 산업적 불모지, 기반시설 주변, 빈 터, 버려진 땅과 동일시되며 경제적 문제로 인한 산업공동화나 영역 분열로 생겨난 '도시 질환'으로 간주된다. 미시적 차원에서는 구축 환경의 변화 과정에서 생겨난 인구 과밀 도시 영역 안의 작은 틈과 구석진 곳으로 볼 수 있다.

사용자들에게 이처럼 과도기의 버려진 지역은 일시적으로 전용할 수 있는 공간이 된다. 감상주의나 향수(nostalgia) 보다는 실용주의를 더 중시하는 사용자들은 그 지역 중에서 자신에게 필요한 공간을 찾아낸다. 빈 땅에서 일어나는 활동은 어느 정도는 합법적이지만 항상 일시적이다. 사실 빈 공간이 전용되는 정도는 구축되고 계획 중이며 공적인 주위 환경에서 느껴지는 억압의 정도와 밀접한 관련이 있다. 그러나 체계적 기능, 프로그램, 규칙이 전혀 없는 '사이 공간'은 해방감을 준다. 우리는 그러한 공간을 자유롭게 해석할 수 있다.

홍콩의 사이 공간은 좁은 뒷길이다. 이 뒷길은 과밀한 도시 형태 안에서 사이 공간으로 된 망을 형성한다. 이 좁은 뒷길은 공식 지도에는 나타나지 않는다. 그러나 전부를 합치면 최근 반환된 서카오룽 문화지구의 네 배인 150 km에 달한다. 크기가 작더라도 여러 도시 영역에서 반복적으로 나타나는 사이공간은 우리가 도시를 인식하고 경험하는 데 커다란 영향을 끼친다. (온갖 금지 사항이 적용되는) 공식적 공공공간보다 규제가 심하지 않고 불모지 같은 (그래서 거주자에 의해 계획 없이 전용되는) 잔여 공간은 도시 전역에서 중요한 완충 지대 역할을 한다.

Behind the legitimate city, for which we can easily draw the outlines, lies the city of 'in-betweens' – a city in a constant state of transformation, but which inspires precisely because it has not been finished or defined; a city that leaves room for the imagining of a future or a past. At a macro-scale, morphological 'in-betweens' are identified as industrial wastelands, infrastructure surroundings, vacant plots, waste grounds; so-called 'urban diseases' resulting from deindustrialization or territorial fragmentation pushed by economic agendas. At a micro-scale, they are small gaps and recesses within the dense urban areas, resulting from the mutational process of the built environment.

For users, these areas in transition, abandoned, offer a momentary space to appropriate. Motivated more by pragmatism than sentimentalism or nostalgia, they find in them a space that answers a need. Activities taking place in these vacant lands are more or less legal, but always temporary. In fact, like concomitant vases, the intensity of the appropriation of voids is tightly linked to the extent of the repression felt in the built, the planned, and the official surrounding them. Yet, devoid of systematic functions, programmes, and rules, 'in-betweens' bring a feeling of liberation. We feel free to interpret them.

In-betweens in Hong Kong are narrow back lanes that form a network of interstitial spaces in the dense urban morphology. These narrow back streets are not registered on official maps; however, put together, they represent an area of 150 km, four times the area of the recently reclaimed land that makes up the West Kowloon Cultural District. Although tiny in size, their recurrence in many urban areas has a strong impact on the way we perceive and experience the city. Less regulated and sterile than official public spaces (where one is welcomed by an arm-long list of interdictions), these residual spaces – unplanned and appropriated by their inhabitants – function as

공간이 부족할 때 넘쳐나는 사람들은 이런 형태학적 사이 공간에 모이게 된다.

사이 공간 환경에서는 소유권이 불분명하다. 홍콩의 뒷길에서는 공공공간에서 사적인 활동을 한다. 중고 소파, 화분, 벽에 걸린 거울, 옷걸이에 걸린 티셔츠는 생활 전체를 보여준다. 소유권의 애매함을 최대한 활용하며 순간적인 필요에 맞게 구성되는 사이 공간 네트워크는 도시의 거대한 거실처럼 보인다. 규제가 있는 공식적 공공장소와 달리 이 공간은 공공장소에서 할 수 없는 활동 전부를 할 수 있는 지대가 된다. "그리고 기능이 정해져 있는 공간이 아니기 때문에 용도 전환의 여지가 있다."[1]

불문율에 따라 만들어진 이 영역 안에서 사용자와 행인은 사적 영역의 한계를 변경한다. 그 형태는 다양한 사용자 사이에서 지속적으로 타협된다. 공공장소에 대한 과도한 계획과 규제는 당국의 통제 필요성뿐만 불명확함을 반영한다. 이 경우에는 소유권의 모호함이 합의를 촉구하는 듯 보인다.

잔여공간의 관찰과 분석은 공백이 필요하다는 인식을 낳는다. 즉 공백이 공간을 형성하는 과정에 필수적인 요소라는 것이다. 렘 콜하스가 공간적 틈을 통해 도시를 연구한 '보이드에 대한 전략'에서 언급했듯이 우리는 단단하고 눈에 보이는 것만 다루며 마주치는 모든 공간을 채우고 싶어 하는 건축에 이의를 제기하고자 한다.

우리는 공공공간과 개인 공간, 내부와 외부라는 양극성을 강요하지 않은 채로 다양한 해석 방법이 있는 공간을 설계할 수 있을까? 사람들은 군중이 아닌 개개인의 합으로 간주될 때 타인과 환경을 직관적으로 좀 더 세심하게 대하는 경향이 있다.

인구밀도 높은 도시에는 이러한 길들이기가 사회를 위협하기는커녕 사회 결속력과 안정성을 높여준다. 사실 '누가 자신의 거실을 스스로 망가뜨리기를 원하겠는가?'[2]

important buffer zones across the city. When lack of space is an issue, these morphological in-betweens collect the overflow of life.

Inherent in the in-between condition is the ambiguity of ownership. In Hong Kong's back lanes, private activities overlap with public ground. A second-hand couch, potted plants, a mirror on a wall or T-shirts on a hanger are indicators of a whole living system. Making the most out of the ambiguity of ownership, the network appears as a giant urban living room, configured to react to instant needs. Compared to official public spaces that are over regulated, it offers ground to all kinds of activities that would not be allowed in them: "and because the spaces are not saturated with function, there is room for appropriation".[1]

Within this territory of unwritten rules, the users and the passers-by are shifting the lines of their personal boundaries. Its forms are in constant negotiation between different users. Over-programming and regulations upon public spaces mirror the authorities' need to control, and of non-definiteness. Here, the lack of clarity of ownership seemed to call for consensus.

Observing and analyzing the leftovers raises awareness of the need for emptiness, a fundamental component in the process of making space. Recalling Rem Koolhaas' *strategy of the void*, the study of the city through the gaps, we are tempted to question an architecture that deals only with the solid and the tangible; that tries to fill in every space it encounters.

Instead of forcing the polarities between public and private space, indoor and outdoor can we design spaces that provide multiple ways of interpretations? When humans are not considered as a mass of people but as a sum of individuals they tend to intuitively behave with more sensitivity to the others and the environment.

Far from being a threat, this sense of domestication in a dense city encourages social coherence and stability: indeed, "who would want to destroy one's own living room"?[2]

1 Géraldine Borio and Caroline Wüthrich, *Hong Kong In-Between*, Hongkong: Park Books: MCCM Creations, 2015.
2 Idem.

1 Géraldine Borio and Caroline Wüthrich, *Hong Kong In-Between*, Hongkong: Park Books: MCCM Creations, 2015.
2 Idem.

밀도 높은 도심지역의
뒷골목은 정의가 불분명한
완충지대다. 열린 공간이자
사적인 공간이고 엄격한
규율이 존재하나 불법행위가
난무하기도 하다.

The interstitial network
of back lanes within the
dense urban fabric are
ambiguous buffer zones
that oscillates between the
classical binary opposition
of public/private, legal/
illegal, inside/outside.

홍콩　　Hong Kong

시간과 공간을 넘나드는
이들의 개입 방식은
파리를 하나의 생태계로
인식함으로써 시민의
주도권에 초점을 둔다.

By considering the city
as an ecosystem, their
ways of intervention cross
the extent of space and
time, all to offer a place
on the initiatives of the
populations.

파리

Paris

선전

Shenzhen

우리는 지금의 도시 구조를
대체할 새로운 공간적
내러티브를 그려내고자 한다.

We hope to choreograph
a new spatial narrative
to superimpose onto the
existing urban structure.

한편, 몇몇 이익 단체는 이 쇼핑몰들을 일상의 유산으로 보존하자는 운동을 벌이고 있다.

On the other hand, several interest groups are campaigning for the conservation of these buildings together with their everyday heritage.

싱가포르

Singapore

로테르담

Rotterdam

미완성 프로젝트는
기존주민의 이탈을 다룬
젠트리피케이션과 도시
단조로움이라는 난제를
넘어서는 역동적인 청사진을
제시하는 살아있는
성명서이다.

Incomplete & Unfinished
is a living manifesto
for a dynamic blueprint
that moves beyond
gentrification and
monotony.

도시성이 방대한 경관을 차지함에 따라 도시라는 개념을 서술하기가 점점 더 어려워지고 있다. 현재의 도시성, 미래 도시에 대한 전망, 방법론에 관한 대화는 우리에게 '도시'라는 용어의 현대적 개념을 논의하는 데에 새로운 시각을 제시한다. 이러한 시각은 공간 개념을 철학적으로 해석하여 도시 현상의 연구에 필요한 과학적 방법론을 만들어낼 수 있다. 예를 들어 주거 양식을 연구한 프라이 오토(Frei Otto)의 작품은 매개변수적 어바니즘(parametric urbanism)의 방식으로 진화했다. 그뿐만 아니라 새로운 스마트 기술로 새로운 분석 도구와 데이터를 파악할 수 있게 되었다. 이러한 분석 도구와 데이터는 21세기 도시 이해의 방식이 되었다. 새로운 분석적 기구와 데이터를 알게 되었고, 이것은 도시를 이해하기 위한 21세기 방법이 되었다. 전 세계 도시의 양적인 특징을 비교하기란 용이하다. '사물인터넷(IoT)' 시대에 환경이 급속도로 디지털화됨에 따라 빅데이터로 도시를 파악하는 일이 가능해지면서 이용 패턴이 드러난다. 이는 도시 관리자들의 도시 운영에 도움을 준다. 데이터는 '스마트 도시'라는 개념으로 잘못 이해되는 일이 많다. 그러나 앙트완 피콩(Antoine Picon)이 언급했듯이 데이터는 인공두뇌나 인공지능에 의해서도 해석될 수 있다. 데이터는 새로운 가상현실이 도시 구성과 집합 공간에 어떠한 영향을 미치는지 보여줄 수 있다. 이 주제로 모인 도시들은 현대의 분석과 정량적 수치와 정성적 환경을 연결하는 연구 방법을 제시한다. 미래 예측과 방법론을 통해 이들은 과거의 모델을 가로질러 미래 현실을 투시하는 목표를 향해 나아가고 있다.

나이로비,
방콕,
바르셀로나,
거대한 계획,
인천,
베를린 + 뭄바이,
타이산,
TYOLDNPAR,
홍콩

The concept of the city is becoming increasingly harder to describe as urbanity takes over an expansive landscape. To have a dialogue among the current urbanities, speculative cities and methodologies disclose new viewpoints for how we can discuss the contemporary conception of the term "city." These can be philosophical interpretations of the idea of space, to scientific methodologies for researching urban phenomena. Research such as Frei Otto's work on settlement patterns has evolved to methods for parametric urbanism, for example. New smart technologies have also been able to reveal new analytical instruments and data, which has become the method of choice in the 21st century for understanding cities. The quantitative qualities of a city can easily be compared across the globe. Out of the era of the "internet of things," there has been a rapid digitalization of our environments that allows understanding the city through big data as a way to reveal patterns of use, helping city managers operate better. Data has often been misconstrued with the concept of "Smart Cities," but as noted by Antoine Picon, can be interpreted through cybernetics or intelligence. Data can demonstrate how the new cyber realities are affecting the urban composition and collective space. The cities collected under this theme present contemporary analysis, and research methods that link quantitative figures to qualitative environments. Through speculations and methods, these cities aim at traversing previous models to project future realities.

Nairobi,
Bangkok,
Barcelona,
Big Plans,
Incheon,
Berlin + Mumbai,
Taishan,
TYOLDNPAR,
Hong Kong

나이로비 시민들은 이러한 이러한 유동성의 흐름을 낳은 케냐의 정치상황에 대해 종종 논쟁을 벌이곤 한다.

The politics that have created Nairobi's mobility flows are often debated amongst the city's residents

THE AUTOMATION MODEL

KATE NANCY WANJURU
uber driver

NELSON MAGWALI
boda boda driver

AWERA NAITORE
matatu conductor

나이로비   Nairobi

Project

17069 USOS
EIXAMPLE
2014-2017

https://300000kms.net/

Author

300.000 KM/S

The main objective of this study is to comprehensively describe urban uses and their impact within the Eixample district.

BARCELONA
POSTS

바르셀로나　Barcelona

정보 전달이란 사물에 형태를 부여하는 일이다. 도시는 인류가 만들어낸 것 가운데 가장 큰 정보 구축물이다. 역사를 통틀어 다양한 체계가 도시 영역을 측정하고 도시의 지도를 제작하며 도시를 조직화하고 계획하며 건설하는 등 도시를 형성하는 데 사용되었다. 건축가, 도시계획 전문가, 공학자의 설계는 공간적 관념을 기하학적으로 표현한 것으로서 결과적으로는 도시 건축물로 구체화된다. 과거 메소포타미아에서 그랬고 현재도 그러하다. 중요한 사실은 이러한 생각이 표현되고 소통되는 도구가 지구상의 문명과 더불어 변화했다는 점이다.

우리는 현실을 더 제대로 이해하고 조직화하며 바꾸기 위해 어떠한 정보든 알고리즘을 통해 디지털로 전환하고 저장하며 활용할 수 있는 시대에 살고 있다.

그렇다면 이러한 사실은 도시를 계획하고 운영하며 바꾸는 과정에 어떠한 영향을 줄 수 있을까?

도시는 도시 통제에 필요한 지도를 제작하고 소유한 사람에 의해 지배된다. 역사적으로는 정부가 그 역할을 했다. 하지만 디지털 시대의 도래와 함께 도시 거주자의 행동, 그들의 이동, 부동산 가치, 주민들의 기호, 다국적 민간 조직에 의해 관리되는 각종 사안 등 다양한 유형의 정보가 발생했다. 이로 말미암아 도시 조직의 공공의식이 사라지고 있다. 이러한 상황 때문에 도시의 암묵적인 의미가 위험에 처할 수 있다. 공동 규칙을 통해 생활, 일, 사회화를 결정하는 물리적 공동체 구조가 사라질 수 있다는 뜻이다. 도시를 정보를 관리하는 다국적 기업의 사업 공간으로만 여긴다면 우리는 시민이라는 자격을 잃고 소비자나 데이터를 생산하는 행위자에 그치게 되어 스스로의 운명을 결정할 수 없게 된다.

우리 시대의 도시가 직면한 큰 도전과제는 우리가 살고 있는 환경에 시민의식을 심는 것과 도시의 빅데이터에 다수를 위한 긍정적인 의미를 부여하는

To inform is to give form to things. Cities are the largest informational constructions that humanity has ever made. And to shape them, throughout history, various systems have been used to measure the territory, map it, organize it, plan it or build it. The plans of architects, urban planners or engineers are geometric representations of a spatial abstraction that finally solidifies to build the structure of cities. So it was in Mesopotamia and it is today. The point is that the tools with which these ideas are represented and communicated have changed throughout the civilizations that have inhabited our planet.

We live in an era where any type of information can be digitized, stored and activated through algorithms to better understand reality, to organize or transform it.

But what effects can it have on the planning, operation or transformation of cities?

The city is dominated by the one that cartographies it, who owns the maps that allow it to be controlled. And historically this has been done by governments. But with the arrival of the digital world there are multiple types of information like the behavior of people in cities, their movements, the value of real estate, the preferences of their inhabitants, and many other issues which are managed by private global organizations, which makes this public sense of the city's organization disappear. And this can put at risk the meme sense of cities, such as the physical construction of a community that decides to live, work and socialize through shared rules. If the city is only seen as a business place, organized by global corporations that manage our information, we cease to be citizens and become either consumers or data producing agents who are unable to lead our own destiny.

This is the great challenge facing cities in our era: being able to give a civic sense to the environment where we live and give a positive meaning to the majority of the urban big data. The

것이다. 도시를 조직하는 법률은 도시 조직을 지배하는 알고리즘이다. 과거에는 시장이 웹마스터 역할을 했으며 현재는 공동체가 웹마스터다.

도시에서 디지털 정보가 제공하는 기회를 창출하는 측면에서 내가 더 흥미롭게 여기는 것은 다음과 같다.

1   도시 기본 계획의 미래는 디지털 시뮬레이터, 즉 매개변수를 통한 지도 제작이다. 이제 도시는 느린 속도로 도시화되다가 정지된 사건이 아니다. 구축된 도시를 운영한다는 것은 모든 사람과 모든 사물이 정보를 내뿜고 소비하는 체계에서 작업을 한다는 뜻이다. 다시 말해 도시의 변화는 더 이상 도시 개혁에 의해서만 이루어지는 것이 아니라 (주택 임대나 자동차 공유 등의) 애플리케이션이나 플랫폼의 실행과 같은 행동에 의해서도 이루어진다는 뜻이다. 따라서 아무런 예고 없이 변화가 발생할 수 있다. 그러므로 도시는 프로젝트와 그 영향 사이의 관계를 평가할 수 있는 공공 도구로 도시의 시뮬레이터를 구축해야 한다. 이를 통해 프로젝터의 혜택과 그 경제적, 사회적, 환경적 비용을 기능적, 공간적, 환경적 차원에서 분석해야 한다. 소수의 집단이 직관적으로 도시 투자를 결정하는 것만으로는 더 이상 도시를 제대로 운영할 수 없다. 도시 건설은 미래를 시각화하고 공동체의 이익을 결정할 권한이 있는 수많은 행위자가 하는 공동 창조 작업이다.

2   데이터는 창출하는 사람이 소유한다. 또는 공개적이고 투명하게 옮긴 사람들의 필요에 따라 이용된다. 이것이 이 시대의 큰 도전과제라는 점에는 의문의 여지가 없다. 이를 위해서 우리는 도시 데이터의 관리를 규제하고 조직할 수 있는 국제적 합의를 새로운 형태로 이루어내야 한다. 오늘날 우리는 정보가 편의에 따라 관리되는 경우 선거가 조작되고 부동산 가격이 조정되며 에너지 가격이 결정되고 관광지가 뜨거나 침체되는 광경을 목격하고 있다. 따라서 도시가 민간 조직들이 시민의 행동이나 미래의 삶을 (거의 항상 원격으로) 은밀히 거래하는 중심지로 변모하는 것을 막으려면 새로운 형태의 정보 관리 체제를 만들어낼 필요가 있다.

3   정부도 기업처럼 혁신적이어야 한다. 그렇지 못하면 정부는 산업이 제공하는 것의 소비자에 그치게 된다. 대기업은 일개 도시보다 더 강력하지만 세계

laws that organizes the urban are the algorithms that govern the organization of cities. Before the web master was the mayor, now it is the community.

The aspects that I believe are more interesting to develop the opportunities offered by digital information in cities are:

1   The urban Master plane of the future is a digital simulator, a parametric cartography. The city is no longer a static event that transforms at the speed of slow urbanization processes. Operating on a built city implies working on systems where everyone and everything emit and consume information. And this means that the changes of the cities are no longer only produced by urban reforms but also by the behaviors, the implementation of applications or platforms (for rent, to share vehicles, etc.) that can change everything without prior notice. Therefore, cities should build urban simulators as public tools to assess the relationship between projects and their repercussions, analyzing the benefits and economic, social or environmental costs of decisions that can be taken at functional, spatial or environmental levels. Intuitive decisions made on urban investments by a small group of people are no longer sufficient for the good governance of cities. The construction of the city is an act of co-creation among many actors, who must be able to visualize possible future and decide for the benefit of the community.

2   The data is own by those who produce it. Or in their case by those who are openly and transparently transferred. This is undoubtedly one of the great challenges of our time. And for this we must make a new type of international agreements that regulate and organize the management of urban data. Today we see how elections are manipulated, the price of real estate can be modified, the price of energy is decided or tourist destinations can be boosted or sank if information is managed conveniently. Therefore, it is necessary to create new forms of information governance that prevent the transformation of cities into centers where private organizations traffic (almost always remotely), with the behavior of citizens and the future of their lives.

3   Governments must be as innovative as corporations. If they are not, they become mere consumers of what the industry offers them. Large corporations are more powerful than an isolated city, but if ten leading cities in the world decide to create standards related to tourism, housing or

10대 도시들이 관광, 주택, 교통 이동 관련 기준을 정할 경우에는 기업들이 그 요구를 충족하기 위해 노력할 것이다. 그러나 도시가 대중의 요구를 규명하지 않는다면 기업이 자신의 이익에 맞게 발전 속도를 정하게 된다. 바로 이러한 이유에서 우리는 전 세계 국가 조직이 시민들의 현실과 동떨어진 이익 집단으로 변모하는 듯한 현재에 전 세계 도시로 구성된 조직을 만들어 스스로의 운명을 체계화해야 한다.

4   도시를 디지털화하면 광범위하고 투명한 도시 모델을 만들 수 있다. 도시에 적용된 인공 지능은 집합 생활에 위기도, 기회도 될 수 있다. 그러나 도시에서 다층적인 디지털화가 개발되고 서비스로 제공됨에 따라 이를테면 '에너지 블록체인' 같은 기술이 탄생하여 에너지 효율이 높은 동네를 개발할 수 있게 되었다. 이러한 곳에서는 모든 건물이 투명한 전용 플랫폼을 통해 에너지를 생산, 소비, 저장, 공유한다. 도시에 배치된 정보의 층과 데이터 관리는 더 큰 목표를 위한 수단이 되어야 한다.

5   기술은 기후 위기를 해결하는 데에 사용되어야 한다. 도시는 현실을 관리하는 데 안주해서는 안 되며 앞으로 다가올 미래에 대한 구상도 갖춰야 한다. 그렇지 않으면 '현상 유지'만을 하는 상태로 운영되고 보다 집합적인 미래를 향한 걸음을 내딛지 못하게 된다. 우리 시대에 가장 중요한 문제는 지구의 기온 상승을 불러와 불행한 결과를 초래할 가능성이 있는 기후 위기에서 비롯된다. 도시 빅데이터는 도시 변화 과정을 평가하고 무배출(Zero Emissions)을 기반으로 한 도시 모델을 도출하는 주요 사명을 이미 수행하고 있다.

6   도시에 필요한 것은 느린 속도, 중간 속도, 빠른 속도로 생산되는 정보가 다양한 민간 행위자나 공공 행위자에 의해 통합되는 도시 운영 체계다. 이러한 공공 정보 플랫폼은 도시 생활과 연관된 각 부문이나 부서가 각자의 정보만 관리하는 수직적 사일로(silo, 원통형 창고 또는 부서 간의 장벽)를 방지하기 위해 통합된 방식으로 구축되어야 한다. 정보가 도시와 별개로 운영되는 클라우드 형태를 띠어서는 안 된다. 그보다는 현실에 맞게 다시 현지화되고 시민 생활의 지속적인 개선에 도움을 주어야 한다.

mobility, companies will work to meet that demand. But if there is no definition of demand from the public, it is companies that set the pace of progress based on their interests. That is why we must create global organizations of cities that organize our destiny, once the global organizations of countries seem to have transformed themselves into interest groups far from the concrete realities of the citizens of the world.

4   The digitization of cities allows developing distributed and transparent urban models. Artificial intelligence applied in cities can be a risk or an opportunity for collective life. But thanks to the development of multiple layers of digitization in cities and their services, it is possible to develop, for example, energy-efficient neighborhoods thanks to the "blockchain of energy", where all buildings produce, consume, store or share energy through transparent platforms created for that purpose. The informational layer and data management that we have deployed in cities must be a tool for greater objectives.

5   Technology must be put at the service of the climate crisis. Cities can not live only to manage reality, but must have a vision of the future to know what to do tomorrow. If not, they would only work based on "business as usual" and not oriented to the progress towards a better collective future. The most important challenge of our time is related to the climate crisis that causes the planet's temperature to rise with potentially dire consequences. Urban Big data of cities already has a key mission that is to evaluate the process of transforming cities and progress towards an urban model based on Zero Emissions.

6   The cities need a City Operating System, in which information produced at low, medium and high speed is integrated bymany private or public agents operating in a city. This public information platform should be structured in an integrated way, overcoming the vertical silos where each field or department linked to urban life manages its own information. The information should not become a cloud that operates separately from the cities, but should be re-localized, down to earth, and be oriented to the constant improvement of citizens lives.

도시설계는 유토피아적, 남성적, 권위주의적인 것으로 여겨지며 한때 배척되어온 적이 있다. 그러나 우리는 이에 동의하지 않으며 이에 대한 담론을 중요시한다. 도시는 개발과 발전수준에 대해 저마다의 출발점과 판단기준을 갖고 있기 마련이다.

The design of cities in their all-at-once has become anathema, deplored as "utopian" (a concept itself, sadly, in bad odor), masculinist, authoritarian, etc., etc. We disagree and think this discourse is vital, not only imaginatively but on the ground: every city has a point of departure, a set of governing ideas that limn the terms – however constricting or loose – for its development and maturation.

HOUGUAN
(WUHAN)

거대한 계획

Big Plans

인천   Incheon

장소 기반의 '인터미디어' 이야기

Site-based intermedia narratives

Mumbai's two worlds

베를린 + 뭄바이

Berlin + Mumbai

385 ELEVATION

INCREMENTAL URBANISM
EVOLVING COLLECTIVE FORMS IN TAISHAN, CHINA

GONGYI XU

SANBA XU

TINGJIANG XU

Traditional Dwellings 三間兩廊式民居

Traditional "3 room-2 corridor house" is a typical residential type in Taishan region, with a mix structure including brick walls and wooden beams.

三間兩廊式民居是臺山地區的典型住民形制，為木屋型磚石土牆承重的混合結構。

Datong Bridge 大同橋

Built in 1930, Datong Bridge exerted a significant part in connecting the Datong market and Nanbian farm-village and booming the economy.

大同橋約1930年建成，是連通大同市集與南邊農村活在重要交通樞紐並幸繁榮了經濟的橋梁。

타이산의 집합적 형태를 들여다봄으로써 우리는 이 도시가 겪은 역사적 우여곡절을 살펴볼 수 있으며, 거시적 차원과 미시적 차원, 유형과 무형, 과거와 미래를 아우르는 기억의 조각들을 하나의 큰 과정으로 이해할 수 있다.

The mapping of the collective form exhibits the wear and tear of the Taishan's historical narrative and bridges fragmented memories spanning across the macro to the micro, tangible and intangible, back and forth, in a dialogue that informs the whole process.

386.

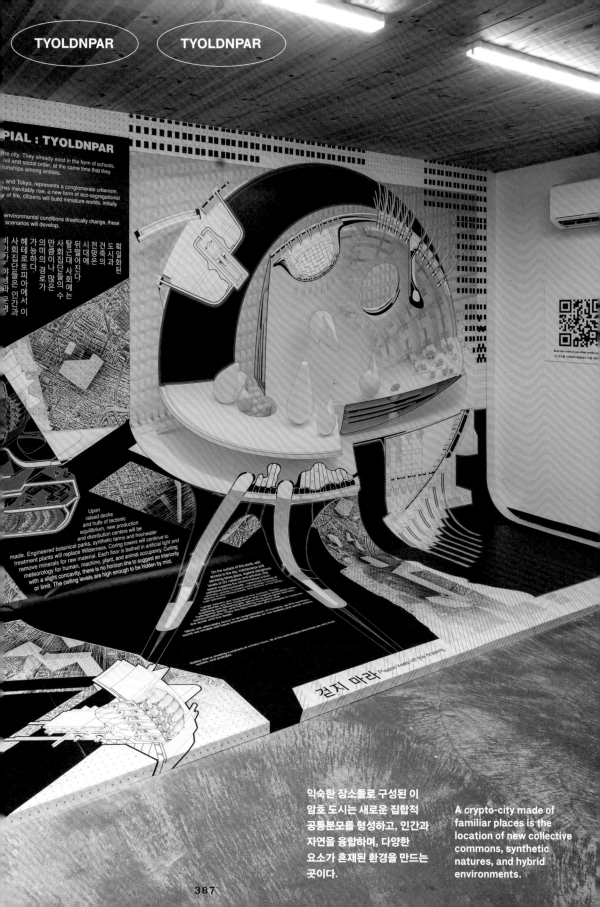

PIAL : TYOLDNPAR

...e city. They already exist in the form of schools,
...ivil and social order, at the same time that they
...onships among entities.

... and Tokyo, represents a conglomerate urbanism.
...res inevitably rise, a new form of eco-segregationist
... of life, citizens will build miniature worlds, initially

...environmental conditions drastically change, these
... scenarios will develop.

Upon
raised decks
and hulls of tectonic
equilibrium, new production
and distribution centers will be
made. Engineered botanical parks, synthetic farms and freshwater
treatment plants will replace Wilderness. Coring towers will continue to
remove minerals for raw material. Each floor is bathed in artificial light and
meteorology for human, machine, plant, and animal occupancy. Curling
with a slight concavity, there is no horizon line to suggest an interiority
or limit. The ceiling levels are high enough to be hidden by mist.

익숙한 장소들로 구성된 이
암호 도시는 새로운 집합적
공통분모를 형성하고, 인간과
자연을 융합하며, 다양한
요소가 혼재된 환경을 만드는
곳이다.

A crypto-city made of
familiar places is the
location of new collective
commons, synthetic
natures, and hybrid
environments.

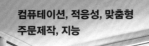

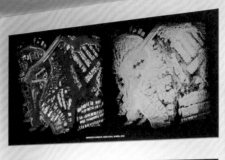

홍콩　　　　Hong Kong

집합적인 도시기억과 레이어

[10]

다양한 시기를 거쳐 진화한 도시들은 다양한 정치 제도, 산업 방식, 이민 유입, 수십 년의 성장을 통한 기반시설 발전 등을 경험했을 것이다. 이러한 속성들은 도시의 네 가지(사회적, 경제적, 정치적, 기술적) 측면으로 나타난다. 이러한 측면들은 도시를 경험하고 이해하는 다양한 방식을 펼쳐 놓는다. 학생이 도시를 대하는 태도는 관광객이나 공무원과는 다를 것이 분명하다. 인류학적인 역할을 상정했을 때 해당 범주 내에서 전시되는 도시들은 도시의 독특한 특징과 집합적 기억을 만들어 내는 다양한 층을 드러내 보인다.

로스엔젤레스,
하노이,
아바나,
프랑크푸르트,
암스테르담,
잔지바르,
코펜하겐,
밀란,
상트페테르부르크

Cities that have evolved through different periods may have experienced different political regimes, industrial patterns, an influx of immigration, and infrastructural developments over decades of growth. These attributes represent four different dimensions of a city (social, economic, political and technical). These dimensions unfold multiple ways of experiencing and understanding a city. A student will address the city differently than a tourist, or a civil servant will. Assuming an anthropological role, the cities presented in this category unearth the different layers that generate unique urban characters and collective memory.

Los Angeles,
Hanoi,
Havana,
Frankfurt,
Amsterdam,
Zanzibar,
Copenhagen,
Milan,
St. Petersburg

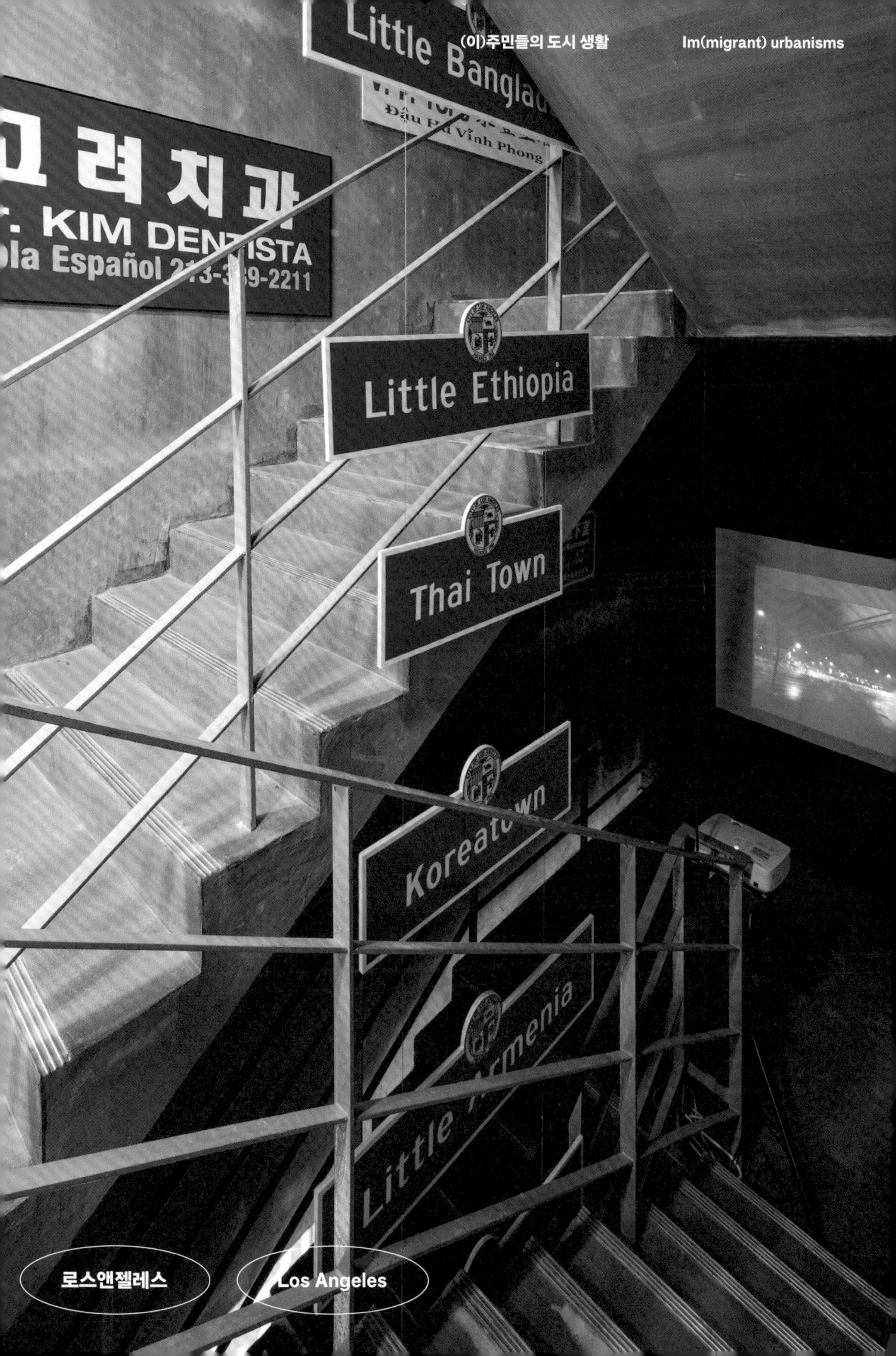

로스앤젤레스

Los Angeles

도시에 대한 기억은 시간과
끊임없이 이 도시로 몰려드는
사람들의 물결과 함께 소멸해
간다.

Memories of the city
keep vanishing with time,
with the flood of people
immigrating tirelessly into
this city.

하노이 Hanoi

재현 작업이 그저 수동적인
반영에 그치지 않는다는
사실을 보여주며, 도시의
재현과 도시의 새로운
모습을 적극적으로 창조하기
위한 노력 간의 상관관계를
제시한다.

Representations are
complicit in not only
passively representing a
city but also in actively
creating one.

아바나

Havana

건축의 기원으로서의 도시

피터 트러머

The City as Origin of Architecture

Peter Trummer

도시는 우리가 알고 있는 가장 큰 유물이다. 도시는 가장 위대한 예술 작품이다[1]. 도시는 질베르 시몽동이 제시한 개념인 네 겹으로 이루어진 개체화 세계(물질[The Physical], 생물[The Biological], 정신[The Psychic], 집합[The Collective])에서 가장 마지막 층으로 볼 수 있다. 태초에 지구는 달과 더불어 물리적 실체로서 태양계에 나타났다. 물질세계가 탄생한 후에는 식물로 이루어진 생물세계가 출현했다. 인간종의 출현과 더불어 생물세계에서 정신세계가 탄생했으며 그 이후 4가지 개체화 과정의 마지막 단계인 집합세계가 나타났다. 마누엘 데란다는 집합이 건물, 거리, 유통을 위한 다양한 행동이 존재하는 상황에서 사람, 네트워크, 조직체들의 조합인 도시를 뜻한다고 말했다.

거대객체로서의 도시

도시는 인간에 의해 일방적으로 만들어진 것이 아니다. 그리스 문명이 그리스 도시 국가를 재창조했다는 아리스토텔레스의 명언과 같이 도시는 우리 인간을 재창조하기도 했다. 아리스토텔레스는 "... 따라서 이러한 사실은 국가가 자연의 창조물이고, 인간은 본질적으로 정치적 동물이라는 증거다..."[2] 라고도 말했다. 도시가 존재함에 따라 인간은 비로소 자연계의 동물에서 정치적 동물로 변화한다. 도시는 정치적 주체인 인류의 발상지이자 정치적 삶(biopolitikos) 또는 한나 아렌트가 말했듯이 실천적 삶(vita activa[3]), 즉 공공 정치 사안에 전념하는 삶에 할애된 공간이다. 도시는 인간과 관련된 모든 실체를 만들어냈다. 여기에는 언어, 법률, 정치, 문화,

The City is the largest human Artefact we know. The City is the greatest work of Art.[1] The City can be seen as the last layer of Gilbert Simondon's fourfold process the individuation of the world: "The Physical The Biological, The Psychic and The Collective". First our planet emerged as a physical entity with its moons and within our solar system. On top of the physical world rose the biological world, with its faunas and floras. Out of the biological world emerged the psychological one, with the emergence of the human species and the last stage of the fourfold process of individuation is the emerging of collectives or as Manuel De Landa describes cities, "assemblage of people, networks, organisations, as well as a variety of infrastructral components, from buildings and streets to conduits for matter and energy flows."

The City as a Hyperobejct

The City is not only made by humans, but also made us, humans, as mentioned by the very well-known passage of Aristoteles in which the Greek Civilisation reinvented the Greek City State he states: "... hence it is evident that the state is a creation of nature, and that man is by nature a political animal..."[2] With the City man is turning from a natural animal into a political one. The City was not only the Birthplace of us as a political subject, a place of "bios politikos" or as Hannah Arendt calls it "vita activa"[3], a life devoted to public-political matters. The City invented all kinds of human entities like Language, Laws, Politics, Cultures, Capitalism, Double entry bookkeeping, Tragedy, Drama, Gambling, Banking, Planning, Real estates, Properties, Plans, Geometry, Bureaucracy, Processions, Tax, Credits, Tolls, Armies, Newspaper, Luxury, Traffic, Manners,

1 Lewis Mumford, *The Culture of Cities*. New York: Harcourt, Brace & Co., 1938, p. 5.
2 ibid.
3 Hannah Arendt, *The Human Condition* (2nd ed.), Chicago: University of Chicago Press, 1998, p. 7.

1 Lewis Mumford, *The Culture of Cities*. New York: Harcourt, Brace & Co., 1938, p. 5.
2 ibid.
3 Hannah Arendt, *The Human Condition* (2nd ed.), Chicago: University of Chicago Press, 1998, p. 7.

자본주의, 복식부기, 비극, 드라마, 도박, 금융, 기획, 부동산, 자산, 계획, 기하학, 관료제, 행진, 세금, 신용, 통행료, 군대, 신문, 사치재, 교통, 예의, 속어, 매독, 빈민가, 위생, 매춘 알선, 소매치기, 인쇄, 복제, 표준화, 사생활, 협동적 소유, 전력망, 심리학, 교육, 보육, 보건, 독점, 협동, 폭력단, 사회, 직업 조합, 기관, 생산 방식, 경찰, 치안판사, 행정, 매춘, 혁명, 프롤레타리아, 교도소, 노동의 사회적 분업, 이방인, 소시민, 상인, 장인, 강도, 관중, 가게 주인, 노동자, 직원, 거지, 한량, 도둑, 원자적 개인, 그 밖의 셀 수 없이 많은 것들이 포함된다[4].

우리 인간은 도시를 만들었고, 도시는 우리를 만들었다. 우리가 도시에 대해 알면 알수록, 도시에 대해 이해할 수 없는 것도 많아진다. 도시는 티머시 모턴이 말한 거대객체(Hyperobejct)[5]가 되어가고 있다.

하지만 도시는 인간뿐만 아니라 그 이외 모든 존재에게도 거대객체다. 도시는 건물로 구성된 물리적 실체이지만 건물 역시 도시를 통해 만들어진다. 도시는 온갖 새로운 유형의 건축물이 만들어지는 장소이다. 건축물 유형으로는 극장, 신전, 법원, 학술원, 실내경기장, 은행, 호텔, 인술라(insula, 고대 로마의 다세대 주택), 방어벽, 박물관, 돔과 첨탑, 병원, 대중목욕탕, 사우나, 비축창고, 감옥, 정신병원, 궁궐, 국고, 도박장, 학교, 조합집회소, 시청, 대성당, 양로원이나 고아원, 대학교, 소극장, 쇼핑몰, 성채 도시, 요새, 수용소, 창고, 콘서트홀, 무도회장, 스포츠클럽, 체육관, 금융시장, 증권거래소, 도서관, 기록 보관소, 검역소, 폐기물 공장, 철도역, 축구 경기장, 주차시설, 병영, 무기고, 창고, 정박 시설, 항구, 공항, 도시 별장, 마천루, 위니테 다비타시옹, 베르사유 사랑의 신전, 데이터 센터, 시장, 묘지, 거리, 골목, 도로, 시민 광장(piazza del popolo), 광장, 오지, 미술관, 통로, 정원, 테마파크, 대로, 공공녹지, 공원, 유원지, 동물원, 볼링장, 축구장뿐만 아니라 풍경, 건축물, 아파트, 원룸, 사무 공간, 백화점, 응접실, 거실, 복도, 개인용 화장실, 유리창(중세의 스테인드글라스),

Vulgarism, Syphilis, Ghettos, Sanitation, pimping, pick pocketing, printing, copying, standardisation, Privacy, Cooperative ownership, Grids, Psychology, Education, Child care, Health care Monopolies, Cooperations, Mobs, Society,Guilds, Institutions, Modes of Production, The Police, Magistrates, Administration, Prostitution, Revolutions, Proletariats, Slams, Social division of Labor, the Stanger, the Burgher, the merchant, the Craftsman, the Robber, the Spectators, the Shopkeeper, The worker, The Employees, the Beggars, the Flaneur, the Thieves, the Atomic individuals and endlessly more.[4]

We humans not only made the City, the City also made us. As more as we know about the City as more the City withdraws from our understanding of it. The City has turned for us into what Timothy Morten calls a Hyperobejct.[5]

But the City is not only a Hyperobejct to us humans. It is a hyperobject to all non-human entities as well. As a physical entity, the City is not only made out of buildings, buildings are made through the City. The City is the birthplace of all kind of new forms of Architecture like: the Theater, the Temple, the Courthouse, the Academies, the Gymnasiums, the Bank, the Hotel, the Insulae, its Defence Walls, the Museums, the Dome and Spire, the Hospital, the bath house, the therms, Storage buildings, prisons. the Mad house or psychiatric, the Palace, the Exchequer, the Casinos, Schools the Guild halls, Town halls, Chatedrals, Asylums for ages and orphaned, Universities, the Play house, the Shopping malls, the Burg, the Fortress, the camp, Warehouses, The Concert hall, The ballroom, the Athletic Club, The Gym, the Bource, the Stock Exchange, Libraries, Archives, Quarantines, Waste plants, Railway stations, Football Stadiums, Parking garages, Barracks, Arsenals and Magzines, Storehouses, Docks, Harbours, Airports, Urban Villas, Skyscrapers, Unite d'Habitations, Temple of love, Data centers, The market, Graveyards, The street, The Alley, The Boulevard, The piazza, The Campo de Popolo, The Square, The hinterland, The Galleria, the Passages, the Gardens, Theme Parks, Grand avenues, the Public green, Parks, Pleasure gardens, zoological garden, Bowling ground, Football field, Vistas and architectural entities like

4   Lewis Mumford, *The Culture of Cities.*
5   I refer her to Timothy Morton, *Hyperobjects: Philosophy and Ecologyafter the End of the World* (Posthumanities, Band 27), University of Minnesota Press, 2013.

4   Lewis Mumford, *The Culture of Cities.*
5   I refer her to Timothy Morton, *Hyperobjects: Philosophy and Ecologyafter the End of the World* (Posthumanities, Band 27), University of Minnesota Press, 2013.

보관함, 기숙사, 분수 등을 들 수 있다.

건축은 문자 상으로뿐만 아니라 개념상으로도 우리 사람과 도시 사이에 있다. 내가 건물 안에 있다고 생각해보자. 말 그대로 벽을 통해 세상을 내다 볼 수 있다. 건축은 세상에서 공간을 잘라내서 안과 밖을 구분해준다. 건축은 실내에 벽, 지붕, 바닥, 창문, 문 등의 모든 건축학적 요소를 압축해 새로운 세상을 창조한다. 인간 중심적인 세상에 사는 우리는 인간적 관점으로 도시 건축을 해석해왔다. 아니면 개인적 관점, 작가적 관점, 건축가와 프로젝트의 관점뿐만 아니라 광범위하게는 폴리스(Polis, 고대 그리스의 도시국가), 집합체, 사회를 통해 만들어진 구체적 추상화의 관점에서 이해하려고 노력했다. 그러나 우리가 도시의 건축을 인간의 주체적 관점이 아니라 인간 이외의 객체적 관점, 즉 도시 그 자체의 형식적인 특징이라는 관점에서 보는 것은 어떠할까?

건축의 기원

기원전 2~3세기에 메소포타미아에서 도시가 나타나기 시작했을 때 도시의 안과 밖을 가르는 벽은 도시의 특징이 되었다. 도시와 시골의 경계는 단순히 세상을 둘로 나누는 역할을 했을 뿐만이 아니라 그 이전에는 존재하지 않았던 종류의 지식인 건축을 생산했다는 점에서 더 큰 중요성을 띤다. 베네볼로(Benevolo)는 유럽 도시에 관한 저서에서 다음과 같이 썼다.

> "도시는 담장으로 둘러싸인 곳이다. 또는 그러한 장소의 연속체다. 그 안에서 중간 거리와 짧은 거리를 조작하는 기술(이후에 '건축'으로 표현되는 기술)이 발달한 반면에 무한한 경관을 차지하고 변형하는 데 사용된 옛 기술은 서서히 잊혔다"[6]

베네볼로는 도시의 탄생과 새로운 지식인 건축이 도시 성벽 내부에서 새로운 공간을 창조해낸 반면에 영역과 도시 외부의 공간에 대한 옛 지식은 서서히 사라졌다고 주장한다.

실제로 베네볼로는 더 나아가 도시, 성벽, 성벽

the flat, the one room apartment, the office space, Department store, the drawing room the salon, the corridor, the private toilet, the Glass window (Rippen window Middle Ages), the Storage box, Dormitories and even Fountains.

Architecture lies literally and conceptually between us, humans, and the City. Imagine you are inside a building. You literally can see through the wall into the world. Architecture cuts out a space from this world and divides the inside from an outside. Architecture creates a new world as an interior, encapsulated by all its architectural elements: the wall, the roof, the floor, the window and its doors. Within the anthropocentric lifetime we have tried to read the Architecture to the City from the viewpoint of us humans: either from the Individual, the Author, the Architect and his Projects or from concrete abstractions given through the Polis, the Collective or the Society at large. But what if we look to the Architecture of the City not from the viewpoint of a human subject, but from a non-human object, namely from the formal properties of the City itself.

The Origination of Architecture

When the City came into existence in the third and second millennium BC in Mesopotamia it was characterised by the a wall which divided the world between an inside and an outside. The boundary between the City and the Country side, not only divided the world in two parts, but more importantly produced a kind of knowledge, one which didn't exit before, called Architecture. In his book on the European City, Benevolo writes:

> "The City is an enclosure, or a series of enclosures, in which the art of manipulating medium and short distances - that which from then on would be described as 'architecture' — reached maturity, while the older art, which aimed at the occupation and modification of the unlimited landscape, was gradually forgotten."[6]

Benevolo argues that with the birth of the City, this new knowledge called architecture was given to the new space created within the City-walls, while the older knowledge of the territory, the space outside

---

6   Paul M. Hohenberg, Leonardo Benevolo, *The European City*, (Trans.) Carl Ipsen, Cambridge: Blackwell, 1993.

6   Paul M. Hohenberg, Leonardo Benevolo, *The European City*, (Trans.) Carl Ipsen, Cambridge: Blackwell, 1993.

사이의 공간이 도시 내 건축의 형식적인 기원이라고 주장한다. 베네볼로는 도시 내에의 주택, 궁전, 사원은 제각각 담장으로 둘러싸여 있고 서로 얼마만큼 떨어져있는지에 따라 중요성이 가감된다고 말한다. 베네볼로가 주장하려는 바는 건물의 형식적 구상이 건축가나 좀 더 광범위하게 사회에서 나온다기보다 다른 객체의 형식적 특성인 도시 자체의 형식적 특성에서 비롯된다는 것이다. 한마디로 도시의 탄생과 더불어 건축의 형식적인 기원이 시작되었다.

the city walls slowly disappears.

Benevolo goes actually even a step further in his argument and claims that the city, its enclosure walls with its space in-between, acts as the formal originator of the Architecture within the City itself. He states that within the City, each house, each palace and each temple are enclosures in themselves and become important due to their degree of segregation from each other. What Benevolo seems to argue, is that the formal idea of buildings do not emerge from either an architect nor from the society at large, but from the formal properties of another object, namely from the formal properties of City itself. With the birth of the city, architecture found its formal originator.

Image credits:
The Architecture of HyperCities
Project by Peter Trummer Architect
Credits: Peter Trummer with Lida Badafreh, Mehrshad Atashi and SAC.

HyperCity

Frankfurt

하이퍼시티는 인류가 가진 가장 거대한 유물이며 이 유물은 다양한 모습의 도시들로 이뤄진다.

HyperCity - The City is the largest human Artefact. This Artefact is formed by multiple Cities.

프랑크푸르트　　　Frankfurt

Building in a brownfield site or
the city, at the height of a real e
impossible but also an opportu

도시의 인기 없는 쪽에 위치한 오
짓는 다는 것은, 특히 부동산 위기
불가능해 보였지만 그러나 일종으

I was successively a develope
area promoter and now reside

나는 개발업자로 시작해서, 단계
연구가, 지역 프로모터를 거쳐서,
(Buiksloterham) 지역의 주민입

Beforehand, residents could b
that they could have a say in th
building.

이전에는, 주민들이 건물의 디자
표출할 권리를 가질 수 있다고 믿

Buiksloterham is not top-dow
The self-builders and institutio
through an intermediate layer
professionals.

바우크슬로터함 (Buiksloterham
상향 방식도 아닙니다. 자가주택-
공동집단체 및 전문가란 중간 계
작업합니다.

Residents' involvement in thei
into a sense of neighborhood,
fun initiatives.

주택 소유자인 거주자들의 참여는
공동체 감정으로, 그리고 영리하고
구상안들로 나타납니다.

암스테르담   Amsterdam

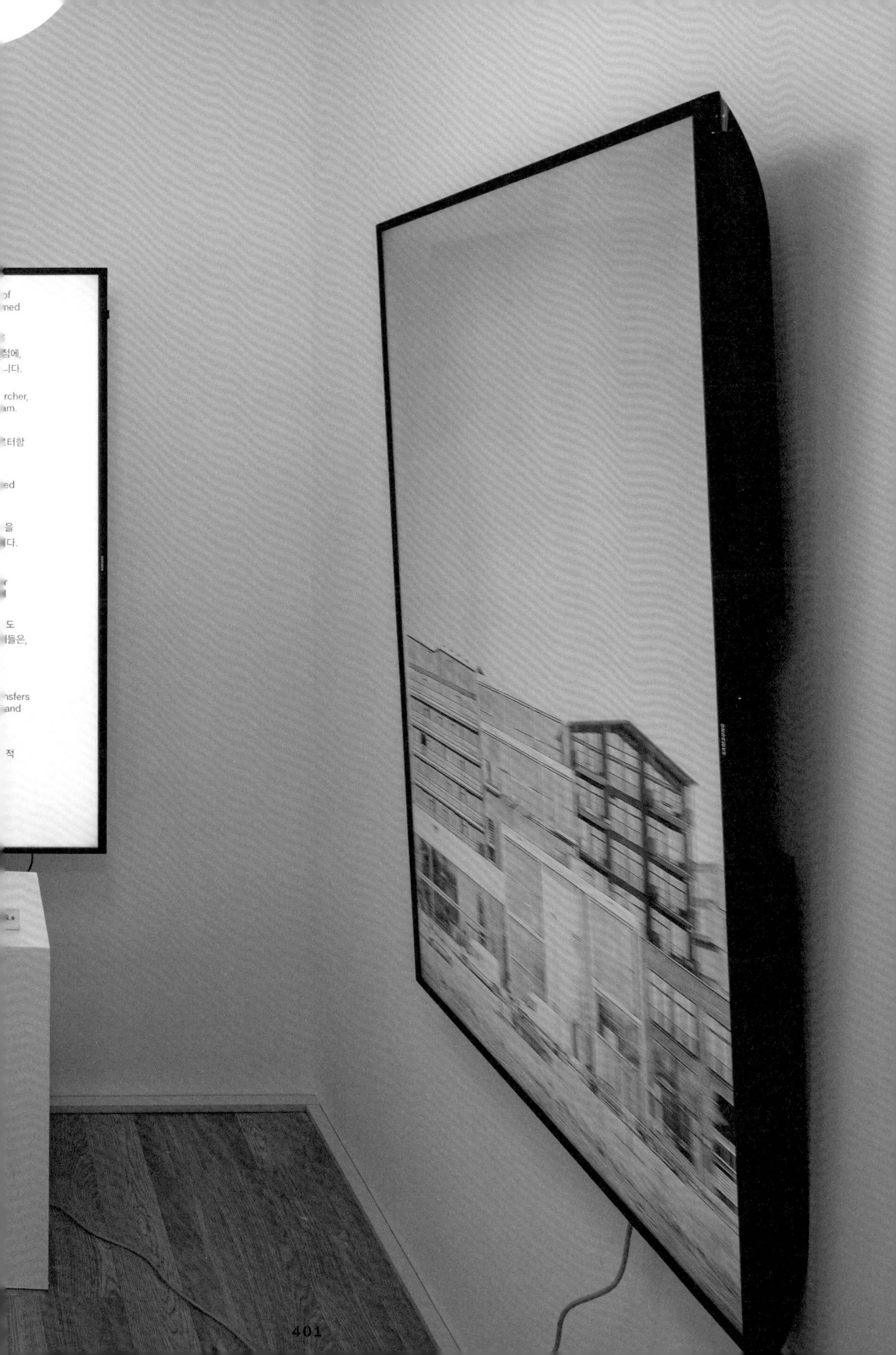

한때는 잔지바르 타운의 역동적인 문화 중심지였으며 '행복한 거리'라는 별명이 붙었던 응암보 지역은 스톤타운 만큼이나 역사적 의미가 있는 곳이다. 응암보 지역에 대한 지식을 보다 많이 연구하여 이곳을 새로운 도심으로 만들기 위한 미래 전략의 기초를 다질 수 있을 것이다.

Once a culturally vibrant heart of Zanzibar Town, known by some as 'the happy streets', Ng' ambo has proven to be as historically as significant as Stone Town. Knowledge that is now considered the basis for the future plans for the area, the new city centre.

잔지바르

Zanzibar

코펜하겐   Copenhagen

Noise Pollution Zoom Map

PLANNING
STRIP SYSTEM

Water Zoom Map

PLANNING
ROOF GARDEN

Water Zoom Map

PROJECT

PROJECT

Infrastructures Zoom Map

ROUTES

Ground Surface Zoom Map

River Zoom Map

MORPHOLOGY

Milan exemplifies several features characterizing modern urbanism and architecture. The natural, social, cultural and economic milieu – as well as the dramatic historical events of the 20th century – have promoted the implementation – albeit an experimental one at times – of ideas whose span stretches from the second half of the 1800's the end of the 1900's. This work suggests various outlooks for describing such phenomena; every interpretation here hinted at enables a description of the modern city as well as being a link to its history. Thus, every reading becomes a distinct representation of the city, sometimes already offered by example, cartographies, studies and designs here only partially mentioned. These descriptions example of how one might consider Milan as a city embodying "modernity". Despite ning uniformity, Milan – due to its stratified complexity – is suitable for experime iptive techniques focusing on how the modern city arises from an ongoing overla nsition of parts and urban fragments. It is noteworthy how these elements hav alized independently from plans and projects considering the city as a

AN. BUILDING CONTEX

우리는 또 다른 층을 만드는 것이 아니라 현재의 도시를 개선해야 한다.

At the present time we do not want to create another layer, but instead improve the city we already have.

HISTORICAL CENTRE
8% OF ST. PETERSBURG TERRITORY

Pre-revolutionary historical quarters. Dense urban environment with some empty sites due to war bombing, partial demolition. Potential to transform blank walls and fenced off spaces into active areas.

Buildings are protected objects of cultural heritage
Yards of various sizes
Old buildings, which need to be reconstructed

BEFORE

AFTER

INDEX BEFORE   INDEX AFTER

GREY BELT
11.8% OF ST. PETERSBURG TERRITORY

Mostly non-residential industrial and warehouse areas surrounding the historical core of the city. Mostly inefficiently used. Potential for mixused quarters with unique identity.

Close to the city center
Pronounced industrial identity of architecture
A large number of rules and regulations for new construction

Unique planning structure for each district
Large unstructured territories
Vast unused land

BEFORE

AFTER

INDEX BEFORE   INDEX AFTER

«TEMPORARY CITY»
5.8% OF ST. PETERSBURG TERRITORY

Individual houses and townhouses: urban villages within the city th
Potential to save unique morphotype.

Hidden within urban fabric
Low-rise buildings. Human scale environment
Mainly private houses for one family

Green, but underd
Poorly developed
Low density

BEFORE

AFTER

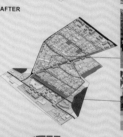

INDEX BEFORE

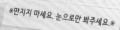

※만지지 마세요. 눈으로만 봐주세요.※

상트페테르부르크   St. Petersburg

집합도시 | 도시수집: 도시의 기억, 상상의 도시

이승택

Collective City | Collectable City: Memory of City, Imaginary City

Peter Lee

"건축과 도시라는 거대한 주제가 일반 시민들에게 다가갈 수 있을까?"
"80여 개의 도시가 '도시전'이라는 하나의 전시 경험이 될 수 있을까?"
"도시라는 거대한 시스템 속에서 개인이 자신만의 집합을 찾는 것이 가능할까?"

2019 서울도시건축비엔날레 도시전을 찾는 방문객의 경험을 설계하기 위해 풀어야 했던 세 가지 숙제였다. 도시전은 80여 개 나라가 참여한 만큼 다양한 전시가 펼쳐진다. 그것만으로도 충분한 볼거리가 있지만 방문객이 도시전을 경험하며 건축과 마주하는 방법, 나아가 도시와의 관계를 다시 생각하게 만드는 작업이 필요했다. 집합도시라는 비엔날레의 주제를 개인의 이야기로 끌어오는 것, 건축과 도시를 하나의 전시로 꾸려낸 시스템 속에서 개인이 뚜렷하게 도드라지는 방법을 고민하기 시작했다.

사람들은 도시와 건축물 안에서 살아간다. 그러기에 도시에서 벌어지는 일상은 특별할 것 없는 일상이다. 그러나 건축과 도시라는 단어를 언급하는 순간 전문가만이 다룰 수 있는 거대한 개념으로 변신한다. 평범한 사람들의 일상에서 멀어지는 것이다. 나 역시 건축물 안에서, 도시 안에서 살고 있지만 건축과 도시를 스스로 설명할 수 없었다. 그저 내 주변의 환경일 뿐 구체적인 대상으로 생각하지 않았기 때문이다. 도시와 건축 속에서 살고 있지만 없는 것이나 마찬가지였다.

도시가 당신의 흥미를 끄는 것은, 그 도시의 일곱 혹은 칠십 가지 불가사의 때문이 아니라, 당신의 질문에 대한 그것들이 만드는 대답이다. — 이탈로 칼비노, 『보이지 않는 도시』 중에서

"Can the extremely heavy topics of architecture and urbanism reach the general public?"
"Can 80 or so cities come together as one for the 'Urban Projects' to create a unified experience?"
"Can individuals find their own examples of collectivism within the massive systems of urbanism?"

In order to design a meaningful experience for visitors of the "2019 Seoul Biennale of Architecture and Urbanism's "Urban Projects," there were three main obstacles to overcome. This project entails the participation of approximately 80 different countries, which contributes to the expansive variety of the exhibits. Just the sheer number of participating countries is expected to provide more than enough opportunities for visitors to enjoy the event. However, the question of how individuals would get a first-hand look at architecture and rethink its relationship with urbanism was still unanswered. In other words, we still had to figure out how we could create a personal experience for visitors through the biennale's theme, which involved clearly demonstrating the role and impact of the individual within the massive context of architecture and urbanism.

Cities and architectural structures serve as the setting for everyday life for the majority of the world's population. As such, the events that take place in these locations are seen as nothing more than just that: everyday life. However, to experts, architecture and urbanism span across a seemingly endless number of concepts and possibilities. In fact, they transform into something that is remarkably distant from our common perceptions of everyday life. Although we all live in buildings and cities, it is difficult to describe the two concepts themselves on our own. Most of our explanations end up being relative to our surrounding environments, for we hardly ever take

도시전의 경험 설계는 방문객 한 명, 한 명에 집중하는 것에서 시작한다. 수동적으로 전시를 받아들이는 것이 아니라 능동적으로 참여하게 만들고 싶었다. 도시전에는 80여 개의 도시가 참여한다. 방문객이 80여 개의 도시를 관람하는 것은 한 번에 80여 명의 사람을 만나 이야기를 나누는 것처럼 느껴진다. 언어도 목소리도 생김새도 다른 사람을 만나는 것은 혼돈으로 남을 가능성이 크다. 그러나 한 가지 주제 혹은 관점을 지닌다면 무의미하게 스쳐 가는 것이 아니라 선택할 수 있고 맥락이 생긴다. 관점은 우연을 인연으로 만들 수 있는 힘이 있기 때문이다.

　놀공이 제시하는 관점은 '기억'과 '상상'이라는 키워드에서 출발한다. 도시는 사회, 정치, 경제 등 모든 요소가 집합적으로 모인 곳이다. 여러 요소가 뒤엉키면서 도시 안에서도 기억과 망각이 일어나고 새로운 상상이 나타난다. 모든 도시는 기억의 집합이고 도시의 상상이 모인 것 역시 도시가 될 수 있다.

　　사람들은 기억도 해야 하고 잊기도 해야 해요. 모든 걸 기억해서는 안 돼요. 왜냐하면 내 작품에 나오는 푸네스처럼 모든 것을 끝없이 기억하면 미쳐 버릴 것이기 때문이에요. 물론 우리가 모든 걸 잊는다면, 우린 더 이상 존재하지 않게 될 거예요. 우린 우리 과거 속에 존재하기 때문이죠. 그렇지 않으면 우리가 누구인지, 이름이 무엇인지 알지 못할 거예요. 우린 이 두 가지 요소가 뒤섞인 상태를 지향해야 해요. 안 그래요? 이 기억과 망각을 우린 상상력이라 하지요. 아주 거창한 이름이에요. ―『보르헤스의 말 – 언어의 미로 속에서, 여든의 인터뷰』중에서

상상력이 기억과 망각 사이에서 생겨나는 것이라고 한 보르헤스의 생각을 빌어 도시의 기억과 도시의 상상을 관람객이 경험하도록 하는 할 수 있는 설계를 준비했다. 놀공은 도시전을 방문하는 사람들이 전시를 바라보는 관점을 만들고자 한다. 언어는 관점을 제시하는 방법 중 하나다. 방문객이 지니고 있는 개인적이고 추상적인 감정과 경험을 도시전 관람을 통해 끌어내기 위해 구체적인 행동을 설계했다.

a detailed look at these two topics individually. In some regards, although we live within a world characterized by cities and structures, we have become so used to them that we are not consciously aware of their existence.

　"You take delight not in a city's seven or seventy wonders, but in the answer it gives to a question of yours." — Italo Calvino, *Invisible Cities*

Our goal in designing the urban projects was to provide a unique experience for each visitor. This included creating something that encourages people to actively engage in the event and what it has to offer. 80 cities from around the world are participating in the urban projects; through a single visit, people will feel as though they just met and listened to the stories of 80 individuals from different places. Of course, there is always the possibility of confusion and hesitation arising when meeting people who speak different languages or look and sound unfamiliar. However, providing a variety of options for visitors makes this event that much more meaningful and allows individuals to choose their own areas of interest, creating their own context rather than taking a mere shallow look at a single topic or perspective. After all, perspective has the power to turn chance into destiny.

　The perspective that NOLGONG aims to provide is rooted in the concepts of memory and imagination. Cities are the result of the collective interactions between society, politics, and economics. These factors constantly intertwine with one another throughout urban settings, undergoing an endless cycle of remembering and forgetting to result in new imaginations. All urban spaces are the result of collected memories, and the imagination of these spaces can become new cities themselves.

　　It is the responsibility of all people to remember and forget. Not everything can be remembered, for, as Funes demonstrated, attempting to remember everything indefinitely can only lead to insanity. By the same token, we cannot forget everything either, for that would only lead to the loss of our very existence. This is because we exist as part of our own pasts; to lose such memories would mean to forget who we are and what our names are. Our lives are in a constant flux between the actions of remembering and

도시수집 관람객 체험 구조

1. 관람객은 관람 전 사전 질문에 답한다. 사전 질문은 관람객이 도시수집: 도시의 기억, 상상의 도시를 자연스럽게 받아들이고 도시에 대한 관점을 상기시키는 것으로 구성되어 있다.
(예시: 당신은 어떤 도시에서 살고 싶나요? 당신이 꿈꾸는 도시를 선택하세요.)

2. 방문객은 사전 질문에 따른 결과로 입장 팔찌를 발급받으며 첫 번째 관람 포인트로 이동한다. 입장 팔찌는 사전 질문에 관람객이 답한 것에 따라 총 세 가지로 나뉜다.
(예시: 도시의 기억, 상상의 도시, 상상과 기억 사이.)

3. 각 도시의 전시물에는 해시태그(#)로 표기된 '도시키워드'가 2개씩 부착되어 있다.

4. 관람객은 스마트폰에서 2개의 해시태그 중 하나를 골라 입력한다. 입력 후 등장하는 2개의 문장 중 마음이 가는 쪽을 선택한다.

5. 관람객은 '도시키워드' 입력과 문장 선택을 반복하여 도시를 수집하고 나만의 도시 컬렉션을 만든다.

'도시의 기억'은 도시가 주인공인 관점이다. 도시가 건설되어온 과정과 도시의 형태, 요소에 대한 이야기를 담고 있다. '상상의 도시'는 도시 속에서 살아가는 사람들이 주인공이다. 도시가 어떠한 방향으로 나아가길 바라는지, 어떤 도시를 설계하고 싶은지에 대한 고민이 '상상의 도시'에는 녹여져 있다. '도시의 기억'이 '도시의 의지'라면, '상상의 도시'는 '사람의 의지'다. '상상과 기억 사이'는 도시를 기억하고 상상하는 것, 모두를 의미하는 것으로 도시를 바라보는 관점이 도시의 기억과 상상의 도시 중간 어디쯤이 된다는 것을 가리킨다.
　도시전에 참여한 도시들은 도시적 경험의 주요 콘텐츠들을 제공한다. 하지만, 무엇보다 중요한

forgetting, and this relationship gives rise to our imaginations, something that is utterly extravagant. — *Borges, Language and Reality: The Transcendence of the Word*

According to Borges, imagination is borne from the constant battle between memory and forgetfulness. We designed the urban projects based on this concept, so that visitors can experience the memories and imagination attributed to cities. NOLGONG aims to create a new perspective for visitors of cities through language, designing a unique series of activities that each individual can relate to, based on their own abstract interpretations, emotions, and personal experiences.

Collectable City Visitor Experience Overview

1. Visitors answer the provided questions prior to viewing the exhibit. The questions are meant to help visitors better understand the Collectable City: Memory of City, Imaginary City project and rethink their perspective of cities.
(e.g. What type of city do you want to live in? Select your ideal city.)

2. Based on their answers to the questions, visitors receive a wristband and are led to the first stop of the exhibit. There are three different wristbands, dividing visitors into three groups based on their answers to the questions provided.
(e.g. Memories of cities; City of imagination; and In between imagination and memory).

3. Each exhibit is marked by two keywords beginning with hashtags (#).

4. Visitors select one of the two keywords using their smartphones and then select the sentence that best describes how they feel.

5. Visitors repeat this process to create their own collection of cities.

The city itself plays the main role in "Memories of cities," including the way it was designed and constructed and its unique characteristics. As for "City of imagination," the people living in the

것은 우리가 원하는 도시전의 경험을 실제로 만들어가는 주체는 도시들이 아닌 방문객이라는 것이다. 집합도시가 도시에 대한 전시라면, 도시수집은 방문객들이 직접 참여해 집합도시를 만들어가는 주체로의 전환이다.

도시수집: 도시를 '수집'하는 행위

'수집'은 자기중심적 행동으로 도시전 전체의 경험이 주관적/개인적 경험에서 도시적 경험으로 연결되는 가장 중요한 출발점이다. 아무것이나 다 모으는 것을 수집이라 하지 않는다. 수집은 어떤 가치 기준을 근거로 선택을 하는 행위이고 그 개별 선택들이 모였을 때 '수집'의 의미와 가치가 완성된다. 이러한 구조는 개별적인 다양한 쌓임과 연결의 결과물인 집합도시가 만들어지는 과정과 유사하다. 이렇게 수집 활동은 '도시수집'의 구조이며 동시에 집합도시의 추상성을 참여하는 행동과 과정을 통해 구체화하는 콘텐츠의 역할을 한다.

1   스마트폰으로 도시 수집

'도시수집'은 방문객이 자신의 스마트폰을 사용해 참여한다. 디지털 시스템은 '도시수집'이 개인적이고 도시적 규모의 경험이 되는 데 중요한 기능을 한다. 스마트폰을 활용한 디지털 시스템은 도시전의 방문객들의 주관적 선택을 디지털 입력을 통해 구체화하고 그에 대한 지속적인 피드백을 제공해 참가자들이 몰입 상태를 유지 할 수 있도록 돕니다. 대중교통이 중요한지 자전거 전용도로가 중요한지는 사회적이며 기능적인 이슈이다. 동시에 개인의 상황에 따라 다른 개인의 가치 선호도 문제이기도 하다. '도시수집'의 수집 과정은 주관적이고 추상적일 수 있는 선호도를 '버튼'을 눌러 선택한다는 과정을 통해 매우 단순하면서 구체화한다.

2   참여자의 선택을 기록, 데이터화

디지털 기기를 통한 선택의 중요함은 디지털 기기가 선택을 인식하고 기억하기 때문이다. 이렇게 생성된 일종의 빅데이터는 "파리에서 ○○를 선택했군요"와 같이 참가자의 활동을 기억하고

city are considered to be the main characters. This group is defined by the ideal direction that residents imagine and hope a city will take in the future , as well as these individuals' perception of an ideal urban design. Whereas "Memories of cities" is based on the will of the city, "City of imagination" is based on the will of the people. Lastly, "In between imagination and memory" covers both the acts of remembering and imagining cities, focusing on the point in the middle at which these two perspectives meet.

The cities participating in the urban projects are the ringleaders that provide visitors content characteristic of a truly urban experience. However, despite the important role that the cities themselves play, the visitors were of primary concern throughout the process of designing these projects and ensuring a valuable experience. In that regard, while Collective City is an exhibition focused on cities, Collectable City is an exhibition that is focused on visitors and their direct participation in creating their own collective city.

Collectable City: Collecting Cities

The act of collecting is a self-driven activity that serves as the core for both the subjective and objective ways that visitors interact with the urban projects. The important thing to note is that "collecting" does not simply refer to the gathering of just anything; rather, it refers to the gathering of something that is specifically selected apart from other things based on its perceived value. This concept is similar to the process through which a collective city is created through the diverse connections and composition of its individual parts. To emphasize the unique and important concept of collection, the Collectable City project aims to play an important role in conceptualizing the abstract ideas and characteristics of collective cities through detailed actions and interactive content.

1.   Collecting Cities Using Smartphones

Visitors of the Collectable City exhibit participate in the project using their smartphones. This digital system plays an important role in providing a personal and urban experience. Using their smartphones, visitors are asked a series of subjective questions, and their answers are then used as important data for feedback, all while establishing an immersive and interactive environment

반응해 주는 피드백을 통해 개인적 경험을 강화하고, 또 "○○님은 B를 선택하셨는데, 같은 40대들은 A를 주로 선택했습니다. 20대 방문객들이 B를 많이 선택하셨네요" 처럼 개인의 선택을 다른 참가자들과 연결하여 개인적이며 거시적인 피드백을 제공한다.

3. 다른 참여자의 선택을 보여주는 것, 거시적 경험

디지털 시스템과 참가자들의 선택을 통해 생성된 빅데이터는 도시수집이 지향하는 개인적이며 도시적인, 미시적이며 거시적인 경험을 가능하게 해준다. 관람 과정에서 방문객들의 스마트폰을 통해 참여하는 수집 활동의 기록이 매 순간 데이터로 저장되고 빅데이터는 방문객들이 만든 집합도시의 결과로 다시 공유되게 된다.

도시가 시간적, 공간적, 사회적 집합의 결과라면, 도시를 수집하는 행위는 개인이 집합을 만들어내는 적극적인 움직임이다. 관람객이 도시수집을 통해 도시를 기억하는 사람인지, 상상하는 사람인지를 확인하고 도시 혹은 관람객 자신의 의지로 도시 수집을 이루어나가길 바란다.

throughout the duration of their visit. For example, the relative importance of public transportation versus that of designated bicycle lanes is a social and functional issue for debate. There is no definitive answer to this question, as one's opinion is highly dependent on one's personal situation and perspective. Throughout the Collectable City exhibit, visitors will be asked similar abstract, subjective, and relatable questions, and our digital system allows these answers to be inputted and recorded in a simple yet detailed manner.

2. Recording and Gathering Data on Participants' Selections

The importance of making selections through digital devices for the purpose of this project is because these digital inputs are used to create a set of data. For example, this might include information like "Visitor A selected option B at Paris," serving as a way for each individual to remember what selections he or she made and give any relevant feedback. Also, other inputs might include "Visitor A selected B, but other visitors in their 40s selected A, while visitors in their 20s selected B." This links the selections of visitors with those of others to provide a broader scale of data analysis and evidence-based feedback.

3. A Macroscopic Experience: Presenting the Selections of Other Visitors

Through digital systems and big data on the selections made by visitors, the Collectable City project provides an experience that is both personal and urban on both the microscopic and macroscopic levels. Visitors participate in the exhibit by using their smartphones to create their own collection of cities, the data of which is recorded and collected to provide a complete set of data on the Collective City that they have created.

Considering that cities are the collective result of temporal, spatial, and social factors, by participating in this project, visitors are actively playing their part in creating the collective identity of cities. We hope that through the Collectable City project, visitors can complete their collection and use the results to determine whether they are the type to remember cities or imagine cities.

글로벌스

GLOBA

STUDIO

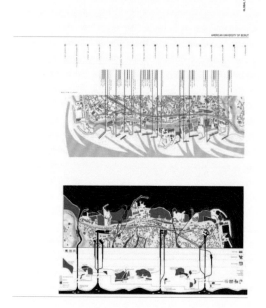

베이루트아메리칸대학교

American University of Beirut

아키텍처럴 어쏘시에이션 건축대학교

Architectural Association School of Architecture

어썸션대학교 + 서울시립대학교 + 호치민 건축대학교

Assumption University + University of Seoul + University of Architecture Ho Chi Minh City

바틀렛 건축대학교

Bartlett School of Architecture

SEOUL BIENNALE BLUEPRINTS
HANYANG UNIVERSITY ERICA

Global Studio 2019
Seoul Biennale of Architecture and Urbanism
SHARED CITY: SEOUL AND LONDON

STUDIO SYNOPSIS

[studio synopsis text]

Studio Members

Studio Leader:
Su Young Kim
Associate Professor, Hanyang University ERICA

Teaching Assistant:
Youngjoon Kim

한양대학교 에리카   Hanyang University Erica

캘리포니아예술대학교   California College of The Arts

SUBVERTING AQABA

컬럼비아대학교   Columbia University

홍콩 중문대학교   Chinese University of Hong Kong

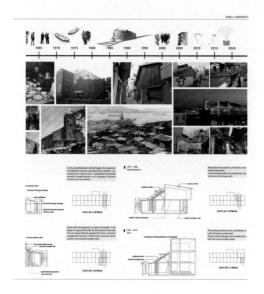

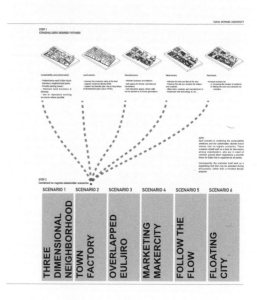

동아대학교　　　　　　　Dong-A University

이화여자대학교 +　　　　　Ewha Womans University
라드바우드대학교　　　　　+ Radboud University

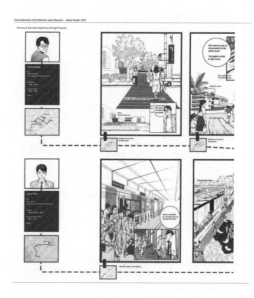

한양대학교　　　　　　　Hanyang University　　　　하버드대학교　　　　　Harvard University

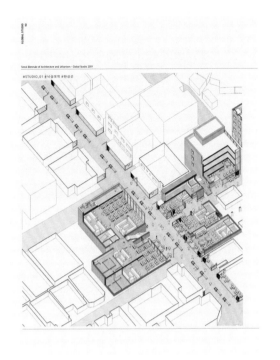

#STUDIO_01 홍익생태역 #한상군

홍익대학교   Hongik University

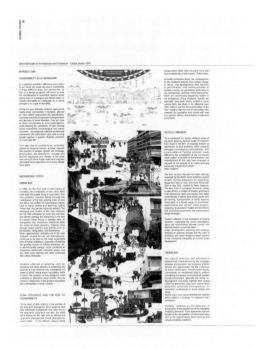

출랄롱코른대학교 INDA 프로그램   INDA – Chulalongkorn University

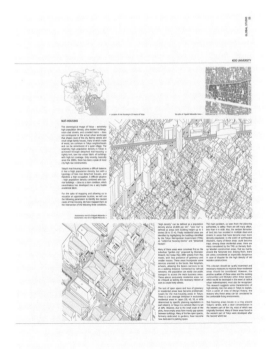

게이오대학교   Keio University

국민대학교   Kookmin University

SEOUL BIENNALE BLUEPRINTS

Sep.2019

Global Studio 2019
Seoul Biennale of Architecture and Urbanism

KUWAIT UNIVERSITY

Mecca Studio: An Ongoing Manifesto

DEEMA ALWUGAYAN-SARA ALHASHASH

STUDIO SYNOPSIS

Studio Members

Studio Leader:
Shaikha Almubaraki

Students:
Bedour Alhoorawe
Fajer Alzayed
Maryam AlNajjar
Sara Alaboduoh
Bedoor Alboloon
Yasmeen Alshaijakh
Deema Alakhdar
Sara AlSeer
Latwa Alboloon
Waitha Al Sabeel
Rada Abu Qoqa
Deema Al Wugayan
Banau Al Khalaf
Sarah Alkhamoni

SEOUL BIENNALE BLUEPRINTS

Sep.2019

Global Studio 2019
Seoul Biennale of Architecture and Urbanism

NATIONAL UNIVERSITY OF SINGAPORE

HDT 445: The Equatorial City and the Architectures of Aggregation / Ho Chi Minh, Vietnam

STUDIO SYNOPSIS

Studio Members

Studio Leader:
Erik L'Heureux   GIA, LEED AP BD+C
H.Arch, BA Arch, NUS B-1008 Carnilton, RA New York, RI, USA

Students:
Cai Tong
Lee Dong Eun
Leh Astrid Mayankleia
Maaya Binte Mohd Yusof
Mesto Lam Jia Chieu
Nicholas Tak Kan Hern
Aswald Hergen Valen Huang
Sio Hui Tarah Binte Osama 1
Chanlyn Hwang Yu Jane
Loh Xusing

Seoul Biennale of Architecture and Urbanism - Global Studio 2019

AD53 – OFFSETTING the OFFSHORE: On the Illusion, Delusion and Dilution of Waterfronts

Tutors: Daniel Fernández Pascual, Alon Schwabe
Teaching Assistant: Guillermo Ruiz de Teresa

Seoul Biennale of Architecture and Urbanism - Global Studio 2019

쿠웨이트대학교　　Kuwait University

싱가포르국립대학교　　National University of Singapore

영국왕립예술대학교　　Royal College of Art

서울대학교　　Seoul National University

SEOUL BIENNALE BLUEPRINTS Sep.2019

Global Studio 2019
Seoul Biennale of Architecture and Urbanism   SINGAPORE UNIVERSITY OF TECHNOLOGY AND DESIGN   Urban Hinterlands / Guangzhou

STUDIO SYNOPSIS

Studio Leader:

Calvin Chua Tzu Jie
Adjunct Assistant Professor
Singapore University of Technology and Design

Students:

Seoul Biennale of Architecture and Urbanism - Global Studio 2019

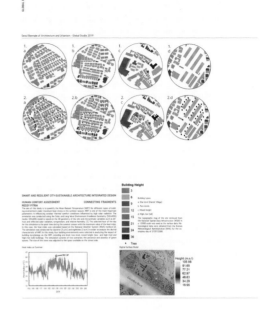

SMART AND RESILIENT CITY-SUSTAINABLE ARCHITECTURE INTEGRATED DESIGN

HUMAN COMFORT ASSESSMENT   **CONNECTING FRAGMENTS**
RESSY FITRIA

싱가포르과학기술대학교   **Singapore University of Technology And Design**

성균관대학교   **Sungkyunkwan University**

Seoul Biennale of Architecture and Urbanism - Global Studio 2019

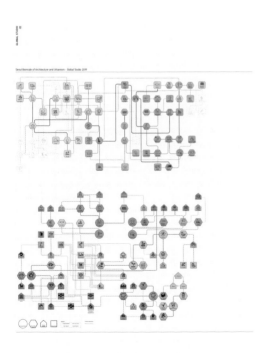

SEOUL BIENNALE BLUEPRINTS Sep.2019

Global Studio 2019
Seoul Biennale of Architecture and Urbanism   TECHNISCHE UNIVERSITÄT BERLIN   COLLABORATIVE RESEARCH CENTER "RE-FIGURATION OF SPACES"

맵핑의 재구성
이민자 거주민
방문객 콜렉-
티비티의 형태

STUDIO TITLE   Studio Members

MAPPING RE-FIGURATIONS
MIGRANTS
RESIDENTS
VISITORS
FORMS OF COLLECTIVITY

시라큐스대학교   **Syracuse University**

베를린공과대학교   **Technische Universität Berlin**

SEOUL BIENNALE BLUEPRINTS
TU WIEN
The Book of Boundaries
Hanstolten, Vienna | Jamsil, Seoul

STUDIO SYNOPSIS

THE BOOK OF BOUNDARIES

Studio Leader: Seong H. Sang, Mladen Jadric, TU WIEN

SEOUL BIENNALE BLUEPRINTS
TEXAS TECH UNIVERSITY
Capacity of Seoktye

STUDIO SYNOPSIS

Studio Members

빙공과대학교   **Techische Universität Wien**

텍사스테크대학교   **Texas Tech University**

SEOUL BIENNALE BLUEPRINTS
UNIVERSITY OF CAPE TOWN
Space of Good Hope Design Research Studio
Delft, CT

STUDIO SYNOPSIS

Studio Members

SEOUL BIENNALE BLUEPRINTS
THE UNIVERSITY OF HONG KONG
Inhabitable Territories: Seoul Urban Mountains
a Collective Infrastructure for the City

STUDIO SYNOPSIS

Studio Members

케이프타운대학교   **University of Cape Town**

홍콩대학교   **University of Hong Kong**

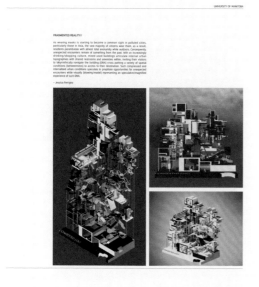

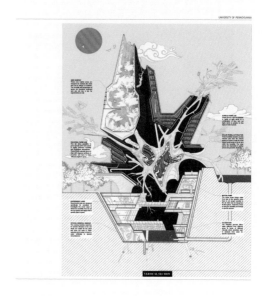

매니토바대학교 + 칼턴대학교 + 토론토대학교

University of Manitoba + Carleton University + University of Toronto

펜실베니아대학교

University of Pennsylvania

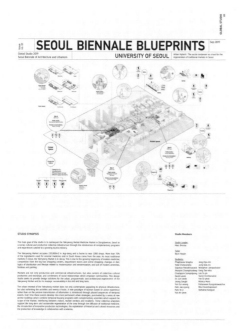

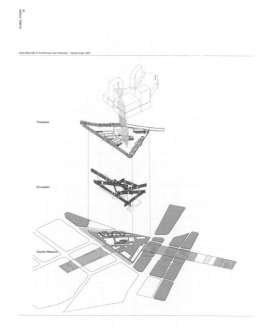

서울시립대학교

University of Seoul

시드니공과대학교

University of Technology Sydney

SEOUL BIENNALE BLUEPRINTS Sep.2019

Global Studio 2019
Seoul Biennale of Architecture and Urbanism
UNIVERSITY OF TEXAS AT ARLINGTON

STUDIO SYNOPSIS

Studio Members

SEOUL BIENNALE BLUEPRINTS Sep.2019

Global Studio 2019
Seoul Biennale of Architecture and Urbanism
URBAM - EAFIT Center for Urban and Environmental Studies

STUDIO SYNOPSIS

Studio Members

텍사스대학교알링턴   **University of Texas at Arlington**

EAFIT대학교 URBAM   **URBAM - Universidad Eafit**

Seoul Biennale of Architecture and Urbanism - Global Studio 2019

건축을 전시한다는 개념 자체가 이미 큰 도전이지만, 건축 연구과제를 전시의 형식으로 바꾸기 위해서는 유효한 취사선택의 결정이 따를 수 밖에 없다. 글로벌스튜디오는 전 세계의 43개 대학이 모여서 '집합도시'라는 공동의 주제를 연구하는 드물고 귀한 기회였다. 시각적으로 의미있는 부분만을 모아서 전시할 수 밖에 없었던 현실에 대한 보완책으로 연구과정 전체를 컬렉션으로 보여주는 블루프린트 출판물이 다음과 같이 기획되었다.

While the idea of exhibiting an architectural work is already a big challenge, it requires immense curatorial decisions when it comes to research works of architectural projects. Global Studio exhibition was a rare and valuable opportunity to gather research projects from 43 universities around the world all working under the common theme of 'Collective City.' As a supplement, the publication of 'BLUEPRINTS' was pursued as a collection to archive the vast portions of research works that had to be edited out from the exhibition that were composed of only the visually compelling portions of the ressarch.

연세대학교   **Yonsei University**

서울도시건축비엔날레 글로벌 스튜디오 큐레이터 메모

최상기, 글로벌 스튜디오 큐레이터

2019 서울도시건축비엔날레 글로벌스튜디오에는 '집합 도시'라는 주제에 따라 전 세계 43개 건축 학교에서 34개 팀이 참여한다. 집합 도시라는 주제는 사회적, 물리적 측면을 토대로 하는 다양한 연구·설계 프로젝트의 촉매제 역할을 하며 학술적 사안과 전문적 관심사를 연계하고 있다. 그러한 의미에서 이곳에 전시된 작품들은 우리의 건축 환경을 구성하는 사회적 역동성을 깊이 있게 이해하도록 도울 뿐 아니라 강력한 물리적 존재감을 전달한다. 이 책자는 글로벌스튜디오 전시회에 출품된 작품들에 대한 다양한 해석과 접근법을 제공하는 연구 플랫폼 역할을 함으로써 건축을 매개로 한 집합적인 활동이 어떻게 해서 도시의 활력에 기여하며 사회에 변화를 가져올 수 있는지 탐구하는 데 도움을 주고자 한다.

집합 기반시설

집합 기반시설은 집합 도시 논의에서 빠질 수 없는 주제다. 도시를 인류의 가장 위대한 발명품이라고 한다면 기반시설은 도시 형성에서 가장 효율적이고 큰 영향을 미치는 요소라 할 수 있다. 이런 점에서 도시의 기반시설은 적극적인 집합 참여 활동이 구현된 형태다. 도로, 교량, 교통 거점, 옹벽 같은 기반시설 대부분은 정부 주도의 공익사업으로 건설되지만 결국에는 집합적 운영에 의한 활용을 통해 잠재적 기반시설인 도시 공간이 활성화된다.

모빌리티의 기반시설

캘리포니아예술대학교의 니라지 바티아 교수가 이끄는 스튜디오는 확장된 영역에서 기반시설이 공평한 자원 재분배와 새로운 공동체 관계 형성에 어떠한 역할을 하는지 조사한다. 이러한 디자인 개입은 기존에 방치된 캘리포니아 내륙을 관통할 예정인 고속 철도 건설 계획을 중심으로 발전된다. 주 단위 기반시설에 비하면 교통 환승지에 대한 건축적 개입은 소규모로 보일지도 모르지만 이러한 건축 개입은 집합 참여에 대해 현실성 있는 해결책을 제공한다. 여기에 나타난 다섯 가지 건축 제안은 과거의 외딴 지역을 '영역의 가장자리'로 전환할 뿐 아니라 우리에게 건축이 사회·경제적 상호작용을 생성할

Curatorial Notes on SBAU Global Studio

Sanki Choe, Curator of Global Studio

The Global Studio exhibition of the 2019 Seoul Biennale of Architecture and Urbanism presents a collection of thirty-four entries from forty-three architectural schools around the globe, all participating under the common theme of the "Collective City." The theme calls for a wide range of research and design projects that are firmly grounded on both social and physical dimensions, bridging academic and professional interests, and the disciplines of architecture and urbanism. As such, the works compiled here convey a strong physical presence with deep understanding of the social dynamics that give form to our built environment. This publication is a collection of entries displayed at the Global Studio exhibition, which functions as a research platform harboring various interpretations and approaches on exploring how collective actions contribute to urban vitality and bring difference to society by utilizing architecture as a medium.

COLLECTIVE INFRASTRUCTURE

Infrastructure is one of the favorite subjects inseparable from the discussion of the collective city. If the cities are mankind's single most important invention, we come to understand that infrastructure is the most efficient and influential contributor in its making. In this sense, urban infrastructures are formal manifestations of active collective engagement. While most infrastructures such as roads, bridges, transportation hubs, and retaining walls are built as government-driven projects to serve the public, their adaptation through collective operation is what eventually activates the infrastructures and other leftover urban spaces that can be perceived as latent infrastructures.

Infrastructure of Mobility

Neeraj Bhatia and his studio at California College of the Arts investigates the role of infrastructure in redistributing resources more equitably across expanded territories, engaging new relationships with the collective mass. The design intervention evolves around the opportunity of

영역 도시: 거대도시의 가장자리
캘리포니아예술대학교

The Territorial City: Edge of the Megalopolis
California College Of The Arts

수 있다는 것을 보여준다. 이러한 역할은 건축의 핵심 요소이지만 기존의 탑다운 방식의 계획 모델에서 흔히 간과되어왔다.

기반시설에 대한 개입은 고속 철도의 경우와 마찬가지로 모빌리티 기술의 발달로 한층 더 큰 추진력을 얻고 있다. 하버드 대학교 디자인 대학원의 안드레스 세브추크 교수는 학생들로 구성된 여섯 개 팀과 함께 최첨단 모빌리티 기술이 로스앤젤레스라는 도시의 미래에 미치는 영향을 조사한다. 자율 주행 자동차에서부터 개인 이동 수단과 자동 택배 배송 시스템에 이르기까지 다양한 유형의 모빌리티 기술이 새로운 대중교통 경험을 창출하고 있다. 새로운 형식의 모빌리티 기술은 기존 기반시설과 병행되기도 하겠지만 대체로 별개의 시스템을 통해 운영되면서 과거에는 상상조차 할 수 없었던 새로운 개념의 도시 관계를 만들어 나갈 전망이다.

텍사스알링턴 대학교의 조슈아 네이슨 교수는 학생들과 함께 서울의 공통 기반시설을 지표면으로 설정하여 집합도시의 매개체로 세우고자 한다. 이 스튜디오는 집합 공유지가 성공적인 플랫폼으로 발전할 가능성이 있다는 점에서 지표면에 내재된 기회에 주목한다. 그 같은 기회가 서로 연결되면 집합적인 상호작용의 기반이 될 광활한 공간을 창조할 수 있다. 그 이외에도 지표면에 내재된 방대한 가능성을 활용한 출품작으로는 어썸션 대학교, 호치민 건축대학, 서울시립대학교의 세 학교가 호치민시에서 함께 진행한 공동 워크숍을 들 수 있다. 학생들은 집합적인 패턴뿐만

high-speed rail being planned to pass through the previously neglected inner lands of California. The architectural intervention of the nodes may appear smaller in scale when compared to the scale of the statewide infrastructure itself, but they prove to convey real solutions in engaging the collective. Five architectural proposals presented here transform the once remote regions into 'territorial edges' and redirect our attention to the role of architecture as a generator of social and economic interactions: the key element often missing in the traditional model of top-down planning.

Just like the high-speed rail, the drive towards infrastructural intervention is fueled by advances in mobility technologies. Andres Sevtsuk of Harvard University Graduate School of Design leads six teams of students to investigate the impact of new mobility technologies on the urban future of Los Angeles. The studied trajectories of new mobility options ranging from automated vehicles and personal mobility devices to automated package delivery systems, suggest new forms of collective traffic options which result in interactions that sometimes align with the existing infrastructure, but more often dictate a system that is detached from the existing urban forms, thus generating a new urban relation unimagined in the past.

Joshua Nason of University of Texas at Arlington leads the students to investigate the common grounds of Seoul with aims to reinstate it as a medium for the collective city. The studio emphasizes the inherent opportunities of the ground surface that may grow out into a successful platform for a collective commons, taking cues from the fact that they are interconnected and yield

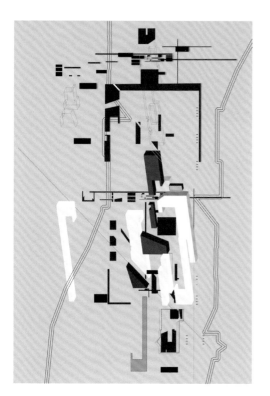

서울의 지표면
텍사스대학교알링턴

Seoul, Skin Deep
University of Texas Arlington

a vast area to serve as platforms for collective interactions to take place. Another entry that taps into the immense opportunities inherent in the ground plane is the joint workshop that took place in Ho Chi Minh City among three universities: Assumption University, University of Architecture Ho Chi Minh City, and University of Seoul. The students observed and analysed the collective patterns and interactions between the community and its built environment. What they discovered was the role of the vast continuous ground plane that often interweaves the interiors of the building and extend out into the flat landscape of Ho Chi Minh City. This ground surface not merely serves the traffic, but also offers a platform that invites massive yet efficient exchange of collective activities delivered by battalions of motorcycles. It is noteworthy to mention how this reappraisal of the ground plane—an important yet underappreciated collective agent—comes at a time when the city is invaded by large scale redevelopments with tendencies to privatize the ground planes under corporate reasoning.

Architecture as Infrastructure

It is not unfair to say that modern architectural developments with their taste for generic, mass-produced, and highly adaptable structures, have pushed aside an autonomous architectural expression and surrendered itself to being transformed into mere 'infrastructures' that are subject to urbanism. While this may sound like a modern curse, there is a trace of blessing if we look into the important role of the architecture/infrastructure hybrid that is capable of accommodating the desires of the community and functioning as a collective agent.

The studio led by Carla Aramouny and the faculty of the American University of Beirut presents a parallel research project on infrastructural intervention applied to the coastal city of Zouk Mosbeh and Beirut where the coastal highway instigates physical segregation between the coastal lands and the inner city. The apparent urban segregation caused by the old infrastructure (highway) is further intensified through bi-polarization taking place in various levels of density, scale, wealth, and pursuit of leisure. If the solution for the coastal land comes from providing new smaller horizontal infrastructures that domesticate the marine landscape into a scale fit for collective use, the inner city employs vertical infrastructures that activate the previously unoccupiable small plots by assigning creative use to the banality of stairs, room layouts, and architectural volumes. The theme of the Texas

아니라 공동체 사이의 상호작용, 공동체의 건축 환경을 관찰하고 분석했다. 또한 연속적인 광활한 지표면이 건축물의 내부를 얼기설기 연결하고 호치민시의 평면적인 지형 속으로 확장된다는 것을 발견했다. 지표면은 단순히 교통수단이 다니는 통로가 아니라 대규모 오토바이 부대의 집합 활동이 효율적으로 교환되는 플랫폼 역할을 한다. 미처럼 그 중요성에도 불구하고 집합적 매개체로서 평가절하된 지표면이 재평가되고 있다. 특히 이러한 재평가가 기업의 이해관계에 따라 지표면이 민영화되고 도시가 대규모 재개발 사업에 지배당하는 시기에 이루어지고 있다는 사실에 주목해야할 필요성이 있다.

기반시설로서의 건축

현대 건축이 개성 없는 대량 생산 형태로 진행되며 지나치게 중성적이고 순응적인 구조에만 초점을 두는 쪽으로 발전함에 따라 자율적인 건축 표현이 위축되고 건축 자체가 도시 계획에 종속된 '기반시설'로 전락하고 있는 상황에 주의를 기울일 필요거 있다. 이렇게 보면 근대화가 저주처럼 느껴질지도 모르지만 건축과 기반시설의 혼합체가 공동체의 욕구를 충족하고 집합 매개체로서의 역할을 할 수 있다는 점을 생각해보면 근대화는 오히려 축복일 수도 있다.

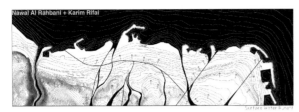

Nawal Al Rahbani + Karim Rifai

ECOLOGIES: FAUNA AND FLORA

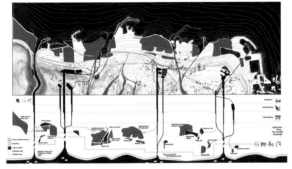

버티컬 스튜디오: 여가 활동과 기반
시설의 생태학에서 소형 대안주택까지
베이루트 아메리칸대학교

Vertical Studio: From Ecologies
of Leisure and Infrastructure to
Small Alternative Housing
American University of Beirut

카를라 아라모니와 교수를 포함한 베이루트
아메리칸 대학교의 교수진이 이끄는 스튜디오는 해안
도시 주크 모스베와 베이루트에서 시행된 기반시설
사업을 비교한 연구 프로젝트를 제안한다. 이곳의 해안
고속도로는 해안 지대와 내륙 도시를 물리적으로 나눈다.
이처럼 오래된 기반시설(고속도로)에 의한 도시 분리는
인구밀도, 규모, 소득, 여가활동 등 다양한 측면에서
일어나는 양극화를 한층 더 심화하고 있다. 해안 지대에
대한 해결책이 소규모 수평적 기반시설을 공급하여 해안
지대를 집합적 이용에 적합한 규모로 전환하는 것이라면
내륙 도시에 대한 해결책은 수직적 기반시설을 마련하는
것으로 나타나고 있다. 이러한 수직적 기반시설을 통해
상투적인 계단, 방, 공간 배치, 건축 용적을 창의적으로
재구성하면 이전에는 사용할 수 없었던 자투리땅을
활용할 방안이 생긴다. 박건 교수와 임리사 교수가 이끄는
텍사스 테크 대학교 스튜디오의 주제도 서교동이라는
인구밀도 높은 동네의 공간 사이 자투리땅을 파악한
다음에 물류 거점, 산책로, 계단 탑, 화분, 등의 기존 도시

Tech University studio led by Kuhn Park and
Lisa Lim also is aligned with this approach on
identifying the residual in-between spaces existing
in the dense urban fabric of Seogyedong and then
transforming the strategic points by means of
installing architectural scale infrastructure such
as logistic hubs, trails, stair towers, planters, and
corridors that penetrate and interconnect the
existing urban boundaries.

Erik L'Heureux and his studio at the National
University of Singapore have conducted an
extensive research and design project on Ho Chi
Minh City discovering an unusual junction between
the equatorial environment and the languages
of modern architecture, and how it starts to
respond to the collective formations of urban
living. While the research initiated from the study
of the climate and the environmental specificities
of the equatorial city, it is interesting to view how
the main focus of the design was channeled into
architectural infrastructures such as open frames,
scaffoldings, brise soleil, and buffer zones. These
architectural features characteristic to the modern
architectural adaptation to equatorial living, not
only negotiate for the extreme climate but also
opens up opportunities for open social interaction
that leads to the inclusion of the collective in the
domain of architecture.

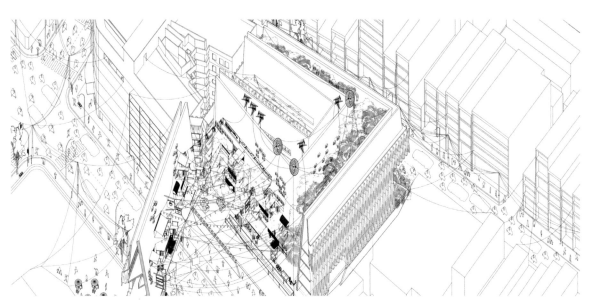

열기: 적도 도시의 집합 건축
싱가포르국립대학교

**HOT AIR: The Equatorial City and
The Architectures of Aggregation
National University of Singapore**

영역을 관통하고 연결하는 회랑 등과 같은 건축 규모의
기반시설을 설치하여 전략적 지점을 전환한다는 점에서
이러한 접근법과 일맥상통한다.

　　싱가포르 국립대학교의 에릭 르뢰 교수가 이끄는
스튜디오는 호치민시에서 적도 환경과 현대건축 언어
사이의 특이한 접점을 찾아내고 이러한 접점이 도시
생활의 집합적 형성에 어떠한 영향을 미치고 있는지에
대해 광범위한 조사와 디자인 프로젝트를 진행했다. 해당
연구는 적도 도시 특유의 기후와 환경에 대한 조사에서
시작되었지만 디자인의 초점이 오픈 프레임, 비계, 차양,
완충영역과 같은 건축 기반시설에 집중되는 양상을 보는
것도 흥미롭다. 적도 지역생활에 맞게 적용된 현대 건축의
특징적인 건축학적 요소는 극단적인 기후 환경에 대한
절충안을 제시하는 동시에 건축 분야의 집합성 포용으로
이어지는 사회적 소통의 기회를 열어준다.

　　자연적 기반시설
도시 기반시설을 집합 대중에게 자원을 전달하는
시스템이라고 넓게 정의하면 도시 기반시설은 도로,
다리, 통신망처럼 인간이 만들어 낸 시스템에 국한되지
않고 산이나 강 같은 자연계 구성요소까지 포괄하게
된다. 홍콩대학교와 EAFIT 대학교의 URBAM 연구소는
이러한 접근법으로 천연 자원을 새로운 시각으로 보고
심도 있게 연구한다.

　　제럴딘 보리오 교수가 이끄는 스튜디오는
홍콩대학교 학생들과 함께 서울의 4대문을 둘러싼 네

　　Natural Infrastructure
By expanding the definition of urban infrastructure
as a system that delivers resources to the
collective mass, the term can also be applied to
natural features such as mountains or rivers rather
than being confined only to man-made systems
of roads, bridges, and data highways. University
of Hong Kong and **URBAM** at **EAFIT** University
carry this approach into further depth and invites
us to look into our natural resources with a new
perspective.

　　The studio led by Geraldine Borio takes
students from the University of Hong Kong to
explore the four mountains that surround the old
capital of Seoul by expanding the extent of the
infrastructure by including how the mountains play
an important role in forming the city and engaging
various user groups that confront the mountains
as important sources in their urban living. An
important lesson learned from this investigation
also sheds light on the open attitude of the city
government that collectively invites the citizens to
participate and contribute as planners in moulding
this organic system through self-initiated activities
and facilities. The result according to their
research proves how such collective interaction
along the mountains can get accumulated to
reformulate the natural infrastructure.

　　Invisible Infrastructure
Infrastructure takes on a more poetic twist in
Yonsei University's entry led by Sangyun Lee. The
studio walks the fine line that traces the ambivalent
desires of our society trying to grab onto the
visible and tangible, while being serviced by many

군데 산을 탐색한다. 이들의 연구는 기반시설의 범위를
확장하여 도시를 형성하고 도시생활의 활력소로 산을
바라보는 다양한 이용자 집단을 수용하는 과정에서 산이
담당하는 중요한 역할을 탐색한다. 이러한 조사를 통해
드러난 서울시의 열린 태도는 유익한 교훈을 제시한다.
서울시는 자기주도형 활동과 시설을 통해 산이라는
유기적 시스템의 형성에 시민들의 집합적 참여를
유도하고 있다. 이 같은 조사 결과는 산을 매개로 한
집합적 상호작용이 축적되면 자연 기반시설이 재구성되는
결과로 이어질 수 있다는 것을 보여준다.

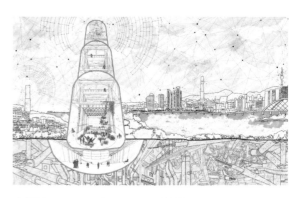

더 파일론
연세대학교

The Pylon
Yonsei University

보이지 않는 기반시설
이상윤 교수가 이끄는 연세대학교의 출품작에서
기반시설은 한층 더 시적인 반전을 취한다. 해당
스튜디오는 볼 수 있고 만질 수 있는 것을 움켜쥐려고 하는
동시에 보이지 않는 (따라서 통제 불가능한) 수많은 도시
기반시설에 의해 돌아가는 우리 사회의 상반된 욕구를
추적하여 그 모호한 차이를 작품으로 보여준다. 전시된
'파일론(Pylons)'이 복고적인 구상의 형식으로 표현되는
동시에 고대 기념비의 형태를 띤다는 사실이 눈길을
끈다. 이처럼 양면적인 특성은 신생 기반시설 파일론의
연결망이 통제 불가능하고 대항 불가능한 상태로까지
발전한 무형적, 기술적 진보에 대한 반작용으로서
우후죽순처럼 증가하고 있다는 사실을 시사하는 듯하다.
그뿐만 아니라 물리적, 건축학적 기반시설이 현대
도시의 참된 집합적 상호작용을 촉진하는 환경의 조성에
필요하다는 사실을 일깨워준다.

집합 생활

'공유 도시'를 주제로 한 제1회 비엔날레에서 탐구된
바와 같이 도시생활의 전통적인 영역이 급격하게
변화하는 까닭은 신자유주의 경제체제 안에서의 한정된
자원 때문이다. 자율경쟁 사회에서 자원은 정보기술과
소셜미디어의 확산으로 고갈될 상황에 처했다. 해당
주제는 연구자들에게 가정에 대한 전통적 관념을
해체하는 새로운 사회적 관계에 기반을 둔 새로운 집합
주거 형태를 모색하고 그러한 주거 형태를 도시 생활의
패턴을 재규정하는 하이브리드 주거 형태로 승화시킬
것을 제안한다.

공유 생활
집합 주거에 대한 담론을 시작하려면 집합성 개념의
핵심인 주택 건축 계획으로 주의를 전환할 필요가 있다.
아키텍처어쏘시에이션건축대학의 샘 자코비 교수는 집단

invisible—and therefore uncontrollable—urban
infrastructure. It is interesting to notice how the
proposed 'Pylons' are rendered in a nostalgic
representational method, and also take the shape
of an archaic monument. This appears to be an
implication suggesting that the network of the
new infrastructural Pylons have mushroomed out
as a counter-reaction to all the intangible and
technological advances that have grown to be
uncontrollable and overwhelming. It also brings
our attention to the necessity of a physical—
and therefore architectural—infrastructure in
preparation of an environment that promotes true
collective interactions in our contemporary cities.

COLLECTIVE LIVING

As explored in the previous biennale under the
theme "Imminent Commons", the traditional
boundaries of our urban living conditions have
drastically been altered due to limited resources in
the neo-liberal society exploited by the proliferation
of information technology and social media. The
theme encourages the researchers to continue
to seek out new forms of collective living based
on emerging social relations that dissolve the
traditional sense of domesticity and infuses them

투사된 도시
아키텍처럴 어쏘시에이션 건축대학교

Projective Cities
Architectural Association School
of Architecture

Scenario 1 - Meeting points

Smart parking system

Sport activities

Crossing Bridge

경계의 기록
빈 공과대학교

The Book of Limes
Techische Universität Wien

into a hybrid that redefines our patterns of urban living. The new patterns of urban forms rooted on the collective exchange of goods as well as ideas are capable of merging function and social practices into hybrid living patterns.

Shared Living

To open up conversations on collective living, we turn our attention to the program of housing, the very architectural program that is centered on the notion of collectivity. I think it is worthwhile to share a provocation by Sam Jacoby of Architectural Association who stated that "housing has been weaponised" in a way that it becomes a tool for liberating as well as oppressing its collective subjects. The studio presents extensive historical research on how modern housing projects have supported the construction of a nuclear family and how they still functions as an efficient agent for developing new forms of collective living. The research focuses on the recent developments of co-living prototypes that expand beyond the family, thus constantly reinstating its relation to the socio-economic transformations in our contemporary society.

Mladen Jadric of Technische Universität Wien leads a studio dealing with housing projects in Vienna and Seoul as parallel design interventions of refurbishing a typical "Dormitory Town" where social interactions do not happen the way it was devised. In order to understand the actual and active opportunities of social interaction among tenants, the students invest most of their research efforts in extracting the patterns of hidden boundaries prevalent in the area. The subsequent design phases utilize design proposals dubbed as "social prostheses" where physical apparatuses added in retrospect, initiates and intensify the collective interactions in the community.

If there is one studio that challenges us to view architecture as a mere background to be altered and reformulated by the actions that reflect the true patterns of collective living, it comes from the University of Cape Town led by Fadly Isaacs and Melinda Silverman. The studio focuses on the political nature of space where sometimes the normative spatial models can transform the living patterns of the residents but also vice versa. The studio displays how the inverse conditions developed in post-apartheid Delft where the residents collectively transformed government housing with new spatial dynamics.

Production City

The studio led by Calvin Chua at the Singapore University of Technology and Design explores the

주체성을 해방하기도 하고 억압하기도 하는 방식으로 "공통주거가 무기화되고 있다"는 문제를 제기한 바 있는데 나는 그가 제기한 문제에 모두가 관심을 기울일 필요가 있다고 생각한다. 해당 스튜디오는 광범위한 고증을 바탕으로 현대 주택 건축이 어떻게 해서 핵가족화를 촉진했으며 아직도 새로운 집합 주거 형태의 발전에 효율적 매개체로서 작용하는가를 보여준다. 해당 연구는 코리빙(co-living, 주거지 공유)의 원형이 발전하면서 가족의 범위가 확대되고 그에 따라 가족 구조가 현대 사회의 사회, 경제적 변화에 따라 지속적으로 변화하고 있다는 사실에 주목한다.

빈공과대학교의 믈라덴 야드릭 교수가 이끄는 스튜디오는 서울과 빈에서 동시 스튜디오 진행을 통해 사회적 상호작용이 원래 계획대로 이루어지지 않은 전형적 '베드타운'을 재단장한다. 세입자들 사이의 실질적이고 적극적인 사회적 상호작용의 기회를 확인하기 위해 학생들은 해당 지역 안에 숨겨진 영역의 패턴을 찾아내는 데 전념했다. 그런 다음 디자인 단계에서는 '사회적 보형물'로 불리는 디자인 제안을 활용했다. '사회적 보형물'은 과거 형태로 추가된 물리적 장치로서 공동체에서 집합적 상호작용을 촉진하고 확대하는 역할을 한다.

파들리 아이작스 교수와 멜린다 실버먼이 이끄는 케이프타운대학교의 스튜디오는 건축이 집합 주거지의 실제 패턴을 반영하는 활동에 의해 바뀌고 재구성될 뿐인 배경 요소라는 관점으로 기존 건축계획의 관습에 도전장을 던진다. 이 스튜디오는 공간의 정치적 속성에

생산적 주변부
싱가포르기술디자인대학교

Productive Peripheries
Singapore University of
Technology and Design

주목하여 규범적 공간 모델이 거주자들의 생활 패턴을
바꿀 수도 있지만 거주자들의 생활 패턴이 규범적 공간
모델을 바꾸기도 하는 양상을 탐구한다. 이 스튜디오는
후자의 사례로 남아프리카공화국 델프트 지역의
거주자들이 아파르트헤이트(인종차별정책) 이후 시기에
정부에서 제공한 계획주택을 집합적으로 변형함으로써
새로운 역동성의 공간을 창조해낸 것을 예로 든다.

생산 도시
싱가포르기술디자인대학교의 캘빈 차 교수가 이끄는
스튜디오는 농촌과 도시, 생산과 생활, 건축과 도시화
사이의 중간지점이라 할 수 있는 틈새 규모의 개발을
탐색한다. 이 스튜디오는 도시화가 급속도로 진행 중인
광저우 외곽에서 성행하고 있는 소규모 제조업과 도시
농업에 초점을 맞춘다. 집합적 주거 패턴에 따른 새로운
공간 유형은 주거, 공유, 생산의 건축적 하이브리드에서
생성된 시너지 효과 덕분에 집합적 주거 패턴에 맞는
새로운 공간 유형이 탄생하고 있다. 이러한 공간 유형은
집합 주거의 대안 모델이 될 것으로 기대된다.

런던과 서울을 실험 장소로 삼은 공동 스튜디오의
작품에서는 도시 생활의 대안 모델이 한층 더 심층적으로
다뤄진다. 바틀렛건축대학교의 사빈 스토프 교수와
패트릭 웨버 교수는 새롭고 창의적인 도시 거주 방식을
모색하기 위해 한양대학교의 김소영 교수와 함께
학생들을 지도해 인구밀도 높은 도시 환경에서 나타나는
새로운 공유 문화에 주목한다. 그뿐만 아니라 일반적인
생산 활동을 체험하도록 함으로써 건축적 하이브리드라는
전환이 일련의 예기치 못한 물리적 행동을 통해 집합 주거
환경을 재구성한다는 것을 입증한다.

성균관대학교의 토르스텐 슈처 교수는 독일과
한국의 7개 대학, 52명의 참여자로 구성된 진정한 '집합'
워크숍과 함께 집합 도시의 잠재적인 모델로 대구 '혁신
지구'에 대한 디자인과 분석 프로젝트를 제안한다. 이들은
진정한 집합과 혁신을 동시에 충족하기 위해 혁신적 건축
해법을 제시할 뿐 아니라 지속 가능한 사회 시스템을

in-between scale developments that claim the
middle ground between rural and urban, production
and living, and between architecture and urbanism.
The studio focuses on micro-manufacturing and
urban agriculture that happens around the urban
peripheries of Guangzhou that is under pressure
of rapid urbanization. New spatial typologies that
respond to collective living patterns are enabled
by the synergy generated from the architectural
hybridization of living, commons, and production,
and are expected to project alternative models of
collective living.

Alternative models of urban living are further
explored in the joint studio of two universities
that takes London and Seoul as their laboratory.
In search for creative new ways to inhabit the
city, Sabine Storp and Patrick Weber of Bartlett
School of Architecture together with So Young
Kim of Hanyang University, lead the students to
focus on the new culture of sharing in the dense
urban environment, and combines it with the
prevalent activities of production to prove how such
permutations of architectural hybrids are capable
of redefining collective living conditions through a
series of unexpected physical behavior.

Thorsten Schuetze of Sung Kyun Kwan
University leads a true collective workshop
comprised of seven universities from Germany and
Korea involving fifty-two participants to propose
a design and analysis project of an "Innovative
District" in Daegu as a possible model for the

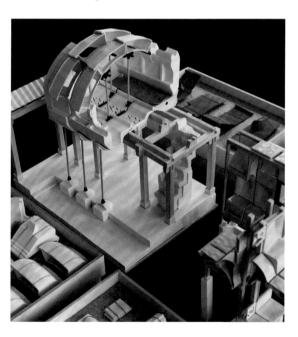

공유도시
바틀렛건축대학교 + 한양대학교 에리카

Sharing Cities
Bartlett School of Architecture +
Hanyang University Erica

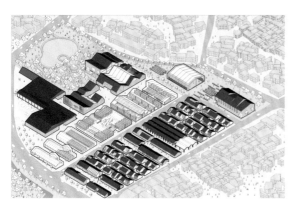

우암동 소막마을 재생
동아대학교

Country for Old Men
Dong-a University

조직하여 공동체 간의 공유, 조율, 대화, 적극적 참여의
기회를 촉진하는 데 초점을 맞춘다.

고령화 시대의 주거
한국의 도시들은 사망률과 출산율 하락으로 말미암아
유사 이래 최초로 마이너스 인구 성장을 경험하고
있다. 2개 스튜디오가 인구밀도가 높은 도시에서 빠른
고령화로 인한 도시 축소와 생활환경 변화라는 주제를
다룬다. 동아대학교 차윤석 교수가 이끄는 스튜디오는
도시 구조의 밀도에서 부산 소막골의 집단 상호작용을
활성화할 단서를 찾는다면, 한양대학교의 라파엘 루나
교수가 이끄는 스튜디오는 다양한 유형의 '공유지'
주위에 마련되고 하이브리드 기반시설의 형태를 취하고
있는 서울의 대안적 도시 주택 유형을 탐색한다. 두 가지
사례에서 공동 주거지의 집합적 행동은 기술 발전으로
보강된 새로운 도시 생활환경의 실험에 중요한 지침이
된다.

집합 도시와 시장
우리말에서 '도시'와 '시장'이라는 단어에 시장을 뜻하는
같은 한자어가 들어간다는 사실을 생각해보면 시장에서
이루어지는 적극적 집단 상호작용이 도시의 형성과
확장에 중추적인 역할을 해왔다고 해도 과언이 아니다.
따라서 현대 시장의 환경에 대한 연구는 건축 환경에
새로운 형태를 더하며 동시대 집합 문화와 공존하는
집합적 활동의 실질적 패턴을 반영할 수밖에 없다.
서울시립대학교의 마크 브로사 교수는 시장이
단순히 상업적 교환의 장소일 뿐만 아니라 집합적 문화
교류의 중심지이며 공동체에 힘을 부여하는 사회적
관계가 응집되는 장소라고 밝힌다. 국민대학교와
서울시립대학교는 이런 생각을 한 차원 높은 수준으로
끌어올리고 집합 도시를 형성하는 촉매제로서 시장의

collective city. In order to be truly collective and
innovative at the same time, the emphasis is
placed not only on the innovative architectural
solutions but also in organizing a sustainable social
system that provides a stable foundation fostering
opportunities of sharing, cooperation, dialogue
and active participation among the collective
community.

Housing the Aging Population
Korean cities are experiencing negative growth
of population for the first time in history due to a
combination of low mortality rates and low birth
rates. Two studios deal with the topic of shrinking
city and changing living conditions caused by a
rapidly aging population in the dense urban areas.
If Cha Youn-Suk's studio at Dong-A University
looks for clues from the density of the urban fabric
to facilitate collective interactions in the Somak
village of Busan, Rafael Luna's studio at Hanyang
University explores alternative typologies of urban
housing in Seoul that are proposed around various
types of 'commons' which take the shape of
hybrid infrastructures. In both cases, the collective
behaviors of cohabitation are taken as key
directions in guiding the experiment for new urban
living conditions that are augmented with advances
in technology.

Markets in the Collective City
When we contemplate on the fact that the
Korean words 'city' and 'market' share the
same etymological root, it is not a far stretch to
speculate that the active collective interactions
prevalent in the markets have played a vital
role in formulating and expanding the urban
presence. Therefore, the study of the conditions
of contemporary markets will reflect the actual
patterns of collective operations that constitute
new forms of architectural conditions that are
aligned with the concurrent collective culture.

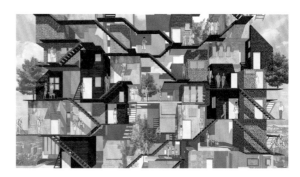

슈퍼마켓 : 초집단 주거
국민대학교

Supermarket : Seeking Super-
Collective Living
Kookmin University

가능성이 무엇인지 탐색한다.

최혜정 교수를 비롯한 국민대학교의 교수진은 학생들과 함께 실제 거래와 교환이 가상 온라인 공간에서 이루어지는 새로운 시장 문화 분석을 바탕으로 고도의 집합 생활 전략을 탐구했다. 학생들은 '슈퍼마켓'을 통해 전통적인 시장의 물리적인 성격을 대체하는 교환 경험을 재창조함으로써 학생들은 건축을 도구로 활용하여 아무것도 없는 상태에서 집합성의 형식적 관계를 새로이 구축하는 실험을 했다. 이와 마찬가지로 서울 시립대학교의 마크 브로사 교수와 변효진 교수가 지도하는 학생들은 온라인 쇼핑, 현대 의학, 기타 도시 공동체의 '기술 발전'으로 인해 어려움을 겪고 있는 서울의 약령시장을 탐색한다. 학생들의 작품은 침을 놓듯이 빈 공간을 전략적으로 관통하는 한편 재래시장의 전통적인 교환 패턴을 응용하여 집합 공동체 차원에서 다양한 교환 방식을 제공할 수 있는 새로운 활동, 행사, 프로그램을 제안한다.

마찬가지로 홍익대학교의 김주원 교수와 임동우 교수가 이끄는 스튜디오는 서울 주거 지역에서 복잡하게 얽히고설킨 거리들이 독특한 도시 집합성을 생성하는 역할에 주목한다. 학생들은 다양한 공공 주도 건축 프로그램들을 조밀하게 연결된 거리들과 나란히 보여주는 작품을 제시한다. 거리가 만들어내는 연결망은 결국 사회·경제적 생태계 내에 유기적인 관계를 만들어낸다. 이러한 상호 연계 개발 모델은 집합 도시의 요구와 가능성을 간과하는 상업 주도 재개발 모델에서는 찾아볼 수 없는 집합적 전략을 제시한다.

집합적 운영

집합 도시는 상호작용, 개입, 사회적 영역 교류 등 다양한 창의적 전략을 통해 운영되고 진화한다. 이 같은 집합 활동의 양상은 시장, 길거리, 주택 등의 구조물과 같이 다양한 물리적 모델이나 전통적 또는 현대적 기반시설에서 이루어진다. 그뿐만 아니라 도시는 혁신적인 통치 모델과 적극적인 집합 활동에 힘입어 규칙과 사회적 합의의 균형을 맞출 수 있고 그에 따라 집합성은 공익을 위한 도시 재구성으로 이어진다. 우리는 도시 환경에 대한 연구에서 어떠한 전략을 도출할 수 있을까? 또한 건축은 좀 더 적극적인 집단 교류의 촉매제로서 어떠한 역할을 할 수 있을까?

집합 의례
2019 서울 비엔날레의 주요 주제인 집합 도시는 집단에 의해 주도되고 건축이라는 인터페이스 위에서 수행되는

Marc Brossa of University of Seoul states that markets are not merely for commercial exchange but "centers of collective cultural exchange" and "condensers of social relationships that empower communities." Kookmin University and University of Seoul takes this speculation to a further level and explore the possibilities of markets becoming catalysts in making a collective city.

The faculty members including Helen Choi at Kookmin University invite students to seek super-collective living strategies based on their analysis of new market culture where the actual transactions and exchange now happen remotely in a virtual online space. By reinventing the experience of exchange that supersedes the physicality of traditional markets,—thus, 'supermarkets'—the students are encouraged to utilize architecture as a tool for setting up new formal relations of collectivity from scratch. Similarly, Marc Brossa and Hyojin Byun led their students at University of Seoul to explore the old herbal medicine market in Seoul that are deteriorating from online shopping, modern medicine and other forms of 'technological progress' in the urban community. The works of students attack the vacant voids like acupuncture, while taking advantage of pre-existing patterns of exchange in the traditional market to inject new activities, events, and programs that form different kinds of exchange in the dimension of its collective community.

Likewise, Juwon Kim and Dongwoo Yim of Hongik University lead the studio in adopting the condensed street network of Seoul's residential neighborhood to act as the generator for a unique urban collectivity. The students propose works that juxtapose various public-driven architectural programs onto the tight network of streets, which in turn, yields an organic relationship within the socio-economic ecosystem. These developmental models of inter-connectivity suggest a collective strategy that breaks away from the commercial-driven renewal model that bypasses the demands and opportunities for a collective city.

COLLECTIVE OPERATION

The collective city operates and evolves through many creative strategies of interactions, intrusions, and exchange of social boundaries. The patterns of such collective actions can be extracted from various physical models of marketplaces, public streets, housing, and other structures or infrastructures both traditional and modern. Cities also benefit from innovative governance models and active collective operations that balance out

현대의 의례
영국왕립예술대학교

Contemporary Rituals
Royal College of Art

운영 전략을 중심으로 진화한다. 이 사실을 알고 나면 이처럼 적극적이고 집합적인 상호작용에 기여하는 요소가 무엇인지에 주목하게 된다. '도시 의례'만큼 이러한 현상을 정확하게 설명하는 용어는 없을 것이다. 도시 의례는 다비드 사코니, 잔프란코 봄바치, 마테오 콘스탄초, 프란체스카 델랄리오 등 교수진들의 지도 아래에 영국왕립예술대학교가 출품한 스튜디오 작품의 주제다.

영국왕립예술대학교 교수진과 학생들은 건축을 사회적 의례의 물리적 구현체로 보는 참신하면서도 설득력 있는 접근법을 제시한다. 여기에서 의례는 '공간 창조 도구' 역할을 한다. 이들의 작품에서 '현대적 의례'는 다양한 사회적 상호작용의 흔적을 건축학적 특징으로 구현하는 일련의 과정으로 해석된다. 의례의 기록물들은 몰개성적이고 타성적이어서 집단의 사회적 형성을 구현하는 데 실패한 건축물과 극명한 대조를 이룬다. 그런 이유에서 해당 프로젝트는 동시대 우리 사회의 행위와 활동에 초점을 맞춤으로써 새롭게 떠오르고 있지만 실제로는 '현시대'에 이미 존재하고 있었을지도 모르는 유형의 집합성을 모색한다.

건축에 숨어있는 '의례'에서 도시의 촉매제를 찾기 위한 해당 접근법은 출라롱콘대학교 건축학부 INDA 프로그램의 알리시아 라차로니 교수가 이끄는 스튜디오 작품에서 한층 더 구체적으로 표현된다. 해당 스튜디오는 멋진 분석적 도면을 취합하여 보여주는데, 이러한 도면에는 건축학적 장치에 드러나는 특수한 문화적 순간이 확대되어 나타난다. 우리는 '의례'로 불리는 그러한 순간이 방콕의 집합적 정체성을 구성하지만 도시환경의

the rules and social agreements, which in turn, collectively contribute to reshaping the city for the public good. What strategies can we extract from the studies on urban settings and how can architecture redefine its role as a generator for a more active collective engagement?

Collective Rituals

As we come to understand that the main theme of the 2019 Seoul Biennale evolves around the operational strategies initiated by the collective and performed on the architectural interface, our attention is naturally driven to defining the agents that constitute such active and collective interactions. No other term explains this phenomenon more clearly than 'urban rituals,' the term used to describe the studio works presented by the Royal College of Arts, led by Davide Sacconi, Gianfranco Bombacci, Matteo Constanzo, and Francesca dell'Aglio.
The faculty and students of Royal College of Art propose a novel yet highly agreeable approach to looking at architecture as a physical embodiment of social rituals, where rituals take on the role of 'space-making devices.' Their entry, "Contemporary Rituals," invites us to a collection of various processes of transforming the traces of social interactions into an architectural physicality. The collection of documented rituals function as a direct critique of the generic and inert architecture that have failed to represent the social forms of the collective. The project therefore focuses on the actions and activities of our contemporary society to search for the new emerging forms of collectivity that may indeed prove to be 'contemporary.'

This approach of trying to find an urban catalyst from the 'rituals' embedded in architecture is presented with further embellishments in the studio works led by Alicia Lazzaroni of INDA program at Chulalongkorn University. The studio presents a collection of beautiful analytical drawings that leads us to observing the magnified view of special cultural moments manifested in architectural devices. We learn that these moments dubbed 'rituals' constitute the collective identity of Bangkok, yet remain hidden and vulnerable beneath the physical surfaces of the urban environment. The students analyse, extract and amplify the found vulnerabilities in a way that begins to collectively celebrate the hidden drivers of the collective city, which in turn propose a new perspective in understanding the operations of our urban environments.

Jorge Almazan of Keio University delivers an accurate diagnosis on how our contemporary collective city has surrendered to market forces

물리적 표면 아래에 취약한 상태로 숨어있다는 것을 알게
된다. 학생들은 집합 도시의 숨겨진 원동력을 집합적으로
기념하기 시작함으로써 숨어있던 취약성을 분석하고
추출하며 확대한다. 그 같은 원동력은 우리 도시 환경의
운영을 이해하는 데 새로운 관점을 제시한다.

게이오대학의 호르헤 알마산 교수는 현대 집합
도시가 어떻게 해서 신자유주의 경제 체제에서 거세어진
시장의 힘에 굴복했는지를 정확하게 진단한다. 해당
연구는 기업 주도형 도시화에 안주하거나 모더니즘
모델인 전체주의적 도시계획에 의존하기보다 도쿄의
비공식적 도시 구조에서 볼 수 있는 자연스럽고 즉흥적인
건축 표현을 영감의 원천으로 하는 방향으로 초점을
돌린다. 계획되지도, 기업화되지도 않은 도시 구조에서
찾아낸 소규모 개입은 집합 도시 거주에 대한 대안 전략을
제공할 것으로 기대된다.

집합 도시의 거버넌스

이화여자대학교의 클라스 크레세 교수와
라드바우드대학교의 에르빈 판 데어 크라벤 교수는 도시
재개발이라는 정치 현장에 뛰어들어 도심 재개발에서
건축이 담당하는 역할을 재정의하는 과정에 관여한다.
이들은 지역의 젠트리피케이션과 기존 주민의 추방으로
이어지는 전통적인 재건축 모델에 대한 비판으로

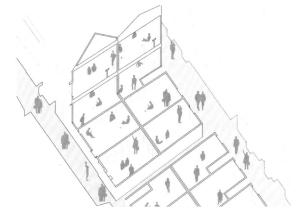

fueled by the neoliberal economy. Instead of
residing within corporate-driven urbanism
or resorting back to the modernist model of
totalitarian urban planning, the research diverts its
focus towards the spontaneous and opportunistic
architectural expressions found in the informal
urban fabrics of Tokyo as sources of inspiration.
The small-scale interventions collected from
the urban patterns that are neither planned
nor corporate, are expected to deliver alternate
strategies of inhabiting the collective city.

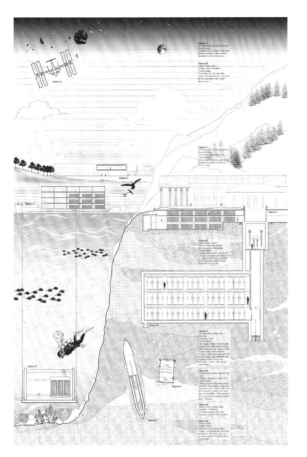

스페큘러티브 시티:
위기와 혼란의 건축적 투사
컬럼비아대학교

Speculative City: Crisis, Turmoil,
and Projections in Architecture
Columbia University

PILaR(+): 을지로의 미래를 위한 후회
없는 시나리오
이화여자대학교 + 라드바우드대학교

PILaR(+): No-Regrets Scenarios
for the Future of Euljiro
Ewha Woman's University +
Radboud University

토지 소유주, 사용자, 임차인 등 이해 관계자 각각이
개발 과정에서 담당하게 될 역할을 재할당하거나
재규정함으로써 미래의 재개발을 위한 청사진을
제시한다. 정치적 정책으로 들릴 수도 있지만 이들의
아이디어는 다양한 이해 관계자의 조그만 역할 변화로
판도가 바뀔 수 있는 게임의 형태를 취한다.

집합 도시에 대한 우리의 이해가 사람과 사람
사이의 사회적 상호작용 양상에 치우쳐 있다면
베를린공과대학교의 교수진과 학생들이 수행한 지도
제작과 연구 프로젝트만큼 강력한 설득력을 지닌
출품작은 없을 것이다. 사회학, 건축학, 도시 설계가
혼합된 다학제간 공동 연구는 지도 제작을 매개로
활용하여 베를린과 인천 송도의 세 지역에서 사회적
영역과 물리적 영역 사이의 숨어 있는 관계를 조사한다.
해당 연구는 평범한 관찰로는 보이지 않는 예상치 못한
발견을 도출하고 집단 공동체가 어떻게 해서 힘을 되찾는
행위자 역할을 하는지 보여준다.

Governance in the Collective City
Klaas Kresse of Ewha Womans University and
Erwin van der Krabben of Radboud University
plunge into the political battlefield of urban
redevelopment and get their hands dirty in the
process of redefining the role of architecture in the
redevelopment of inner urban sites. Equipped with
critiques on conventional redevelopment that often
ends up in gentrified neighborhoods and displaced
communities, they propose new scenarios for
future developments by re-assigning, or re-defining
the role of each stakeholder in the development
process: the landowner, city hall, renters and
architects. As political as it may sound, it actually
takes the form of a game where a slight shift in the
roles of different stakeholders can become a game
changer.

If our understanding of the collective
city weighs highly on the patterns of social
interactions among people, no other entry speaks
stronger than the mapping and research project
conducted by the faculty and students of the
Technische Universität Berlin. This interdisciplinary
collaborative research between sociology,
architecture and urban design, utilizes the medium
of mapping to investigate the hidden relations
between the social and physical spheres of three
sites located in Berlin and Songdo. The research
yields some unexpected findings that are not
visible just from plain observation but prove to
illustrate how collective communities can act as
agents for reclaiming power.

Architectural Speculations
David Moon leads the students of Columbia
University GSAPP into an extensive research
project that refreshes our attention on how
social, political, and economic context influences

몬트리올, 감각의 증강
펜실베이니아대학교

Montreal, Sensate and
Augmented
University of Pennsylvania

건축적 추론

데이비드 문 교수는 컬럼비아대학교 학생들과 함께
사회적, 정치적, 경제적 맥락이 건축과 도시화에 어떠한
영향을 미치는지에 대한 광범위한 연구를 진행함으로써
우리의 주의를 환기시킨다. 학생들은 도시 한 곳만을
사례로 제시하지 않는다. 이들의 작품이 보여주는 다양한
건축학적 고찰에는 동시대 환경을 만들어낼 정도로
중요한 사회·정치적 순간에 대한 강력한 시각적 표상이
포함되어 있다. 펜실베이니아대학교의 사이먼 김 교수는
학생들이 인간 중심 디자인의 전통적인 체계에서 벗어나
인간 아닌 주체를 고려하도록 유도함으로써 이 주제를
또 다른 차원으로 옮겨 놓는다. 이러한 접근법은 공기,
물, 계절과 같은 환경 요건의 행동 양상을 반영하는 설계
과정을 도입하는 데 도움을 준다.

샤이카 알 무바라키 교수가 이끄는
쿠웨이트대학교의 스튜디오는 기업 도시로 변화할 위기에
처한 집합도시의 사례로 메카를 제시한다. 하지만 메카가
처한 위기는 집합적 활동의 부족보다 과잉에서 비롯된다.
전 세계 이슬람교도에게 추앙받는 성지 메카에서는
방문객 급증뿐만 아니라 도심과 변두리 격차로 인해
공동체 활동이 순례자들의 발길이 끊이지 않는 탑과
상점가가 있는 알카바 중심 구역에서 밀려나 갈수록
상황이 악화되어가는 변두리 정착촌으로 옮겨가고 있다.
해당 스튜디오는 지속 가능하고 공정한 발전 가능성을
찾고자 메카의 건축과 공동체의 역사적, 종교적, 정치적
의미를 조사하면서 메카에 대한 심도 있는 연구를
진행한다.

제3의 공간의 집합적 역할

구역, 지구, 동처럼 비교적 큰 규모의 도시 단위 통치
체계를 통한 집합 도시 운영이 갈수록 일반화되고 있듯이

architecture and urbanism. Instead of dealing with
a single city as an example, the students present
a collection documenting many architectural
speculations with strong visual representations
that pertain to the seminal socio-political moments
that have shaped our contemporary environment.
Simon Kim at University of Pennsylvania carries
this theme to another level by encouraging
students to break from the classical hierarchy of
human centric design and allow for a non-human
authorship. This approach helps introduce a design
process that respects the behavioral patterns of
environmental requirements such as air, water, and
seasons.

Kuwait University studio led by Shaikha Al
Mubaraki presents the case of Mecca, another
emblematic collective city under the risk of
transformation into a corporate city, but this time
not by the lack of collective activities but by its
excess. The Holy City adored by all Muslims in
the world, has drastically increased in footprint,
and the disparity between the central district of
Al Kabba with its towers and malls fueled by the
pilgrims, further pushes communal activities out
into the deteriorating settlements in the outskirts.
The studio searches for possibilities that allow
for a sustainable and equitable development by
conducting in-depth research of the Holy City
through mapping its historical, religious, and
political connotations on the architecture and the
community in Mecca.

Third Space as Agents for the Collective

While it is more common to observe collective
operations occurring through the system of
governance applied to larger scale urban units
such as districts, zones, or the neighborhood,
the same principle can also be tested on the
architectural scale seen in the renovation of
industrial heritage. Seoul National University and
the University of Technology at Sydney test out
such patterns of collective operations on their
renovation project for a factory structure from the
1930s. The keyword used in their intervention is
'third space' which Gerard Reinmuth and Andrew
Benjamin of UTS consider as actual 'openings'
that is defined by its "relations of indetermination."
The students of UTS explore and identify the
previously under-used interstitial 'openings' within
the existing structure to assign new programs
such as threshold, mixing chamber, and circulation
that promote active exchanges of events that in
turn, help connect the site back to its collective
urban context. The students of SNU in John
Hong's studio take the idea of 'third space' further
by assigning familiar programs that are directly

도시 리노베이션과 제3의 공간
서울대학교

Urban Renovation and the Third
Space
Seoul National University

산업 유산의 재생에서도 그러한 원리가 건축학적 규모로
실행될 수 있다. 서울대학교와 시드니공과대학교는
1930년대 공장 구조물의 재생 프로젝트를 통해 그 같은
유형의 집단 운영을 시험한다. 이들의 개입에서 가장
중요한 부분은 시드니공과대학교의 제라드 라인무스
교수와 앤드루 벤저민 교수가 '제3의 공간'이라 부르며
'불확실성의 관계' 로 정의한 틈새 공간이다. 시드니
공과대학교 학생들은 이전에는 기존 구조물 내에서
제대로 활용되지 않던 사이 공간으로서의 틈을 탐색하고
밝혀내어 문턱, 혼합실, 동선 등의 새로운 역할을
부여한다. 이로 말미암은 적극적인 이벤트 교환은 틈
공간이 집합적 도시 맥락과 다시 연결되도록 돕는다. 존
홍 교수의 스튜디오에 참여한 서울대학교 학생들은 우리
일상생활과 직접 관련이 있는 익숙한 프로그램을 정하여
'제3의 공간' 개념을 확대한다. 이런 식으로 공간을 설득력
있는 건축 도면으로 변형시킴으로써 보는 사람이 집합
도시 운영을 시각화할 수 있도록 한다.

집합 공동체
홍콩중문대학교의 피터 훼레토 교수는 탄탄한 연구가
되기 위해서는 '실제 환경'에 대한 적극적인 조사를 토대로
한다는 사실을 잘 알고 있다. 홍콩중문대학교 연구진은
급격한 도시화의 뒤안길로 빠르게 사라지고 있는 중국
농촌 마을에 집중한다. 집합적 상호작용의 양상은 이러한
'실제 농촌 환경'과 그 이후 공동체의 '실제' 피드백을

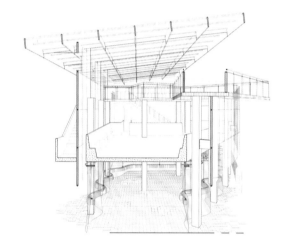

제3의 공간과 집합도시
시드니 공과대학교

Third Spaces and the Collective
City
University of Technology Sydney

relevant to our daily life. Thus, they transform
them into persuasive architectural drawings, which
help the viewers visualize the operations of the
collective city.

Collective Communities
Peter Ferretto at the Chinese University of Hong
Kong understands very well how strong research
is founded on active involvement in investigating
the "actual conditions" even before any researcher
can come up with an analysis. The research unit
of CUHK submerge themselves into one of the
hakka villages of China that are disappearing

실제 마을의 상태: 다시 생각해보는
중국 농촌
홍콩 중문대학교

CONDITION / REAL VILLAGES:
Rethinking China's Countryside
Chinese University of Hong Kong

설계 | 에너지 | 미래: 중국, 슝안 지구
시라큐스대학교

Design | Energy | Futures:
Xiong'an, China
Syracuse University

at an alarming rate under the shadow of rapid urbanization. The patterns of collective interactions are learned from these 'actual rural conditions' and the subsequent design/build project of a communal facility is based on the 'actual' feedback from the community. The project delivers a clear and strong case on how and why the collective communities are going through such transformation.

Fei Wang's studio at Syracuse University looks into Xiong'an as a laboratory ground for the experimentation of urban typologies of the future. This setting becomes possible due to the city's proximity to Beijing and other industrial areas, and from the projection that it will soon grow to accommodate the overflowing urban features already saturated in the neighboring megapolises. The studio takes this opportunity to create a collective prototype that occasionally takes the form of workplaces, living spaces, community gardens, public facilities, and even villages, while allowing open opportunities for collective exchanges to happen.

Jae-Sung Chon brings to the Global Studio curated entries from three Canadian universities (University of Manitoba, University of Toronto, and Carleton University). The entries are the results of a collective investigation on the formal and informal characteristics of Seoul under the common topic of "Block Mutations." The work is a collection of urban impressions on how collective desires mutate and become transplanted into an architectural hybrid, which in turn, constitutes our collective city. It is particularly interesting to see the entries from the three schools together under the same theme because each seems to present varying phases of the hybridization process: While Adrian Pfiffer's studio (Toronto) focuses on interpreting the results of such urban hybridization, Jae-Sung Chon's studio (Manitoba) seizes the clashing moments of collective desires that infuses the familiar with unfamiliar urban conditions.

토대로 한 공동 시설의 설계/건설 프로젝트에서 찾을 수 있다. 이 프로젝트는 집합 공동체가 이런 변화를 겪게 된 과정과 이유를 명확하고 확실한 사례로써 전달한다.

시라큐스대학교의 페이 왕 교수는 미래 도시 유형의 실험장으로 슝안 신구를 조사한다. 해당 연구는 슝안 신구는 베이징 및 다른 산업 지역과 가까이 있기 때문에 가능했으며 이미 포화상태가 된 주변 광역시가 더 이상 수행하지 못하는 도시 기능을 수용할 수 있을 정도로 이 지역이 성장할 것이라는 예측에 근거를 두고 진행된다. 해당 스튜디오는 이러한 기회를 이용해 일터, 주거 공간, 공동체 정원, 공공시설은 물론 마을의 역할까지 할 수 있는 집합 공동체의 원형을 만들어내는 한편 집합적 교류가 일어날 수 있는 기회를 창출한다.

전재성 교수의 스튜디오는 캐나다의 3개 대학(매니토바대학교, 칼턴대학교, 토론토대학교)의 출품작으로 구성된다. 해당 스튜디오의 작품은 '블록 돌연변이'라는 공통주제 아래에서 서울의 형식적, 비형식적 특징을 집합적으로 조사한 결과물이다. 작품은 집합적 욕망이 어떻게 변이하고 건축적 하이브리드에 이식되어 집합 도시를 구성하는지를 도시에 대한 인상으로 보여준다. 무엇보다도 흥미로운 점은 세 학교의 출품작이 같은 주제에 따라 하이브리드화 과정의 다양한 단계를 보여준다는 점이다. 애드리언 파이퍼 교수의 스튜디오(토론토)는 그러한 하이브리드화의 결과와 해석에 초점을 맞추는 한편 전재성 교수 스튜디오(매니토바)는 낯선 환경이 익숙한 환경으로 스며드는 집합적 욕망의 충돌 순간을 포착한다. 오자이어 살루지 교수의 스튜디오(칼턴)는 매우 '일반적'이며 집합적 욕망의 하이브리드화가 이루어질 수 있는 건축학적 토대를 찾아내는 과정에서의 체계적으로 조율된 시도를 보여준다.

2019 서울도시건축비엔날레 글로벌 스튜디오에 참여하는 34개 작품을 준비하고 조직하면서 새로운 생각을 접하는 귀한 경험을 했다. 이 모든 것은 전시를 준비하는 과정에서 계속해서 아이디어를 공유한 저명한 스튜디오 리더들과 재능있는 학생들 덕분이었다. 글로벌스튜디오는 전 세계 여러 명문 대학의 스튜디오 작품이 모인다는 점에서 드물고 값진 기회이다. 해당 대학들은 공통 주제에 따라 디자인과 연구 성과물을 보여주며 그러한 작품들은 건축과 도시화라는 공통 분야에서 다양한 사회적 접근법, 디자인 전략, 그래픽 표현법에 대한 우리 이해를 넓혀줄 것으로 기대한다. 이번 전시가 동시대 도시 환경의 분석에 지침을 제시할 뿐 아니라 건축과 도시 분야의 미래 전략에 영감을 주는 시기적절하고 의미 있는 집합적 노력으로 자리매김 하기를 기대해본다.

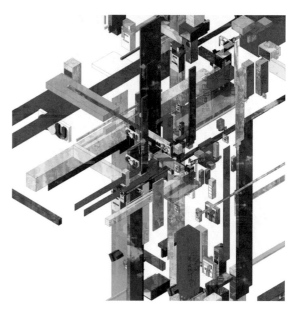

블록 돌연변이: 도시 블록 다섯 곳의 투사적 진화
매니토바대학교 + 칼턴대학교 + 토론토대학교

Block Mutations: Projective Evolutions of Five Urban Blocks
University of Manitoba + Carleton University + University of Toronto

Ozayr Saloojee's studio (Carleton) presents a well-coordinated effort in trying to identify the very 'generic' architectural base upon which all such hybridization of collective desires can be mounted.

Working on the curation and organization of thirty-four entries for the Global Studio exhibition of the 2019 Seoul Biennale of Architecture and Urbanism has brought wonderful and enlightening experiences thanks to many illustrious studio leaders and talented students who made themselves available for the continuous exchange of ideas in the process of preparing for the exhibition. The Global Studio truly is a rare and valuable opportunity to bring together studio works from many inspiring schools around the world. The schools all worked under a common theme to present a collective body of design and research studios, which in turn, allows us a broader understanding of many different social approaches, design strategies and graphic representational methods within the common discipline of architecture and urbanism. I hope this exhibition will be marked as a meaningful collective effort in time, not only for being informative in analysing our current urban conditions, but also for inspiring future strategies in the discipline of architecture and urbanism.

WORKSHOP   DONG

RE-
ACTIVATE

2016

TIMBER
PROBING

2017

PROTO-
TYPE

2018

MASTER-
PLAN

2019

WORKSHOP

TEACHING WORKSHOP

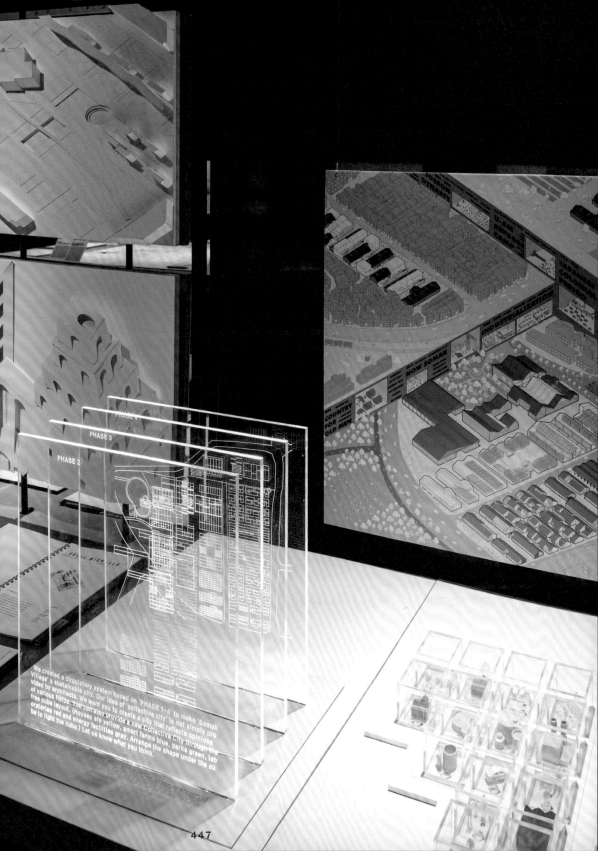

We created a circulatory system based on 'PHASE 1-5' to make 'Somak Village' a sustainable city. Our idea of 'collective city' is not simply provided by architects. We want you to create a city that reflects opinions of various fields. You can also provide a new 'Collective City through the free cube layout. (Houses are yellow, smart farms blue, parks green, laboratories red and energy facilities gray. Arrange the shape under the cu be to light the cube.) Let us know what you think.

PHASE 4
PHASE 3
PHASE 2

2019. 9. 8. 일요일
12:00-18:00
DDP 동대문디자인플라자
디자인랩 3층 아카데미 홀

1부
집합 기반시설

발표자
니라지 바티야,
　캘리포니아예술대학교
조슈아 네이슨,
　텍사스알링턴대학교
카를라 아라모니,
　베이루트아메리칸대학교
에릭 루뢰, 싱가포르국립대학교

토론자
최상기, 서울시립대학교 (진행)
라파엘 루나, 한양대학교
패트릭 웨버, 바틀렛건축대학교

최상기 도시를 인류의
가장 위대한 발명품이라고
한다면 인프라스트럭쳐(사회
기반시설)는 그것을 작동시키는
원동력으로 볼 수 있어요.
왜냐하면 이는 자원을 위에서
아래로든 아래에서 위로든
대중들에게 분배할 수 있는
효과적인 수단이니까요. 따라서
저는 교통이나 자원 등 외에 자연
지형 또한 사회 기반 시설이 될
수 있다고 봅니다. 지형과 밀접한
관계를 가진 조슈아 네이슨의
텍사스알링턴 대학의 스튜디오나
싱가폴 국립대학의 에릭 르뢰
스튜디오 학생들의 작품들을
보면 알 수 있죠. 대지 그 자체뿐
만 아니라 그 위에서 일어나는
행위에 집중하는 것도 같은
결론을 내리는 데 도움을 줍니다.
또한, 오래된 사회 기반 시설이
해변 지역을 나누고 작은 단위의
인프라스트럭쳐들이 그 분절에
의해 발생한 문제들을 완화시키는
베이루트 아메리칸대학의
스튜디오의 작품들과 마찬가지로,
작은 단위의 구조물들이 분리된
두 지역을 잇는 텍사스 알링턴
대학의 작품들은 공통 분모를
갖고 있다 하겠습니다. 제 생각에
스튜디오들 간 겹치고, 공유하는
내용들이 많은 것 같습니다.

라파엘 루나 그렇다면 저의
질문은 이러한 상황에서 건축가의
역할은 무엇인가예요. 니라지
바티야의 스튜디오에서 보여준
거시 영역 규모의 프로젝트에서
우리는 어바니즘 보다는 도시로
회귀하고 있다고 말하는 것
같은데 그 안에서의 건축가의
역할은 무엇일까요?

패트릭 웨버 전 발표된
프로젝트들의 디자인이 각각의
다른 스케일로 적용되는 것이
흥미롭다고 생각해요. 동시에
도시차원의 디자인 요소들을
사람 눈높이로 적용할 수 있는지
궁금해요. 사회 기반 시설들이
사람들의 상호작용을 품을 수
있는 구조물로써 적용된 에릭
스튜디오의 프로젝트와 고속
열차가 커뮤니티를 연결하는
니라지 스튜디오의 프로젝트는
무척 흥미로웠어요. 니라지 씨께
질문을 드릴게요: 사회 기반
시설이 보여주신 과정들을 통해
어떻게 휴먼 스케일로 문제를
확장하고 있다고 보시나요?

니라지 바티야 첫 번째 질문에
대답부터 할게요. 저희
스튜디오의 목표는 학생들이
보통 건축가들에게 보이지
않는 도시화의 규칙 혹은
경향들을 스스로 해석하는
것이었어요. 그 것들은 보통
자본이나 토지 소유권 등에 의해
움직이기 때문이지요. 또한
배경을 개방함으로써 배경을
조직하는 사회 기반 시설과 그
과정에서 발생하는 문제들을
이해하는 것이죠. 스튜디오를
진행하면서 왜 이러한 집중된
노드들이 더 많이 분포되어
있는지 깨닫게 되었는데, 그 것은
조경적인 요소들이 많은 배경을
강조하고 그 것을 많은 자본과
무수한 결정들이 이루어지는
앞의 무대로 끌어오려 하기
때문이었어요. 샌트럴밸리에는
다양한 계층의 사람들이 살아요.
불법 노동자들이 위태롭게 살고
있고 캘리포니아에는 아까
말씀하신 님비현상이 팽배하지만
여기서는 그렇지 않지요. 그래서
우리는 해당 사회 기반 시설을
노출시키거나 전복시키는 등의
과정을 통해 이러한 도시화
과정을 해석하고자 했지요.

Sunday, September 8th,
12:00- 6:00PM
Dongdaemun Design Plaza
(DDP), Academy Hall

Session 1
Collective Infrastructure

Presentation
Neeraj Bhatia, California
　College of the Arts
Joshua Nason, University of
　Texas at Arlington
Carla Aramouny, American
　University of Beirut
Erik L'Heureux, National
　University of Singapore

Discussion
Sanki Choe, University of
　Seoul (Moderator)
Rafael Luna, Hanyang
　University
Patrick Weber, Bartlett School
　of Architecture

Sanki Choe If we agree with
the saying that cities are
mankind's best invention,
we are also admitting
that infrastructure is what
makes it possible, because
it distributes resources to
mass public and works in
both directions of top-down
and bottom-up. If you think
of the infrastructure in that
broader definition, it does not
just apply to traffic and civil
structures, but the natural
terrains can also become an
infrastructure. This is why
the works of UTA in Joshua's
studio can be closely related
to Erik's NUS studio on how
they are looking at the ground
plane as a common field. The
focus on the ground plane
and its activities helped us
conclude that the ground can
become an infrastructure.
One can also find similar
approaches in the AUB
studio where the smaller
scale infrastructures start to
mitigate the old infrastructure
that segregates the coastal
landscape from the inner
lands. I think this resonates
with UTA studio that also
deals with smaller structures
that bind the two areas
segregated by wealth, density
and wetlands. I think there
are a lot of similarities that
are intertwined and correlated
among the studios.

Rafael Luna I guess my
question is what is the role
of the architect in this urban
condition, by going back to
that relationship between
city and urbanism which
architects have been left out
of the project of the city. The
projects that Neeraj showed
go back to more of a city
form rather than urbanism
so I actually had a question
regarding that. These
territorial infrastructures… the
projects you presented were
about city form by almost
countering the urban project.
The projects like prison one
and agrarian city were quite
interesting. So what is the role
of the architect in this new
condition?

Patrick Weber It was quite
interesting for me to see
bringing design to very
different levels. So I have a
lot of questions for you all.
How would you make sure
to bring down urbanism to
the human level. Seeing the
infrastructure just a scaffold
of human interaction and
connection in Erik's studio
was quite interesting. And
for Neeraj's studio, the first
high-speed rail, and how
that connection connects
community. Actually
what I love is translated
or transformed projects/
problems from one place,
as it extends, to another
place. To me the question
for Neeraj is this: how do
you see infrastructure as
something that can actually
extend problems giving a very
different human aspects due
to that progress?

Neeraj Bhatia I will answer
that first question. The goal of
studio was to try to encourage
students to decode some
of the rules of urbanization
which are often more invisible
for architects to see, driven
by things such as capital,
landownership and so forth.
And through opening up the
background and making sense
of the infrastructure that is
organizing that background,
asking how these problems
can be addressed through
architecture. Perhaps the
reason for having more
concentrated nodal forms
is to ask if even within the
confines of an architectural
footprint there is agency to

라파엘 루나 저는 그 프로젝트가 도시화의 해석이라고 보지 않아요. 만약 이 프로젝트가 어바니즘의 해석이었다면 산호세부터 샌프란시스코까지 기차 노선을 따라 이어지는 도시화 지역에 관한 것이어야겠죠. 하지만 제가 본 것은 각각의 기차역에 집중된, 즉 도시 형태에 관한 프로젝트였어요.

니라지 바티야 음식과 농산품의 생산, 관리 및 유통을 해석하는 것은 도시화에 관한 중요한 질문이라고 생각해요. 질문에 대한 대답은 아마 해당 과정을 앞서 말씀 드린 집중된 지점들에 잘 배분해야 하는 것이겠죠. 집중되어 있지만 분배의 기능도 하는.

라파엘 루나 우리는 심지어 도시가 무엇인지 정의를 내리지 못하는 환경에 살고 있는지도 모르겠습니다. 예를 들어, 죠슈아와 에릭의 스튜디오 내용을 보면 건축가는 도시를 최적화의 속도로 보아야 한다고 말씀하시는 것 같습니다. 모더니스트들의

유산을 기반으로 삼아서요. 저는 니라지 스튜디오의 작업을 거점 중심적인 도시의 형태를 제안하면서 도시 공간을 무분별하게 발전하게 방관해선 안 된다고 이해했어요. 그리고 카를라 스튜디오의 작업을 보았을 때는 도시화의 과정이 수자원과 레져 산업, 그리고 환경 생태적인 측면의 새로운 이미지를 주는 기존과 다른 사회 기반 시설의 확산, 즉 앞서 말씀 드린 최적화의 과정으로 보신 것 같은데 여기서도 질문이 있어요: 왜 앞의 기능들이 다 분산되어 있나요? 수자원 자체가 레져 산업, 사회 기반 시설적 가치, 그리고 환경 생태적인 시스템도 다 갖춘 하나의 요소로 볼 수는 없었을까요?

카를라 아라모니 첫 번째 질문에 대답을 하자면 건축가는 건축의 범위가 아닌 여러 문제들을 해결하고자 하는 책임감을 갖고 있어요. 전 학생들이 이러한 이슈들을 분석하고 디자인을 통해 어떻게 우리가 대안 점을 제시할 수 있는지 배우길 바랬어요. 지적하신 기능의 분리에 대해서는

affect the territory. In Central Valley, you have very different classes of people, a lot of undocumented workers who live very precariously, and this causes particular forms of exploitation. We saw the role of decoding those rules of urbanization and somehow exposing those tensions and leveraging or subverting those infrastructures as a way of problematizing the power relationship in itself.

Rafael Luna I think it is not just about decoding urbanization because it would have been continuous urbanized field from San Jose all the way to San Francisco along the train line. But what I saw was proposing each train station which was about city form.

Neeraj Bhatia You can understand urbanization as the continuous mat of sprawl, but it also includes logistical networks. For instance, decoding food logistics is an urbanization question: how agricultural fields are organized, how food is gathered, moved and distributed has particular

rule sets. Unpacking these, we were able to be more precise with an architectural intervention that rewires these networks.

Rafael Luna Going back to the other question, what are you teaching your students in terms of what is the role of the architect in this new setting that we can't even define what a city is. For example, Joshua and Erik, architect should be one that takes the city as a place for appropriation, and modernism becomes the platform for the city to grow within itself. I read yours (Neeraj Bhatia) as we shouldn't even keep on growing in an urban environment, but we still could propose a city project that is nodal and has a city form responsive to one specific point. And yours (Carla Aramouny) is almost like an appropriation, sort of spreading within the urbanization that reevaluates the infrastructure in order to give new image to the city which brought up water, leisure or environment and ecological aspects, which I actually have a question

전략상 프로젝트마다 집중하는 시스템이 달라서였어요: 첫 번째 프로젝트는 결정적 요소들에 대해서, 두 번째는 모든 지점을 연결하는 몸체에 대해서, 그리고 마지막은 해상 동식물 종들에 관한 보존이에요. 하지만 사람이 만든 자연적 구조물을 통해 기존의 환경을 다르게 바라보는 것에 대해서는 저도 동의해요.

죠슈아 네이슨 전 앞서 라파엘과 패트릭이 언급한 두 가지 질문의 융합이 상당히 흥미로워요. 저희 스튜디오의 경우 이러한 상황에서의 건축가의 역할을 명확하게 정의하진 않았지만 작업을 진행하면서 깨달은 건축가의 역할은 건물과 도시를 휴먼스케일로 변환하는 거였어요. 사람들의 다양한 활동들도 개개인의 스케일로 낮춰져야 한다는 것도 말이죠.

에릭 르뢰 건축가의 역할은 우리가 생각하기에 다소 과장되어 온 것 같습니다. 우리는 우리가 생각하기에 무엇을 알고 하는지, 우리의 범위가 어디까지인지 안다고 생각해왔으니까요. 보수적인 환경에서 온 저로서는 아직까지 건축가의 역할은 사적인 영역에서 공적인 영역, 작은 스케일에서 큰 것까지 무척 중요하다고 생각합니다. 하지만 앞서 다뤄진 영토 영역의 스케일에서 건축가가 시민이나 정치적 운동가의 역할을 하는 것에 회의를 느낍니다. 저는 개인적으로 건축가의 역할을 두 개의 영역 중 어느 곳에 더 중점을 둬야 할 지 늘 어려움을 겪어 왔습니다. 아마 둘 다일 수도 있겠지요. 그러나 여전히 저는 거대 규모의 도시계획적 접근에는 점점 더 의심이 들어요. 작은 규모가 아닌 이러한 거대 스케일의 개발은 이제 더이상 의미가 없어요.

라파엘 루나 사실 이 질문들은 당신께서 보여주셨던 1927년 즉, 르 코르뷔지에나 힐버자이머와 같은 건축가들이 도시에 대해서 진지하게 고민하던 시절의 자료들로부터 기인했어요. 또 보여주셨던 상황들은 버나드 루도프스키의 "Architecture without Architect"에서 잘

묘사된 것과 같은 형식이고, 반복된 박스 형태의 주거들로 점철된 현대 서울의 모습 역시 공통적인 내용을 공유하고 있어요. 즉, 결국 이 모든 이미지들에서는 건축가의 역할이 사라졌죠. 그렇다면 에릭 씨는 모더니스트들이 남겨놓은 것들을 우리가 재해석하며 건축가의 역할을 회복해야 한다는 말씀이신가요?

에릭 르뢰 네. 호치민의 경우 지금 엄청난 개발을 겪고 있지요. 한국의 삼성이나 쌍용 같은 회사들이 들여온 자본력으로 타워들을 마구 지어대고 있죠. 많은 도시들이 그렇듯 새로운 개발 모델이 들어서고 기존의 것들은 지워지죠. 그래서 우리는 지금과는 다른 개발 프레임워크를 디자인해야 해요. 하지만 모든 것을 다 정하는 것이 아니라 사람에 의한 변화나 밀도 조절의 가능성 같은 여지는 남겨두어야 하구요.

최상기 이 대화는 이틀 전 주제전 심포지움에서 나온 내용을 환기시킵니다. 안드레아 루비는 과연 건축이 완성되었다고 말할 수 있냐는 질문을 던졌죠, 마치 아기가 갓 태어났을 때 아기가 완성되었다고 말할 수 없듯이 말이죠. 이제부터가 시작인 거죠. 저는 이번 글로벌 스튜디오에 출품된 많은 프로젝트들에서 모더니즘의 건물들을 중성적인 프레임으로 치부하는 접근 방식을 많이 봤습니다. 아마도 건축이 사회 기반 시설로서 인식된다면 그 안에서 이루어지는 사용자들의 모든 집합적 행위들이 "건축가 없는 건축"처럼 조명될지 모르겠습니다. 하지만 동시에 그러한 건축이야말로 바로 우리의 도시를 더 흥미롭게 만들어주는 도구 역할을 합니다.

에릭 르뢰 하지만 그 프레임조차도 엄청난 지적 능력이 수반되어요. 즉, 그 것들에는 여전히 구조적이고 공간적인 열정이 부여되어 있고 그 틀 자체를 어떤 특정한 거주의 기능을 하게 하기도 하죠. 그리고 그 곳에는 우리의 지식을 가져와 융합할 수 있는 가능성도 있고요.

about: why are they separate? Shouldn't they be just water? And then students understand water could be an infrastructure per se but it provides leisure industry, infrastructural value, and environmental and ecological system at the same time?

Carla Aramouny To answer your first question, as architects we have a responsibility to respond to and tackle the issues that are often assumed to be not part of our scope. In the studio, we wanted students to actually look into these infrastructural / ecological issues and try to propose through design and strategic macro visions potential alternatives. As for separation between the themes, I would say each student project had their own concentration in terms of how they strategically engaged with the site: The first project defines water as the main factor that informs the design scheme; the second project focuses on leisure and the connective spine as the main strategy, and the last one focuses on an ecological approach to sustain coastal species. But I completely agree with this observation, that there are overlaps between the three themes, which can all be framed under natural/man-made environments.

Joshua Nason I think the conflation of the initial two questions by Rafael and Patrick are quite interesting. And I will be the first to admit that we did not explicitly talk about what the role of architect is in these environment, but one thing that came out of that is, the study that affected the role of the architect in these settings were actually humanizing scale of a building and a city. The scale of human activity needs to be drawn to the individual projects.

Erik L'Heureux The role of the architect is always expansive; we often like to imagine having powers that we might not really possess – we see this in the presentations by Neeraj and Carla which I am a bit suspect of. As I get older I am becoming

more conservative. I strongly believe that the agency of the architect still has a really important role at human scale: at a scale of the building or a scale of a small cluster of buildings. I know this sounds boring but street, edge, envelope and architecture still have an important role to play today. Decisions here are not dump or overthought, or without intelligence – this is a super relevant domain of architecture and the city. Though I appreciate the works that are proposing to operate at a very large territorial scale, I wonder if it is beyond the agency of an architect. Territorial issues are important to understand as a citizen or as a political activist, or even as a planner – but not necessarily as within the capabilities of architect to affect significant or meaningful change across territory. But we can implement change within a neighborhood, or on a street corner, or on a rooftop. We shouldn't be distracted by the hidden megalomania of large territorial projects.

Rafael Luna Actually the questions came from the references you pulled like Le Corbusier and Hilberseimer from 1920's where the architect was thinking of the city, and then you showed this conditions what was best described as "Architecture without Architect" by Bernard Rudofsky, along with the contemporary conditions of Seoul that is just a repeated cookie cutter of housing. So in all these images the role of the architect is gone, you are proposing a new recuperation that we should take all this as infrastructure. In such case, the role of the architect is to reevaluate the setting that we have inherited which is basically a leftover from the modernists' agendas.

Erik L'Heureux Ho Chi Mihn City is under extreme pressures of development and rapid population growth from Vietnamese pouring in from the countryside. The city is radically transforming as Korean capital flows into Vietnam as Vietnam's single largest investor is South Korea. Towers are sprouting

그래서 저는 그 것들이 특히 진부하다고 생각하지는 않습니다.

라파엘 루나 조슈아 씨는 서울을 공부하는 목적은 더 좋은 서울을 제시하려는 것이 아니라 배운 것들을 가지고 스케일이나 밀도의 측면에서 텍사스에 적용시키기 위함이라고 하셨습니다.

조슈아 네이슨 네 맞습니다. 흥미로운 사실은 텍사스 주의 달라스 지역과 다른 작은 도시들은 미국에서 가장 빠르게 성장하고 있는 도시들이고 수십 년 내에 거대한 대도시 지역이 될 거라는 것입니다. 물론 속도는 아시아의 도시들에 비하면 무척 느리지만요. 제 학생들이 있는 도시들은 아직 빠른 속도로 성장한다는 사실을 인지하지 못하고 있는 것 같습니다. 그 것들은 계속해서 영역적으로만 확장해 나가고 있죠. 그래서 학생들은 서울과 같은 도시들을 보면서 도시가 시간이 흐르며 어떻게 성장하는 지에 대해 놀라울 만한 양의 지식과 정보를 얻었습니다. 특히 도시가 몇 백 년의 세월이 지나며 수많은

사용자와 건축가와 계획가의 목소리들을 통해 변해왔다는 사실을 배우는 것은 특히 유익했습니다. 이제 우리 학생들은 소수의 마스터플랜으로 도시가 바뀌면 안 된다는 것을 깨닫고 이를 몸소 실천해 나갈 것입니다.

2부
집합 생활

발표자
믈라덴 야드릭, 빈공과대학교
사빈 스토프, 바틀렛건축대학교
캘빈 촤,
　싱가포르기술디자인대학교
토르스텐 슈터, 성균관대학교
페이 왕, 시라큐스대학교

토론자
마르크 브로사,
　서울시립대학교(진행)
김소영, 한양대학교

마크 브로사 샘 자코비가 블루프린트에서 언급한 "공동주거는 무기화되었다"라는 인용구를 중심으로 주거에 대해 이야기해 보고자 합니다. 그리고

like crazy. The results of this developmentalist agenda is occurring in real time, but based on a real estate model that is entirely foreign to HCMC. In many ways its breaking the neighborhood bonds that make a collective city. There is an irony of course that in Seoul we are promoting "The Collective" while in Vietnam, Korean developers and builders are promoting a vision of Hilberseimer. As an architect, we can identify frameworks and design those frameworks in a very different way than this large scale tower model. Such a counter proposal needs to keep the loose fit, operate in closer proximity to the exiting city, and work directly with its inhabitants to empower them. We no longer need to micro-managing every detail, rather allow a loose fit inviting intelligent solutions that accommodate density and high demographic pressures without losing the agency of the people of HCMC. This can still be great architecture—it just looks a bit more messy. And this aesthetic shift is just as important as the structural

shift required in our discipline.

Sanki Choe This conversation mirrors something that we discussed a couple of days before at the thematic symposium. It was Andrea Ruby who mentioned that architecture cannot be 'finished' the same way a baby is not 'finished' upon birth; it is really just a beginning. I have seen a lot of approaches in the global studio entries that takes on the modernists' buildings as generic frames that invites people to build upon. Architecture is becoming an infrastructure shaped by the collective activities, that leads to "architecture without architect." Perhaps this is what makes architecture an agency for making our cities more collective and interesting.

Erik L'Heureux The frame requires tremendous intelligence; structural, spatial, materially. The frame has a grammar and the architect is the one who can establish this grammar – this is not generic – its hyper local and contextual that works

저는 공동주거는 애초부터 무기화되었다고 상정하고 대화를 시작하고 싶습니다. 다들 아시다시피 1872년 엥겔스의 여러 글을 보면 그는 산업도시에서 노동자들의 주거가 사실은 자본주의의 산물이라고 결론을 내립니다. 그리고 자본주의에 의해 환경이 나아질 것이라는 데에 회의감을 갖지요. 또한 저는 최근 데이비드 매든(David Madden)과 피터 머큐즈(Peter Marcuse)의 주거에 관한 책을 소개하고 싶습니다. 그들은 요즘의 주거난은 시스템의 잘못에서 비롯된 것이 아니라 사실은 제대로 작동되고 있는 데서 오는 당연한 결과라는 결론을 내놓습니다. 공동주거가 무기화되었다는 것은 많은 정치적 긴장감을 함축하고 있습니다. 저는 이러한 긴장감이 여러분께서 스튜디오를 진행하실 때 어떤 프레임워크로 작용되었는지 궁금합니다.

캘빈 촤 네, 공동주거라는 테마는 저희 스튜디오에 중요한 촉매제가 되어 왔고 그래서 저희는 단순히 "주거"이상의 혹은 이외의 다른 기능을 생각해왔지요. 또한

공동주거는 찰스 푸르니에의 "Phalanstère"와 같이 이상적이고 공공의 주거라는 독특한 역사를 갖고 있지요. 비록 2000년대에 들어서 우리는 조금 회의적인 시각을 갖고 바라보고 있긴 하지만요. 그 것은 또한 다른 삶의 방식에 대한 가능성뿐만 아니라 현실적인 부동산의 관점에서도 우리에게 시사하는 점이 있는 것 같습니다. 저는 모든 학생들의 프로젝트가 단순히 주거의 여러 유형들을 분석하기보다 더 큰 차원의 힘 즉, 사회적, 경제적인 역학관계를 다룰 수 있는 도구가 되어가고 있다고 생각합니다.

사빈 스토프 저희는 공동주거에 한정하기 보다는 거주행위라는 개념에 촛점을 맞추어 왔습니다. 거주에 집중하면 갖가지 주거 유형들 이상의 것들로 연구를 확장할 수 있고, 사실 저희 학교에서 제 교육 경험에 비추어보았을 때 하우징은 다른 프로젝트들처럼 "섹시"했던 적이 별로 없었던 것 같습니다. 특히, 주거난에 당면해 있는 런던을 보았을 때 학생들은 점점 더

with the material and structural realities of a place as well as the aspirations and capabilities of the neighborhood that it should support.

Rafael Luna Johsua, you mentioned that your studio was not about "studying Seoul to provide a better Seoul!" but more like a 'Made in Tokyo' type of approach, where we can learn from the condition and bring it back to… say, Texas in terms of scale and density.

Joshua Nason Yes absolutely. I think one thing that it is really interesting is how Dallas region and mini cities in Texas are the fastest growing cities in the US which is incredibly slow compared to most city growth in Asia, but still potentially the cities that will become the largest metropolitan region in the US within just a couple of decades. Of course there are different ways to view that in just a different sets of numbers, but the reality is that the environment my students are working in, is completely ignoring the fact that it is growing quickly and unnoticed: it is sprawling and sprawling. So looking at a city like Seoul, it actually imbues my student with incredible amount of intelligence and information about how a city can grow specifically over time. What we are talking about now in many US cities particularly in Texas happened instantly, and everything is going to be master planned. Realizing that cities grow over hundreds of years and through different voices of users and architects and planners in the city, is quite helpful for my students in particular. So now they don't think they have to come up with all of these master plans, and instead they can drop down this learning in our region or wherever they go and work for.

Session 2
Collective Living

Presentation
Mladen Jadric, Technische Universität Wien
Sabine Storp, Bartlett School of Architecture

Calvin Chua, Singapore University of Tech & Design
Thorsten Schuetze, SungKyunKwan University
Fei Wang, Syracuse University

Discussion
Marc Brossa, University of Seoul (Moderator)
So Young Kim, Hanyang University

Marc Brossa In the introduction text, Sanki quotes Sam Jacoby in saying "housing has been weaponized". I would like to use this provocation to frame the discussion on collective living, and to take it further by adding that housing in the modern period has not just been weaponized at some point: it has actually been a weapon from the beginning. Already in 1872 in a series of articles entitled 'The Housing Question', Friedrich Engels denounced the housing conditions of the working class in German industrial cities, and wondered what could be done about it. He concluded the housing problems of the working class were part and parcel of the capitalist mode of production and thus, of class struggle. He doubted they could be solved under the capitalist system. In a more recent book entitled 'In Defense of Housing', David Madden and Peter Marcuse follow up on this idea by stating the housing crisis affecting populations around the world is not a result of a system failure, but rather of the system working properly. Housing is political. The crisis is systemic. And the idea of housing as a weapon encapsulates a series of inherent dialectic tensions. I wonder whether and how this understanding of housing as a weapon, conflict or tension has been addressed in the studios included in the 'Collective Living' session.

Calvin Chua Yes, housing is indeed a provocation of our studio and that is the reason why we limit the settings and design to something other than habitat, and it has its own history looking back to Charles Fourier's "Phalanstère", This is an ideal, communal and collective living model but of course in our

새로운 모델을 발견해 나가는 데에 오히려 흥미를 느낍니다. 5년 동안 저는 많은 학생들이 다양한 방법으로 프로젝트들을 진행하는 것을 보아왔습니다. 그래서 저는 "무기"라는 것은 어찌 보면 새로운 기회로 받아들여지지 않을까 생각합니다.

토르스텐 슈처 저희는 같은 주제를 한국인의 관점에서 보려고 노력했습니다. 한국에서의 젠트리피케이션을 보시면 경제의 논리에 이끌려 온 것이 사실이고 그 안에는 기존의 건물들을 부수고 밀도가 높은 하우징 타워들이 등장하는 시나리오가 있지요. 그래서 아파트 유닛을 구입해 렌트를 주며 돈을 벌던 사람들은 재개발 후에도 그 활동을 지속할 수 있지만 원래 그 곳에서 거주했던 분들은 더 이상 돈을 벌 수가 없게 되지요. 이러한 유형의 재개발은 사회적 분리를 야기합니다. 하지만 저희 "재il도화" 모델은 유럽에서는 무척 보편적인 방식이고 기존의 건물들을 제거하는 것 없이 진행할 수 있어서 기존에 거주하는 사람들도 계속 생계를 이어갈 수 있습니다.

페이 왕 저희 학교 3년은 모두 하우징 스튜디오를 듣게 되어있어요. 저희가 작업해왔던 프로젝트들은 삶의 방식들에 관한 것이었어요. 최근 수 십 년간 중국에서는 많은 것들이 바뀌었지요: 1950년대-1970년대는 공공주거, 1990년대에는 주상복합이 화두였습니다. 요즘에는 많은 사람들이 1970년대의 공공이라는 개념으로 회귀하고 있지요. 저희는 그 움직임에 초점을 맞추고 최근 2년동안 선전의 도시 주거에 대해 연구해왔습니다. 도시화가 사회적, 정치적, 경제적 그리고 문화적 맥락에 종속되어 있고 그 것들은 상호 밀접한 관계가 있다는 사실은 정말 흥미롭습니다. 주거라는 것은 저희가 다뤄야 하는 것들 중 무척 기초적이고 중요한 프로젝트에 속하는 것 같습니다.

김소영 하우징에 대한 대화의 연계선상에서 서울, 런던, 빈 그리고 대구 등에서 국제 협력스튜디오를 진행하신 분들께 질문을 드리고 싶습니다. 각기 다른 백그라운드에서 온 학생들이 프로젝트들을 진행하면서 마주하는 이슈들을 어떻게 다뤄왔는지 궁금합니다.

믈라덴 야드릭 우리는 우리가 평행 현실에 살고 있다는 것을 인정할 필요가 있습니다. 이 세계엔 아마 하나가 아니라 네다섯 개의 현실들이 존재하고 있는지도 모릅니다. 정치적, 문화적 갈등이 있는 빈의 접근 방식을 다른 곳에 적용할 수는 없죠. 하지만 저는 국제스튜디오와 같은 공동 작업이 필요하다고 생각합니다. 왜냐하면 그 과정 속에서 해결책이나 경험을 공유할 수 있기 때문이죠. 싱가포르나 홍콩, 인도, 인도네시아 혹은 중국의 건축가들이 공동 작업을 하면 이상적일 것 같습니다. 급속도로 성장하고 있는 중국의 도시 문제들을 다른 나라의 그것들과 비교하긴 힘들겠지만요.

토르스텐 슈처 국제적 공동 작업은 퀄리티를 반영한다는 점에서 꼭 필요한 것 같습니다. 유럽에서 온 학생들은 기존의 도시 조직을 보존하는데 힘을 쓰고 공간의 질과 같은 요소들을 생각하며 논의하기 시작합니다. 그래서 저는 한국학생들에게 있어 문화적으로 기존의 것들을 다르게 보거나 부모님 세대에게서 물려받은삶의 방식에 대해 새로운 질적 관점을 적용시켜보는 이 배움의 과정이 정말 중요하다고 생각합니다.

마크 브로사 저는 사빈 씨가 언급하셨던 "공동주거는 섹시하지 않다."로 되돌아가보고 싶습니다. 저는 이 것이 현재의 우리 상황을 잘 반영하고 있다고 생각합니다. 노동자를 위한 집의 제공을 위해 공동주거는 모더니즘의 중요한 기초가 되었지만 1980년대의 신자유주의 이후 건축가들은 주거이론 등에서 배제되었고 공동주거는 시장논리에 남겨졌습니다. 하지만 2008년 경제위기 이후 도시화나 주거는 자본의 종속에서 벗어나 오히려 중요한 동력이 되었습니다. 저는 이 상황이 우리의 공동주거나

case of 21st century, we look at it in a much more cynical manner. And it is a way to speculate for a possible way of living and also speculate it in terms of real estate. So that is the numbers in this case is no longer utopian ideal but rather the reality that we have to deal with. And I think all of the students' projects on housing itself becomes a tool to discuss the greater forces like socio-economic dynamics rather than just designing a housing in terms of typology.

Sabine Storp Our point of view is that we tend to name it 'habitation' more than 'housing.' The habitation allows you to speculate more than just a typology and from the point of the teaching experience from our school, it is always not as sexy as any other kind of things, and students tend to look away from the housing. Especially in London which is facing housing crisis, students more and more discover that actually it is not just really important but it is quite interesting to start to think new models. Dealing with this for five years intensively have shown that a lot of students are now trying to investigate in very different ways. So I think "the weapon" in that sense, kind of returns to new opportunities.

Thorsten Schuetze Our project was trying to address collective living from the Korean perspective. When you think about urban marginalization in Korea, so far it was mainly driven by economic forces like deconstructing existing buildings and increasing the density. More housings were created when you look at the urban typologies in Seoul. The other point is that people who own multifamily buildings earn their living by renting spaces. When there is an urban redevelopment to apartment buildings, then the building owners get one apartment unit assigned but they cannot earn their living anymore by renting out smaller spaces. This type of redevelopment by nature generates segregation. In contrast, the urban regeneration and densification model applied in

our studio is very common in Europe. It is possible to create more space for collective living in the existing fabric without demolishing the environments, and building owners can still earn their living by income from rent and lease.

Fei Wang In my school our third year undergraduate studio requires everyone to work on housing issues. The projects we have been working on were about ways of living. During past decades in China, a lot of things have changed: 1950-70 social housing, and 1990's and on: commercial housing. Right now a lot of people go back to the communal idea of 1970s which shows how people can live all together rather than having everyone so isolated. We tried to take a closer look at that, and during the past two years we were working on the urban villages in Shenzen; it was quite phenomenal to see urbanization in social, political, economic, and cultural context as they are all inter-related. Housing and collective living are really something fundamental we have to deal with.

So-Young Kim In continuation of housing as a platform, I am going to direct questions to presenters who did a joint studio where students were working in Seoul, London, Vienna and Daegu. I wonder how students from different background observe, propose and deal with issues of collective living that they encounter.

Mladen Jadric I think we have to accept that we are living in parallel realities. There is not only one but four or five worldwide. You cannot take one recipe which is working in Vienna where there are political issues and cultural clash of society that resulted in a political agreement based on hundreds of such experiences. But I think it is still necessary to work together because some recipes of goods, solutions, and experiences can be shared. Countries could be like Singapore, Hong Kong partly, Doshi in India who is doing amazing things,

더 나아가 건축에 대한 접근에 어떻게 영향을 끼쳤나 궁금합니다. 최근에는 단지 주거 유형학적인 접근뿐 만 아니라 사회적인 요소와 같은 새로운 관심사항이 대두되는 것 같기 때문이죠.

사빈 스토프 저희 스튜디오는 주거 자체에만 초점을 맞췄습니다. 학생들이 공동주거를 다루고 나서는 꼭 왜? 라는 질문을 초기부터 던져왔죠. 결국에 학생들은 그러한 질문들이 얼마나 중요한지 깨닫게 됩니다. 왜냐하면 그 누구도 우리가 당연시 여기는 공동주거의 요소들에 대해 왜? 라는 질문을 던지지 않기 때문이죠. 왜 우리는 주방이 필요하지? 왜 방들은 싱글 혹은 더블 룸으로 분류가 되지? 모든 가족은 서로 다 다릅니다. 따라서 우리는 다르게 살아야 합니다. 이러한 일련의 질문들은 학생들로 하여금 공동주거나 주거 유형들을 기존과는 다르게 생각하도록 합니다.

믈라덴 야드릭 빈에는 많은 젊은 사람들이 살고 있습니다. 그들 중 37%는 혼자 살고 있는데요. 이는 그들이 필요한 공간은 50제곱미터를 넘기지 않는다는 것을 뜻합니다. 그 친구들은 때때로 공간을 주거나 일 혹은 그 둘 모두를 위해 공간을 공유하기도 합니다. 즉, 젊은 친구들은 전통적인 삶의 방식을 고수하는 이전 세대들과는 완전히 다르게 살고 있는 것이죠. 저는 아직 이 모순을 어떻게 해결해야 할 지 잘 알지 못합니다.

캘빈 촤 말씀하신 논점에 살을 더 붙이자면, 저는 효율을 고려한 주거 유형을 다루는 것이 바람직한가 생각하게 돼요. 왜냐하면 우리가 개발자들을 이길 방법은 없기 때문이죠. 그들은 무척 효율적인 주거 모델을 영국 심지어 싱가포르에서 계속 발전시켜왔기 때문이죠. 우리가 방의 타입에 대한 고찰을 중단한다면 "주거"는 아마 경계의 공간 즉, 복도나 계단 등에 존재하고 있을지도 몰라요. 그래서 저는 사람들이 모이고 교류를 하는 이러한

공간들을 더 발전시키는 것이 19세기적 생각인 방의 유형들에 대해 생각을 하는 것보다 더 중요하다고 생각해요.

마크 브로사 토론을 마무리하기 전에 저는 캘빈 씨가 보여주셨던 맨 마지막 슬라이드의 "우리가 자본주의에서 벗어날 길은 없다."에 대해 이야기를 하고 싶어요. 그리고 기존 사람들의 사는 방식을 보고 우리가 배우는 것, 맹목적으로 숭배하는 것, 하지만 결국 우리는 그 것들에서 배제되는 것에 대해서도요. 그렇다면 우리는 어떻게 이 것들을 캘빈 씨가 말씀하셨던 경계의 공간에 적용시켜야 할까요? 스튜디오를 가르치는 방법론에 대해서 이야기했던 것들도 무척 흥미로웠어요. 오늘 나누었던 대화들은 정말 유익했습니다. 모두들 감사합니다.

———————————————

3부
집합 운영

발표자
호르헤 알마산, 게이오대학교
알리샤 라차로니, INDA,
　출랄롱코른대학교
전재성, 매니토바대학교
제라드 라인무스,
　시드니공과대학교
파들리 아이작스,
　케이프타운대학교

토론자
존 홍, 서울대학교
클라스 크레스, 이화여자대학교
프란체스카 로마나 델랄리오,
　왕립예술대학교

존 홍 저는 나오신 모든 분들의 말씀에 동의해요. 저도 보여주셨던 굉장한 스튜디오 같은 작업들을 진행하고 싶네요. 하지만 제가 그러지 않는다면, 저는 무엇을 할까요? 우리는 건축가이므로 우리는 항상 옳아요. 하지만 만약 완전히 옳지는 않다면? 한 분씩 돌아가며 제가 크리틱을 해볼게요. 먼저 호르헤 씨. 마지막 슬라이드는 제게 특히 공격적이었어요. 큰 스케일의 개발과 작은 스케일의

Indonesia and China partly. But it is very hard to compare problems in China where cities are doubling every five or ten years.

Thorsten Schuetze I think this international collaboration is very helpful to reflect on the quality of living, because in Korea the approach in general is that it's better to be in apartment buildings, and that the densely grown neighborhoods are generally less attractive. However, the students from Europe become enthusiastic about preserving existing urban structures. By discussing the qualities of space and all that can be reactivated, I think it becomes a very important cultural self-reflection process on seeing things in different ways, and revisiting what the younger generation had been learning from their parents in terms of living quality.

Marc Brossa I would like to go back to what Sabine mentioned about "housing not being sexy", since it reflects where we are as a profession. Housing projects became test beds for the new urban and architectural ideas that gave birth to the Modern Movement since the early 20th century, but since 1980s the neoliberal turn brought increasing deregulation and the decadence of the welfare state. The provision of housing shifted largely to the private sector. Architects were alienated from the process and the discipline deserted housing theory. Since the 2008 it has become increasingly evident that housing and urban development are not just byproducts of capitalism, but actually some of the main forces driving it. If we acknowledge that, What critical approaches to the housing issue can be used to re-conceptualize the architecture of dwelling for a post-neoliberal period? Is the collective the answer?

Sabine Storp Our studio is just very strict. They can only do housing and interpretation which brought a lot of questions like why? at the beginning, and at the end they all realized how important it

is. But it is all based on the realization that people think what is housing? And nobody actually questioned if we really need them. Do we need kitchens? Why room has to be double or single bedrooms? Families are different. So we could live together differently. These are all questions that are forcing students to think housing typologies or habitation differently.

Mladen Jadric There are a lot of younger Viennese in the city. The percentage of singles is 37% already in Vienna so a lot of people chose to be alone, which means they are not looking for anything bigger than 50sqm. They sometimes share their space for living or working or sometimes both so I think youngsters are living in a totally different way whereas older generation still wants to preserve conventional way of living and therefore housing. I am not sure how to solve this conflict in a long term.

Calvin Chua In order to build on that point, this makes me question whether it is still fruitful to discuss about typology in terms of efficiency. Because there is no way we can outgain the developers. They have been developing much more efficient layouts in UK or even in Singapore. If we abandon the typology of the room, perhaps the act of living itself will exist in the spaces of threshold in a much more elemental dimensions such as corridor and staircases. They are the places for exchange and places for people to gather. Maybe it makes more sense to spend more time investigating these spaces rather than traditional 19th century typology of rooms.

Marc Brossa To wrap it up, I would like to turn Calvin's solemn last slide which read "no escape from capitalism" into an open-ended conclusion by suggesting that, rather than looking for an unlikely escape from capitalism, the role of architects should be to facilitate alternatives to the present housing system. There were other issues I wanted to discuss, such as the fetish for informal practices,

분석의 대결 구도였죠. 둘을 완전히 양분되면서요. 저는 호르헤 씨가 그 둘 사이에 중간 스케일의 공간을 놓치지 않았나 생각해요. 다음 알리샤 씨. 저는 "취약한 상태"라는 단어가 참 좋았어요. 하지만 저는 알리샤 씨께서 보통은 눈에 잘 보이지 않는 일상의 요소들을 엄청난 색들과 시각적 제스처들을 통해 페티쉬화하지 않았나 싶어요. 재성 씨가 다음이었나요? 재성 씨는 학생들을 고문했어요! 최저 시급을 받는 노동처럼 당신은 학생들의 노동력을 착취했지요. 제라드 씨께서는 제3의 공간의 생성에 대해서 말씀하셨는데, 그렇다면 우리는 클라이언트를 속이며 교묘하게 제 3의 공간을 집어 넣어야 하나요? 마지막으로 파들리 씨. 발표하시는 내용 중에 전원 도시를 보았는데 이는 기본적으로 스스로 생존하고 발전해나가는 민주적인 조건을 가지고 있어요. 하지만 파들리 씨는 아이콘이 될 수 있는 권리를 제거했어요. 왜 우리는 기념비적이고 특이한 건축을 할 수 없죠? 왜 우리는 항상 획일적이고 지루한 건축만 해야 하나요? 파들리 씨는 그들은 아이콘을 가질 수 없다며 사람들의 열망에 한계를 부여했어요.

<u>프란체스카 로마나 델랄리오</u>
단어 "collective"의 한국어 번역은 "집단"이 맞죠? 저는 이 "collective"라는 단어의 역할이무척 미묘하다고 할 수 있는데 "공동의," "공통적인," "집단의"과 같은 단어에 대체될 수 있죠. 또한 수식 받는 명사를 순식간에 "공공 장소"로 만들어 버리거나 공공 장소를 "집단의 장소"로 바꾸기도 하지요. 저는 제시된 질문들에 약간 덧붙이고 싶은데요, 제 생각에 이번 발표자 분들의 작업들은 오늘 다루었던 많은 주제들을 포함해 다루고 있는 것 같아요. 첫 번째 주제는 건축가나 건축의 역할이었고 마지막 패널 분께서 문화인류학적으로 다뤄주신 부분은 제가 재직하고 있는 영국 왕립예술대학의 스튜디오와 개인적인 연구분야와 일치하는데요, 저는 도시를 문화인류학의 한 형태로 볼 수 있다고 생각합니다. 이야기가

and methodology of the studio which I thought that was very interesting from all of you. I appreciate all the conversations we had. Thank you so much.

Session 3
Collective Operation

Presentation
Jorge Almazan, Keio University
Alicia Lazzaroni, INDA, Chulalongkorn University
Jae-Sung Chon, University of Manitoba
Gerard Reinmuth, University of Technology Sydney
Fadly Isaacs, University of Cape Town

Discussion
John Hong, Seoul national University
Klaas Kresse, Ewha Womans University
Francesca dell'Aglio, Royal College of Art

<u>John Hong</u> So I agree with everybody. I would love to do studios like you all did. It was quite incredible but, if I wasn't to do a studio what would I do? We are all incredibly correct here because we are architects. But what if I was not completely correct? How would it be seen? So let me just go back one at a time on how you took the bait. First, Jorge, so the last slide was very offensive to me. You split it up, basically between large scale developers versus small scale analysis, and you did it through very beautiful drawings. And you put this dichotomy between the two. I'm wondering if there is a spatial condition or in-between spaces around those large and small scales that you might have missed. Next Alicia, I love the term "vulnerable states." One of the things that you did was you took the daily practices and you fetishize them in this book you made with incredible graphics and bring them forward to look at. But at the end of the day this is kind of fetishization of the everyday. It is usually invisible "everyday" but through the colors and visuals, I think it is fetish. Was Jae-Sung next? Wow you tortured your students! What you did was you made

them labor like underpaid labors of that whole area! You made them suffer slum. As for Gerard, you talked about third space, but then you talked about the developer's brief. So basically what you were saying is to come in and deviously put it in this third space, just carve them in but we don't tell the client. Is that what we do? Tricking clients? And finally Fadly, I saw garden city, this drawings that were basically democratic field conditions of self-sustaining and self-made additions. But what you have taken away from these is the right to iconography. Why can't we have something monumental, singular architecture? Why do we always have to have kind of flat things. You were also putting the ceiling on aspirations of people. You are saying that they don't deserve iconography.

<u>Francesca dell'Aglio</u> This translation of "collective" into Korean is "communal", right? So I would like to add into conversation that the role of the word "collective" which everyone uses and is interchangeable with "communal," "commonable," "collectiveness", is quite a delicate noun which immediately defines a public space, and sometimes substitutes public space with a "collective space." I think that is quite an interesting discussion of the biennale in general especially from the perspective of educators. I would add to all these very straightforward questions, what I think that these panels did somehow summarize a lot of topics that emerged throughout the day. One of the first topics was certainly the role of the architect and architecture, but the last panel also brought forward the anthropological aspect which has direct connection with our studio at Royal College of Art, and my personal research in general. It is parallel with the architecture looking up to anthropology, so we can see the city as an anthropological form. In order to answer the initial question, I think first of all architects are citizen. And where does the agency of the architect sit? I find it very interesting that the idea of the

잠시 다른 곳으로 흘렀는데요, 첫 번째 질문에 대답하자면, 전 건축가 또한 시민이라고 봐요. 건축가의 활동 범위가 어디까지 확장되든, 최대한의 활동 범위로서의 프레임이라는 생각은 참 흥미로워요. 왜냐하면 그 것은 유연하고 변할 수 있어서 우리가 영향력을 끼칠 수 있거든요. 저는 고정된 프레임을 가진 프로젝트들을 주로 다뤄왔는데요, "어떻게 건물이나 공간이 도시를 모을 수 있을까?" 라는 물음은 공동성의 의미를 생각하는데 도움을 주어 왔습니다. 제가 드리고 싶은 질문들은 이 것들이에요: 우리의 교육 방식이나 모델은 무엇이고 어떻게 이 것들이 공동체나 공공프로젝트에 녹아들 수 있을까요?

호르헤 알마산 첫 번째 이슈는 공공 공간이로군요. 저는 우리가 퍼블릭 스페이스에 대해 말하는 것이 이제는 지루하다고 생각하실 걸 알아요. 하지만 제 경험에 비추어 보았을 때 적어도 동아시아에서는 공공 장소에 대한 담론이 중요해요. 예를 들어 한국에서의 공공 공간은 무척 적절한 논쟁의 대상이죠. 첫째 식민 시대 때 유입된 "공공"이라는 단어, 일본어로는 "코쿄"의 첫 글자는 제국과 연관이 있어요. 당연히 이 말은 한국 사람들로 하여금 일제강점기를 떠올리게 하기 때문에 환영 받지 못하죠. 두 번째 글자의 뜻은 공산주의를 연상시키는데 이 또한 문제를 갖고 있고요. 따라서 동아시아에서의 공공 공간의 생성 및 발달은 한 순간에 이루어진 것이 아니에요. 그래서 저는 당연히 국가 간 문화의 비교 및 대조를 통해 이 개념을 연구해야 한다고 생각해요. 결국엔 우리가 의견의 일치를 이끌어낼 지라도 사회 정의란 무엇인지에 대한 논의를 수반할 거에요. 저는 공공 공간이나 사회 정의 그리고 지속 가능성과 같은 개념들은 우리가 계속해서 연구하고 발전시켜 나가야 할 주제들이라고 생각합니다. 두 번째 포인트는 중간 스케일 공간인데요. 저는 일부러 그 두 개념을 양분했음을 인정해야겠어요. 사실 제가 한국,

중국 그리고 일본 건축가들을 초대해서 심포지움을 개최한 적이 있습니다. 저희가 낸 결론들 중 하나는 최근 한국의 건축가들은 중간 스케일의 공간을 발전시켜 가고 있다는 건데 이는 일본에서는 일어나지 않습니다. 그래서 일본 건축가들은 한국 건축가들에게 배워야 한다고 생각해요. 도쿄 시는 독립된 건축가들을 통해 현대의 서울에서 일어나는 것처럼 중간 스케일을 발전시켜야 해요.

알리샤 라차로니 저는 확실히 일상 생활의 요소들을 페티쉬화 했습니다. 하지만 그 것이 문제가 되나요? 그 과정은 일상과 직접적인 관계를 맺는 것이고 일반적인 디자인 스튜디오에서 주로 다루는 환경의 변두리를 탐구하는 것이지요. 막상 학생들의 작품을 리뷰해 보면 단순히 주변 사물을 보고 듣는 것이 아니라 그들 만의 흥미로운 접근 방식으로 이야기를 풀어 나갑니다. 제 생각에 가장 중요한 점은 어떻게 그 요소들을 마주하고 관계를 맺는 방식들을 정의하는 지라고 봅니다.

전재성 노동이라는 개념은 일종의 구성 방식 혹은 시스템입니다. 세워진 시스템 안에서 학생들은 사이트에서 노동을 반복적으로 하는 것입니다. 그리고 쉴 수 있는 유일한 시간은 상점에 간다거나 잔다거나 밥을 먹을 때이고 따라서 학생들은 단순히 한국 신문지를 무한히 자르는 것이 아니라 노동 행위 그 자체에서 사는 것입니다. 그들은 스스로 필요한 도구들을 구하기 위해 직접 을지로의 많은 상점들을 방문하기도 했죠. 저는 교육자로서 구성 방식을 생각할 때 두 가지를 생각해야 한다고 봅니다: 청중과 청중 속의 작가지요. 저는 또한 교육 모델로서 "변이"라는 개념을 중요시합니다. 모든 학생이 변이 과정을 통해 바뀌거든요. 그리고 오늘날 우리는 때때로 오리지날 곡을 커버해서 원곡보다 많은 히트를 받는 유튜버들을 심심찮게 볼 수 있는데요. 이는 건축가라는 집단 외의 영역에는 우리만큼 잘하는, 심지어

frame becomes the utmost space of agency because it is something malleable that we can work on. For most of the time I was working on projects that have very rigid frame. "How does a building or an enclosure collect the city?" and this also allowed us to share this idea of collective and commonality. These are the questions that I want to address: what is our pedagogical process or project, and how this can be translated into our unities and unit project?

Jorge Almazan The first issue is public space. I know we are bored about it, maybe it is better to just give up and talk about other things, right? But in fact, that might be true to some areas on this planet, but at least not in East Asia in my experience. For example the concept of public space in Korea is very controversial. First in colonial period, the Japanese word that was used is "Gonggong" or "Kokyo" in Japanese, and the first character has connection to imperialism. Of course Koreans are against that concept of public space because it was referring to imperial Japan. And the second character "Gong" or Kyo" in Japanese is the same as the one meaning communism which is also problematic. So in East Asia, the rise of public space was not an immediate process, I think it is still worth working on it, exploring it, and trying to find meanings through cross-cultural comparison. At the end even if we achieve a certain consensus, then it will obviously prompt a discussion of what it means for social justice which is important. I believe notions like pubic space, social justice and sustainability are things that we should be continuously talking about. Second point about middle scale, I should admit that I deliberately dichotomize them. Actually I organized once this symposium where Korean, Japanese, and Chinese architects were attending, and one of the conclusions was that it is interesting to see Korean architects are trying to develop this middle scale space between large and

small, which is not happening in Japan. So Japanese architects should learn from Koreans. I believe Tokyo needs to develop middle scale by independent architects as it is happening now in contemporary Seoul.

Alicia Lazzaroni I was certainly fetishizing everyday modes of life. Why not? I mean fetishizing everyday life is direct engagement and in a certain way spout like statues of peripheral environments in design studio, why not? Then obviously there is a time to critically deal with work of students, and when you see them they are not just naively appreciating these but they are interestingly approaching the projects on their own ways. I guess the main point is how you enter and define this moment of engagement within it.

Jae-Sung Chon The idea of labor is kind of a setup if you think about it. You set it up in such a way that you have to work in-situ, and you have to repeatedly work on it. And the only time we have to rest is to go to shops, to sleep, and to eat, so you have to live the whole thing. I think that is the pedagogical model of this modeling process. And not only are you cutting through Korean newspapers, but also you are living it by doing it. They searched out the metal, bolts and nuts shops by drawing it and engaging it. So the setup I think as an educator you have to think about two things: audience and authors within the audience. I also think as a pedagogical model, the idea of mutant is very important for me because it is a neutralizer. Everybody was shifted through it. And today's world, we are maybe thinking about great architects but as we know the Youtubers are converting the songs and they are getting much more hits than the original sometimes, so we realized we are professionally educated authors but there are other authors out there who are as good as we are or even better than us. As soon as we accept this, then we start to engage more of a public and collective authorship.

우리보다 실력이 좋은 작가들이 있다는 것을 뜻합니다. 우리가 이를 받아들이는 순간 공공의 작품 활동에 더욱 참여할 수 있게 되죠. 그래서 저는 공공의 도시에서는 다중의 저작권을 고려하는 게 중요하다고 생각합니다. 우리가 질문에 대답할 수 있는 유일한 방법은 저작하고 재 저작하는 것입니다. 이 개념을 받아들일 수 있다면 한 완성된 작품은 계속 다중의 작가들에 의해 재 저작되어가며 다중 저작권을 획득해 나가겠지요. 이는 역사에게 맡겨버리는 즉, 제 생각에는 포기해버리는 상태보다 바람직한 방식입니다.

클라스 크레스 제 생각에 모든 프로젝트들에 최상기 큐레이터님께서 논의 초반에 말씀하신 집단 활동의 중요성이 각기 다른 형태로 존재하는 것 같습니다. 건축 활동을 보는 시각은 동경에서의 그 것처럼 자본이나 큰 빌딩들에 휘둘리지 않고 문화나 삶을 보존하는 수단인 문화적 활동으로 보는 시각과 개발자들에 이루어지는 경제적인 시각으로 나눌 수 있습니다. 개발자들은 재정 관리 등의 현실적인 문제들을 잘 다룰 줄 알기에 우리가 그들의 방식을 이해하지 못하면 그들을 이기기 힘들지요. 이 점은 패널 분들이 보여주셨던 프로젝트들에는 조금 미비하지 않았나 생각이 듭니다. 제가 하고 싶은 질문은 앞서 마크 씨께서 말씀하신 무기화 즉, 개발의 무기화가 필요하다면, 건축가들이 경영이나 재정 관리 등을 더 배우고 능숙하게 해야 하지 않냐는 것입니다. 만일 우리가 그럴 수 있다면, 우리의 제안 및 디자인이 받아들여지고 결국 개발자들이 원하는 것이 아닌 우리 건축가들이 원하는 것을 지을 수 있지 않을까요?

제라드 라인무스 제 생각에는 집합 도시에 대한 존 홍 씨의 크리틱은 일리가 있는 것 같습니다. 신자유주의는 개발 등과 같은 행위들을 그 안으로 희석시켜 버리기 때문에 공공 도시는 딜레마입니다. 우드스톡 세대는 모든 것을 가졌죠. 그래서

결국 교육자의 역할이 그들의 사고 방식에 맞추어져 들어간다는 것은 참 흥미롭습니다. 건축가의 역할을 다른 분야로 확장하는 것은 정말 힘듭니다. 왜냐하면 모든 건축가들은 기후 변화에 대해서 전문적으로 할 수 있는 일이 적기 때문이죠. 매우 복잡한 맥락 속에서 건축가의 역할에 질문을 던지는 것은 중요하고도 어렵습니다. 그래서 우리가 무엇을 해야 하는 지에 대해 물어보는 것은 항상 흥미롭지요. 따라서 존 홍 씨의 질문에 대해서는 "Yes"가 저의 대답입니다. 왜냐하면 우리가 긋는 선 하나로도 10,000명의 삶을 풍족하게 할 수 있으니까요. 제가 만약 건물 스케일의 프로젝트를 진행한다면 저는 아마 공공의 기회를 "교묘히 끼워 넣을" 겁니다. 그리고 다른 종류의 스케일에서의 문제를 풀면 그제서야 모든 딜레마들을 마주친다는 점에서 이는 건축에 있어서 가장 어려운 과제들 중 하나입니다. 아마 우리가 어디로 나아가고 있는지를 진지하게 고려하며 다음 세대들을 가르치는 것 또한 중요한 과제가 되겠습니다.

파들리 아이잭스 존 홍 씨께서 말씀하신 부분은 일리가 있습니다. 우리는 디자인은 항상 단순해야 한다는 생각을 자주 하는 것 같습니다. 제 생각에 이 분야에는 디테일이나 문화적 상정보다는 더 중요한 것들이 있습니다. 우리는 형태에서 디테일이라는 개념을 잃어버리고 있고 사회와 문화의 복잡한 층위들을 잘 인지하지 못합니다. 우리는 글로벌 경제라는 컨텍스트에서 상기의 모델을 받아들여야 합니다. 그리고 저는 학계와 실무 분야가 빠른 속도로 뒤섞여 가는 것을 압니다. 이 둘은 각자의 세계에 서로의 차이를 위해서 존재해야 합니다. 우리가 만약 실무를 무시하기 시작한다면 그 날은 학계가 또한 그들의 기반을 잃는 날이 될 겁니다.

I think it is important to consider multiple authorship in collective city. And the only way we tried to answer the question was to author and re-author. If you can accept that, your output could be re-authored, re-sampled, and re-mixed, and then we can get into the multiple authorship rather than just impasse where we leave it to history to take care of it, which I think is just giving up. That's why I would rather go the other way accepting that it could be re-authored.

Klaas Kresse As a reaction to your provocation, and Sanki's statement in the beginning, where he was addressing the importance of collective operation. I think that was something there in all the project in different ways. If you look at architectural practice, the nature of architecture is also debated like some people see it as a cultural practice like the one in Tokyo where there is an idea of preserving culture and life, otherwise they will be overwhelmed by capital and big buildings. Others see it as an economic practice which is mainly taken over by the developers. As they know how to finance, get things done and so on, it is difficult to win against them if you are not aware of the tools that they use. So that is something that was lacking a little in the presentation and maybe in the approaches. The question I want to raise is whether we need a weaponization (which was previously mentioned by Marc) of development finance. Shouldn't architects be more smart in terms of management and financing. If we know how to organize and get finance for them, can't we just make them like our designs as well, and in the end build what we want instead of what developers want?

Gerard Reinmuth I think you are right with a collective city. So the collective city is a dilemma because we know that neoliberalism folds these actions on to itself. Woodstock generations out-own everything. So in the end what is interesting is that our role as educators is also flipping onto their thinking and I

was wondering if we weren't radical enough, the 16-year-old-dude will be really pissed off as I should be, because anyone is going to have vague working knowledge of climate change research. It is really hard to get into that discourse. I think it is a big question of what is the role of the architect in a very complex context. So it is interesting to ask about what we should be doing. So the answer is yes to your provocation. If you draw a line, and you are convincing a state to do something, you may improve 10,000 people's lives. If our work is scale of building, maybe I can "sneak in" public opportunity. And I think this is one of the great challenges to the discipline by the fact that all the dilemmas we have after we solve all the different scales. So I think you are right. I think the clicking things perhaps has been a distraction because otherwise it is maintaining setups, and perhaps even for all of us I think it is interesting mulling over how are we educating next generation if we really take seriously the trajectory where we are going.

Fadly Isaacs The points you have made make sense. I think that too often our design imagination relies on too singular references. . If you look within self-sustaining and self-made collective fields, one finds layered palettes of everyday details and diverse cultural assumptions. We lose the notion of detail in singular form and I don't think we appreciate much more complex layers of society and culture. We need to accept the model of cities as diverse collectives in the context of a global economy. And I think what has become conflated is the role of academia, the discipline of architecture and the practice of the profession. These spaces are conflated too quickly. Their difference should be recognised and appreciated. . I think that the day that the profession defines academia is the day it has lost the plot.

현 장 프 로

LIVE

PROJE

젝트

CTS

을:地 공:存

2019
서울도시건축비엔날레
SEOUL BIENNALE
OF ARCHITECTURE
AND URBANISM

주최 I·SEOUL·U
주관 Seoul Design 서울디자인재단
기획 YOUNHYUN TRADING 玧呟商材 URBANPLAY

1차 사람의 취향 9/21 ~ 22
2차 사물의 스펙트럼 10/26 ~ 27

윤현상재 보물창고

SEOUL Cl x

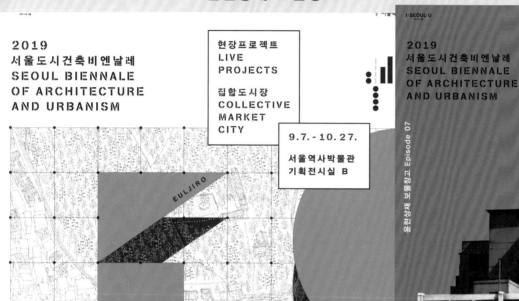

2019
서울도시건축비엔날레
SEOUL BIENNALE
OF ARCHITECTURE
AND URBANISM

현장프로젝트
LIVE
PROJECTS

집합도시장
COLLECTIVE
MARKET
CITY

9.7. - 10.27.
서울역사박물관
기획전시실 B

을 地 공 존

윤현상재 보물창고 Episode 07

EULJIRO
CHEONGNYANGN
NAMDAEMUN
DONGDAEMUN

세운상가
SEWOON PLAZA

Makercity
SEWOON

현 장
프 로 젝 트
LIVE
PROJECTS

2019
서울도시건축비엔날레
현장프로젝트 서울도시장

주최 서울특별시
주관 서울디자인재단
기획 윤현상재

서울도시건축비엔날레
SEOUL BIENNALE
OF ARCHITECTURE
AND URBANISM

집합도시
COLLECTIVE
CITY

2019년 10월
오전 11시
무

2019
서울도시건축비엔날레
SEOUL BIENNALE
OF ARCHITECTURE
AND URBANISM

윤현상재 보물창고
EPISODE 07 을地:공存

시민이 다 함께 즐길 수 있는
공간 조성을 위하여 윤현상재와
코오롱글로텍이 함께 만들었습니다.

PICNIC ZONE

2019.8.27. TUE -
2019.11.10. SUN
다시세운광장

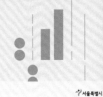

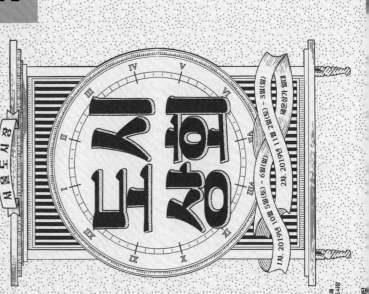

현장프로젝트

장영철, 현장프로젝트 큐레이터

Live Projects

Young-chul Jang, Curator of Live Projects

시장은 집합으로서 도시의 형태를 가진다. 도시는 교환의 장소로 부터 시작되었으며, 그 시장이 밀도가 높아지고, 시장 주위로 다양한 기능들이 부가 되면서 도시의 형태로 진화하였다. 도시의 형태는 산업 혁명 이후의 철도와 도로의 발달, 분업화된 산업구조, 집약된 자본 등은 동력원으로 변화하고 있지만, 그 안에서 원초적인 집합도시 형태로서의 전통시장은 어느 도시나 아직도 건재하다.

전통시장의 시간을 거슬러 쫓아가다 보면, 일상적이며 격동적인 서울 저자거리의 이야기를 만나게 된다. 예를 들어, 서울에서 최초로 조선의 민간자본으로 설립된 배오개(현재 종로4가)에 설립된 동대문시장(현재 광장시장)은, 전태일의 분신, 정치깡패 이정재의 축재(蓄財), 1899년 서대문-동대문 경성전차의 개설, 임진왜란 이후로, 어영청등의 군인들이 월급으로 받는 미곡을 다른 물건으로 바꾸던 자연발생적 난전(亂廛)의 등장등, 일상적이고 굴절된 우리의 역사를 투영할 수 있다. 또한 남대문 시장은 조선시대 재정기관인 공물세로 현물 대신 대동미와 포·전을 받기 위해 설립한 관청인 선혜청이 자리잡고 있었던 곳이다.

변화하는 도시지리적 상황에 가장 민감하게 반응하는 곳도 시장이다. 경강(京江), 포구, 신작로, 전차, 도시계획, 교통망, 재개발등 시간과 당대에 필요한 역할에 따라 재편되는 도시상황에 따라, 시장은 도시생태계에 재빨리 적응하고, 생태계 자체를 재편하기도 한다. 시장이 집합의 힘을 가지는 것은, 그것의 네트워크에 있다. 많은 상가들과 도시 공장들은 거미줄처럼 연결되어 있으며, 정비되지 않은 자연발생적인 도시 조직 내에서 언뜻 불가능해 보이는 연결방법으로 최적화되어 있다. 정교한 기계장치나 생명체처럼, 개별로의 분해는 쉽지만, 그 개별요소들을 다시 조립하려면 어려운 것 처럼 전체가 개별들의 합보다 크다. 이것이 집합이 필수조건이 되는 시장의 힘이며 가치가 된다.

이 원초적인 집합도시인 전통시장의 다양한 관점을

The market is a collective taking on the form of the city. The city began from a place of exchange followed by density of the market and then evolved into the form a city as various functions surrounded the market. As we witness the expansion of railroads and roads, as well as the development toward capital intensive and specialized industries, the market has maintained a consistent presence and typology in many cities across the globe, serving as important socio-economic infrastructure for our dense urban cores.

As you trace the history of traditional markets, you come across the stories of everyday yet dynamic shopping streets of Seoul. For example, Dongdaemun Market (now Gwangjang Market) was Seoul's first market to be established with Joseon's private capital in Baeogae (now Jongro 4-ga). The market reflects our ordinary yet distorted history: from Jeon Tae-il's self-immolation to political bully Lee Jeong-jae's accumulation of wealth, the 1899 establishment of Seodaemun-Dongdaemun Gyeongseong streetcar, *Imjin-waeran* (the Japanese Invasion of Korea), and *nanjeon* (small markets), which was naturally formed as soldiers of Eoyeong-cheong traded rice—their salary—with other goods. Moreover, where Namdaemun Market exists today used to be the office of Seonhyecheong, a tax office of the Joseon Dynasty, where rice, cloth and money were collected for tax instead of in-kind goods.

It is also markets that are the most sensitive to urban-geographic changes. Markets adapt accordingly and even reshuffle the ecosystem, as the city reorganizes to assume the role required by the time—be it Gyeonggang, ports, new roads, streetcars, urban planning, transportation networks or redevelopment. The collective power of markets comes from their network. Stores and urban factories are intertwined like spider webs and optimized with the seemingly impossible connection within the naturally forming, unorganized urban frame. Much like delicate machines or organisms, markets can be broken down into parts, but cannot be put back together

통해서 외적으로는 글로벌 하지만 내적으로는 단절된,
미시적으로는 집합적이나, 거시적으로는 파편화된,
겉으로는 조화롭게 보이지만, 내부적으로는 갈등이
산재한, 우리의 도시문제를 다시 한번 들여다 보고, 함께
도시적 대안을 찾아보고, 의논하며, 서로 배우고 알아가며,
체험하며 즐기는 과정을 만드는 것이다.

또한, 자연발생적으로 발생한 집합적 도시 조직을 가진
전통시장이 다양한 사회적, 문화적 행위를 포함하며,
시민들이 더욱 좋아할 수 있고, 경제적으로도 풍요로운
장소가 되기위해서 건축도시 디자인이 어떻게 기여를
할 수 있을 것인가라는 질문은 프로젝트를 이끌어가는
중요한 화두가 될 것이다.

so easily. As such, markets add up to exceed
the total and create synergy. This, indeed, is the
power and value of markets, with collectivity as an
indispensable condition.

The Live Projects exhibition explores the
challenges currently facing our cities from this
important and universally recognizable urban
form. While Seoul is a global and international
city, it grapples with issues of social isolation and
fragmentation that is reinforced through physical
separation and boundaries. The exhibition explores
urban alternatives in Korea that can strategically
and innovatively address these contradictory
problems of the contemporary condition. Situated
within the heart of the city itself, it actively
engages with the local citizens of Seoul, tackling
the most pertinent questions of the city.

The marketplace serves as an example and case
study of how questions of the collective have
evolved and the socio-political, economic, and
cultural impact of these market models. These
spaces also offer the opportunity to study and
understand how architecture and urbanism can
make a significant contribution to these traditional
spaces of trade and exchange and for them
to continue to be enjoyed by everyone. Public
engagement in these spaces and the exhibition is
central to the project.

집합도시장
집합도시장을 펴내며

유아람

Collecitve City Market
Publishing Collective City Markets

Aram You

"도시는 교환의 장소로부터 시작되었다."

서울에는 수많은 물건, 사람, 자본, 그리고 정보가 끊임없이 움직이고 교환된다. 이들은 개인의 주변에서부터 거대한 도시까지 아우른다. 우리를 둘러싼 교환이 모여들어 물리적 환경을, 지역을, 장소를, 그리고 이 도시를 이룬다.

서울 도성 안 세 개의 시장—남대문, 동대문, 그리고 을지로일대—과 청량리 시장은 조선 후기부터 오늘날까지 한 자리에서 교환의 장이자 도심의 중심지 역할을 해왔다. 모든 물건이 모이고 흩어지는 시장백화점 남대문시장, 독특한 도시산업생태계로 진화한 세운상가와 을지로, 의류산업중심지 동대문시장, 그리고 계절마다 색을 달리하는 청량리시장은 독립적이되 서로 연계된 서울상업공간이다.

집합도시장은 네 개의 재래시장이 갖는 역사를 살펴보는 것에서 출발한다. 도시가 다양한 교환의

"Cities began as spaces for exchange."

In Seoul, there are endless movements and exchanges of goods, people, capital, and information, not only at the local level in an individual's surrounding area, but also at the larger citywide level. The various exchanges that surround us come together to create the physical environment, local regions, spaces, and the identity of Seoul as a city.

The markets of Namdaemun, Dongdaemun, Euljiro, and Cheongnyangni have, since the late Chosun Dynasty, served as spaces for exchange and as Seoul's urban center and continue to do so today. Dongdaemun is known as a sort of department store of marketplaces, where anything and everything can be found. Located here are Sewoon Plaza and Euljiro Market, which have evolved into urban industrial centers; Dongdaemun Market, which is the center of the clothing industry; and Cheongnyangni Market, whose colors change

장이라는 것을 "서울장 연표"에서 시장과 서울의 역사를 통해 살펴본다. 사람과 물건, 자본이 모이는 결절점인 네 개의 전통시장은 100년 넘게 독립적으로, 또 서로의 역학관계 속에서 존재했다. 지금은 마치 고정된 듯 각자의 위치에서 각자의 물건을 파는 것이 당연하게 느껴지지만, 크고 작은 사건들 속에서 이들은 견제하고 또 상생해왔다. 시장의 역사는 서울 역사의 커다란 한부분이다.

과거의 시간을 간직한 채, 네 개의 재래시장은 지금의 모습을 우리에게 전한다. 집합체로서 시장은 아주 오랜 시간동안 형성되어 왔기에, 교환의 과정이 축적되어 온 중요한 기록이다. "현재의 서울장"은 미시적인 눈을 통해 시장의 물리적 유형과 사회적 구조를 살핀다. 시장을 샅샅이 추적하다보면 우리는 숨어있는 사람과 물건의 이야기를 찾을 수 있다.

서울의 재래시장은 다양한 물건들이 가득 쌓여서 우리로 하여금 '오감'을 통해 서울이라는 도시를 느끼게 한다. 고유의 냄새, 습도, 온도, 색과 밝기, 그리고 대화소리. 이들은 시장 전체를 이루는 가게들에서, 가게의 매대, 매대의 상품과 작은 물건들이 모였을 때, 거기에 상인과 행인이 함께 만들어내는 미시적 화학작용의 결과다. 시장은 분명 사람이 만들었지만 통제되고 계획되지 않은, 교환 과정에서 의도치 않게 배어난 자연발생적인 총체이다. 서울의 시장은 서울에서 일어나는 교환의 물화(物化)이며, 그 자체로 교환의 장소이다.

with every season. While they each operate independently, these marketplaces are linked in their identity as the heart of Seoul's industries.

Collective City Markets begins with an exploration into the history of these four traditional markets of Seoul. The "Timeline of Seoul's Markets" provides an overview of the history of Seoul's markets, which have and continue to serve as the city's spaces for diverse forms of exchange. These four traditional markets, where people, goods, and money come together, have existed for over 100 years, both independently and, also, as parts of one another. Nowadays these markets seem naturally settled in their places selling their own particular goods, but they have been through thick and thin, ranging from both small and major events. Indeed, the history of these markets makes up an important part of Seoul's overall history.

While harboring the past, these four markets provide us with a reflection of our present. As collective spaces, these markets have been formed over the course of long periods of time, naturally becoming important records of various exchanges built up over those years. "Today's Markets of Seoul" takes a microscopic look at different physical forms and the social composition of markets. By thoroughly tracing back the histories of markets, we can discover hidden stories of various people and objects.

The traditional markets of Seoul are places where different things gather and pile up, giving

경제발전기
ECONOMIC DEVELOPMENT

90년대와 2000년
90S AND 2000S

교환의 장으로서의 시장은 끊임없이 변화하며 매 순간 도시의 시작점이 된다. 그렇기에 "미래의 서울장"은 작가들의 눈과 귀, 그리고 손으로 재탄생한 서울의 시장을 그린다. 우리가 살펴본 오래된 장소는 이 새로운 모습마저 품어낸다. 새로운 교환의 가치를 담아내어 또 하나의 더 큰 집합으로 만들어내고 있는 중이다.

우리가 살아가는 도시도 바로 그렇게 만들어지고 있다.

us an experience of Seoul's urban identity that stimulates all five of our senses. The various smells, atmosphere, temperature, hues, and the sounds of people talking can be felt throughout the storefronts, and their displays of goods. Here, merchants and consumers engage in a special chemical reaction at microscopic levels, resulting in the markets that we recognize today. Although it is true that people established markets, the identity of these markets result from the uncontrolled and unplanned exchanges that occur in these spaces. Seoul's markets are not only the unit of exchange within the city, but also the space where such exchange occurs.

As platforms for exchange, markets undergo nonstop change and become new starting points of urban identity with every passing moment. Against this backdrop, "Seoul's Markets of the Future" depicts the rebirth of Seoul's markets through the artist's eyes, ears, and hands. The spaces that we have observed over such a long period of time are incorporated in this new image of Seoul's marketplaces. Here, we can see signs of the value of new exchanges and the process of creating yet another large collective space.

The cities in which we live are also undergoing this same process of change.

INTERVIEW 1 : 남대문시장
"재래시장의 특성이 뭐냐면 신속성이에요. 내가 디자인을 구성해서 원단을 들여 가지고 그 주 안에 빨리 소도매를 해야 하는 거에요. 메이커나 대기업 같은 경우엔 6개월 전, 1년 전에 기획 생산에 들어가지만, 재래시장은 흐름을 빨리 읽어서 일주일 단위, 한 달 단위로 신상이 계속 나오면서 유지를 해야 하니까." - 포키 아동복, 아동복 사장님

INTERVIEW 2 : 동묘 벼룩시장
"결혼하면서 남편을 따라 장사를 시작했고 제주를 제외한 전국을 대상으로 공장 3곳에 종업원도 3명이지만 모두 축소된 상태고, 1990년 전후에는 소비력이 좋아서 생산도 하고 공장도 여러 곳에 있었지만 지금은 경기가 좋지 않아서 산지에서 공급을 받아 판매하고 있지..." - 청량리 인삼상인

INTERVIEW 3 : 을지로 세운상가
"청계천이라는 게 하나의 기업이나 마찬가지야. 예를 들어서 말이에요, 탱크를 만든다고 해보자구. 그럼 선반, 칠, 주물 등 여러 업종이 필요하다고. ... 부품이 필요한 것 같으면 몇 개의 협력업체가 성립한다고 봐야 해요. 혼자 할 수 있는 건 작은 거고, 협력하지 않으면 할 수 없어. 왜 바닷가에 모래알 같은 거 있잖아. 멀리서 보면 하나같지만 가까이 자세히 보면 다 다르다구요. 청계천도 마찬가지야. 청계천도 청계천이라고 하는 큰 것으로 보일 뿐이지 자세히 보면 개개인이 다 달라요. 협력하지 않으면 안 돼." - 입정동 소재 기계공장(I-115)

INTERVIEW 1: Namdaemun Market
"Traditional markets are all about being quick. I design, get fabrics, do retailing and wholesaling and finish the same week. Those big brands and companies need to plan and manufacture 6 months to one year ahead. But here, you've got to always read the trend, introduce new items every week, every month to keep up with the pace." – Owner of Poki Children's Clothing

INVERVIEW 2 : Dongmyo Flea Market
"I started the business with my husband after we got married. We target all parts of the country except for Jeju Island. We have three factories and three employees but we've recently scaled down. We had more factories and customers around the 1990s, but the economy isn't doing well now, so we purchase directly from the producers." – A ginseng seller in Cheongnyangni

INVERVIEW 3 : Euljiro Sewoon Plaza
"Cheonggyecheon itself is like a big corporation. Let's say we are making a tank. We need shelves, paint, casting and other materials from different industries. If you need parts, you need to find several contractors. If it's small, you can do it on your own. You can't do it unless you collaborate. Think of a sand grain on the beach. They look the same from afar but when you look closely, they are all different. Same with Cheonggyecheon. It looks like one big thing but every business is unique. You gotta work together." – A machinery factory in Ipjeong-dong (I-115)

서울 2045
오영욱

가까운 미래의 서울을
상상한 입체지도 작업이다.
실제적인 물품 거래가
이루어지기보다는 기억과
인간 본성을 주고받는 장소로
살아남은 재래시장과 그
주변에 몰려 사는 사람들과
인구감소와 기술 발전으로
많은 곳이 숲으로 돌아가게 된
도시를 그렸다.

Seoul 2045
Young Wook Oh

This is a three-dimensional
map imagining Seoul in
the near future. We see
people flocking to these
areas while other parts
of the city are reclaimed
by forests owing to
both sparse population
and technological
advancement.

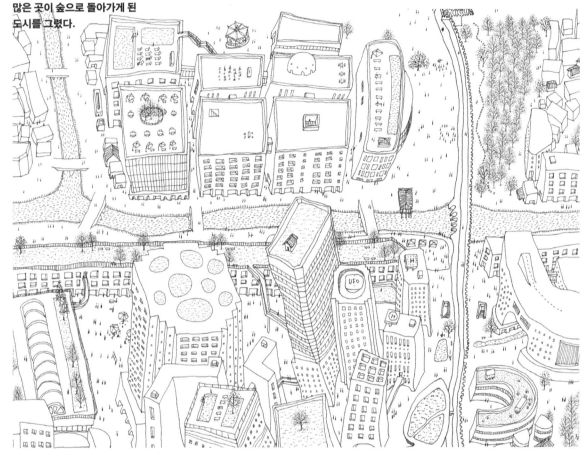

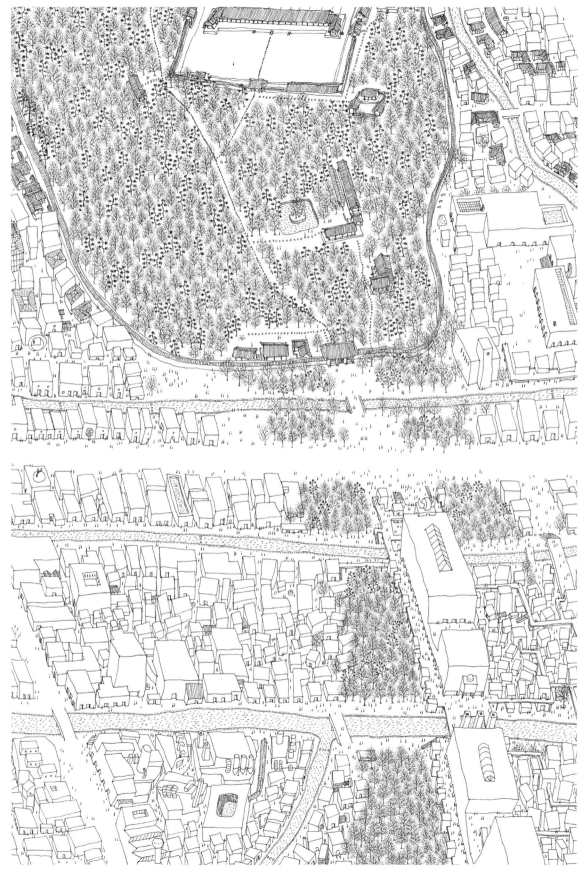

2019 DISPLAY_02
토마즈 히폴리토
공간과 그 안에서 발생하는
활동에 주목하며, 이로 인해
공간이 어떻게 변모해가는지
추적한다.

2019 DISPLAY_02
Tomaz Hipolito
This project addresses the
questions of space, the
activities occurring within
that space, and how the space
transforms as a result.

동대문 시장과 배후기지
000간
동대문시장과 인근
생산기지들의 관계도와 함께
지역의 문제를 해결하는
'지속가능한 디자인'을
통해 지역의 자산이 새로운
가능성으로 변화해 나가는
실험을 보여주고자 한다.

Dongdaemun Market and
the Backstreets
000gan
This exhibition shows
this intricate network of
production sites that make
up Dongdaemun Market in
seeking solutions for local
issues through creating a
'sustainable design' that
transforms local assets in
to future possibilities.

시장의 초상
노경
광장시장의 현 모습을 기록한
영상 작업이다. 광장시장에서
우리가 경험하지 못한 시간을
상상해보고, 앞으로 남겨질
시장의 모습이 어떠할지
생각해 볼 수 있는 기회가
되었으면 한다.

Portraits from the Market
Kyung Roh
Portraits from the Market
is a video documenting
the current scenes of
Gwangjang Market.
This project provides an
opportunity for everyone to
imagine what we missed
in Gwangjang Market and
how the market will look
like in the future.

무엇이 가만히 스치는 소리
오재우
시장은 사람들이 모이는
곳에서 시작해 사람들이
빠져나가는 곳에서 끝난다.
… 시장에서는 사람들이
주고받는 소리가 그 안을
채우고 있는 것이다.

The Sound of gently
brushing by
Jaewoo Oh
Markets open where the
crowd is and close when
and where people leave.
… Markets are filled with
sounds of exchange.

데이터스케이프: 서울장의
형태
방정인, 둘셋
상품, 자본, 유통, 소비, 관광
등 현대 전통시장의 모든
것들은 '숫자'를 갖는 데이터로
존재하고, 이는 곧 시장의 또
다른 모습을 가시화한다.

Datascape: Form of Seoul
Market
Jeongin Bang, Studio
Twothree
Everything in contemporary
traditional markets such
as commodities, capital,
distribution, consumption,
and tourism exist as data
with 'numbers', which
makes another aspects of
the market visible.

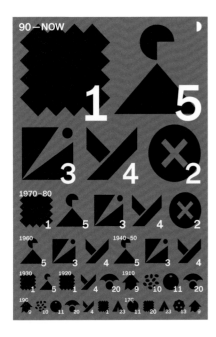

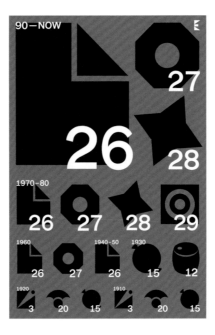

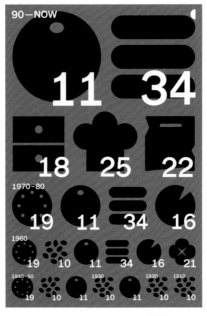

서울도시장 1
을地:공존
윤현상재 보물창고 에피소드 7

최주연

Seoul City Market 1
Eulji Coexistence
Younhyun Trading
Bomulchanggo Episode 7

Jooyeon Choi

윤현상재는 현장프로젝트가 진행될 을지로 지역의 시간과 사람이 만들어낸 공존의 실체를 '을지공존(을地:共存)'이라는 타이틀 아래 마켓으로 재현하고자 한다.

원초적인 집합 도시인 전통 시장의 개념을 다양한 관점을 통해서 들여다보고 시민들이 함께 즐길 수 있는 플랫폼을 만들고자 다양한 기획을 시도한다. 도시는 교환의 장소에서 시작 되었음을 인지할 때 시대에 따른 시장의 변화가 현시대의 도시 문화를 투영 시켜 주기도 한다. 마켓은 거대 집합 도시의 작은 축소판이라고 해도 과언이 아니다.

과거와 현재가 공존하고 있는 사이트 '을지로_세운상가, 청계·대림상가'를 중심으로 디자이너와 상인이 모이고 시민들이 함께 할 수 있는 프로그램을 기획한다. 이것은 공간과 사람을 엮는 스토리 기획을 통해 더불어 살아가는 집합 도시의 한 풍경을 기록하는 과정이다. 을지로는 너무나 많은 것들이 공존한다. 청계천이 흐르고, 우리나라 최초의 주상복합 주택인 세운상가가 빈티지를 머금으며 존재하고 있고, 또 한쪽에서는 개발이 이루어지기도 한다. 도시는 이렇게 다양성이 공존한 체 시간이 흘러가며 어떤 방향으로건 변화해간다. 윤현상재는 이러한 공존의 사이트를 토대로 사람이 모이게 하는 마켓을 기획함으로 이 시대의 한 단편을 기록하고자 한다. 윤현상재의 보물창고 마켓 '을지공존'은 물건을 사고파는 상업적 마켓을 넘어 다양한 사람들이 만나 삶의 무늬를 만들어가는 무형의 오프라인 플랫폼 기획이다.

윤현상재 보물창고 을지공존은 두 번의 마켓으로 진행되는데 서로 다른 주제어를 가지고 있다. 기획의 중심에 사람이 있기를 바라는 마음과 사물을 바라보는 관점의 스펙트럼을 넓히기 위한 시도이다.

The Younhyun Trading Live Project attempts to recreate the Eulji Coexistence, a shared space created as a result of the time and efforts of local residents.

This project was developed under the concept of taking marketplaces, the fundamental model of collective cities, and interpreting them from various perspectives to create a platform that the public can enjoy. Coupled with the awareness of the fact that cities began as spaces of exchange, one can notice how aspects of the present-day generation are reflected in and help transform the image of modern marketplaces. In this sense, marketplaces can be considered as small models of large, collective cities.

As part of this project, we have developed programs that enable market vendors and the public to interact with each other in Euljiro, Sewoon Plaza, and Cheonggye, Daelim Plaza, areas where the past and the present coexist. By creating a story that ties together spaces and the people in them, we were able to create a piece that records an image of cities as a collective entity. There are many things that coexist in Euljiro: Cheonggyecheon; Sewoon Plaza, the first ever collective residential space and symbol of Seoul's vintage history; and sites of urban development. Cities are centers for diversity, eventually changing in a certain way as time passes. We decided upon Younhyun Trading as the location for this live project because it is a place of coexistence, thus making it the perfect place for showing a snapshot of this generation in the form of a marketplace where people gather. The marketplace in this Eulji Coexistence is not only a place where goods are bought and sold, but it is also an offline platform where people from all different backgrounds come and leave their own unique marks behind.

This idea behind this space encompasses a number of different themes, stretching across the spectrum of different perspectives, from a people-centered one to an object-centered one.

시민共ZONE
다시세운광장

Picnic (A)
A01 코오롱글로벌
A02 솔로우
A03 슥쿠리
A04 피티스커피
A05 빅벳이프로

Giver Market (B)
B01 운현상제

부산스러운 사람들 (C)
C01 이중타일
C02 이윤순한복
C03 레모블랑
C04 유제까바늘
C05 리포래쉬텔
C06 손뜨개
C07 모루과자점
C08 테이블엔테이블
C09 피타피부

브랜드共ZONE
세운상가 청계대림상가

살림생활 (D)
D01 살림하는다지
D02 모벨체이
D03 무진
D04 내추럴빈스
D05 솔로우포타
D06 곤숙스튜디오
D07 자반
D08 플엔폴리나
D09 피티스커피
D10 바디로
D11 노고추음식공방
D12 외촌시움
D13 조선유기공방
D14 뭘
D15 여인
D16 민너롬

F&B (E)
E01 오대장
E02 카페린정성
E03 오별이당
E04 동이강정
E05 무한신단발
E06 빅벳이프로
E07 단물가게
E08 시골여자의 바른먹거리
E09 취니스커피
E10 부지막이솜
E11 메도로전비
E12 베이킹센터
E13 클라스틱밥초
E14 상영아이사
E15 식화테이블
E16 블로동구우소
E17 다리망메도릴리(라빙)

리빙&소품 (F)
F01 콘스호이테
F02 SWSW(W센)
F03 오롤디에이
F04 쿠린
F05 모나키람
F06 재우노나가카(메션)
F07 OFR SEOUL (아토스트)
F08 로메티포
F09 취그레이더밀
F10 에이브문
F11 플라이충
F12 쉭히
F13 프리다릴람초
F14 폼간뎁(F&B)
F15 크로
F16 모노모노
F17 단에드웨드페인트
F18 햅뉴어네임
F19 BZB
F20 핑크마켓
F21 코이코이스튜디오
F22 평소관망
F23 에이룸

리빙&소품 (G)
G01 레아로우
G02 부부1206
G03 온혜직물
G04 한이조
G05 재우노나가카(메션)
G06 피피미스튜디오
G07 부보웍스
G08 촌리강가
G09 드름드름드롬
G10 이강가
G11 이플플라리마켓
G12 TWB
G13 콜라이프스튜디오
G14 티노셀렉트
G15 파인드스타일
G16 스파라우트
G17 호시노엑쿠카스
G18 세톤드모잉
G19 아나피스케어
G20 미함당(F&B)
G21 에이슬판드
G22 다이노탭
G23 아웃도어박스
G24 하이브로우

아티스트 (H)
H01 바바선
H02 시윌벤스테디워크
H03 유트
H04 선겨선분
H05 서운정회사
H06 히어리
H07 조언예
H08 오우글라스워크
H09 도침(라빙)
H10 에플카인드(F&B)
H11 폴라위베리
H12 마이콤아

패션 (I)
I01 아이에이트데이
I02 에아오브운
I03 오디너리
I04 언더른톰
I05 엘라엉리
I06 SGJ
I07 러프
I08 엘로우스톤

디자이너共ZONE
다시세운

인테리어디자이너 (J)
J01 8I7디자인스페이스
J02 디자인부토
J03 비하우스
J04 삼플라스디자인
J05 본미라이디자인
J06 셀라드보올스튜디오
J07 라무디자인
J08 스튜디오에소오
J09 씨앗
J10 코니예블
J11 한크영
J12 얼연스타일
J13 폴딘트
J14 마담가
J15 오롯디자인
J16 디자인불

패션디자이너 (K)
K01 서울패션레디오
K02 오츠스
K03 예포티크레이
K04 두발디스하우스
K05 나세
K06 엘롬라이프
K07 와이엠지스페이스
K08 포크라노스
K09 이긴
K10 한송인
K11 크레비
K12 디레니모
K13 린병환
K14 슬로우마마세(라빙)
K15 윌테이블(라빙)

범례
● 인내소
◑ 화장실
◐ 시민共ZONE
◐ 브랜드共ZONE
◐ 디자이너共ZONE
○ 파벨리온

시민共ZONE
다시세운광장   창신   國민대 ⑪   ⓐ ⓘ

브랜드共ZONE
세운상가   D01 중앙대 ⓓ   E01 연세대 ⓔ ⑪   D16   E17

디자이너共ZONE
다시세운교   J01 ⓙ   K01 선문대 ⓚ ⑪ 가톨가릴가릴 K15   J16
○ 선문대
○ 한양대

브랜드共ZONE
청계 대림상가   F01 ⓕ   G01   I01 ⓘ   I08 ⓗ H01 H12   H01   ⓖ   F23   G11 ⓖ ⑪ 서울시립대 ⓘ   G24

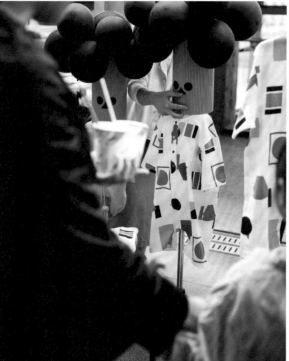

을地:공존 doc. 01 사람의 취향
(디자이너, 상인, 건축가, 예술가, 브랜드 and 시민)
2019. 9. 21 - 22, 11:00 - 17:00

첫 번째는 다양한 사람들의 공존을 마켓의 키워드로 끌어온다. 사람 사는 이야기를 마켓을 통해 펼쳐 나가고자 함이다. 디자이너들이 소장하고 있는 소장품 혹은 브랜드의 제품 등 '호기심'의 스토리를 사물보다는 사람으로부터 먼저 시작해 보면 어떨까? 가 이번 기획의 시작이다. 또한 을지로에서의 에피소드를 참여 하는 기획자, 디자이너, 혹은 브랜드 속 사람들을 통해 들어 보며 온라인에 기록으로 남기고자 한다. 윤현상재는 마켓의 스토리텔러가 되어 참가자들의 취향을 이야기하며 시민의 참여를 유도 다 함께 즐길 수 있는 의미 있는 마켓을 만들고자 한다.

을地:공존 doc. 02 사물의 스펙트럼
(재화, 예술 수공예품, 주문제작 상품, 기타)
2019. 10. 26 - 27, 11:00 - 17:00

두 번째는 다양한 사물의 스펙트럼을 이야기하고자 한다. 특히 스토리가 있는 사물을 소개함으로 사물을 더 깊이 들여다 볼 수 있는 기회가 되기를 바라고 이것은 만드는 과정, 만드는 사람 등에 귀를 기울이기 바라는 마음이다. 을지로에서 생산되는 제품, 을지로에서 활동하는 작가, 또 을지로가 아니어도 스토리가 있는 사물은 두 번째 마켓의 주인공이 된다. 전시 콘텐츠, 아티스트 참여 프로그램, 다양한 브랜드의 재미난 이야기들이 기획의 중심으로 두 번째 마켓이 형성된다.

시민共존, 디자이너共존, 브랜드共존: 조닝을 통한 윤현상재의 기획은 스토리를 만들기 위함이며 이번 마켓에서는 3개의 카테고리(시민 공존, 디자이너 공존, 브랜드 공존)로 구성된다.

1 시민共존 - 시민을 위한 윤현의 플레이스 메이킹: 종묘를 바라보는 세운광장의 사이트에 시민들이 즐길 수 있는 피크닉 공간을 구현한다. 코오롱의 협찬으로 인조잔디가 깔리고, 행사 기간 동안은 슬로우 브랜드의 매트리스가 깔린다. 우리가 상상하는 그림은 시민들이 이곳에 누워 종묘를 바라보고 하늘을 바라보며 도심 속 쉼의 시간을 가지는 것이다.

2 디자이너共존 - 디자이너스 플리마켓: 디자이너의 취향이 묻어나는 아이템과 그 안에 담긴 스토리에 집중하며 디자이너들의 애정 담긴 중고 물품과 디자이너들이 운영하는 브랜드의 제품을 플리마켓 형식으로 구성한다. 디자이너 간의 소통은 물론 소비자와 오프라인에서 소통할 수 있는 플랫폼을 형성한다.

3 브랜드共존: 윤현상재가 큐레이션한 다양한

Eulji Coexistence Zone doc.01
Personal Taste (Designer, marketer, architect, artist, brand, and citizen)
Sep 21 – 22, 2019, 11am – 5pm

The first key concept of this live project is its representation of a market where people from various backgrounds coexist. The lives of these people are shared as personal stories through the market's operations. Our approach here was to consider how the stories behind the products and different brands are not attributed to the objects themselves but, rather, to the people involved. Also, we attempt to record online the experiences of the coordinators, designers, and brand representatives participating in the various episodes in Euljiro's history. Participants become the story tellers for Yoonhyun Bomulchanggo Market and share their own personal preferences, before encouraging everyone to enjoy their time together and help create a newfound marketplace.

Eulji Coexistence Zone doc.02
The Spectrum of Objects (Material items, art crafts, customized goods, various items)
Oct 26 – 27, 2019, 11am – 5pm

The second market of this project is to share the story of the diverse spectrum of objects in this marketplace. This is done out of the hope that, by introducing objects that have deep background stories, visitors can better understand the significance of not only these objects, but also of the creators and their creative process. The main characters of this second market are the objects created in and the artists who are active in Euljiro, as well as those who are active elsewhere but have a meaningful story to share. The second market is comprised of exhibits, an artist participation program, and stories about various brands.

The Public's Coexistence Zone, Designer Coexistence Zone, Brand Coexistence Zone: Different zones were created in Yoonhyun Bomulchanggo as part of this project to create different stories falling under three different categories: the public, designers, and brands.

1 The Public's Coexistence Zone Yoonhyun's Place Making for the public: As part of this project, a picnic space was designed and established at Sewoon Plaza, overlooking the Dongmyo area. Thanks to sponsorship by KOLON and SLOU, there are plans to install fake grass and place mattresses in this space. We hope that by creating this space visitors can come and lie down, relax, and enjoy the view of Dongmyo and the open skies.

2 Designer Coexistence Zone Designer's Flea Market: Here, we focused on creating a flea

브랜드를 만나볼 수 있는 가장 큰 조닝의 카테고리다. '리빙', '소품', '아티스트', '패션', 'F&B' 비롯해 살림의 고수가 직접 추천하는 브랜드들을 모아 만든 '살림生活'까지 다양한 카테고리 속 브랜드를 구성했다. 살림生活은 28년 차 살림의 노하우 및 살아가는 평범한 이야기로 많은 사람들과 소통하는 인플루언서 '살림하는 여자' 김연화 씨가 평소에 애정 하던 브랜드 15개를 큐레이션했다.

프로그램
시민들의 참여를 적극 유도하기 위해 직접 체험하며 즐길 수 있는 다양한 프로그램들을 준비했다.

1. 기버마켓
윤현상재는 시민과 브랜드들이 선뜻 내어주신 아이템들을 기부받아 다시세운광장에서 '기버마켓'을 운영한다. 기부의 문화를 만들기 위한 기획으로 기부의 과정에서부터 시민의 참여를 온라인으로 유도, 선한 영향력이 확산되기를 의도했다. 판매 후 수익금 전액은 기부된다.

2. 나의 모습
을지로는 변화가 많고 사라지는 풍경이 유난히 많은 곳이다. 이곳을 배경으로 곳곳에 '나의 모습'을 사진으로 찍을 수 있는 공간을 만들고 시민들의 참여를 유도, 기록을 남기는 프로그램이다.

3. 클래스
일러스트 작가 마마콤마 스튜디오의 '실크스크린 클래스', 윤현상재 '타이포타일 클래스', 틸테이블의 '식물, 잘 심고 가꾸는 법', 일상 속 즐거운 배움의 장을 만들었다.

4. 퍼포먼스
소소한 움직임부터 큰 움직임까지 각기 다른 성격을 가지는 다채로운 퍼포먼스를 준비했다. '차돌 솥밥 짓기', '김치 담그기', '퍼레이드' 등 곳곳에서 벌어지는 퍼포먼스는 누구나 함께 즐길 수 있다.

market where items that have their own stories and that local designers are fond of are sold. It is meant to be a space where designers can communicate with each other and with consumers through an offline, flea-market-like platform.

3   Brand Coexistence Zone: This is the biggest zone of the three, where various brands curated by Younhyun Trading are gathered. Experts in various categories, including "living", "decorations", "artists", "fashion", "food and beverages", share their recommended brands and create their own, collective lifestyle space. Social media influencer and lifestyle coach Kim Yeon-hwa, who for the past 28 years has shared everyday lifestyle techniques and tips, curated this space to introduce 15 of her favorite brands.

Program
We have prepared various programs to encourage the public to actively participate in experiencing the live project first-hand.

1. Giver Market
At Sewoon Plaza, we plan to operate the "Giver Market," where items donated by people and brands from Younhyun Trading are put up for sale. The objective of this market is to create a culture of giving, encouraging the public to participate online to do something good through the act of donations. All of the profits from this market will be donated to a good cause.

2. Portrait
Euljiro is known for being a place that witnesses constant change and the loss of several, unique images and landscapes. Based on this concept, we have created several spaces throughout the area where participants can take portrait photos and leave behind traces of their memories and time spent there.

3. Class
In collaboration with illustrator MamaComma Studio, we are providing classes on silk screens, Younhyun Trading typo-tiles, and how to plant and care for plants by Teal Table. These programs help participants learn and have fun as part of their daily lives.

4. Performances
We have prepared a number of diverse performances characterized by a wide range of motions, from small to big. Everyone is welcome to enjoy these performances, which are being held everywhere including making chadol sotbab, preparing kimchi, and joining in a parade.

UAUS 파빌리온 프로젝트

UAUS Pavilion Project

21개 대학생 건축과 연합회인 UAUS는 대학생들로만 이루어진 연합회이다. 매년 그들만의 전시와 축제를 즐기는 것으로 이미 건축계에서는 꽤나 유명한 작가팀이다. 이번 해는 특별히 "시장"의 주제에 맞춰 진행되었고, 그 중 선별된 7개팀이 이번 비엔날레에 참여하게 되었다. 그들이 다루고자 했던 "시장"은 단순히 고정관념의 시장이 아닌 변화하고 있는 개념의 "시장"에 대해 다루었다.

UAUS is an alliance of students from twen-one universities who gather every year to create exhibitions and festivals. Seven universities were chosen to join the Seoul Biennale of Architecture and Urbanism this year with the theme of 'markets'. The market as a place of change.

가천대학교
유시연, 강유림, 김다솔, 김희수, 박준석, 양재욱, 오승원, 윤수빈, 이원식, 임도진, 최수연

국민대학교
이동우, 심주용, 김동하, 박준방, 이오랑, 진효원, 권법호, 김지한, 정세현, 이수연, 김세인, 남수빈, 도희원, 서희준, 최예인, 이광혁, 이영주, 이유진, 조주나

서울시립대학교
정욱호, 강정원, 김경환, 김동욱, 박소은, 장혜림, 전지호

선문대학교
김무현, 김동민, 최경민, 강지혜, 김정민, 김재진, 맹승주, 이경준, 정창희, 제이슨 브란드너, 함기훈, 이종민

연세대학교
정영제, 강수빈, 김재욱, 나영수, 남궁찬우, 박준수, 박찬호, 박초원, 이성원, 이윤서, 이정수, 이진경, 정진우, 조우현

중앙대학교
김정근, 공승환, 김가은, 김소미, 김수환, 도상혁, 박창현, 안성환, 유하연, 이자윤, 장지연, 정민섭, 정재성, 최기열, 최지유

한양대학교
윤경익, 강동형, 고성준, 구예림, 권규리, 김동윤, 김재연, 류진혁, 문지연, 박태인, 배장우, 성호석, 오기백, 이광진, 이성민, 이성호, 이승훈, 이영웅, 이종승, 이지형, 전충민, 정건희, 정동준, 조민기, 최성욱, 허승학, 황영록

Gachon University
Si Yon Yu, Yu Rim Kang, Da Sol Kim, Hee Su Kim, Jun Seok Park, Jae Wook Yang, Seung Won Oh, Su Bin Yun, Won Sik Lee, Do Jin Lim, Soo Yoen Choi

Kookmin University
Dong Woo Lee, Ju Yong Shim, Dong Ha Kim, Jun Bang Park, Oh Lang Lee, Hyo Won Jin, Ho Beob Kwon, Ji Han Kim, Se Hyun Jung, Su Yeon Lee, Se Een Kim, Su Bin Nam, Hee Won Do, Hee Jun Seo, Ye In Choi, Gwang Hyeok Lee, Yeong Ju Lee, Yu Jin Lee, Ju Na Jo

University of Seoul
Wook Ho Jung, Jeong Won Kang, Kyung Hwan Kim, Dong Wook Kim, So Eun Park, Hye Lim Chang, Ji Ho Jeon

Sunmoon University
Mu Hyeon Kim, Dong Min Kim, Gyeong Min Choi, Ji Hye Kang, Jeong Min Kim, Jae Jin Kim, Seung Ju Maeng, Keung Jun Lee, Chang, Hee Jung, Jason Brandner, Ki Hoon Ham, Jong Min Lee

Yonsei University
Young je Jeong, Subin Kang, Jae Wook Kim, Young Soo Na, Chan Woo Namgung, Jun Su Park, Chan Ho Park, Cho Won Park, Sung Won Lee, Yoon Seo Lee, Jeong Soo Lee, Jin Kyung Lee, Jean Woo Jeong, Woo Hyun Cho

Chungang University
Jeong Geun Kim, Seung Hwan Kong, Ga Eun Kim, So Mi Kim, Su Hwan Kim, Sang Hyuk Do, Chang Hyun Park, Seong Hwan An, Ha Yeon Yoo, Ja Yoon Lee, Ji Yeon Jang, Min Seop Jeong, Jae Seong Jeong, Ki Yeol Choi, Jee You Choi

Hanyang University
Kyung Ik Yun, Dong Hyung Kang, Sung Jun Go, Ye Lim Goo, Gyu Ree Kwon, Dong Yoon Kim, Jae Yeon Kim, Jin Hyeog Ryu, Ji Yeon Moon, Tae In Park, Jang Woo Bae, Ho Seok Seong, Ki Baek Oh, Gwang Jin Lee, Seong Min Lee, Seong Ho Lee, Seung Hoon Lee, Young Woong Lee, Jong Seung Lee, Ji Hyeong Lee, Chung Min Jeon, Gun Hee Jeong, Dong Jun Jeong, Min Gi Cho, Mally Choi, Seung Hak Heo, Young Rok Hwang

감각 場
UAUS: 국민대학교
시장에서 느낄 수 있는 다양한 감각들에서 시장의 냄새, 색깔 등을 요소로 추출하였고 이를 개별주제인 향의 판매상황에 녹였다.

WEAVING DOME
UAUS: Kookmin University
We focused on the various ways that it stimulates our senses, through different smells and a wide array of colors, and incorporated these aspects into the selling of everyday items.

풍경재생
UAUS: 연세대학교
사선으로 얽힌 각재의 다양한
패턴은 평면적이지만 서로
연결되며 입체적인 지붕을
이룬다.

PLAY-SCAPE
UAUS: Yonsei University
Although the interwined
diagonal lines of the
wooden beams are flat,
they link together to create
a three-dimensional roof.

내가 만드는 키오스크
UAUS: 중앙대학교
누구나 판매자가 될 수 있고
쉽고 빠르게 자신만의 공간을
창조하고 재구성할 수 있다.

IKIO
UAUS: Chung-Ang
University
Everyone has the
opportunity to create
and reorganize their own
spaces, as they desire, in a
quick and easy manner.

받히다, 바치다
UAUS: 가천대학교
아치 형태의 지붕은 기존
재래시장의 지붕을 현대화 한
것이다.

CRATER: CREATE WITH
CRATE
UAUS: Gachon University
The arch roof is a
reinterpretation of the
rooves of the traditional
markets.

501

컵플라워
UAUS: 서울시립대학교
버려지는 일회용 컵들이 모여
쓰레기더미가 되는 대신 도심
속의 작은 정원과 같은 공간이
될 수 있지 않을까?

CUPLOWER
UAUS: University of Seoul
What if there was a way
to turn these disposable
cups into little pieces of
nature by turning them
into makeshift plant pots?

플로트폼
UAUS: 한양대학교
시장을 드러내기 위해서
시장에서 가장 외면되는 선을
사용했다.

FLOATFORM
UAUS: Hanyang University
For this exhibition, we used
these neglected lines to
recreate the true essence
of the marketplace.

리: 커버
UAUS: 선문대학교
시장에 가면 차가 다니지
못하는 좁은 골목을 자유롭게
다니며 다양한 물품들을
판매하는 리어카들을
움직이는 하나의 작은
시장이라고 해석하였다.

RE: COVER
**UAUS: Sun Moon
University**
Since the market alleyways
are too narrow, people use
pushcarts to sell a variety
of goods. With this image
in mind, we decided to
express the marketplace
by reinterpreting the image
of pushcarts moving about
within a small market.

도시상회

홍주석

City Markets

Jooseok Hong

고성장 시대의 기술 혁신은 도시를 고효율의 상업적 도시를 만드는 원동력이 되었다. 하지만 4차 산업혁명 시대의 기술의 발전은 사람의 도시 행태를 변화시키고 이제는 사람들이 도시를 변화시키고 있다. 도시가 더 이상 오프라인 공간의 효율과 수익의 논리에 의존하지 않고 경험과 커뮤니티, 온라인 서비스와의 접점 등 다양한 요소들에 의해 재구성될 것이다.

누구나 크리에이터가 될 수 있는 시대적 흐름 속에 오프라인 도시는 더 이상 랜드마크나 메가스트럭쳐가 아닌 다양한 개인 콘텐츠가 만들어내는 작은 점들로 재구성된다. 이 작은 점들은 이벤트 도시로서의 성격을 내포하고 있어 언제 어디서 어떤 콘텐츠들이 모일 수 있는가가 매우 중요한 이슈가 될 것이다.

이제 시장은 교환의 장소를 넘어 사람과 사람이 만나는 곳, 나아가 지역 콘텐츠의 집합 생산소로서의 역할로 진화할 가능성을 가진다. 이번 현장프로젝트 도시상회에서는 재생의 가능성을 기반으로 다양한 크리에이터가 모여 만드는 창발하는 도시의 잠재적 가치에 집중하여 가치의 부활, 취향의 부활이라 두 가지 마켓 실험을 진행한다.

Technological innovations in periods of high growth were the driving factor in transforming urban areas into highly efficient, industrial cities. However, with the onset of the 4th Industrial Revolution, technological innovations transform the behaviors of those living in cities, and these populations, in turn, transform the cities in which they live. Now, cities no longer depend on the logical basis of efficiency and profitability as defined by offline spaces. Rather, cities have been reconstructed through the incorporation of communities, online services, and other various factors.

In today's era, where anyone can become a creator, offline cities are no longer comprised of landmarks and mega-structures but, rather, by points created by various, individual content. These small points include the characteristics of an event city, thus placing importance on when, where, and what type of content can collectively meet.

Now, markets have gone beyond areas where simple exchanges take place, becoming spaces where people meet and that display the potential to evolve into platforms where amalgamations of local content is created. The live event "City Markets" displays a collection of works by various creators based on the potential for urban regeneration. These pieces are presented through two market experiments that focus on the potential of cities for the revival of values and the revival of personal taste, respectively.

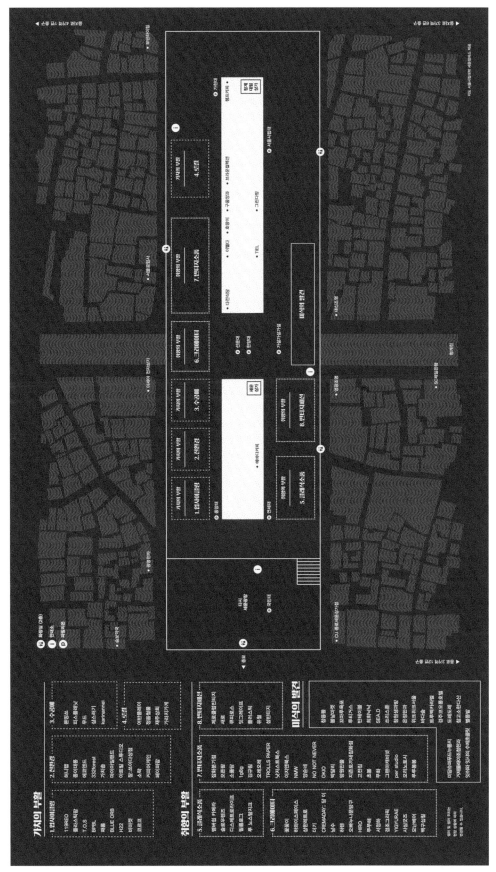

가치의 부활

'가치의 부활'에서는 도시 생활을 기반으로 그 쓰임과 역할이 재정의 되어 재화와 콘텐츠가 재생산되는 가치에 집중한다. 가치 재생산의 기지 역할을 해온 세운상가에서 단순히 헌 것이라는 개념을 넘어 누군가의 콘텐츠와 가치의 재발견을 통해 새로운 가치를 만들어 내고 있는 현장의 경험을 공유할 수 있는 도시문화의 장을 열고자 하였다.

취향의 부활

'취향의 부활'에서는 온라인 커머스가 마켓을 장악하며 현대인들의 라이프스타일의 변화를 가속화 시키는 현 상황속에서 도심 마켓의 의미를 재해석해 보고자 한다. 판매 행위가 오가는 마켓의 기능을 넘어 사람과 사람이 만나 콘텐츠와 커뮤니티가 생성되는 마켓의 원리를 기반으로 다양한 취향 커뮤니티로서의 마켓의 실험을 하고자 하였다. 전통 5일장이 도시의 이벤트 역할을 해왔듯이 현대도시에서의 취향 커뮤니티로서의 마켓 실험을 통해 다양한 방식의 커뮤니티 중심의 도시 재구성의 가능성을 경험할 수 있는 장을 만들고자 하였다.

Revival of Values

The "Revival of Values" market experiment focuses on the value of recreating goods and content based on the newly defined use and role of urban life. The purpose of this market experiment is to show that Sewoon Plaza, which has served as a sort of home base for the revival of values, is not a space characterized by old ideas and goods but, rather, a space that is very well alive. This project shows that Sewoon Plaza is platform for urban culture where individuals can rediscover and create new content and values and share their experiences with others.

Revival of Personal Taste

The "Revival of Personal Taste" market experiment attempts to reinterpret the meaning of urban markets in modern society, which is characterized by lifestyles that are rapidly transforming under the influence of the dominant online commerce markets. Through this project, we go beyond the understanding of markets as spaces where simple exchanges, buying, and selling takes place and focus on them as spaces where people can meet and create things as a community with a wide spectrum of individual tastes. Against this backdrop, just as traditional daily markets have served as spaces for urban events over the years, this project reflects the personal tastes of modern communities to provide a market experience that demonstrates the potential for community-driven reorganization of cities.

서울시장산책

이희준

Discovery Seoul Market

Heejun Lee

IT와 도시문화적인 요인들로 인해 사람들의 도시 속 행태가 경험 위주로 재구성되어가고 있다. 이색적인 콘텐츠에 기반으로 하여 수시로 이벤트가 일어나는 골목이나 시장을 중심으로 사람들이 모이고, 이는 다시 온라인을 통해 더 많은 이들에게 새로운 콘텐츠로 전달된다. 특히 다양한 콘셉트를 자랑하는 큐레이션 마켓에 대한 관심은 갈수록 높아지고 있다.
이에 따라 '서울시장산책' 프로젝트를 통해 한국만의 역사와 콘텐츠가 축적된 전통시장에 대한 의미를 재해석하여 현대 도시 속 전통시장의 가치와 지속 가능성에 대한 새로운 고민을 던지고자 한다. 전통시장에서만 만나볼 수 있는 상점과 특화된 상품, 자기만의 철학을 가지고 장사를 해온 상인들을 소개하며 한국만의 콘텐츠로 이를 둘러싼 도시 생태계에 대한 이해도를 높이고 그 잠재 가능성에 대해 논하고자 한다.

Urban activities are reorienting to various participatory experiences due to advances in IT and current urban cultures. People are drawn to contents based events happening frequently in streets and markets. Who then quickly share their experiences online creating another experience for others. Curation markets are especially growing popular known for its diverse contents.
This program, "Discovery Seoul Market," takes into consideration the increasing interest in the curation of different markets. The goal of this project is to reinterpret the significance of traditional markets, which are collections of the local characteristics and history, and rethink their value and potential in the context of the cities today. This program offers opportunities for a closer look to learn about each stores, their specialized goods and the philosophies behind the traders who work there. We hope for a better understanding of the local contents produced within the urban ecosystem that surrounds the traditional markets, while also increasing awareness and interest in their potentials.

통인시장

한국전쟁 이후 경복궁의 서쪽 마을인 서촌의 인구가 급격히 증가하면서 노점과 상점이 현재의 시장 안팎으로 넓게 확장되었고 지금의 통인시장의 형태가 갖춰졌다. 점포 수는 약 70여 개로 아담한 규모이며, 확실히 일본인들이 이용하던 시장이었기 때문에 건축, 역사학적으로 시장을 바라보면 골목이 작게 형성된 점, 상점과 상점이 거리를 두지 않고 붙어 있는 점, 생활형 시장에 맞게 일상생활을 위한 생필품 상점이 주요 상점인 점 등이 새롭게 인지될 시장이다. 식당과 반찬가게 등 요식 관련 업소와 식료품 상점, 옷과 가방, 구두 등을 수선하는 가게까지 다양한 상점이 입점해 있다. 최근 통인시장은 생활형 시장에서 전환하여 관광시장으로서의 변모를 갖춰가고 있다. 오전 11시부터 오후 4시까지 즐길 수 있는 '도시락카페 통'은 상인들의 협종와 엽전이라는 전통적 아이템의 결합으로 국내 전통시장 상인들에게는 문화관광형 시장이 가야 할 길에 대해서 인사이트를 주고 있고 국내외 관광객들에게는 관광명소로서 다양한 체험을 먹거리와 함께 즐길 수 있다는 기회를 제공하고 있어 주목된다.

망원시장

망원시장은 망원동 사람들을 위한 마을 시장으로 시작해 2000년대 이후 망원동 주민뿐 아니라 서울시민들이 사랑하는 시장이 되고 있다. 뿐만 아니라 망원역 2번 출구부터 망원시장 입구까지 이어지는 약 1km의 주택 골목은 현재 시장이 태동하는 모습을 확인할 수 있다는 점에서 도시 건축학적 관점에서 주목할 만한 사례라 하겠다. 특히, 1인 가구와 젊은 층이 거주하는 주거지역의 앵커 시장으로서 기능하고 있는 망원시장은 채소와 과일 그리고 생필품 가격이 저렴하여 더 많은 팬을 늘려가고 있다.

광장시장

"현존하는 가장 오래된 전통시장은 어디일까요?" (…) 사실 빈대떡은 1905년 시장이 생겼을 당시 전국으로 광장시장의 포목을 날랐던 봇짐장수들의 요깃거리로 시작했다. 이들에게 포목을 넘겼던 포목 거상들은 육회 비빔밥을 먹었기에 광장시장의 먹거리는 풍부해졌다. 하지만 광장시장을 먹거리 시장으로만 봐서는 안 된다. 서울 4대문 안에서 쌀을 사려면 줄을 서서 사야 했던 미곡시장으로서의 기능을 이해해야 하고 먹자골목 뒷골목의 포목시장은 우리나라 포목시장의 뿌리임을 잊으면 안 된다. 광장시장 포목 가게들은 현재 짧게는 2대, 길게는 4대까지 대를 잇고 있다.

경동시장

경동시장은 누가 뭐라 해도 '서울의 부엌'이다. 일본 교토에 '니시키 시장'이 '교토의 부엌' 임을 자청하고 네덜란드 암스테르담의 '더 마켓 홀 시장'이 '암스테르담의 부엌'이자 '유럽의 부엌'으로 불리는 것처럼 우리는 잘 모르지만 적어도 식자재를 다루는 요리 연구가나 셰프들에는 없어서는 안 될 존재가 바로 경동시장이다. 경춘선, 경의선, 경원선, 중앙선 철도를 이용해 경기도와 강원도의 농부들이 각종 특산품을 내다 팔았던 그야말로 파머스마켓이었다. 한국 전쟁이 끝난 이후에는 상설시장으로 자리 잡으며 현재의 모습이 갖춰지기 시작했다. 경동시장에서 구할 수 없는 제 철식재료는 서울 안에 없다'는 명제는 지금도 유효하다. 30여 개의 상점이 몰려 있는 기름 골목, 각양각색의 약재를 취급하는 100여 곳의 약재상이 모여 있는 약령시, 전국 제철 과일이 도매 거래되는 청량리 청과물시장, 인삼과 도라지의 도매 거래가 이뤄지는 인삼도라지 도매 상가 등 한 품목당 10개 이상의 상점이 모여 영업을 하는 대형 시장에 속한다.

Tongin Market

After the Korean War, the population in Seochon (located west of Gyeongbokgung Palace) dropped instantly, which led to the expansion of street vendors and stores both inside and outside of the market, leading to the Tongin market today. Now, Tongin market is quite compact with about 70 stores. As this market was mainly used by Japanese residents, it possesses unique characteristics from architectural and historical perspectives: With narrow alleyways; stores sit tightly side by side; selling daily necessities befitting a daily life-oriented market. There are diverse stores in this market from restaurants to grocery stores to bag and shoe repair shops. Now, Tongin market has transformed itself from a daily life-oriented market to a market oriented to tourism. "Lunch box TONG," combining the collaboration of local food stands and the use of yeopjeon(traditional brass coins), may be the ideal direction for these old markets to sustain as a cultural place. This program, which you can enjoy from 11 am to 4 pm, offers diverse experiences as well as foods to local and foreign visitors.

Mangwon Market

Although Mangwon market was initially set up for the local neighborhood, it has started to attract urban tourists. In addition, the route from the subway station to the market hasn't changed. which is another special quality from the urban architectural perspectives. In particular, Mangwon market, which functions as an "anchor market" in a residential area where single-person households and the younger generation are living, is selling fruits, vegetables and daily necessities at affordable prices, and thus getting more popular these days.

Gwangjang Market

The most famous is their *bindaetteok* (mung bean pancake) which originated as a snack for peddlers who traded linen and cotton nationwide from Gwangjang market in 1905 at the time when the market was set up. Meanwhile, the rich merchants who also sold linen and cotton enjoyed *Yukhoe-bibimbap* (beef tartare bibimbap). Likewise, the foods in this market became diverse and affluent. However, Gwangjang market is not necessarily just a food market. It also functioned as a rice market where people had to queue in a long line to buy rice. Gwangjang market also served as the root of linen and cotton markets in Korea, located in the backstreets. Linen and cotton stores in Gwangjang market have kept their businesses for two to four generations.

Gyungdong Market

Gyungdong market is no doubt the "kitchen of Seoul," just like the Nishiki market calls themselves the "kitchen of Kyoto" and the Market Hall Market are calling themselves the "kitchen of Amsterdam (or the kitchen of Europe)". Gyungdong market is an indispensable place for food researchers and chefs who procure food ingredients every day. It was a famers' market as well. Farmers went to Gyungdong market from Gyeonggi-do Province and Gangwon-do Province via Gyeongchun line, Gyeongui line, Gyeongwon line and Jungang line, and sold their local products. After the Korean War, it became a permanent daily market. There is a famous saying that "Gyungdong market sells every seasonal food ingredients existing in Seoul," which is still true. You can see not only nationwide fruits and vegetables in season but also daily necessities and industrial products, recently, with changes in lifestyle.

서울마당

SEOUL

MADAN

G

을지로 노가리골목
무교동 유정낙지
북한산 큰숲 사람의 마을 은평
장충동족발
구민이 꿈꾸는 가치, 함께 만드는 광진
미래도시 용산
희망 서울
ItaewonFreedom
GangnamStyle
세계는 서울로 서울은 세계로
잘생겼다서울
자연과 문화 속으로! 힐링도시 노원
한강고수부지컵라면
사람중심 명품도시 종로
아름다운 변화, 열린구청, 행복도시 서대문구
서울의 문 꿈을 여는 동대문구
I.SEOUL.U
What is a reason why you like Seoul ?
What public space is essential for Seoul ?
아름다운 Seoul
Welcome to Seoul
Jesus

서울마당

Seoul Madang

서울마당은 서울건축도시전시관 지하3층에 위치하는 공간이다. 서울마당은 2019 서울도시건축비엔날레의 홍보관 역할을 하면서 "서울의 발견" 전시를 통해 서울의 다양한 모습을 담아내는 공간이다.

서울건축도시전시관의 입구를 들어서면 오른쪽 벽에 2019서울도시건축비엔날레를 홍보하는 영상을 접하게 되고 계단을 내려오면서 오른쪽 벽면에 3면의 영상을 통해서 비엔날레를 소개하게 된다. 서울 마당에 도착하면 터치스크린을 통해서 비엔날레에 관한 상세한 정보를 접할 수 있다.

"서울의 발견" 전시는 도시를 시민과 정부 및 지자체 그리고 전문가들이 집합적 노력을 통해서 만들고 그 도시를 시민들이 공평하게 누려야 한다는 2019 서울도시건축비엔날레의 주제인 "집합도시"에 대해 다양한 방식으로 시민들과 소통하려고 한다.

시민들이 직접 제작하여 응모하고 직접 당선작을 뽑은 "서울의 발견: 시민들이 좋아하는 공공 공간" 공모전의 결과를 전시한다. 시민들은 현장에서 집합도시에 대한 의견을 제안할 수 있고 직접 서울시를 디자인에 참여하는 체험을 할 수 있다. 또한 25개 구청이 자랑하는 도시의 공공공간으로 꾸민 "집합도시 서울풍경"을 통해 자신이 속한 구의 아름다운 공공공간은 물론 서울의 공공공간의 전체적인 풍경을 확인할 수 있다.

마지막으로 현재 진행 중인 도시 프로젝트들을 통해 미래의 집합도시 서울의 풍경을 점쳐 볼 수 있다.

Seoul Madang is located on the third lower level of the Seoul Hall of Urbanism and Architecture. It is not only a promotional space for the 2019 Seoul Biennale for Architecture and Urbanism but a home to the "Finding Seoul" exhibition.

Upon entering the space, there are videos introducing the 2019 Seoul Biennale projected to the right along the stairs down to the main hall. "Seoul Madang" begins with three touch-screens where visitors can access more information about the Seoul Biennale in greater detail.

Then "Finding Seoul" communicates with the public on the topic of "collective cities," which is the overarching theme of the 2019 Seoul Biennale. It relays the message that cities should be collective spaces that offer equitable living opportunities. It also emphasizes that they are spaces created through collaboration among the public, the central government, local governments, and experts.

Visitors can participate on site to share their opinions and ideas on collective cities. Also, through the "magic wall," visitors can participate in designing the city. They can see all of the public buildings across twenty-five different districts throughout the city, including one's own neighborhood.

Finally, visitors can take a look into the future of Seoul by experiencing the ongoing projects of experts in different areas.

서울의 발견
시민공모 사진 및 영상 전시

Finding Seoul:
Photography and Videography Exhibition Open to Public Entries

시민들이 직접 서울비엔날레에 참여하는 방식의 하나로 온라인 사진 및 영상 공모전 "서울의 발견: 함께 누리는 도시"를 개최하였다. 총 사진 1,519개와 영상 100개의 작품이 접수 되었고 시민들의 투표를 통해서 최종 수상작이 결정되었다. 이 전시를 통해서 시민들이 생각하는 함께 누리는 좋은 공공 공간이란 무엇인지? 시민들은 어떠한 공공 공간을 꿈꾸고 있는지 확인할 수 있는 좋은 기회가 되었다. 도시의 주인은 시민이고 따라서 시민들은 그 도시를 공평하게 누릴 수 있어야 한다.

"Finding Seoul: Enjoying the City Together" was a pre-Biennale event encouraging public participation before the opening. A total of 1,519 pictures and 100 videos were submitted, followed by a public vote to select the winners. It revealed which public spaces were popular by the masses and what kind of public spaces citizens dream of. After all, the city is of the people and by the people, and everyone should benefit equally from the city.

사진 부문 대상 조민성

Grand medal in photography
Jo Minseong

사진 부문 금상 송상은

Gold medal in photography
Song Sangeun

사진 부문 은상 전지용

Silver medal in photography
Jeon Jiyong

사진 부문 은상 송승욱

Silver medal in photography
Song Seunwook

영상 부문 금상 김수종

Gold medal in moving image
Kim Soojong

영상 부문 대상 하상은

Grand medal in moving image
Ha Sangeun

영상 부문 은상 박새아

Silver medal in moving image
Park SaeA

영상 부문 은상 정훈구

Silver medal in moving image
Jeong Hoonku

집합도시 서울의 풍경

"서울의 발견"이 시민들의 관점에서 본 함께 누리는 도시의 풍경이라면 "집합도시 서울의 풍경"은 서울시 25개 구청이 선정한 시민들이 사랑하는 공공 공간에 대한 전시이다. 서울시의 공공 공간의 풍경은 25개 구에 속한 공공 공간의 집합으로 이루어진다. 서울시는 25개의 파편적 풍경을 하나의 집합적 풍경으로 묶어내는 새로운 전략이 필요하다. 전시는 각 구청 별로 32인치 모니터가 하나씩 주어지며 그 것을 이용하여 사진과 영상전시를 자유롭게 하게 된다. 전시장 바닥에는 서울시 지도가 설치되어 모니터에 전시된 공공 공간들의 위치들을 확인할 수 있다. 관람객들은 먼저 개개인의 특별한 위치를 확인할 것이고 인근에 있는 시민들이 좋아하는 공공 공간들을 확인 할 수 있을 것이다. 모니터에서 펼쳐지는 서울의 풍경도 감상하면서 지도 위에서 펼쳐지는 공공 공간의 풍경도 같이 감상할 수 있다.

While "Finding Seoul" depicts Seoul from the perspective of citizens and the ways in which they can enjoy cities together, the "Collective Cityscape of Seoul" displays public spaces enjoyed by the public as selected by twenty-five district offices throughout the city. These districts come together to comprise the landscape of Seoul's public spaces. A strategy is needed to connect these twenty-five distinct, fragmented components into a single, collective landscape. For this exhibit, each district office is represented by a 32-inch monitor that displays selected pictures and videos. Furthermore, there is a map of Seoul on the floor of the exhibit so that audiences can visualize the locations of the public spaces shown on the monitors. Using this map, not only can audiences see these different spaces and their locations, but also experience other popular spaces that are in nearby areas. Through this exhibition, visitors can enjoy the scenery as shown on the monitors, along with the images shown on the accompanying map.

서울량반은 글 힘으로 살고..

강민선, 이마지나, 전수현 외 공동작업

Textual Collective, Seoul

Collective Work by Minsun Kang, Imagina, Sooh and more

"도시는 장소를 넘어 시대물이다."
— 패트릭 게데스

도시의 모습은 문자로도 그릴 수 있다. 속담, 슬로건, 영화나 드라마 제목, 시와 노래부터 현대의 신조어까지 다양한 매체에서 서울이라는 도시에 대한 개인 혹은 집합으로서의 시간, 기억, 경험, 욕망, 문화, 움직임과 느낌들을 담고 있는 문구들은 무수하다. 길고 섬세한 묘사부터 짧지만 마치 주문처럼 우리의 기억을 바로 소환하는 문구까지, 만들어진 시기와 형식과 쓰임이 달랐지만 다함께 커다란 벽을 장식한다. 어떤 문구들은 여럿이 함께 공유하고 있는 것이 무엇인지 알려주고 또 어떤 문구들은 생소해서 궁금증을 불러 일으켰으면 한다. 관객의 참여(인터랙션)가 더해지면 벽에 뿌려져 있던 문구들이 재조합되어 도시의 특정한 모습을 보여 준다.

"But a city is more than a place in space, it is drama in time."
— Patrick Geddes

The image of the city is also found in texts. Various texts in old and new media like proverbs, slogans, titles of films and TV dramas, poems, songs, and even modern-day slang hold its own time, memories, experiences, desires, and culture. They also reflect sentiments of the individual and the collective city. Some of these texts are long and detailed descriptions, some short and musical or even magical, conjuring up forgotten memories. These endless words, phrases and sentences from different times and places are collected to form a wide wall. And while some texts are shared by the masses, others are rare and foreign, likely to spark the viewer's curiosity. Finally with the visitors' sensual interaction the texts come alive to tell different collective stories.

프로젝트 서울

Project Seoul

"프로젝트 서울"에서는 현재 서울시에서 진행중인 대표적인 도시 스케일의 공공프로젝트 두개를 소개한다. 하나는 백사마을 주거지 보존 프로젝트이고 다른 하나는 강남권 광역복합환승센터이다.

서울시에는 현재 수 많은 주택단지 프로젝트가 진행되고 있는데 백사마을 주거지 보존 프로젝트는 몇 가지 관점에서 차별화된다. 첫째, 대부분의 대규모 단지들이 기존 지형, 기존 마을의 공동체의 흔적을 물리적으로 밀어 버리고 시작하는 반면 백사마을은 기존 지형, 터, 마을의 풍경, 기존 공동체의 문화 및 흔적(삶의 풍경) 등을 최대한 지켜내려고 노력하고 있다. 둘째, 프로그램도 기존의 원주민은 물론 다양한 계층의 사람들을 위한 임대주택이다. 그저 물리적인 공간을 디자인하는 것이 아니고 그 들의 삶을 담을 수 있는 공동체를 디자인하고 있다. 그의 삶을 지원하게 되는 다양한 주민 공동 시설의 프로그램을 주민공동체와 같이 설계하고 있으며 그 운영방식도 같이 고민하고 있다. 마지막 차별점은 설계 방식이다. 도시 스케일의 공공프로젝트는 한명의 계획가가 단지 전체를 설계하는 경우가 많지만 백사마을은 10명의 건축가가 기존의 터를 20개로 나누어 설계하고 있다. 길과 터를 같이 설계하는 방식이다. 더불어 개개 프로젝트의 설계도 중요하지만 각각의 프로젝트가 만나는 경계가 더욱 더 중요하다. 많은 시간을 이런 경계지점에 할애하고 있다. 진정한 집합도시의 실천이다.

강남권 광역복합환승센터는 도심을 관통하는 다양한 교통 인프라들이 만나는 지하도시를 지상의 도시의 맥락과 연결시키는데 의미가 있는 작업이다. 도시의 공공 공간이 수평적으로만 확장되는 것이 아니고 수직과 수평 방향으로 동시에 확장되는 것이 이 프로젝트의 도시적 가능성이다.

마지막으로 "프로젝트 서울"에서는 2013년부터 진행되고 있는 서울시의 공공 프로젝트의 면면을 영상자료로 확인 할 수 있다.

"Project Seoul" introduces two of the most definitive public projects that are currently being conducted in the urban level in Seoul: the Baeksa Village Residential Area Preservation Project and the Gangnam Intermodal Transit Center.

The Baeksa Village Residential Area Preservation Project is different from other ongoing residential projects in Seoul. First of all, while most large-scale areas are built upon a foundation that strays away from local identity and traces of the previous community, Baeksa Village aims to preserve as much of the original characteristics, spaces, scenery, community, local culture, and its traces as possible. Second, this program is to create living spaces not only for local residents, but for incoming residents from all socio-economic backgrounds. And the purpose is not to simply design the physical spaces but to design a community that facilitates the lifestyles of the residents. Most importantly, those implementing this project are collaborating with the local community to develop various programs and facilities that meet local demands, as well as to find the best way to operate such infrastructure. Lastly, unlike most cases where a single individual is responsible for the entire project, ten architects divided Baeksa Village into twenty different sections. This method involves designing the roads and plots in tandem. Furthermore the border where plots meet were as important as each of the plots. On that note, a lot of time and attention are being paid to these boundaries as a real practice of a true collective city.

The Gangnam Intermodal Transit Center aims to take infrastructure models for various modes of transportation that runs through the heart of the underground city and connect them with similar infrastructures above ground. The main function of this project is to help urban spaces break away from horizontal growth and explore ways to facilitate their vertical growth as well.

Lastly, audiences can view video materials for a closer look at the various accomplishments of Seoul's public projects since 2013 as part of "Project Seoul."

시민참여
프로그램

PUBLIC
PROGR

AM

2019 서울도시건축비엔날레
주제강연, 조민석, 밤섬 당인리
라이브

2019 SBAU Lectures,
Minsuk Cho, Bamseom
Danginri Live

시민참여프로그램

김나연

2019 서울도시건축비엔날레(이하 서울비엔날레)는 각각의 도시가 시간적, 공간적, 사회적 요소들에 의해 끊임없이 변화되고 있는 과정에서 시민이 함께 참여하고 누릴 수 있는 새로운 집합유형을 모색하고자 '집합도시'란 주제를 담았다. 그리고 도시가 당면한 문제의 해법을 찾기 위해 세계 도시의 다양한 경험과 연구를 전시를 통해 공유하였고, 이는 곧 문제성을 알리는 수단으로도 새로운 대안을 제시하는 형태로 표현되었다. 이러한 연구와 실험, 그리고 토론을 토대로 보다 쉬운 방식으로 시민들에게 제공 되었고, 시민의 참여로 이루어지는 프로그램으로 구현되었다. 이는 크게 교육, 투어, 영화영상 프로그램의 세 가지로 나뉜다.

교육프로그램은 기존의 학술적인 주제 강연을 비롯해 시민들이 쉽고 전방위적인 관점에서의 도시건축을 이해할 수 있도록 건축가 뿐만 아니라 만화가, 미디어아티스트, 방송작가 등 여러 분야의 전문가들의 이야기를 들을 수 있는 특별강연으로 구성되었다. 더불어 기존 서울비엔날레에서는 볼 수 없었던 게임형, 건축 키트 만들기형, 아이디어를 쏟아내는 토론형의 전시연계 체험프로그램과 어린이를 대상으로 한 건축학교 등 시민이 직접 체험하고 즐길 수 있는 프로그램도 기획되어 운영되었다.

2017년도에 이어 연일 매진이 되었던 투어프로그램은 올해도 역시 더욱 더 특별한 프로그램의 기획으로 예약이 순식간에 마감되어 많은 분들의 참여가 돋보였다. 평소에 쉽게 가볼 수 없는 대사관과 관저를 찾아가는 오픈하우스 서울 프로그램과 서울의 역사와 현재 이슈 및 생활상을 엿볼 수 있는 서울역사투어 및 서울테마투어를 통해 새로운 서울이라는 도시를 발견 했으리라 짐작된다. 올해 투어프로그램의 차별점은 건축가, 지역 전문가를 초빙하여 전문 도슨트가 콘텐츠 내용면에서 더 알찬 내용을 시민들에게 제공할 수 있도록 한 점이었다. 그리고 다소 생소한 도시 프로젝트의 연구 결과물의 전시들을 관람객 스스로 오디오 가이드를 통해 전시에 집중할 수 있었고, 그뿐만 아니라 주말에는 전시 도슨트를 통해

Public Program

Nayeon Kim

The 2019 Seoul Biennale of Architecture and Urbanism (hereinafter, Seoul Biennale), under the topic "Collective City," explores a new, collective structure of cities that welcomes the participation of all citizens throughout the urban changes as a result of the different factors of time, space, and society. Also, in an effort to find a solution to the numerous problems that our cities face today, the Biennale provides a platform for representatives from cities all around the world to share their experiences for further research. By including the public in this program which is largely divided into the three parts, education, tours, and movie screenings, it is much easier to experiment and discuss issues related to architecture and urbanism.

The education programs offer academic lectures to help participants better understand different perspectives on the theme. Led not only by experts in the field architecture, but also by cartoonists, media artists, screenwriters, and many more. Furthermore, there will be hands-on programs, in which participants can play games, build kits, share their ideas, and an architecture school for children.

The Biennale will also hold the Open House Seoul Special Program for the public to be able to visit different architecture sites and also organize tours of Seoul, including History Tours, Themed Tours, and Docent Tours. In particular, the Seoul History Tour will be a fascinating opportunity for participants to listen as a renowned architect takes them on a journey through urban spaces in Seoul that reflect the city's history, while they collectively consider its past and future. The Seoul Themed Tours take a different approach, as participants take a closer look at the living conditions of Seoul and the problems in urban architecture that the city is facing today. At the same time, it is also an opportunity to see and appreciate the remarkable hidden spaces that are often missed in and appreciate the remarkable hidden spaces that are often missed in everyday life.

This year, for an upgraded experience, experts

직접 설명을 들음으로서 관람객의 편의에 집중하였다. 영화영상프로그램은 서울국제건축영화제와의 다시 협업하여 영화제 기간 동안에는 비엔날레 입장권으로 영화 관람도 가능한 연계 방법으로 보다 많은 시민들이 비엔날레와 영화제에 참여할 수 있었다. 올해 시민참여프로그램은 시민이 적극적으로 참여할 수 있는 프로그램으로 다양화하였고, 보다 더 쉽게 서울비엔날레를 이해할 수 있는 프로그램이 되게끔 노력하였다. 앞으로 서울비엔날레가 더욱 더 다양한 시민참여프로그램으로 찾아올 수 있도록 많은 관심과 참여를 바란다.

were invited to lead the docent tours and provide the highest quality content for the themed tours. Additionally, audio guided tours will be provided for the exhibition tours, so that visitors may enjoy the event on their own and at their own pace. Docent tours will be provided on the weekend as well to further accommodate our guests. Lastly, the movie screening programs were prepared in collaboration with the th Seoul International Architecture Film Festival (SIAFF) to grant Seoul Biennale ticket holders access to the SIAFF, encouraging the public to attend both events. For this year's citizen participation program, we made it our top priority to make it easier for the public to engage in the event and better understand the Seoul Biennale.

집합도시 서울투어
역사투어
한양-경성-서울

Collective City Seoul Tour
Seoul History Tour
Hanyang-Gyeongseong-
Seoul

교육 프로그램

2019 비엔날레 주제강연

2019 서울비엔날레에 참여하는 전시 큐레이터, 작가, 디자이너가 직접 서울의 도시·건축, 비엔날레 주제 '집합도시'를 전시 기획의도 및 작품 등을 통해 강연한다. 시민들이 보다 쉽게 이해하고 관심도 증진 및 소통의 기회를 마련하는 자리이다. 총 8회 진행이다.

1. 프란시스코 사닌, 집단성의 시급성과 당위성에 관하여
2. 홍은주와 김형재, 건축가와 함께 일하기
3. 홍주석, 로컬크리에이터가 바꾸는 도시의 미래
4. 최상기, 집합도시를 이해하는 도구로서의 건축
5. 조민석, 밤섬 당인리 라이브
6. 유아람, 서울 시장 이야기: 서울이라는 교환의 장소
7. 장영철, 건축의 기획, 기획의 건축
8. 임동우, 집합적 결과물로서의 도시

특별강연

'건축 + α 분야'의 결합으로 비엔날레 주제 '집합도시'와 전시를 다양한 방면으로 폭넓게 이해한다.미디어아트, 건축영화, 만화, 다큐멘터리 등과 같은 다양한 분야와 도시건축을 접목하여 일상 속 공감과 흥미를 불러일으킨다. 총 6회 진행이다.

1. 이이남, 건축 그리고 미디어 아트의 향기
2. 김일현, 공공과 일상사이의 건축
3. 서현석, 개인의 이상, 국가의 이념: 모더니즘 건축과 아시아 근대 국가의 형성
4. 김소현, 사람이 살고 있습니다: 동네한바퀴와 도시 이야기
5. 박정현, 발전 체제와 건축, 그리고 집합성
6. 최호철, 박인하, 만화, 손과 눈과 발로 그린 공간

전시연계체험프로그램

서울 비엔날레는 전통적인 도시건축 전시의 모습을 넘어서는 종합적인 문화예술의 장을 지향하며, 다양한 분야의 전문가들이 다각적인 매체를 통해 도시건축의 메시지를 전달한다. 전시연계체험 프로그램은 다양한 연령대, 직업, 관심사들을 담을 수 있도록 폭 넓은 프로그램을 구성하여 각계각층 시민들이 각자의 취향에 맞는 프로그램에 참여 할 수 있으며, 이를 통해 보다 쉽게 전시를 이해하고 공감할 수 있도록 하였다.

1. 모두의 비엔날레
2. 나도 건축가
3. 유레카! 서울: 더 나은 세상 만들기

서울시 건축학교

역사적으로 볼 때, 도시는 오랜 기간 다양한 사회 가치와 지배구조를 통해 진화했다. 비엔날레 기간에 운영되는 서울시 건축 학교는 학생들이 이번 도시전에 참여하는 도시를 주제별로 살펴보고, 탐구하는 과정 속에서 도시의 물리적 성장과 변화를 유도했던 사회적 상황과 의사 결정자의 의도, 유토피아적 비전과 그의 현실화 과정 등 다양한 진화 요소들을 통해 깊이 있게 도시를 탐구하고자 한다. 학생들은 개인의 관심도에 기반하여 주제와 도시를 선정하는 서울시 건축 학교의 자기주도형 수업을 통해 도시에 대한 흥미를 더욱 키울 수 있을 것이다.

– 1 아이, 1 도시

투어 프로그램

집합도시 서울투어

서울이라는 도시를 크게 역사투어와 테마투어로 나누어 걸어본다. 반복되는 일상 속에서 단편적으로 지나쳤던 서울의 곳곳을 다양한 키워드와 방식을 통해 비일상화 할 수 있는 기회로, 시민들은 일상과 비일상의 경계에서 마치 여행을 떠나는 설렘으로 서울을 마주하고 비엔날레 주제 '집합도시'를 이해할 수 있을 것이다. 더불어, 현재 도시 서울의 이슈인 재생건축, 재개발과 관련된 현장을 방문하고 전문 도슨트와 시민들의 소통을 통해 나아가야 할 미래 서울의 모습을 그려본다.

1. 서울역사투어
   - [a] 한양-경성-서울
   - [b] 조선-대한-민국
   - [c] 성문안첫동네
   - [d] 세운속골목
   - [e] 그림길겸재
   - [f] 타임슬립
2. 서울테마투어
   - [g] 인스타시티성수
   - [h] 을지로힙스터
   - [i] 서울생활백서
   - [j] 지하도시탐험
   - [k] 서울파노라마

개인자유투어(스탬프투어)

전시장 및 현장 별로 독특한 디자인의 스탬프를 비치하여 방문기록을 남길 수 있다. 안내소에 비치된 비엔날레 지도를 각 스팟에서 스탬프로 완성시킨 후 기념품과 교환할 수 있다.

연계투어 정동역사탐방

2019년은 3.1운동 100주년, 임시정부수립 100주년입니다. 의미가 깊은 해인만큼 정동의 근대역사와 문화에 대한 이해를 높이고자 기획된 투어 프로그램입니다. 투어는 고종의 밀사 역할을 맡은 배우들이 직접 등장해 스토리텔링을 하고, 시민들과 함께 소통하며 진행됩니다. 탐방 중간에 진행되는 거점연극 아관파천, 을사늑약을 통해 자연스럽게 정동의 공간을 이해할 수 있습니다.

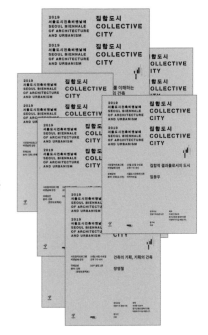

Education Program

2019 SBAU Lectures
Curators, authors, and designers will give lectures on urbanism, architecture, and the topic of "collective cities" by explaining the exhibitions and the intent behind them. These lectures are meant to help the general public better understand and build interest in the topic, while also providing them with a space to communicate with experts and each other.
1　Fransisco Sanin, The Urgency and Agency of the Collective
2　Eunjoo Hong and Hyungjae Kim, Working Together with Architects
3　Jooseok Hong, The Local Creator Changing the Future of Cities
4　Sanki Choe, Architecture as a tool to understand a collective city
5　Minsuk Cho, Bamseom Danginri Live
6　Aram Yoo, The Story of Markets in Seoul: Seoul as A Trading Space
7　Young Jang, The Planning of Architecture, The Architecture of Planning
8　Dongwoo Yim, City as collective consequences

Special Lectures
Taking into consideration that the topic of the 2019 Seoul Biennale is "collective cities," these lectures present audiences with a multi-disciplinary approach to better understand the different exhibitions. Experts in media art, architectural film, comics, and documentaries will present their work in relation to architecture and urbanism to promote interest and understanding of these concepts in everyday life. There are a total of six scheduled special lectures.
1　Leenam Lee,The Scent of Architecture and Media Art
2　Ilhyun Kim, Architecture Between Public Spaces and Everyday Life
3　Hyunseok Seo, Individual Ideals and National Ideology: The Composition of Modernist Architecture and Modern Asian Nations
4　Sohyun Kim, The People Who Live Here: Around the Neighborhood with Kim Youngcheol and Local Stories
5　Junghyun Park, Developmental Regime, Architecture, and Collectivism
6　Hochul Choi, Inha Park Comics: Spaces Drawn by the Hands, Eyes, and Feet

Hands-On Exhibition Program
is a platform that serves as a comprehensive space for arts and culture within the Biennale. In collaboration with experts from various fields, this event presents the underlying message of architecture and urbanism through a wide spectrum of media composed of various activities and exercises to appeal to people of all ages, professional areas, and interests. In this way, more people can better understand and appreciate exhibitions on architecture and urbanism.
1　The Biennale for Everyone
2　I Am an Architect, Too!
3　Eureka! Seoul: Making a Better World

Seoul Metropolitan City Architecture School Program
From a historical perspective, cities have evolved over a long period of time through various sets of social values and governance structures. There are various factors that can impact this process of evolution including the physical development of cities, the social circumstances that lead to changes, the intentions of the final decision makers of such change, utopian visions of change, and the process of achieving that change overall. The Seoul Metropolitan Architecture School will review and select the cities by theme and then proceed to explore and learn more about them in detail. The classes are self-led, with students selecting cities and topics of their interest and then learning about them through the exhibits.
–　1 Child, 1 City

Tour Program

Collective City Seoul Tour
This program is a walking tour of Seoul, focusing on different historical themes throughout the session. The goal is to take the mundane, everyday places in Seoul and presenting them from a different perspective. By blurring the line between everyday life and new adventures, participants will join this exciting trip through Seoul and learn more about collective cities along the way. Furthermore, this tour will give participants a first-hand look at urban regeneration construction and urban redevelopment, which are both important issues currently in Seoul. This program is an opportunity for the public to share their thoughts and ideas on steps forward in Seoul's urban development.
1　Seoul History Tour
　a　Hanyang-Gyeongseong-Seoul
　b　Joseon-Korea-Republic
　c　First Village within the City Gate
　d　Alleyways of Sewoon
　e　Gyeomjae Art Street
　f　Time Slip
2　Seoul Themed Tour
　g　Insta-City Seongsu
　h　Euljiro Hipster
　i　Seoul Living Guide
　j　Exploring the City Underground
　k　Seoul Panorama

Independent Tours (Stamp Tours)
We have designed unique stamps for each site and hall throughout the exhibitions of the Biennale so that visitors keep a record of where they have been. Visitors can receive a map from the Information Desk that they can use to collect these stamps. Upon collecting them all, the map can be exchanged for a special prize as a souvenir.

Jeongdong History Tour
2019 is an important year in Korean history, marking the 100-year anniversary of the March 1st Movement and the establishment of the provisional government of the Republic of Korea. On that note, this program aims to raise awareness and understanding of the modern history and culture of Jeongdong. As part of the tour, actors playing the role of envoys to the Great Emperor Gojong interact and communicate with visitors through storytelling. Also, two plays, Emperor Gojong Seeks Refuge at the Russian Legation and The Korea-Japan Treaty of 1905, are given in the middle of the tour to help visitors better understand Jeongdong and its historical significance.

관광에서 여행으로, 여행에서 일상으로

김일현

From Sightseeing to Traveling, From Traveling to Daily Life

Il-hyun Kim

여행의 추억

태어나서 지금까지 여행했던 장소들을 떠올려보자. 아주 오래 전에 다녔던 여행지들을 갑작스럽게 떠올린다는 것이 쉬운 일은 아니다. 처음은 어렵지만 일단 시작하고 나면 도움을 줄 만한 일종의 단서들이 도처에 널려 있다. 현대인은 인류의 역사상 가장 많은 자료를 만들고 남기는 놀라운 아키비스트가 아니었던가. 옷장위의 상자나 책장의 높은 곳에 까맣게 잊고 있었던 사진첩들이 있을 것이다. 사진을 한 장 한 장 보는 순간 마술과 같이 잊고 살았던 여행의 기억과 그 순간의 감흥이 되살아난다. 보다 최근의 여행과 관련된 자료들은 인화된 사진들보다 컴퓨터 폴더 어딘 가에 저장해 놓은 파일들이 더 많은 것이다. 사진을 찍고 인화하는 시간 그리고 이를 보여주는 과정이 모두 생략되어 여행 때 실시간으로 인스타그램 혹은 페이스북과 같은 곳에 올려놓은 자료들도 등장하기 시작한다. 이를 통해서 점차 촘촘하게 내 인생의 특정한 시기에 살았던 집을 기준으로 이를 떠나 방문했던 곳들을 자세히 파악할 수 있다.

이렇게 재발견한 나의 이동사의 기록을 종이 한 장에 그려보자. 가급적이면 좌표를 고려해서 표기해보자. 다양한 국내외의 여행이 종이 위에 그려지기 시작한다. 어린 시절 자동차나 시외버스를 타고 처음으로 도시 밖을 나간 기억부터, 기차의 덜컹거리는 소리와 펼쳐지는 풍경, 어색한 공항에서 작은 책자를 들고 불과 몇 걸음을 나가면 모든 세계를 접할 수 있을 것만 같았던 긴장감도 생생하게 기억할 수 있을 것이다. 동아리에서 다녀온 답사나 엠티장소, 혹은 비행기를 처음타고 방문한 해외여행지도 다시 떠오를 것이다.

각 여행지에서 각별하게 생각나는 기억들을 목차처럼 적어본다면 위치를 나타내던 하나의 점에 의해 음식, 대화, 풍경 등이 상기되며 순간들이 생생하게 살아나고 확장된다. 같은 장소를 방문한 사람 안에는 각기 다른 지도가 그려진다. 우리가 희열을 느끼는 순간과 찰나를 살아있게 남기는 유일한 방법은 기억하는 것과 기록하는 것이다. 그 반복의 과정을 통해 그 순간을 고증하는 것이 중요하지 않으며 현재의 의의를 느끼는 것이 중요하다는 것을 깨닫게 된다.

Memories of Travel

Take a moment to think about all of the places to which you have traveled. It is not always easy to think back to the places that you visited a long time ago. While difficult at first, once you start, there are hints all around that can help piece the memories together. In the history of all humanity, today's society is unmatched in its ability to create and archive various materials. Many of us have long-forgotten photo albums left in boxes in our closets or on top of bookshelves. Just one look at each of these photographs brings back a rush of memories of our previous trips and the emotions attached to them. For more recent trips, most of us are more familiar with saving our photographs on our computers rather than having them printed out. Nowadays, the entire process of getting photographs printed and then showing them to others is long gone; we live in a world where, when traveling, photographs are taken and shared with the world instantly through social media like Instagram and Facebook. Now, more than ever, we keep closer track of our everyday events, including where we go and what we do.

Let us take a closer look at this newly rediscovered method of recording our daily lives, drawing it out on a piece of paper. We can begin by writing out our travel experiences both inside and outside of the country. This can go as far back as the first time we ever took a car or bus trip outside of the city; the first time we rode a train and heard the clanking of the railcars along the railroad tracks, all while beautiful scenery passed us by; or the first time we found ourselves in the unfamiliar setting of an airport, excited and nervous at the thought of venturing out into the open world. For others, this first memory might be a retreat with fellow classmates, or the first time we rode a plane and traveled abroad.

If we were to list the memories we have of each trip that we have been on, each of the items on that list could be further described by the foods, sights, and conversations that we experienced at that particular destination. Even for those who have traveled to similar locations, no two final images reflecting their memories would be the same.

관광에서 여행으로

방송 프로그램에 나온 여행을 복제한 패키지 투어의
인기는 여전하다. 패키지 여행은 아파트와 비슷하다.
편안하고 효율적이지만 보편적인 것 외의 모든 돌발적인
변수를 차단한다. 여행사가 정해 놓은 유적지와
식당과 상점을 확인하듯 방문하는 관광의 일정에 따라
선험적으로 결정된 경로로 다니며 그 평균적인 경험을
하게 된다. 하지만 무엇보다 현지인 들과의 접촉이나
우연한 만남은 철저하게 배제된다. 집으로 돌아오면
친구들에게 짜여진 프로그램으로 다녀온 여행에 대해
이야기를 하게 될 것이다. 하지만 아무리 많은 이야기를
해도 정작 그 사람의 이야기는 하나도 없을지도 모른다.
그렇다고 해서 이러한 여행 중에 의미를 찾는 것이
불가능한 일은 아니다. 만일 어느 날 숙소나 방문지 근처
동네 주민들이 주로 가는 시장이나 골목길 아니면 가게에
우연히 들어가면, 그 것은 나만의 소중한 경험으로 남는다.
홀로 어느 골목에 눈이 가서 오래된 벤치를 사진찍고
거기에 잠시 앉아있었다고 생각해보자. 혹은 아침에
남들보다 일찍 일어나서 동네 한바퀴를 돌아보았다고
해보자. 바로 그 의자와 거기서 보낸 시간만으로도 그
여행은 값어치를 한 것이 된다. 여행지에서 낯선 사람을
만나고 대화를 나누는 것이 어색하고 두렵기까지 하지만,
그 곳에 사는 사람을 만나거나 대화하는 것은 여행지를
체험하는 것만큼 소중하다. 왜냐하면 그 사람들은
내가 지금 여행을 통해 나의 개인사를 재구성하는 것과
마찬가지로, 그 장소의 역사의 살아있는 증인들이기
때문이다. 특히 우연히 만난 사람들은 다시 만나지 않게
되는 내 인생에서 스쳐가는 특별한 관객이나 조연과도
같다. 물론 그들의 입장에서 당연히 나는 조연으로 기억될
것이다.

지금은 여행보다는 관광이라는 말이 친숙하게
사용되지만, 인간이 자신의 여가를 위해 자유롭게 국경을
넘어서게 된 지가 그렇게 오래되지는 않았다. 하지만
위험을 무릅쓰고 오지를 답사했던 인류의 행보도 그만큼
오래되었다. 생존을 위한 유목과 상업을 위한 여정을
제외하면 여행의 기원으로 생각할 만한 것은 성지순례일
것이다. 특정한 장소를 찾아가는 이 여행의 방식은 모든
종교에서 찾아볼 수 있다. 성지순례는 중요한 장소들을
경유하면서 동시에 내면과 존재에 대한 성찰을 하는
여정이다. 이처럼 은둔과는 정반대의 모습을 띠는
여행이 종교에서 중요한 성찰의 방도였다. 릴케는 이런
맥락에서 "진정하고도 유일한 여행은 결국 나의 내면에서
이루어진다"고 말했다. 그와는 반대로 나는 가만히
있지만, 나를 찾아와서 세상의 소식을 전해주던 사람들이
있었으니 그는 바로 음유시인이다. 음유시인은 이야기

Recording and recalling our memories are the only
true methods we have to feel the excitement of our
past experiences. By repeating these moments of
nostalgia and recollection, it is no longer important
to find the meanings of these memories in relation
to the past; rather, in our reminiscence, we come to
the realization that these memories carry with them
a great importance in the context of our present
lives.

From Sightseeing to Traveling

Television programs that show people going on
group package tours have always been popular.
These package tours are similar to apartment
buildings. While they are comfortable and efficient,
they tend to limit the possibilities of any variables
that stray from the universally accepted itinerary.
Travel agencies select the sites, restaurants,
and stores to try out along their travel route,
and then, for future groups, decide to include
the destinations that ensure travelers to have,
if anything, an average experience. The most
important thing to note is that these package deals
often exclude any interaction with the locals and,
of course, any unexpected meetings or events.
Upon returning home, you will find yourself telling
your friends about your trip, which was perfectly
planned by someone else. However, the details that
you share will most likely lack any information on
or experiences you had with local people you met
while abroad. This is not to say there is no meaning
to be found in these types of trips. After all, if one
day you happen to wander off into a street or store
frequented by locals, that would, in a sense, be your
own, unique experience outside of the previously
planned itinerary. Imagine walking down one of
these streets and stumbling across an old bench,
which you take a picture of before sitting on it for a
moment. Just by sitting and spending time in that
spot, that morning stroll will have become much
more valuable. Although it might be awkward and
even scary at times to meet and talk to strangers
when abroad, these are the very experiences that
make one's travel experience so special. Traveling
changes people in one way or another, and it
is the local people with whom we interact who
act as witnesses to such change as part of that
place's ongoing history. We might never see the
people that we meet in passing while traveling, but
these individuals are the supporting characters
in the performance that is our lives. Of course,
those individuals will think of me as a supporting
character in their lives, in turn.

Although the word "sightseeing" is used
more often than "traveling" nowadays, not that
much time has passed since people began to

보따리이다. 그는 한 장소에서 평생을 보내는 사람에게
먼 곳을 경험하게 하는 세상의 창이다. 사실과 진실은
그의 이야기를 판단하는 잣대가 아니다. 해학과 감동을
통해 잠시나마 우물을 넘어 우주를 품게 만드는 수사학에
묘미가 있다.

궁극의 여행, 일상

여행은 세상을 대하는 태도이다. 우리의 궁극적인 과제는
어떻게 지금까지 했던 여행을 내 자산으로 가치화할
수 있도록 기록할 것인가이다. 망각된 것 같지만 사실
달팽이가 진액을 그 행로에 남겨 놓듯이 우리의 기억 어딘
가에 그 모든 흔적은 남아있다. 단지 소중하게 생각하지
않거나 돌아볼 마음의 여유가 없을 뿐이다. 그리고
무엇보다도 그 여행의 감성을 일상으로 끌어들일 수 있을
것이냐의 여부이다. 왜냐하면 일상도 여행이기 때문이다.
프루스트는 "발견을 위한 진정한 여행은 장소를 바꾸는
것이 아니라 새로운 눈을 갖는 것에 있다."고 말했다.
우리는 각자 지구라는 땅에 엄청난 확률을 뚫고 태어나
자신만의 여행을 하고 있는 중이다. 시간이라는 매번
변하는 기묘한 흐름 속에서 우리는 인생이라는 여정을
보낸다. 지구 위에 흔적을 남기며 지도를 그리고 있는
우리 개개인은 삶의 지리학자들이다. 지구라는 표면위에
일회적인 생애를 통해 여행중인 우리들은 그 위에 보이지
않는 발자국을 남기고 있다. 친숙한 도시에 대해서 각자
새로운 지도를 그리면 그제서야 도시가 새로운 얼굴을
보여준다. 마지막으로 여행을 일상으로 확장하기 위해
비유를 하나 들어보자. 신발에 마르지 않는 물감이 칠해져
있다고 생각해보자. 그 물감 덕분에 태어나서 지금까지의
오갔던 나의 모든 행보가 지면에 기록되어 있을 것이다.
혹은 잊고 사는 어느 평범한 하루의 길목과 먼 곳을 향한
여행에 이르는 그 모든 것이 남겨진 도시의 모습은 나의
개인사를 간직하는 거대한 지도의 모습일 것이다. 만약에
인공위성이나 요즘에 쉽게 구할 수 있는 드론이 있다면
태어나서 이사를 하고 지금까지 살아온 인생의 행보와
그 장소를 한 눈에 바라볼 수 있을 것이다. 하지만 여기에
그치지 않는다. 이 지도는 생각보다 많은 것을 알려준다.
짧은 시간에 좁은 골목이나 놀이터에서 엄청난 양의
거리를 오갔던 어린 시절에 그 잊었던 순간들을 함께 했던
친구들을 만날 수 있다. 한 자리에 머물러서 긴 시간을
보내며 지금으로서는 기억도 나지 않지만 당시로는
심각한 마음고생을 했던 내 모습을 회상할 수도 있다.
혹은 붓으로 글씨를 쓰듯 느릿하게 비틀거리며 길을
걸은 흔적도 읽을 수 있을 것이다. 이 모든 것은 태어나서
지금까지 살아온 여행과 같은 인생과 일상이다. 방문한
장소들과 길 바닥에 남겨진 발자국의 보폭 혹은 모양을

travel across borders for leisure. And yet, there
is a long history of humanity accepting the
dangers of exploring remote areas. Perhaps
the only instance of traveling long distances for
reasons other than survival and trade is religious
pilgrimages. These exist universally as part of
all religions, with believers traveling to a specific
location while passing through other important
sites and observing the ins and outs of urban
areas along the way. This type of journey became
a new method of self-reflection among religious
populations, strikingly different from otherwise
hermetic approaches. In a similar context, poet
Rainer Maria Rilke once said that the truest of
journeys occurs within oneself. However, there were
also groups of troubadours who traveled to various
locations to share stories from around the world.
Troubadours were gold mines of information;
they provided people who stayed in one place
throughout the entirety of their lives with a window
through which they could see the rest of the world.
These storytellers were not concerned with telling
the truth; rather, their tales were characterized
as extravagant rhetoric that combined humor and
emotion to take sheltered individuals on a cosmic
journey to another world.

The Ultimate Trip: Daily Life

Traveling can be defined as one's means of
addressing the world in its entirety. Our ultimate
task is to determine how to keep a record of our
past journeys as valuable assets for the future.
Although there are times when we seem to have
forgotten these experiences, just as a snail leaves
behind a trail as it travels, the traces of all our
moments are found somewhere deep within our
memories. More often than not, we fail to realize
the importance of these memories or lack the
will to look back on them. The most important
factor regarding our travels is whether or not the
emotions that they evoke can be incorporated
into our daily lives; after all, everyday life is also a
journey in itself. Novelist Marcel Proust once said
that true journeys for discovery do not come from
changing one's location but, rather, from changing
one's perspective. Every one of us beat the odds
to enter this world, and every day is a part of our
own journey through life, which constantly changes
against the flow of time. We become geographers
leaving our marks by drawing our own unique maps
throughout the course of this journey, leaving
behind footsteps that are unseen to the naked
eye. It is through the brushstrokes of our different
experiences when visiting familiar cities that we
paint different images of them. To express another
metaphor for how travels can extend into our daily

통해서 인생을 점철해 온 구성하는 순간들을 만날 수도 있을 것이다. 우리가 인생에서 기억하는 것은 날짜가 아니라 순간이라는 어느 문호의 말처럼, 멀리서 그리고 가까이서 살펴보는 이 지도는 우리의 개인사를 기록하는 자서전 그 다름이 아니다. 발터 벤야민은 로지아에서 도시가 시작된다고 말했다. 우리식으로 도시는 현관에서 시작된다고 말할 수 있을 것이다. 덧붙여서 그는 '길을 잃는 법을 알아야 도시의 진정한 모습과 조우할 수 있다.'고 말했다. 이렇게 집을 나설 때 마다 그 누구도 관심을 두지 않겠지만 이렇게 매일 아침 우리 각자의 여행이 시작된다.

lives, let us take a shoe that is covered in ink that has not yet dried. The colored marks represent all of the steps I have taken since birth until now. Or think of a trip that took you from a city to faraway places and seemingly normal alleyways that you had forgotten about, and how these images are all part of your own personal history. Imagine if we could use drones or satellites to track and view all of the places we have visited since birth. Even if we could, this map of our lifetime journey would not be the final picture. In fact, this map tells us more than what first meets the eye. Perhaps the steps we took back and forth between narrow streets and local playgrounds within a short period of time symbolize memories we had with our childhood friends that we had simply forgotten about over the years. Perhaps a point on the map of our life where we stayed in the same place for a while is actually a memory of when we were going through a difficult time. These points and lines retracing our lives dance along this map like text recording our past, representing all of the daily events and moments on our journey since birth. The places we visited and the size and shape of the footprints we left behind show the important events that have shaped our lives up until the present. When we look back on these events, it is not the dates that we remember but, rather, the moments themselves expressed in a literary sense, a map observed from both near and afar that serves as a record of our own personal history. German philosopher Walter Benjamin once stated that "modernity finds its form in cities," or that cities begin from the front door. He also added that one needs to know how to get lost in order to better understand the true nature of any given city. It is in this sense that with every morning, although no one might pay us any special attention, we embark on a new journey that is unique and all our own.

로컬크리에이터가 바꾸는 도시의 미래

홍주석

Local Creators Changing the Future of Cities

Jooseok Hong

자본주의 시장에서 누구나 비슷한 아파트에 살고, 도심 고층 오피스에 모여 일하며, 백화점이나 몰에 모여 소비하는 패턴이 주를 이룬다. 도시의 효율성을 강조하는 하드웨어 중심의 도시 개발은 전 세계인이 비슷한 라이프스타일을 추구하는게 본능이 아닌가 하는 착각을 들게 할 정도로 도시를 획일화 시켜 나갔다. 그렇게 자본에 의해 편리함을 공급받아 생활하는게 익숙해지는 세대는 부동산 불패신화를 장착한 고효율의 도시개발 속에서 자신만의 도시생활을 만들어 나갔다. 그러나 '인구감소'와 'IT혁명'은 우리 도시의 패러다임을 흔들 수 있는 발판을 마련했고 나아가 '어떤 도시가 행복한 도시일까?' 라는 질문을 다시금 던질 수 있는 사회적 공감대가 형성되고 있다.

왜 사람들은 다시 골목을 찾고, 동네 서점에 생겨나고, 사라져 가던 시골 전통시장에 젊은 사람들이 모이는가? 수 십년간 우리 도시를 좌지우지 했던 백화점, 대형마트, 언론사, 방송사들은 이제 미래를 알 수 없는 처지에 놓이게 된 것일까? 급성장의 시대에 대부분의 권력은 정보와 유통에 집중되어 있었다. 자본을 중심으로한 규모의 경제를 통해 유통의 독점권을 획득할 수 있는 시대에 소수 기업의 문어발방식 사업확장은 우리 도시를 다양한 문화가 아닌 공급자에 의해 개발된 획일된 문화가 주도할 수 있는 환경을 만들어 주었고, 덕분에 아주 빠르게 우리 도시는 새로운 옷으로 갈아입 듯이 급변했다.

우리 도시의 문화적 본질이 어디에 있는지 찾기 어려울 정도로 새로운 것들도 지워지고 덧칠해지는 일들이 반복되며 우리 도시의 흔적이 거의 사라지는 사이 인터넷과 스마트폰의 등장으로 대표되는 IT혁명은 권력의 상징이었던 정보와 유통의 흐름을 흔들기 시작했다. 사람들은 TV와 신문이 아닌 스마트폰을 통해 SNS, 비디오 서비스 등을 통해 취향에 맞는 콘텐츠를 소비하고, 왠만한 생활필수품은 온라인을 통해 배송 받는다. 매스미디어는 더이상 여론을 형성시키기 어려워졌고, 프랜차이즈와 리테일 중심의 오프라인 비지니스는 다른 상점과 경쟁하는 것이 아니라 온라인과 경쟁하는 시대가 되었다.

이러한 사람들의 행태를 반영하듯 초고층 건물을

In a world driven by capitalism, it has become the norm for most people to follow similar lifestyle patterns: they live in typical apartments, work at high-rise office buildings in the heart of the city, and flock to local malls or department stores to spend their money. The extent of the similarities across major urban areas around the world gives rise to the misconception that this design, centered on hardware-driven development, represents the type of lifestyle that is universally desired, almost as if it aligns with human nature. Gradually, people began to grow accustomed to the comforts that capitalism provided, utilizing the perceived to be infallible real estate market as a tool for highly efficient urban development and unique lifestyles within those cities. However, now we are faced with the issues of population reduction and the IT revolution, which could very well shake the foundations of the urban paradigms as we know them. This reality has led many members of society to consider in depth what exactly enables a city to be characterized as genuinely "happy."

What is the reason behind people once again seeking out the alleyways and side streets; bookstores popping up here and there; and young populations gathering at local markets that were, at one point, slowly disappearing? Why is it that the department stores, supermarkets, press agencies, and broadcasting stations that had such a tight grip on our cities for decades now find themselves faced with an uncertain future? During previous generations of rapid development, the majority of power was concentrated on information and exchange. In the past, business built their success upon capital-driven economies that they manipulated in their favor to claim monopoly over various areas of trade. A minor group of these businesses expanded their influence however they could, which resulted in major cities losing the freedom to express their diverse cultures, becoming instead little more than a piece of a cultural puzzle designed by their suppliers. As a result of this hard push for growth, our cities experienced an unprecedented, rapid burst of growth, changing its overall identity in the process. However, modern society has gotten to the

지어 도시의 밀도를 높이고 효율성을 강조하던 도시들이 공유 경제를 통해 도시 시스템의 효율화를 위한 새로운 접근들을 시도하고 있다. 소유의 전유물인 자동차는 시간단위로 렌트가 가능하고, 글로벌 대기업만 쓸 수 있었던 대형건물은 코워킹 오피스로 변모하여 다양한 스타트업의 공유 공간이 되었으며, 부동산 투자의 대상이던 주거는 커뮤니티 서비스로의 주거로의 실험이 있다.

특히 골목상권의 변화는 그야말로 창조산업의 가능성을 그대로 보여주는 크리에이터들의 경연장이 되어가고 있다. 동네서점, 맴버쉽 살롱, 공유키친, 식음료 편집샵, 동네 쌀집, 연필 편집샵, 책바, 코워킹 공간 등 이름도 낯선 로컬 공간들이 사람을 끌어당기고 나아가 동네를 변화시키는 원동력이 되고 있다는 사실은 부인하기 어렵게 됐다. 동네서점은 책을 파는 공간을 넘어 사람이 모이고 책을 매개로 사람과 사람이 만나는 '연결의 공간'으로 동네 문화와 커뮤니티 서비스를 제공한다. 20년 전 단골 고객 중심의 동네 장사가 품고 있던 느슨한 커뮤니티가 새로운 로컬 공간을 통해 다시금 재해석되고 그 안에 새로운 사업모델이 탄생하고 있다.

스마트폰을 비롯해 다양한 온라인 플랫폼이 대중화되면서 동네 커뮤니티에서 차별화된 공간 콘텐츠로 성공하면 동네를 넘어 누구나 성공할 수 있는 세상이 되었다. 밀레니얼세대들은 온라인을 통해 본인의 콘텐츠를 알리고 사람들을 끌어 모은다. 전국의 다양한 골목으로 젊은 사람들이 모이고 있다. 사람이 공간을 소비하는 것이 아니라 공간의 콘텐츠가 사람을 모으는 시대, 그리고 그것이 도시를 움직일 수 있는 시대다. 인구감소의 시대 어떤 사람들이 어디로 모이고 어떻게 소통하느냐는 앞으로의 시대에 매우 중요한 이슈가 될 것이며, 이는 우리 도시가 재구성되는 근본적인 물음의 출발점일 것이다.

이제는 우리가 그 동안 가지고 왔던 소상공인에 대한 개념이 바뀌어야 하는 시기다. 동네 비즈니스가 곧 크리에이터 영역으로의 새로운 비즈니스의 잠재력을 내포하고 있기 때문이다. 급변하는 IT기술에 의한 라이프스타일의 변화로 인해 소상공인이 판매자가 아닌 크리에이터로서 건물주는 임대인이 아닌 공간기획자로 작동하거나 공간 매니저가 필요한 시대가 오고 있는 것이다. 4차 산업 혁명시대에 우리가 명심해야할 것은 기술 개발을 넘어 '사람'과 '연결'이다. 사람과 사람과의 연결, 사람과 공간과의 연결, 사람과 지역과의 연결은 지역만의 콘텐츠를 기반으로 지속가능한 미래의 동네로 가는 시작점이 될 것이다.

point where new things are constantly introduced and fade away so quickly that it is difficult to identify the fundamental urban culture of any given city. In the midst of this endless cycle, as the traces of society are replaced and forgotten, the flow of information and exchange is also going through extensive transformations with the rapid onset of the IT revolution, spearheaded by the overbearing influence of smartphones and the Internet. Nowadays, people no longer partake in the consumption of news and other interests through TV or the newspapers like they did in the past but, rather, turn to their smartphones and social networking platforms. Even when purchasing everyday household items, people have an increased tendency to shop online as opposed to offline. In today's generation, mass media is no longer effective in forming public opinions, and offline retail stores and franchises are faced with competition not only from other offline stores, but also from online shopping sites.

Cities that were once characterized by high-rise buildings and high population density to accommodate common lifestyle patterns are gradually shifting toward new approaches that instead boost the efficiency of open economy systems. In the past, cars were once considered to be a symbol of material wealth; now, they are available for rent by the hour for anyone to use. Large working spaces that were only accessible to global conglomerates have now been reinvented in the form of co-working spaces, facilitating the growth of diverse startups. Also, housing properties, which were once prime investment areas, are now being transformed into experimental community service spaces.

In particular, the shift of street shopping centers toward platforms for creative competition demonstrate the potential for creative industries. In recent years, unfamiliar businesses, including local bookstores, membership salons, open kitchens, select grocery stores, local rice shops, select stationary stores, book bars, and co-work spaces, are attracting curious consumers. Most importantly, it is clear that these new creative shops and stores have set the stage for more changes within their respective, local areas. Local bookstores are not only places where books are sold, but also serve as gathering places for people who share an interest in books. Unlike before, these stores are gradually turning into spaces that connect people, creating a sense of communal culture while also providing community services. Whereas twenty years ago local businesses were focused on appealing to regular customers without a solid foundation of community, today's generation is reinterpreting the essence of local spaces and exploring their potential for creating new business models.

With the introduction of smartphones, various online platforms have become more accessible across all members of the public. This means that any successful content unique to local communities has the potential to spread universally. Millennials are characterized by the constant sharing of creative content through online platforms, and young people are gathering all over the streets throughout the country. Now, it is no longer the norm for people to seek out and use spaces; rather, spaces are the operating agents in encouraging people to come together. This phenomenon has the potential to move and transform cities. Some of the crucial factors in this age of small traders and businesses are what types of people are gathering, where they are gathering, and how they communicate. These factors comprise the foundation of the questions we need to ask when restructuring existing cities.

Now is the time for us to change the way we think about small traders and businesses. This is especially true when we consider the fact that local businesses have the potential to expand into new, scalable creative areas of business. Thanks to the lifestyle changes facilitated by increasingly improving IT technology, today's generation is one in which small traders and businesses do not simply sell products but, rather, create them. On a similar note, these individuals and businesses need spatial designers, planners, and managers, as opposed to simple building owners who can rent out their spaces. This is true more now than ever with the onset of the 4th Industrial Revolution, which takes technological development to a new level by emphasizing the importance of people and connectivity. The latter of these two themes refers not only to connectivity among people, but also between people and spaces and between people and regions. Rethinking these linkages marks the starting point for the transition of local regions into spaces that can achieve sustainability through regional content alone.

시민참여부분 전시연계체험 프로그램

권현정

Public Participation in Exhibition Program

Hyun-jeong Kwon

과거에는 도시와 건축을 단순히 물리적인 공간으로 여겼지만 현재는 시민들의 요구가 반영되어 사회적, 문화적 이슈를 담은 공간으로 여겨지고 이를 위한 시민참여가 중요해지고 있습니다. 과거 전문가 집단을 중심으로 이루어지던 수직적인 공급과정에서 벗어나 일반 시민 모두가 함께 만들어가는 도시로 발전되어 도시 조성과정에 능동적으로 참여할 수 있는 기회가 확대되고, 전문가와 일반 시민 공동의 노력과 관심이 수반될 필요가 있습니다.

시민참여 프로그램은 일 방향 지식전달이나 전문가 양성의 측면보다, 스스로 관심을 갖고 과정에서 참여할 수 있는 경험을 통해 도시와 건축에 대해 관심을 고취시키고, 도시공간에 대한 시민의 이해를 바탕으로 더 나은 도시 환경을 이루고자 합니다.

흔히 도시건축은 시민의 삶과 밀접한 관계를 맺고 있음에도 동떨어져 보이고 일부만의 문화, 또는 건설로만 인식되어왔습니다. 도시건축이 일상생활 안으로 들어옴에 따라 도시건축에 대한 인식도 변화하기 시작했습니다. 서울비엔날레를 통해 시민들은 극히 좁은 의미, 즉 건설로만 인식하던 도시건축을, 넓은 의미인 도시 환경 전반에 대한 작업으로 인식하기 시작했으며, 도시환경, 도시 속의 삶, 시민의 생활과 직접 영향을 주고받는 것임을 인지하기 시작했습니다. 서울비엔날레의 건축, 도시연구, 시각디자인, 사진, 영화, 공공미술, 교육, 투어, 워크숍 등 다양한 분야의 전문가들이 함께 준비하여 다각적인 매체를 통해 도시건축의 메시지를 전달합니다. 이러한 다양성을 가진 도시건축의 통합적인 면모를 직관적으로 이해할 수 있도록 시민들을 위한 다양한 프로그램을 구성하였습니다.

'모두의 비엔날레-위기의 지구를 구해라!' 보드게임은 환경이라는 주제를 시작으로, 우리가 함께 살고 있는 도시와 지구에 대한 지속가능성에 대해 생각해 보는 시간입니다. 주사위의 말판을 따라가면서, 우리에게 당면한 문제가 무엇인지 알아가게 됩니다.

'나도 건축가'는 도시를 이루는 최소의 단위인 집을 만들면서, 사적공간의 영역성, 사회적 요소, 일조, 조망, 이웃과 관계 등 시민들이 가상의 마을을 만들면서, 마을

In the past, we considered urbanism and architecture only as physical spaces and nothing more. Now, however, reflecting the demands of the public, these two concepts encompass social and cultural issues, and the participation of the public, in turn, has become increasingly more important. Modern approaches have branched away from the previously expert-heavy processes of creating cities, now characterized by efforts to collaborate with everyday people toward the common goal of developing cities and promoting active participation of the public in urban structuring. More than ever, there is a pressing need to strengthen joint efforts between experts and the general public in developmental efforts in architecture and urbanism.

Rather than implementing a one-way system of conveying information and fostering expertise among those who are already active in the field, the public participation programs that we have planned encourage all individuals to take interest in the exhibited content and provide them with the opportunity to learn more about urbanism and architecture first-hand. The goal of these projects is to gain a better sense of how the public understands urban spaces and later use this information to improve future urban environments.

Throughout the years, despite the fact that urban architecture is closely related to the lives of the public, the majority of people have been quite unfamiliar with this relationship, aside from the more commonly perceived influences of certain cultural aspects of urban spaces or construction projects that take place in these areas. As urban architecture worked its way into different parts of daily life throughout the years, people's awareness of its role in their lives began to change. Thanks to the Seoul Biennale of Architecture and Urbanism, the public has gradually become more aware of the expansiveness of urban architecture and its role and identity as a means of creating a given urban environment and influencing the daily lives of those who live in these urban areas. For the Seoul Biennale, various experts working on urban research projects, visual designs, photographs, movies, public arts, education, tours, and

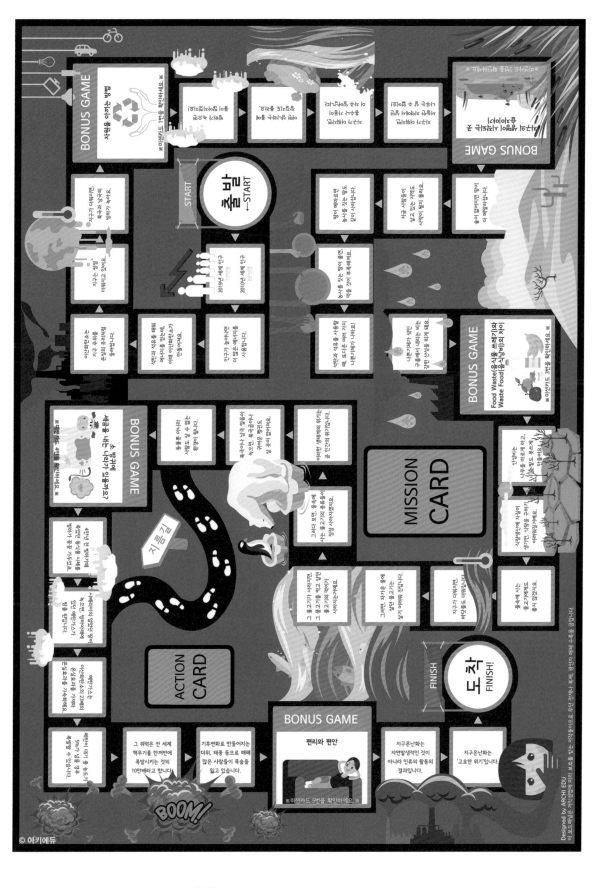

속에서 일어 날 수 있는 다양한 삶에 대해 참여자 모두 '자신의 집'을 만들면서 '더불어 사는 삶'에 대해 진지하게 고찰해 보는 시간입니다.

'유레카! 서울'을 통해 곧 우리사회의 미래의 주인공인 중. 고등학생들과 함께 전 세계의 다양한 학생들의 '더 나은 세상을 만드는 방법'에 대한 실천 사례를 공유하며, 전시의 주요 키워드를 바탕에 두고, ' 더 나은 미래' 에 대해 생각을 모아보는 프로그램으로 다각적인 관점에서 도시건축을 바라보는 수업입니다.

workshops collaborated to help promote the public's understanding of urban architecture. With their help, we prepared various programs to show the public the diversity behind urban architecture in a clear and effective way.

"Biennale for All-Save the Earth from a Crisis" is a board game based on the theme of "environment." When playing this game, participants will have the opportunity to think about the sustainability of our planet and how we contribute to it. As they move around the game board, players will encounter and learn more about the problems that we face regarding the environment.

"I am an architect, too!" is a program in which participants begin by building houses, the basic unit of any given city, before moving on to build an entire imaginary town with its own social elements, relationships among residents, landscapes, concepts of personal space, and social roles. Here, participants have the opportunity to think closely about the various aspects of life within this imaginary city and observe the workings of their own space in the context of urban design and social interactions.

Through the "Eureka! Seoul" program, participants can take a close look at the stories shared by today's middle and high school students, the soon-to-be main characters of future society, from all around the world, on the steps they have and are taking to make the world a better place. This program encourages participants to consider and gather numerous, different perspectives on what can be done to ensure a brighter tomorrow.

서울시 건축학교:
1Child, 1 City

권현정

Seoul School of Architecture:
1 Child, 1 City

Hyun-jeong Kwon

서울시 건축학교는 건축을 통해 개개인의 성장과 더불어 공공의 선한 가치를 함께 학습합니다. 건축을 교육수단으로 학생들의 창의력과 통합적 사고능력을 배양하고 수업을 통해 주변 환경을 가꿀 수 있는 능력과 관심을 유도합니다.

건축을 통한 교육은 교육의 주제가 건축과 도시환경을 대상으로 하여 개개인의 주변 정보와 아이디어를 바탕으로 이를 객관화하는 작업을 통해 창의성과 심미적 감성, 통합적 사고 능력을 향상시키면서, 타인들과 함께 공간을 사용하고 노약자를 배려하는 물리적 공간을 탐구하면서 사회적 공감능력이 향상되고, 공동체 의식이 성장합니다.

아이들이 건축교육을 통해 미래 시민으로서의 소양과 지혜를 습득하는 것은 물론 개인의 창의적 문제 해결 능력까지 향상하도록 도움을 주고 있습니다. 인터넷 검색으로 전 세계의 다양한 지식과 정보들을 손쉽게 얻을 수 있는 현시대에 기존 주입식 교육방식은 체험 위주의 교육방식으로 변화하고 있다. 이러한 변화에 발맞추어 기초건축교육 또한 창의적이고 통합적인 사고능력을 증진시킬 수 있는 체험 위주의 교육방식을 취하고 있습니다.

비엔날레 기간에 이루어지는 '서울시 건축학교'는 돈의문 박물관 마을에서 열리는 '도시전'을 수업의 도구로 삼고 있습니다. 도시는 공간적, 시간적, 사회적 환경의 집합체를 보여줌과 동시에, 경우에 따라서는 계획되지 않은 요소들의 개입으로 끊임없이 변화하는 생물과 같습니다. 수업에 참여하는 학생들은 이러한 도시에 대해, '도시전'에 전시되는 80개의 도시 중 개인적인 흥미로 하나의 도시를 선택하여 전시내용을 분석해 보며 자신이 선택한 도시에 대해 탐구하는 수업입니다.

The Seoul School of Architecture uses architecture as a tool to teach students about personal growth and the good-natured value of a communal public mindset. The lessons taught here foster creativity and integrated thought processes and build capacities for and interest in improving one's surrounding environment.

The curriculum focuses on architecture and urban environments as the main teaching subjects. Here, students are encouraged to share their ideas and objectively implement information from their surroundings with the goal of building creative and aesthetic skills and improving capacities for integrated thinking. Also, in an effort to instill a sense of community and empathy toward others, students are given opportunities to explore physical spaces that are shared by groups of individuals and meant to accommodate those with special needs.

Through the program, students build the wisdom and capacities needed to fulfill their duties as upstanding citizens of the future, while also developing creative problem solving skills. In today's society, where any desired bit of information from anywhere in the world can be easily accessed with a simple Internet search, educational models are gradually straying from the previously preferred models of one-way lectures and input-output based classes. The Seoul School of Architecture aligns itself with this rapidly changing approach to education by implementing hands-on, practice-based lessons that more effectively help students in building their creative, integrated thinking and problem-solving skills.

The Seoul School of Architecture, which will be held during the duration of the 2019 Seoul Biennale of Architecture and Urbanism, is presented to students in tandem with the Cities Exhibition on display at the Donuimun Museum Village. Using these projects as a tool, the school's curriculum aims to show the identity of cities as entities comprised of spatial, temporal, and social components; at the same time, this program demonstrates that cities are a unique type of organism, always changing at the hands of unexpected, outside factors. Participating students will have the opportunity to choose one out of the

80 cities presented as part of the Cities Exhibition that appeals to them, and then analyze their selected exhibit's content and explore it in great depth as part of the class.

I·SEOUL·U

서울특별시

국립민속박물관
The National Folk Museum of Korea

창경궁
Changgyeonggung

국립현대미술관 서울관
National Museum of Modern and
Contemporary Art, Seoul

서울대학교병원
Seoul National Unive

2019
서울도시건축비엔날레
SEOUL BIENNALE
OF ARCHITECTURE
AND URBANISM
9.7.–11.10.

경복궁역
Gyeongbokgung Station ③

창덕궁역 ③
Changdeokgung Station

도시 서울을 크게 역사적, 테마별 주제로 나누어 다양하게
걸어본다. 반복되는 일상 속에서 단편적으로 지나쳤던
서울의 곳곳을 다양한 키워드와 방식을 통해 비일상화한다.
일상과 비일상의 경계에서 마치 여행을 떠나는 설렘으로
도시 서울을 마주하고 비엔날레 주제 '집합도시'를 이해한다.
더불어, 현재 도시 서울의 이슈인 재생건축, 재개발과 관련된
현장을 방문하고 전문 도슨트와 시민들의 소통을 통해
나아가야 할 미래 서울의 모습을 그려본다.

종로3가역
Jongno 3-ga Station

집합도시
COLLECTIVE
CITY
서울투어
SEOUL
TOUR

광화문역 ⑤
Gwanghwamun

종각역 ①
Jonggak Station

①③⑤

세운상가 세운홀
Sewoon Hall

서울도시건축전시관
Seoul Hall of Urbanism &
Architecture

청계천
Cheonggyecheon

세운보행데크 청계대림데크
Sewoon Pedestrian Deck,
Chunggye-Daelim Pedestrian De

덕수궁
Deoksugung

을지로1가역 ②
Euljiro 1-ga Station

을지로3가 ②③
Euljiro 3ga

을지로4가역 ②⑤
Euljiro 4-ga Station

시청역 ①②
City Hall Station

미술관
Museum of Art

명동역 ④
Myeong-dong Station

충무로 ③④
Chungmuro

회현역 ④
Hoehyeon Station

Station

서울 역사 투어 ②③
약 2시간 소요 예정

테마1 [한양·경성·서울]
2019.09.28. 토요일 14:00
경복궁 → 육조거리, 광화문 네거리 → 서울광장
→ 숭례문 → 서울도시건축전시관
(도슨트: 안창모)

테마2 [조선·대한·민국]
2019.10.05. 토요일 10:00
광화문 → 장충공원, 박문사터 → 남산2호터널
→ 유관순동상 → 자유센터, 국립극장 → DDP
(도슨트: 안창모)

테마3 [성문 안 첫 동네]
2019.09.21. 토요일 14:00
돈의문박물관마을 (도슨트: 조정구)

테마4 [세운 속 골목]
2019.10.19. 토요일 14:00
세운상가 일대 (도슨트: 조정구)

테마5 [그림 길 겹재]
2019.10.12. 토요일 14:00
경복고 → 청운초 → 청풍계 → 현대家 → 옥인동
군인아파트, 윤버친가 → 수성동 계곡 →
배화여고 필운대 (도슨트: 임형남)

테마6 [타임슬립]
2019.10.26. 토요일 14:00
드라마센터 → 중앙정보부(구) → 서울예술대학
→ 와룡묘 → 남산신궁 → 회현아파트 → 후암동
적산가옥 (도슨트: 임형남)

서울 테마 투어
약 2시간 소요 예정

테마1 [인스타 시티 성수]
2019.09.08. 일요일 / 10.06. 일요일 14:00
대림창고, 성수연방 → 카페 어니언 → 드림
인쇄소 → 오르에르, WxDxH → 우란문화재단 →
서울숲, 붉은벽돌 재생지역 → 블루보틀 → DDP
(도슨트: 심영규)

테마2 [을지로힙스터]
2019.09.15. 일요일 / 10.20. 일요일 14:00
DDP → 4F CAFÉ, 방산시장 → 금속공장,
N/A갤러리 → 세운상가 → 을지로OF → 만선호프
③④ → DDP (도슨트: 이희준)

테마3 [서울생활백서]
I: 2019.09.22. 일요일 10:00
마장키친 → 풍물시장 → 동묘벼룩/창신완구시장 →
DDP
II: 2019.10.13. 일요일 12:00
마장키친 → 창신동채석장터일대 →
이음피움봉제역사관 → DDP (도슨트: 이희준)

테마4 [지하도시탐험]
2019.09.15. 일요일 / 10.20.일요일 10:00
돈의문 박물관 마을 → 경희궁 방공호 →
서소문 역사공원 → 뮤지스땅스(10.20only)
→ 여의도sema방커 → 서울도시건축전시관
(도슨트: 김선재)

테마5 [서울파노라마]
2019.09.29. 일요일 10:00, 14:00 /
10.27. 일요일 14:00
서울로 7017, 윤슬 → 서소문청사 정동전망대
→ 서울마루 → 서울도서관 하늘書 → 세운,
청계상가옥상 (도슨트: 김나연)

550

전통시장 도슨트 투어

이희준

Traditional Market Docent Tour

Hee-jun Lee

전국의 전통시장을 기록으로 남기는 작업을 하고 있다. 지난 7년간 전국 1,011 곳의 전통시 장을 누볐고, 그중 45개를 골라 2015년 『시장이 두근두근』 1, 2 두 권의 책으로 펴냈다. 전 통시장에서 새로운 기회를 발견해 2016년 7월 서울 구로시장의 청년 상인으로 국내 최초의 참기름 편집매장 '청춘주유소'를 운영했고, 2018년 서울 연남동 연남방앗간의 참기름 소믈리 에로 식문화 기반 동네 커뮤니티 공간 기획 및 운영에 참여했다. 무엇보다 시장의 역사, 철학 있는 상인들의 이야기, 특화된 상품을 기록하고 해설하는 국내 1호 전통시장 도슨트로 활발히 활동 중이다.

Our work involves documentation of traditional markets across the country. Over the past seven years, we visited 1,011 markets countrywide and included the stories of 45 of them in two books, *The Market's Heartbeat* 1 & 2. We discovered the potential for new opportunities in the markets that we visited, which, in July 2016, led us to introduce the first market stall at Guro Market with young adults selling sesame seed oil. In 2018, we designed and operated a community space as "sesame seed sommeliers" at Yeonnam Bangatgan in Yeonnamdong, Seoul, creating an appealing environment for those interested in local food culture. The most important part of our activities is the docent tours we provide to share the history of the local markets, the stories of the local merchants, and explanations of unique, local products.

이번 2019년도 서울도시건축비엔날레의 집합도시 서울투어 프로그램의 전문 도슨트로 참여하여 '집합도시 속의 시장, 시장 속의 집합도시'를 주제로 이야기를 전달하고자 서울의 역사가 깃든 공간들을 잇는 전통시장 연계형 집합도시 도슨트를 기획했다. 현재 서울에는 330여개의 전통시장이 운영되고 있으며 걸어서 찾아갈 수 있는 거리 20분 이내에 1개 이상의 시장을 보유하고 있는 세계에서도 보기 드문 생활형 시장을 가진 도시로 집합도시 속 전통시장 자원에 주목할 필요가 있다.

이러한 서울 집합도시의 전통시장 자원을 기반으로 두 개의 코스를 기획했다. 하나는 축산물 전문시장인 서울 마장동 마장축산시장을 시작점으로 하여 동묘벼룩시장, 창신동 문구완구 도매시장까지 이어지는 '집합도시 속 시장의 역사'를 현장에서 만나보는 기획이다. 서울이라는 도시 안에 있지만 시간을 내 가보지 않으면 만나기 어려운 하나의 품목에 특화된 시장들을 걷 는다. 또 하나의 기획은 콘텐츠 기반 동네 매니지먼트사 어반플레이가 출간하는 아는동네 매거진 '아는동네 아는을지로' 편에서도 언급되고 있는 과거와 현재가 공존하는 레트로 구역, 힙지로라는 별칭을 얻은 을지로 구석구석을 탐방하는 '힙지로 도슨트'가 기획의 주요 골자다.

두 코스의 전통시장과 연계한 집합도시 전문도슨트를 통해 서울 시민들과 서울을 방문한 관 광객에게 전달하고 싶은 메시지는 분명하다. '서울의 집합도시 속 전통시장은 시간이 켜켜이 쌓여 만들어져 왔다는 것'. 총 네 번의 도슨트를 통해 과거로의 여행, 그리고 과거의 흔적을 발견하고 앞으로 나아가는 도시의 역동적인 모습 또한 함께 경험해 볼 수 있는 시간이 될 수 있을 것이다.

As part of the 2019 Seoul Biennale of Architecture and Urbanism Public Programs, we are once again contributing our docent tour services under the theme "collective cities." Specifically "Marketplaces within collective cities and collective cities within marketplaces," with the purpose of sharing stories of the historically rich traditional markets of Seoul and the spaces that connect them. Currently, 330 markets are being operated throughout Seoul, with several markets located within a 20-minute walk of each other. Indeed, Seoul is a city that boasts a marketplace culture within the context of its identity as a collective city that is unique and worth the experience.

We have designed two separate courses to take advantage of the rich selection of marketplaces in Seoul. The first begins with Majang Agricultural Market, one of the most well-known locations for a variety of agricultural products. The next stop to Dongmyo Market, before heading to Changsin-dong Market; through this course, visitors can experience the history of marketplaces within the collective city of Seoul. Although these marketplaces are a part of the city as a whole, they are distinctly characterized by the unique, specialized products sold in each market, most of which cannot be found elsewhere in the city. The second course is to the winding streets of "hip zero" an old neighborhood in Euljiro rediscovered as the retro part of the city where the past and present coexist.

There is a clear message to the locals and visitors of both these docent tours regarding the marketplaces that are closely linked to Seoul's identity as a collective city: that the traditional markets located across the city are the collective product of the passing of time. Through a total of four docent tours we hope the participants can experience a journey to the past where they can discover important pieces of history and the dynamic advancement of the city.

서 울 테 마 투 어

○ 2030세대의 집합놀이문화 키워드별로 능동적 체험 및 주제를 탐구하는 투어 (매주 일요일 운영, 총 11회)

*자세한 시간 및 정보는 맨 뒷면 참조

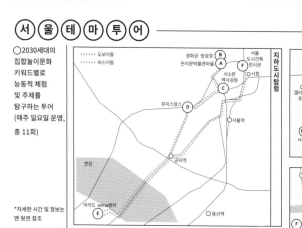

지하도시탐험

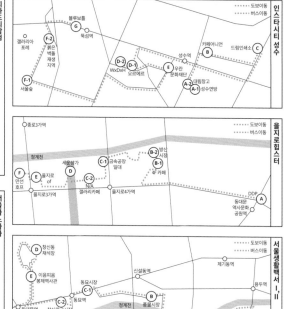

인스타시티 성수

을지로힙스터

서울생활백서 I, II

I : Ⓐ마장동축산시장, 마장키친-Ⓑ풍물시장-Ⓒ동묘벼룩/창신완구시장-ⒻDDP
II : Ⓐ마장동축산시장, 마장키친-Ⓓ창신동 채석장 전망대-Ⓔ이음피움봉제역사관-ⒻDDP

서울파노라마

서 울 역 사 투 어

한양-경성-서울

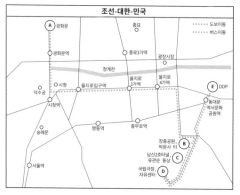

조선-대한-민국

성문 안 첫 동네

● 근, 현대 서울의 도시건축, 한국적 모더니즘 양상 등 도시의 역사를 따라 진행되는 투어 (매주 토요일 운영, 총 6회)

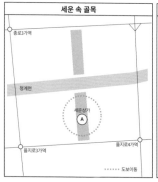

세운 속 골목

그림길 겸재

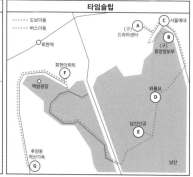

타임슬립

*자세한 시간 및 정보는 맨 뒷면 참조

도시는 시간의 집합이다

조정구

A City is a Collection of Time

Jung-goo Cho

어쩌다 깊은 밤 서촌의 골목길을 걸으면 느껴지는 묘한 느낌이 있다. 곧은가 하면 살짝 휘어져 물처럼 흐르고, 이리 꺾고 저리 꺾다 보면 어느새 '아~' 하고 감탄사가 나오는 골목 마당과 마주한다. 아마 서울이 계획된 도시였다면, 자연의 지세 위에 사람의 삶터가 자연스레 얹어진 도시가 아니었다면 이런 한밤의 산책이 주는 감동은 없었을 것이다.

올해의 서울도시건축 비엔날레의 주제는 '집합도시'이다. 설명문에는 '새롭게 탄생하고 구성될 도시를 연구하여 집합이 실제로 도시의 정치적 활동과 변혁의 주체임을 시사할 것'이라 밝히고 있다. 다가올 미래에 방점을 두고 있지만, 오히려 관심이 가는 것은 '시간'이었다. 도시가 집합으로 이루어졌다면 그것은 '오랫동안 켜켜이 쌓인 시간 위에 새로운 시간이 쌓여 집합을 이룬 결과'가 아닐까. 그래서 도시는 '오래된 것'이고 지금도 축적돼 가고 있는 '시간의 집합'이란 생각이 들었다.

답사의뢰를 받고 서울의 어디를 같이 볼까 고민했다. 십여 년 동안 답사하고 기록했던 동네들과 건축가로 참여해 진행했던 도시재생작업을 되돌아보면서, 사람들이 '시간의 집합으로서 도시'를 느끼고 생각할 수 있는 장소가 어딜까 찾아보았다. 그 과정에서 북촌이나 서촌처럼 오래되었지만 안정된 동네는 제외하고, 가능한 참가자들에게 '앞으로 우리는 이것을 어떻게 해야 할까' 걱정과 질문을 던지는 장소, 더불어 여간해서는 혼자 다니기 어려운 곳을 택하였다. 그 결과 선정된 곳은 서대문 터 바로 안쪽에 자리한 '성문안첫동네'와 세운상가 양쪽으로 펼쳐진 공장과 가게의 밀집지 '세운속골목' 두 곳이었다.

성문안첫동네: 사라진 동네 교남동과 돈의문 박물관 마을

서울의 아름답고 오래된 동네들이 사라지고 있다. 한양 도성과 이웃한 '교남동'도 이 무자비한 파도를 피해가지 못했다. 탑상의 아파트들이 점령군처럼 늘어선 단지에는 이전의 어떤 모습도 비치지 않는다. 구부러져 내려가는 만초천 물길을 따라 둥그렇게 바깥채를 지은 '자전거포

There is a strange feeling when you walk in the alleyways of Seochon deep in the night. The seemingly straight yet slightly curved paths turning corners here and there, will eventually lead you to an urban clearing. Had Seoul been a planned city, and had it not been a city where urbanization naturally settled along the topography, this walk in the night would be less special.

The theme of the 2019 Seoul Biennale of Architecture and Urbanism is "Collective City." The introduction says that the Biennale will "explore the condition of the collective in the city, not as the natural state of the city today, but rather as a condition to be reclaimed and reframed, suggesting that the collective subject is in fact the instrument of political action and transformation of the city." It focuses on the near future; but what really caught my attention was 'time.' If a city is formed of collectives, then maybe a city is the result of the accumulation of time. It made me think that a city is something 'old' with 'a collection of time,' where time continues to accumulate.

I carefully thought about where in Seoul to visit when I received the request for advice. I looked back on the neighborhoods I've visited and recorded, as well as urban regeneration projects I've participated with fellow architects over the last decade or so. I tried to choose places where people can feel and think 'city as a collection of time.' I excluded old but well-recognized neighborhoods like Bukchon and Seochon. Instead, I focused on the ones that make us ask 'what should we do with this neighborhood in the future?' I also looked for neighborhoods that are not easy to visit or travel alone. Two neighborhoods came to my mind: Seongmunan-cheotdongne (Seongmunan First Village) inside the Seodaemun site, and Sewoon-sokgolmok (Sewoon inner alleyways) on both sides of the Sewoon Plaza where factories and shops were concentrated.

Seongmunan-cheotdongne: Gyonam-dong, the Long-gone Neighborhood, and Donuimun Museum Village

Old yet beautiful neighborhoods of Seoul are disappearing. Gyonam-dong, once next to Hanyang

집'이나, 화방벽이 무척이나 아름답던 '할머니 집', 향나무, 모과, 맨드라미, 꽃사과 등 꽃과 과실이 넘쳐나는 골목풍경 역시 사라지고 말았다. 도시는 시간이 쌓여가는 '오래된 것'이라지만 서울은 점점 기억과 정취를 지워가며 '날 것'이 되고 있다.

　　한 편, 같은 재개발구역에 속해 근린공원이 될 운명이었던 정동초입의 작은 동네는 살아남아 '돈의문 박물관 마을'이라는 낯선 이름을 얻었다. 조선 시대와 개화기 그리고 근현대기의 켜가 집과 골목에 온전히 남은 동네는, 주거지에서 상업지로 그리고 지금은 서울시 재생사업의 시험장이 되고 있다. 이번 답사는 기록과 연구작업을 통해 사라진 교남동을 다시 구현한 돈의문 전시장을 중심으로, 돈의문 박물관 마을의 조성과정과 의미 그리고 앞으로의 재생방향을 함께 고민하는 시간이 될 것이다.

세운속골목: 역사도심 속 삶의 터전과 살아있는 시간

종묘 앞에 펼쳐진 피와 땀이 베인 땅을 '세운상가와 주변 블럭'이라 단순히 부르기엔 아쉬움이 남는다. 굳이 말하면 '작은 공장과 가게들이 수백 년 된 골목 속에 마련한 삶의 터전'이란 표현이 맞을 듯하다. 2019년 철거를 멈추게 하고 서울시장은 '노포의 가치'를 말했고, 재생을 생각하는 이들은 지금 '산업생태계'를 말하고 있다. 하지만 노포보다

Doseong, is one of them. Tower-like apartments form endless complexes, completely erasing traces of the past. Gone are the bicycle shops that formed round annexes along the curvy Mancho Stream, Grandmother's House that boasted the prettiest hwabang wall (fireproof wall), and fruits and flowers like juniper trees, quinces, cockscombs and crab apples that once filled our neighborhoods. We say time builds up and makes a city 'antique'. Yet Seoul is erasing its memories and ambience, becoming something rather crude.

On the other hand, the tiny neighborhood at the entrance of Jeong-dong, once destined to be redeveloped and become a neighborhood park, survived and was given the new name Donuimun Museum Village. With periodic layers from the Joseon Dynasty, the time of enlightenment and modern times accumulated in houses and alleyways; the neighborhood now serves as a residential area, a commercial area and the testbed of Seoul's urban regeneration efforts. This visit will center on the Donuimun Exhibition, which reproduces the long-gone Gyonam-dong through records and research, allowing us to consider the formation process and importance of Donuimun Museum Village, as well as the future direction for regeneration.

오래된 공장들이 있고, 생태계보다 생계가 중요하다는
사실, 그리고 주목하는 이가 적으나 골목과 집들이 이루고
있는 '서울이라는 시간의 바탕'을 우리는 간과하고 있는
것은 아닐까?

　　60년대 말 소개공지 위에 미친 듯이 지어진
세운상가는 난파한 배처럼 이 땅 위에 주저앉았다.
종묘와 남산을 잇는 꿈은 요원하지만, 데크를 중간에
두른 상가부터 아뜨리움이 있는 아파트까지 모든 칸들은
가게와 사무실, 창고로 가득하다. 이번 답사는 골목과
속골목, 필로티와 데크, 옥상을 누비며 지층과 구조물을
관통해 갈 것이다. 살아있는 시간의 중심을 볼 수 있는
소중한 기회가 될 것이다.

　　앞으로 우리 도시의 정체성과 관련하여 가장 중요한
주제는 바로 '시간'이라 할 수 있다. 오랜 시간을 두고
현재까지 쌓아온 '시간의 집합'을 어떻게 바라볼 것이며,
어떻게 해야 할 것인가는 피하기 어려운 우리 세대의
과제가 되었다. 보다 많은 사람들이 도시의 시간성을
존중하고 배려함으로써 미래에 적합한 '새로운 시간의
집합도시'를 스스로 주체가 되어 만들어가기를 기대해
본다.

Sewoon-sokgolmok: Life in the Historical Urban Center and Time history

It'd be a shame to call the land of blood and sweat in front of Jongmyo simply the "Sewoon Plaza and nearby blocks." Perhaps "home of small factories and shops built on alleyways of hundreds of years of history" is better suited The Mayor of Seoul halted the demolition of the area in 2019 and emphasized "the importance of old stores"; those who promote regeneration emphasize its "industrial ecosystem." However, what we should ask ourselves is this: are we, in any way, ignoring the fact that life is more important than the ecosystem, and—although not a lot of people seem to notice—the very base of Seoul's history, which is built upon alleyways and houses?

　　Inexorably built on the firebreak during the late 1960s, the Sewoon Plaza landed on the ground like a stranded ship. Its goal of connecting Jongmyo and Namsan seems far away; yet every available corner from the decked Plaza to the apartments with an atrium is filled with shops, offices and warehouses. We will be able to explore the alleyways, inner alleyways, pilotis, decks and rooftops, passing through the ground level and structures. At the same time, we will be able to seize the invaluable opportunity to witness the center of living history.

　　'Time' is indeed the most important subject in regards to the urban identity. We are now faced with the inevitable task of how we will view the collection of time that has accumulated over such a long time, and how we are going to deal with it. I hope more people become active entities in creating a collective city of a new history, one that is more suitable for the future, by respecting and taking into consideration the temporality of the city.

글쓴이

프란시스코 사닌(Francisco Sanin)은 시라큐스대학교 교수이다. 프린스턴대학교, AA스쿨, 킹스턴대학교, 한국예술종합학교의 객원교수로 지내기도 했다. 아시아, 유럽, 남아메리카 지역의 건축설계 프로젝트를 진행하고 있으며 2008 베니스비엔날레 한국관 공동 커미너셔, 2011 광주비엔날레 디자이너 및 2017 서울도시건축비엔날레 큐레이터로 활동한 바 있다.

임재용은 건축사사무소 OCA(Office of Contemporary Architecture)의 대표이다. 그는 사회, 경제, 문화의 전반적인 상황의 변화를 인식하고 그 결과를 건축에 담아내는 새로운 유형들을 찾는 건축가이다. 이번 비엔날레도 도시의 새로운 유형을 찾는 작업의 연장선 상에 있다. 그는 〈Seoul: towards a Meta-city〉 전시의 총괄 커미셔너를 맡았고 2011–2012 한일현대건축교류전의 총감독이었다. 또한 2011년 UIA대회 서울홍보관 커미셔너이기도 하였다. 서울대학교에서 학사학위를 받았으며, 미국 미시건대학교에서 석사학위를 받았다.

베스 휴즈는 주제전의 큐레이터로, 영국 왕립예술대학교 건축프로그램 학장이다. 호주 시드니공과대학교를 졸업했으며 출신 학교에서 교편을 잡기도 했다. 2004년부터 2009년까지 OMA에서 일했다. 2009년에는 그리스 아테네의 포인트 슈프림 아키텍츠에 파트너가 되었고, 2011년에는 런던에 기반을 둔 자신의 건축사무소를 설립하여 현재까지 운영 중이다.

김성홍은 서울시립대학교 건축학부 교수이다. 2007~2010년 프랑크푸르트, 베를린, 탈린, 바르셀로나, 서울에서 열린 〈메가시티 네트워크: 한국현대건축전〉을 총괄기획했고, 2016년 베니스비엔날레 한국관, "용적률 게임: 창의성을 촉발하는 제약"의 예술감독을 맡았다. 〈Megacity Network: Contemporary Korean Architecture〉(2007), 『도시 건축의 새로운 상상력』, (2009), 『On Asian Streets and Public Space』(2010, 공저), 『길모퉁이 건축: 건설한국을 넘어서는 희망의 중간건축』, (2011), 『Future Asian Space: Projecting the Urban Space of New East Asia』(2012, 공저), 『The Far Game: Constraints Sparking Creativity』(2016) 등 도시와 건축에 관한 논문과 저서를 국내외에 발표 했다.

켈러 이스터링(Keller Easterling)은 예일대 건축학과의 교수이자 작가이자 건축가이다. 그녀의 가장 최근 저서인 『Extrastatecraft: 인프라스트럭처 공간의 힘』(Verso, 2014)는 지구상의 인프라를 정치의 매개체로 고찰한다. 최근에 출판된 『Medium Design』(Strelka Press, 2018)은 공간적 문제와 비공간적 문제 모두 혁신적인 생각을 촉진하기 위해 물체와 인물에 대한 강조를 고찰한다. 또 다른 최근 저서인 『Subtraction』(Sternberg, 2014)에서는 건물 철거나 개발 기계를 역방향으로 만드는 방법을 제안하고 있다.

도그마(DOGMA)는 2002년에 피어 비토리오 아우렐리(Pier Vittorio Aureli)와 마르티노 타타라(Martino Tattar)에 의해 설립되었다. 활동 초기부터 주로 도시 설계와 대규모 프로젝트에 집중하여 건축과 도시의 관계에 힘썼다. 지난 몇 년 동안 도그마는 실내 공간의 잠재력과 변형 가능성에 초점을 맞춘 디자인 연구를 하고 있다. 2006년 도그마는 제1회 이아코브 체르니코브 상(Iakov Chernikhov Prize)에서 최고의 건축실무상을 수상하였다. 피어 비토리오 아우렐리는 현재 런던의 AA 건축학교에서 강의를 하고 있으며 예일대학교 초빙교수를 역임하고 있다. 그의 주된 이론적 관심은 건축과 대규모 도시 설계의 관계이다.

Francisco Sanin is professor at Syracuse University. He has taught at Princeton University, Architectural association, Kingston University and KNUA (Korea). He maintains a professional practice with works in Asia, Europe and Latin America. Sanin was co-commissioner of the Korean Pavilion for the 2008 Venice Biennale and the designer of the 2011 Gwangju Biennale 2011, and the curator for the Seoul City Architects Forum 2017.

Jae Y. Lim is a principal architect of OCA (Office of Contemporary Architecture). Based on his perception of social, economic and cultural changes, he strives to reflect them in his architecture through new prototypes. This biennale is also an extension of the task of finding new prototypes of cities. He served as director of *Seoul: towards a Meta-city* exhibition which was held at Aedes Gallery in 2014 and Korea-Japan Contemporary Architecture Exchange Exhibition in 2011 and 2012. Also he was a commissioner of UIA promotion Hall in 2011. He received his B. Arch degree from Seoul National University and M. Arch degree from the University of Michigan.

Beth Hughes is the Head of Architecture at the Royal College of Art. Educated in Australia, she graduated from the University of Technology Sydney where she later taught. From 2004 - 2009 Beth was an Associate at the Office for Metropolitan Architecture, (OMA). In 2009 she joined as partner at Point Supreme Architects in Athens, Greece and in 2011 established her own practice now based in London.

Sung Hong Kim is a professor of architecture and urbanism at the University of Seoul. Between 2007 and 2010, he organized an exhibition, *Megacity Network: Contemporary Korean Architecture*, in Frankfurt, Berlin, Tallinn, Barcelona and Seoul. He curated *The FAR Game: Constraints Sparking Creativity*, for the Korean Pavilion at 2016 Venice Biennale International Architecture Exhibition. He has authored research papers, essays, and books about contemporary Korean architecture and urbanism including *The FAR Game* (2016), *Future Asian Space* (2012), *Street Corner Architecture* (2011), *On Asian Streets and Public Space* (2010), *New Imagination of Urban Architecture* (2009), and *Megacity Network* (2007).

Keller Easterling is an architect, writer and professor at Yale. Her most recent book, Extrastatecraft: The Power of Infrastructure Space (Verso, 2014), examines global infrastructure as a medium of polity. Another recent book, Subtraction (Sternberg, 2014), considers building removal or how to put the development machine into reverse. Other books include: Enduring Innocence: Global Architecture and its Political Masquerades (MIT, 2005) and Organization Space: Landscapes, Highways and Houses in America (MIT, 1999). Easterling is a 2019 United States Artist in Architecture and Design. Her research and writing was included in the 2014 and 2018 Venice Biennales.

스틸스.언리미티드(STEALTH. unlimited)는 도시성을 생산하는 "산업"에 그다지 흥미를 느낀 적이 없는 대신, "폭발 직전의" 공간 문제를 자극하는 데 흥미가 있었다. 그것은 종종 우리가 집단이나 지역사회를 뒤흔들어 미래의 지평선이 무엇인지 다양하게 하는 상황을 보여주기 때문이다. 그와 관련된 오랜 연구들을 토론, 워크숍, 전시 등을 통해 지속적으로 보여주고 있다. 최근에는 '로테르담의 제작과 누가 베오그라드에서 도시를 건설했는가'의 연구를 통해 생산과 (사회적) 재생산의 공간성과 공간을 다루고 있다.

OFFPOLINN(Office for Political Innovation, OFFPOLINN)은 뉴욕과 마드리드를 기반으로 디자인, 연구 및 주요한 환경적 실무를 다루는 국제적인 건축 사무실이다. 이 사무소는 건설 환경의 통합성을 불어넣기 위해 다양한 규모와 매체로 프로젝트를 진행한다. 안드레스 하케(Andres Jacque)는 2003년에 OFFPOLINN을 설립했다. 그는 뉴욕 컬럼비아대학교 건축대학원의 건축설계프로그램의 디렉터이며, 프린스턴대학교와 쿠퍼 유니언의 초빙교수를 역임했다. 또한 그는 마드리드의 The Escuela Técnica Superior de Arquitectura로부터 건축학 석사와 박사학위를 받았다.

알릭 맥린(Alick McLean)은 역사적 도시들이 어떻게 오늘날 건물과 도시의 정체성과 기회를 만들고 유지하는 것에 대한 통찰력을 제공하는지를 연구한다. 그의 출판물로는 『법을 무시하지 말라: 이탈리아 공동체의 부패에 대한 시각 및 공간적 방어』와 『프라토: 투스카나 주의 건축, 경건함, 정치적 정체성』이 있다. 그는 피렌체의 시라큐스 대학에서 약 20년 동안 강의를 하였으며 2017년부터 살렘과 메사추세츠의 중학교와 고등학교 인문학 수업에서 도시학을 소개하고 있다.

임동우는 PRAUD의 공동 설립자이자, 홍익대학교 조교수이다. 그는 서울대학교에서 학사 학위와 하버드대학교에서 석사학위를 받았으며, 2013년 젊은 건축가 상을 수상하였고, 2019 서울도시건축비엔날레 도시전 공동 큐레이터를 맡았다. 그의 작품들은 2014 베니스건축비엔날레 한국 전시관, 뉴욕 MoMA, 베를린 DNA 갤러리 등에 전시되고 있다.

라파엘 루나(Rafael Luna)는 한양대학교 조교수이자, 건축사무소 PRAUD의 공동 설립자이다. MIT에서 건축학 석사 학위를 받았다. 그는 도시 효율성 증진을 위한 시스템으로 인프라-건축 하이브리드에 초점을 맞춘 연구를 진행하고 고, 관련 주제에 관한 에세이를 『MONU』, 『IntAR 저널』, 『Inner Magazine』, 『Studio Magazine』 등에 기고하였다. 2013년 젊은건축가상을 수상하였고, 그의 작품은 MoMA, 베니스건축비엔날레, 서울도시건축비엔날레 등에 전시되었다.

알프레도 브릴렘보그(Alfredo Brillembourg)는 건축사무소 도시건축 연구소(Urban-Think Tank) 원장이다. 뉴욕 태생으로, 컬럼비아대학교에서 1984년 건축학 학사를, 1986년 건축 설계 석사 학위를 받았다. 2010년에서 2019년, 휴버트 클럼프너(Hubert Klumpner)와 스위스 ETHZ에서 건축 및 도시 설계 공동 의장을 맡았다. 2010년 랄프 어스킨(Ralph Erskine)상, 2011년 라틴아메리카대상 골드 홀심 어워드(Gold Holcim), 2012 베니스건축비엔날레 황금사자상 등 다수의 수상 경력이 있다. 2018년, 그의 임파워 쉑(Empower Shack) 주택 프로젝트가 그 우수성과 특별함을 인정받아 RIBA(왕립영국건축가 협회) 수상 최종 후보자 명단에 들었다.

훌리오 데 라 푸엔테(Julio de la Fuente)는 마드리드에 위치한 세계적 건축, 도시 연구기관인 Gutiérrez-delaFuente Arquitectos의 공동 설립자로, 다수의 대회 수상 경력이 있다. 또한, 테크니컬 커미티 오브 유로판 유럽(Technical Committee of Europan Europe) 회원으로 활동하고 있으며, 스페인, 독일, 스웨덴 등지에 위치한 세계 유수의 대학 객원 교수이기도 하다. 그의 작품들은 건축 및 도시학적 관점에서 탈공업화 시대의 도시의 변화 과정을 다루고 있으며, 창조성이 복잡성에 대한 과학적 접근 방식이라는 생각에 기반하고 있다.

라피 시갈(Rafi Segal)은 건축가이자 메사추세츠공과대학교 건축 및 도시학과 부교수로, 건축 도시 석사 과정(SMArchS-Urbanism Program)을 지도하고 있다. 그는 건축, 도시 및 지역 단위에 대한 연구 및 설계 활동 등을 수행하고 있으며, 최근에는 현재 부각되고 있는 공유 및 집단 관념이 건물 및 도시 형태 디자인에 어떻게 영향을 미칠 수 있는지에 대해 중점적으로 연구하고 있다.

Dogma was founded in 2002 by Pier Vittorio Aureli and Martino Tattara. From the beginning of its activities, Dogma has worked on the relationship between architecture and the city by focusing mostly on urban design and large-scale projects. Parallel to the design projects, the members of Dogma have intensely engaged with teaching, writing and research, activities that have been an integral part of the office's engagement with architecture. In the last years, Dogma has been working on a research by design trajectory that focuses on domestic space and its potential for transformation. This work, made of studies and projects, has been exhibited at different venues among which the Tallinn Architectural Biennale (2014), the HKW Berlin (2015), the Biennale di Venezia (2016), the Chicago Architectural Biennial (2017), and the London Design Museum (2018).

New Academy is an education initiative for architecture and urbanism based in Helsinki. Working at the intersection of theory and practice, the goal of New Academy is to operate as a platform for continuing education, with a focus on critical theoretical discourse and its relation to forms of architectural labor. New Academy runs studios and visiting programs for universities and organizes evening lectures and seminars on architecture and the city. Currently New Academy is engaged in developing alternative models of building development based around the training and organization of construction labor. For the 2019 Seoul Biennale of Architecture and Urbanism, New Academy is represented by Leonard Ma and Tuomas Toivonen.

STEALTH.unlimited (2000, Rotterdam/Belgrade) are Ana Džokić and Marc Neelen. Initially trained as architects, their work is equally based in the context of contemporary art and culture. Their recent book Upscaling, Training, Commoning features what it means to distance from a set of economics, architectures and politics devoid of future. Jere Kuzmanić (Split/Zagreb) pioneers as architect, urbanist, and researcher degrowth and new forms of social and environmental responsibility in the field of urban survival. Predrag Milić (Belgrade/Vienna) is a trained architect and urban researcher working between urban development and critical pedagogy, and particularly the oppressed, silenced, and the poor people of Belgrade's urban periphery.

Andrés Jaque/ Office for Political Innovation (OFFPOLINN) is an international architectural practice, based in New York and Madrid, working at the intersection of design, research, and critical environmental practices. The office develops transmedium and interscalar projects, intended to bring inclusivity into the built environment. In 2014, the office won the Frederic Kiesler Prize; and in 2014, the Silver Lion to the best project of the 14 Biennale di Venezia. Andrés Jaque is the Director of Columbia University, Advanced Architectural Design Program; and has been previously architecture professor in Princeton University and Cooper Union.

Alick McLean studies how historical cities provide insight into building and sustaining urban identity and opportunity today. His publications include "Don't Screw with the Law: Visual and Spatial Defenses against Judicial Corruption in Communal Italy" and Prato: Architecture, Piety, and Political Identity in a Tuscan City-State. He taught at Syracuse University in Florence for 20 years. Since 2017 he has been introducing urban studies to middle school and high school humanities classes, at Saltonstall School in Salem, and now at The Academy at Penguin Hall, Wenham, Massachusetts.

Dongwoo Yim is the co-founder of PRAUD and assistant professor at Hongik University. He received his master's degree at Harvard University and a bachelor's degree at Seoul National University. He is the winner of the Architectural League Prize 2013, and the co-curator of the Cities Exhibition in 2019 Seoul Biennale of Architecture and Urbanism. His works have been exhibited worldwide including the award-winning Korean Pavilion in Venice Biennale 2014, Museum of Modern Art in New York, DNA Galerie in Berlin.

Rafael Luna is an assistant professor at Hanyang University and co-founder of the architecture firm PRAUD. He received a Master of Architecture from the Massachusetts Institute of Technology. Luna's research focuses on infra-architectural hybrids as systems for urban efficiency, with essays on the subject published in MONU, IntAR Journal, Inner Magazine, and Studio Magazine. Luna is the award winner of the Architectural League Prize 2013, and his work has been exhibited at the MoMA, Venice Biennale, Seoul Biennale.

샌드라 카지-오그래디(Sandra Kaji-O'Grady)는 호주 퀸즐랜드대학교 건축학과 교수로, 설계를 지도하고 있다. 실험 건축의 과학적 구현에 관한 그녀의 연구는, 크리스 스미스(Chris L. Smith) 및 러셀 휴즈(Russell Hughes)와 공동 편집한 『Laboratory Lifestyles: The Construction of Scientific Fictions』(MIT Press, 2018)과, 크리스 스미스와 공동 집필한 『LabOratory: Speaking of Science and its Architecture』(MIT Press, 2019)에 집대성되어 있다. 그녀는 다음 책에서 애완동물을 위한 도시 공간 설계를 다룰 예정이다.

밀라노폴리테크닉대학교에서 건축학 학사를, 베니스 IUAV에서 박사 학위를 받은 실비아 미켈리(Silvia Micheli)는 현재, 호주 퀸즐랜드대학교에서 강의하고 있으며, 설계를 지도하고 있다. 실비아는 20세기와 21세기의 건축 및 도시를 대상으로 글로벌 건축 및 비교 문화 교류를 연구하고 있으며, 최근 『Italy/Australia: Postmodern Architecture in Translation』(URO, 2018) 책을 공동 편집했다. 실비아는 알바 알토 재단, 비트라 디자인 박물관, 퐁피두 센터, 맥시 뮤지엄 등 여러 문화 기관들과 함께 다양한 국제적 협업을 진행하고 있다.

스위스건축가협회 정회원인 제럴딘 보리오(Géraldine Borio)는, 홍콩에 위치한 독립 연구소로 건축 기반 실행을 추구하는 보리오 랩의 설립자이자 홍콩대학교 건축학과 조교수이다. 보리오는 아시아 도시들 간 차이, 공허함, '그로 인해 생긴 공간'에 관한 책인 『Hong Kong In-Between』(2015)과 The People of Duckling Hill』(2016)의 공동 저자이다. 본 저서의 집필은 보리오가 아시아 도시의 건설 환경 메커니즘을 이해하는 시작점이 되었다.

피터 트루머(Peter Trummer)는 건축가이자 교육자이다. 그는 인스브루크대학교 도시설계 및 계획 연구소 교수이자 소장이며, SCI-Arc(Southern California Institute of Architecture)와 프랑크푸르트 슈테델슐레(Städelschule)의 객원교수이기도 하다. 트루머는 2004부터 2010년까지 로테르담에 위치한 베를라헤 인스티튜트의 연합 디자인 프로그램 책임자였다. 그의 작품은 2006년과 2012년 베니스건축비엔날레에서 전시되었다.

네이란 투란(Neyran Turan)은 캘리포니아대학교 버클리 캠퍼스 조교수이자, NEMESTUDIO의 파트너이다. 그레이엄 재단상을 수상한 투란의 저서, 『수단으로써의 건축』(Actar, 2019)가 최근 출간되었다. 투란은 최근 2020 베니스건축비엔날레에서 터키 전시관 큐레이터 최종 후보자로 선정되었다.

Urbanization.org는 300,000 km/s 및 Actar와의 공동작업을 통해 IAAC가 개발한 프로젝트로, 카탈로니아 자치 정부, 바르셀로나 메트로폴리탄, 바르셀로나 시의회, 블룸버그 재단 등의 지원을 받았다. IAAC는, 디지털 혁명으로 등장한 도시 및 건축에 관한 연구 수행이라는 설립 비전에 따라, 도시 조성을 위해 새로운 과학을 발전시키고, 21세기 더 나은 도시 건설 및 개혁을 추진하기 위해 노력하고 있다. 본 프로젝트는, IAAC 공동 설립자이자, 이전 바르셀로나 시 의회 대표 건축가인 빈센트 구알라트(Vicente Guallart)가 감독을 맡아 진행하고 있다.

오우아라로우 + 초이(OUALALOU + CHOI)는 린나 초이(Linna Choi)와 타릭 오우아라로우(Tarik Oualalou)가 설립하였다. 이들의 작품은 건축가들의 전통적 한계(및 제약)에 대해 의문을 제기한다. 주요 프로젝트로는 메크네스(Meknès) 로마 유적지에 설립된 볼루빌리스(Volubilis) 박물관, 밀라노엑스포 모로코전시관, 마라케시의 COP 22 캠퍼스 등이 있다. O + C 프로젝트는 파리와 카사블랑카 스튜디오를 비롯하여 전 세계에서 전시되고 있다.

마이클 파이퍼(Michael Piper)는 토론토대학교 건축 도시 설계 조교수이다. 그의 연구는 도시계획 정책, 부동산 관행, 도시 형태 간 관계에 초점을 두고 있다. 파이퍼는 로스앤젤레스와 토론토에 위치한 더브 스튜디오(Dub studios) 대표로 도시 설계 작업을 주도하고 있다. 더브 스튜디오는 계약 전 도안의 샘플 혹은 완성품을 고객에게 공개하는 작업과 커미션 작업을 진행하는 한편, 다양한 공공시설 건설프로젝트 입찰에 초청받고 있다.

Alfredo Brillembourg is Principal of the architecture firm Urban-Think Tank. Born in New York, he received his Bachelor of Architecture in 1984 and Master of Science in Architectural Design in 1986 from Columbia University. From 2010-2019, together with Hubert Klumpner he has held the Chair of Architecture and Urban Design at ETHZ, Switzerland. He received the 2010 Ralph Erskine Award, 2011 Gold Holcim Award for Latin America and the 2012 Venice Biennale of Architecture Golden Lion. In 2018 the Empower Shack Housing project was shortlisted by RIBA as an exceptional housing project.

Julio de la Fuente is the co-founder of Gutiérrez-delaFuente Arquitectos, a Madrid based architecture, urbanism and research firm of international scope, awarded in numerous competitions. He is a member of the Technical Committee of Europan Europe, and regular guest professor at universities in Spain, Germany, and Sweden. His works address the post-industrial urban processes from an architectural urbanism point of view, understanding creativity as a scientific approach to complexity.

Rafi Segal is an architect and Associate Professor of Architecture and Urbanism at the Massachusetts Institute of Technology (MIT) where he directs the SMArchS Urbanism program. His work involves design and research on the architectural, urban and regional scale, currently focusing on how emerging notions of sharing and collectivity can impact the design of buildings and the shaping of cities.

Sandra Kaji-O'Grady is Professor of Architecture at the University of Queensland, Australia where she teaches design. Her research on the expression of science in laboratory architecture culminated in two books: Laboratory Lifestyles: The Construction of Scientific Fictions, edited with Chris L. Smith and Russell Hughes (MIT Press, 2018) and LabOratory: Speaking of Science and its Architecture, co-authored with Chris L. Smith (MIT Press, 2019). Her next book examines the design of urban spaces for companion animals.

Silvia Micheli (B Arch Milan Polytechnic; PhD IUAV, Venice) is Lecturer at the University of Queensland, Australia, where she teaches design. Silvia's research investigates global architecture and cross-cultural exchanges in the 20th and 21st century architectural and urban context. She has recently co-edited the book Italy/Australia: Postmodern Architecture in Translation (URO, 2018). Micheli has a range of international collaborations with cultural institutions, such as the Alvar Aalto Foundation; Vitra Design Museum; Centre Pompidou and MAXXI Museum.

Géraldine Borio is a Swiss Registered Architect and founder of Borio Lab, an independent research laboratory and architectural-based practice-based in Hong Kong. She is as well an Assistant Professor at The University of Hong Kong, Department of Architecture. Co-author of the books Hong Kong In-Between (2015) and The People of Duckling Hill (2016), the gaps, voids and so-called 'resultant spaces' of Asian cities have been her entry points to understand the mechanisms of the built environments.

Peter Trummer is an architect and educator. He is a professor and head of the Institute for Urban Design & Planning at the University of Innsbruck. He is a visiting professor at Southern California Institute of Architecture (SCI-Arc) and Guest Professor at the Städelschule in Frankfurt. He was head of the Associative Design Program at the Berlage Institute in Rotterdam from 2004 to 2010. Trummer has been exhibited at the Venice Biennale in 2006 and 2012.

Neyran Turan is an Assistant Professor at the University of California-Berkeley and a partner at NEMESTUDIO. Turan's authored book titled Architecture as Measure, which has been awarded a Graham Foundation grant is released by ACTAR Publishers in Fall 2019. Turan has recently been shortlisted to curate the Turkish Pavilion at the Venice Architecture Biennale in 2020.

이승택은 "아직 이름 지어지지 않은 것을 만든다" 라는 모토로 교육문화 영역의 새로운 경험을 만드는 게임 디자인 아티스트, 놀공의 공동대표. 서울을 기점으로 유럽과 세계를 무대에서 활동중이며, 독일문화원과 협업한 〈Being Faust: Enter Mephisto〉는 전세계 15개국에 공연되었으며, 한국과 독일의 70년 분단과 평화 프로젝트 〈월페커즈: DMZ에서 베를린장벽까지〉는 2019년 1월 베를린과 한국 도라산에서 선보이고 현재 한국과 독일 학교로 확산을 진행중이다. 뉴욕 활동 시 gameLab(1999), Institute of Play(2006), Come Out & Play Festival(2006)을 공동 설립했으며, 2009년 디지털세대를 위한 학교 Quest To Learn 설립에 참여했다.

임애련은 놀공의 공동대표이다. 교육 공학 전문가로서 삼성화재, NHN 등에서 활동했으며, NOLGONG PLUS의 대표이사로 놀공의 교육 프로젝트들을 이끌고 있다. 'Playing Ground', 'Story Playing' 등의 교육 프로젝트들을 개발하고, 독일문화원과 협업한 〈Being Faust: Enter Mephisto〉, 〈월페커즈: DMZ에서 베를린장벽까지〉 등의 프로젝트를 선보였다.

최상기는 서울시립대학교 건축학부의 프로그램디렉터이며 학부장을 역임하였다. 하버드대학교와 연세대학교 졸업 후 노스이스턴대학교와 UC버클리에서 교수 및 연구원으로 경력을 쌓았고, 뉴욕주 등록 건축사로서 서울, 뉴욕 및 캠브리지에서의 건축 실무를 바탕으로 현재는 SCA설계연구소를 운영하고 있다. 서울시건축상 수상자이며 서울특별시 공공건축가로 활동하며 건축과 도시의 상관관계에 대한 연구 및 설계활동을 진행하고 있다.

장영철은 홍익대학교를 졸업하고, UC버클리에서 수학하였다. 현재는 전숙희와 함께 WISE 건축을 운영하고 있다. 공공예술프로젝트를 기획하고 실행하며, 여러 집단과 연계되어 건축 놀이 활동을 지속하고 있다. 2011년에 대한민국 젊은 건축가 상을, 2012년과 2015년에 '전쟁과 여성인권 박물관'과 '어둠 속의 대화'로 서울시건축상 최우수상을 수상 하였고, 2015년 코리아디자인어워드 공간부문 대상을 수상하였다. 2017년 빼빼한 막대나무로 가구를 만드는 '가라지가게'를 시작하였다.

최주연은 이화여자대학교를 졸업하고, 홍익대학교에서 장식 미술 및 실내 디자인을 수학하였다. 현재는 윤현상재의 갤러리 스페이스비이(Space B-E)와 아트숍 윤현핸즈의 디렉터이자 건국대학교 건축학부 겸임교수를 맡고 있다. 갤러리 스페이스비이의 총괄기획자로서 소재에 집중한 작품들에 대한 전시를 이어가고 있다.

홍주석은 한양대학교를 졸업하고, 카이스트 문화기술대학원에서 수학하였다. 2013년에 어반플레이를 창업하여, 현재는 아카이브를 기반으로 다양한 도시문화콘텐츠를 창작한다. 기획부터 디자인과 개발 역량까지 갖춘 어반플레이는 로컬 콘텐츠와 디지털 미디어의 융복합적 접근을 통해 도시해프닝, 공간&전시, 미디어, 웹&디자인, 지역콘텐츠제작이라는 형태로 기존에 없었던 새로운 형태의 문화콘텐츠 경험을 제시한다.

유아람은 서울대학교에서 건축을 전공, 동대학원에서 현대건축에 대한 이론을 연구하여 석, 박사 학위를 받았다. 건축이론을 기반으로 대학에서 강의를 하고 있으며 새로운 기술과 현대 건축에 대한 연구를 진행하고 있다. 건축적 저변을 넓히기 위해 현재 숭실대학교 NRF 전임연구원으로서 도시재생 정책에 대한 연구를 진행 중이다. 도시와 건축에 대한 사유가 비단 건축가만이 아니라 모든 사람들에게 스며들었으면 하는 희망을 품고 있다. 현대 건축이 사람들 사이에서 많이 회자되는 시대를 위해 연구와 교육에 뜻을 두고 활동 중이다.

김나연은 건국대학교 영문학과를 졸업하고, 경희대학교 문화예술경영학 석사를 취득하였다. 백해영갤러리, 갤러리인, 부산비엔날레에서 현대미술전시 큐레이터로 경력을 쌓았고, 이후에 총 4회의 시카프 행사의 기획과 홍보를 총괄하였다. 또한 2014년 만화 '열혈강호'의 20주년 특별전 큐레이터로 활동한 바 있다. 현재 2017년 1회의 홍보 총괄을 거쳐 2019년 2회 서울도시건축비엔날레사무국에서 사무국장으로 근무하고 있다.

김일현은 서울대학교와 베네치아국립대학교에서 건축을 공부했다. 이탈리아 현대건축 연구로 박사학위를 받았고, 다수의 논문과 공저를 이탈리아와 한국에서 출간했다. 국립현대미술관 서울관과 용산국립공원의 전문위원(PA)을 역임했고, 도시의 공공성에 관심이 있다. 현재 '개인 속의 인류사'라는 주제로 연구를 진행 중이고, 경희대학교 건축학과 교수로 재직 중이다.

Urbanization.org is a project developed by IAAC in collaboration with 300,000 km/s and Actar, with the support of the Generalitat de Catalunya, the Metropolitan Area of Barcelona, the Diputacion de Barcelona and Bloomberg Philanthropies. IAAC, following its foundational vision of researching the city and the architecture that emerged from the digital revolution works to build a new science to make cities and promotes the construction and reform of better cities for the 21st century world. The project is directed by Vicente Guallart, IAAC co-founder, and former Chief Architect of the Barcelona City Council.

OUALALOU + CHOI was founded by Linna Choi (USA, S. Korea) and Tarik Oualalou (Morocco, France). Their works serve to question the traditional limits (and limitations) of the architectural profession. Significant projects include the Volubilis Museum on the site of the Roman ruins in Meknès, the Morocco Pavilion for EXPO Milan, and the campus for the COP 22 in Marrakech. With studios in Paris and Casablanca, O+C projects have been exhibited around the world.

Michael Piper is an Assistant Professor of Architecture and Urban Design at the University of Toronto, Daniels Faculty of Landscape Architecture and Design. His research focuses on the relationship between planning policy, real estate practices, and urban form. Piper leads the urban design work as a principal at Dub studios in Los Angeles and Toronto. In addition to its speculative work and commissions, the firm has been invited to prepare proposals for different public entities.

Peter Lee is co-founder of NOLGONG and a game design artist who creates new experiences in the educational and cultural sectors with the motto of "creating something that has yet to be named." He has been working actively on a global stage, especially in Europe while based in Seoul. His work "Being Faust: Enter Mephisto" in collaboration with the Goethe Institut has been performed in 15 countries around the world. Also, his "Wallpeckers: from DMZ to Berlin Wall," a project that addresses the 70 years' division and peace of Korea and Germany, has been spread to Korean and German schools since it was first introduced in Berlin and Dorasan in Korea January 2019. While working in New York, he co-founded gameLab (1999), Institute of Play (2006), and Come Out & Play Festival (2006). In 2009, he participated in the establishment of "Quest To Learn," a school for the digital generation.

Aelyun Lim has worked with Samsung Fire & Marine Insurance and NHN as a specialist in education engineering, and is a co-founder of NOLGONG where she leads its educational projects. She has developed educational projects such as Playing Ground and Story Playing; introduced "Being Faust: Enter Mephisto" and "Wallpeckers: from DMZ to Berlin Wall" in collaboration with the Goethe Institut.

Professor Sanki Choe, curator of Global Studio, is the Director of the Architecture Program at University of Seoul and had served as the Dean of Architecture. He is a New York licensed architect running SCA Design Lab based on his practical architectural experience in New York, Cambridge, and Seoul. He built his career in this field as a professor and researcher at Northeastern University and UC Berkeley after graduating from Harvard and Yonsei University. He was awarded the prestigious Seoul Architecture Prize and appointed as Public Architect by the Metropolitan Government of Seoul. Currently, he is focusing on research and design projects regarding correlations between architecture and cities.

Young-chul Jang received an M.A. at the University of California-Berkeley after graduating from Hongik University. He currently runs WISE Architecture with Sukhee Jeon. He has been active in planning and carrying out public art projects and continues to be active in architectural play in association with various groups. He won the Young Korean Architect Award in 2011; the Seoul Architecture Awards in 2012 and 2015 with the War and Women's Human Rights Museum and "Dialogue in the Dark," respectively; and the Korean Design Award in 2015. In 2017, he launched Garagegage where furniture is made of thin wooden sticks.

Aram You majored in architecture and received M.A. and Ph.D. from Seoul National University for her research on theories of modern architecture. She teaches architectural theories at university and carries out research on new technologies and modern architecture. In addition, in order to expand her architectural base, she also conducts research on urban restoration and policies as a researcher in NRF, Soongsil University. She hopes that not only architects but everyone will be able to ponder over cities and architecture. Dreaming of an era when modern architecture is actively discussed among ordinary people, she is engaged in education and research activities.

홍주석은 KAIST 문화기술대학원 석사, 한양대학교 건축학 학사 학위를 취득하였고, 2013년부터 도시콘텐츠 크리에이티브 그룹 어반플레이를 창업하여 소프트웨어 중심의 도시를 위한 다양한 도시실험을 해오고 있다. '연희걷다'와 같은 로컬 프로젝트와 '연남방앗간' '연남장'과 같은 로컬 공간 프로젝트를 총괄하였으며, 로컬 아카이브 매거진 '아는동네 매거진' 발행인을 맡고 있다.

권현정은 파리 라빌레뜨 건축학교를 졸업, 프랑스 건축사(D.P.L.G.)를 취득하였다. 현재 서울시공공건축가와 서울시 도시디자인 심의위원으로 활동 중이다. 2018년부터 UIA(국제 건축가협회)의 Architecture & Children Work Programmes에서 한국 대표를 맡고 있으며, 교육부 학교공간혁신추진단의 자문 위원이자, 경기도교육청 학교공간혁신사업의 실행총괄로도 활동 중이다. 어린이를 위한 '어린이 건축수업'과 서울시 건축학교 교재를 집필하였다.

이희준은 과거 어반플레이 콘텐츠 디렉터였으면 현재 국립식량과학원 현장명예연구원, 더로컬프로젝트 대표, 연남방앗간 참기름 소믈리에, 전통시장 도슨트이다. 전통시장 도슨트로서 전국 1,500개 전통시장의 역사, 상인, 상품을 해설하고 전국 100여개 전통시장에서 진행하고 책 시장이 두근두근을 출간했습니다. 뿐 만아니라 참기름을 착유하는 참기름 소믈리에로 활동하고 있으며, 전국 로컬 장인과 생산자의 이야기를 수집하고 이야기하는 전국 로컬 식재료의 콘텐츠 아카이빙과 상품 기획, 판매를 하는 더로컬프로젝트의 크리에이티브 디렉터로 운영하고 있습니다.

조정구는 1966년 서울 보광동에서 태어나 자랐다. 서울대학교 건축학과와 동 대학원을 졸업하고, 일본 도쿄대학교 박사과정을 거쳤다. 2000년 구가도시건축 사무소를 만들어, '우리 삶과 가까운 보편적인 건축'에 주제를 두고, 지속적인 도시 답사와 연구, 설계작업을 하고 있다. 2019년 현재 840여 회를 진행한 '수요답사'를 통하여 서울의 동네와 사람들이 사는 모습을 찬찬히 관찰하고 기록하고 있다.

Jooyeon Choi studied decorative art and interior design at Hongik University after graduating from Ewha Womans University. She is the director of Gallery Space B-E and Art Shop Younhyun Hands, as well as an adjunct professor of the School of Architecture at Konkuk University. As the Chief Director of Gallery Space B-E that emphasizes materials, she continues to exhibit works focused on materials.

Jooseok Hong studied at KAIST Graduate School of Culture Technology after graduating from Hanyang University. Since founding URBANPLAY in 2013, he has created diverse urban culture contents based on archives. URBANPLAY, with capabilities of planning, designing and developing, offers totally new cultural contents and experience in the forms of urban happening, space & exhibition, media, web design and local content creation through the approach of convergence between local contents and digital media.

Nayeon Kim majored in English and Literature from Kunkuk University and received MBA from Kyung Hee University. She built her career in curating contemporary art exhibitions at Paik Hae-young Gallery, Gallery-in, and Busan Biennale, and she served as the team leader of the planning and promotion team of the Seoul International Cartoon Animation Festival four times and also served as curator of the special exhibition for the 20th anniversary of *The Ruler of the land*. She is currently serving as the general manager of Seoul Biennale after serving as the manager of public relations for the first Seoul Biennale.

Ilhyun Kim studied architecture at Seoul National University and the University of Venice. He received Ph.D. with his research on Italian modern architecture and published numerous papers and books in Korea as well as in Italy. He served as a Professional Advisor (PA) for the Seoul Branch of the National Museum of Contemporary Art as well as the Yongsan Family Park. With an interest in the public characteristics of cities, he conducts research under the theme of the "human history of individuals" as a professor of architecture at Kyunghee University.

Kwon Hyun-jeong graduated from ENSA Paris-La Villettee, and received the D.P.L.G. She is a member of the Seoul Public Architect and Seoul Urban design deliberation committee. Since 2018, she has represented Korea in the Architecture & Children Work Programmes of the International Union of Architects (UIA). She also works as an advisor of the "Space Innovation Team" under the Ministry of Education that promotes innovation in school spaces, while leading a project for innovating school spaces conducted by the Gyeonggi Provincial Office of Education. She published the book Architectural Class for Children and developed teaching materials for Seoul city's architecture school for children.

Heejun Lee was a content director of URBANPLAY, and now an honorary researcher of the National Institute of Crop Science; a CEO of the Local Project; a sesame oil sommelier of Yeonam Bangagan; and a docent of traditional markets. As a traditional market docent, he published Fluttering Markets in which 1,500 markets' histories, merchants and products are introduced. In addition, he works as a sesame oil sommelier who extracts sesame oil and as a creative director of the Local Project, in charge of managing the contents archive of local food ingredients, local masters and producers as well as product planning and sale.

Junggoo Cho was born in 1966 and grew up in Bogwang-dong, Seoul. He received a B.A. and M.A. in Architecture of Seoul National University, and a Ph. D. from the University of Tokyo. Since founding the guga Urban Architecture firm in 2000, he has been carrying out urban exploration, research and design works under the theme of "universal architecture close to our lives." Through his "Wednesday Urban expeditions," which has been conducted 840 times as of 2019, he thoroughly observes and records the people and neighborhoods of Seoul.

비엔날레를 만든 사람들

Credits

주최/주관

서울특별시
박원순(시장)

총감독
임재용, 프란시스코 사닌

서울특별시 도시공간개선단
김태형(도시공간개선단장)
최원석(도시공간개선반장)
안종연(도시건축교류팀장)
윤정두, 안정연, 김은향,
유연경(도시건축교류팀)

서울디자인재단
최경란(대표이사)

서울도시건축비엔날레 사무국
김나연(총괄), 금명주, 박상현, 구아람,
우윤지, 정윤지, 정혜린, 천주현, 최예지

서울도시건축비엔날레 운영위원회
승효상(위원장), 김영준(부위원장),
서정협, 유연식, 강맹훈, 권기욱, 류훈,
김태형, 이경선, 고병국, 최경란, 김승회,
배형민, 임옥상, 이창현, 조민석, 노소영,
이영혜

주제전

큐레이터
베스 휴즈

협력큐레이터
김효은

보조큐레이터
리비아 왕, 이자벨 옥덴,
제프리 킴, 이유진

코디네이터
황희정

영상상영코디네이터
안나 리비아 바슬

참여자

더불어 사는 일상
이엠에이건축사무소(주)(이은경)

SNS 집합도시
(주)건축사사무소SAAI(이진오)

두레주택
(주)조진만 건축사사무소(조진만)

건축의 공적 역할
정기용

난민 헤리티지
DAAR(알레산드로 페티, 산디 힐랄)

두 개의 변신 전략
알레한드로 에체베리, 호르헤 페레스-
하라미요

건물들과 그 영역
토니 프레튼 아키텍트 (토니 프레튼)

기후변화대응조치 2.0: 건축학적 연대
카담바리 백시

멕시코 주거 도시화
엘 시엘로(아르만도 하시모토, 수렐라
세구)

디지털 광장
페드로 앙리크 데 크리스토(+D)

멀티플라이도시
아틀리에 얼터너티브 아키텍처(인유준)

Host/Organizer

Seoul Metropolitan Government
Won-Soon Park (Mayor)

SBAU 2019 Co-directors
Jaeyong Lim, Francisco Sanin

Seoul Metropolitan Government
Urban Improvement Bureau
Tae Hyung Kim (Director-general),
Won Suk Choi (Director),
Jong Youn An (Team Leader),
Jeong Doo Yun, Jung Yon Ahn,
Eun Hyang Kim, Yeonkyoung Yoo

Seoul Design Foundation
Kyung-ran Choi (CEO)

Seoul Biennale Division
Nayeon Kim (General Manager)
Myeongju Keum, Sanghyun Park,
A Ram Ku, Yoonji Woo,
Yoonjee Jeong, Hye Rin Jeong,
Ju Hyun Cheon, Ye Ji Choe

Seoul Biennale Steering
Committee
H-Sang Seung (Chair),
Young Joon Kim (Vice Chair),
Jeong Hyup (Thomas) Seo,
Yeonsik Yoo, Maeng Hoon Kang,
Kie Wook Kwon, Hoon Ryu, Kyung
Sun Lee, Byung Kook Ko,
Kyung-ran Choi, Seunghoy Kim,
Hyungmin Pai, Ok Sang Lim,
Chang-hyun Lee, Minsuk Cho,
Soh Yeong Roh, Young Hye Lee

Thematic Exhibition

Curator
Beth Hughes

Associate Curator
Hyoeun Kim

Assistant Curator
Livia Wang, Isabel Ogden, Jeffrey
Kim, Yoojin Lee

Coordinator
Heejung Hwang

Film researcher
Anna Livia Vørsel

Participants

Collective Form of the
Everyday
EMA architects & associates
(Eunkyung Lee)

SNS Collective City
Architects Office SAAI (Jinoh Lee)

Du-Re House
Jo Jinman Architects (JJA)
(Jinman Jo)

The Public Role of Architecture
Guyon Chung

Refugee Heritage
DAAR (Alessandro Petti and Sandi
Hilal)

Two Transformation Stratgies
Alejandro Echeverri & Jorge Pérez-
Jaramillo

Buildings and their Territories
Tony Fretton Architects (Tony
Fretton)

Climate Actions 2.0: Architectural
Solidarities
Kadambari Baxi

Housing Urbanism Mexico
El Cielo (Armando Hashimoto and
Surella Segu)

DIGITAL AGORA
Pedro Henrique de Cristo (+D)

MULTIPLICITY
Atelier Alternative Architecture
(Yujun Yin)

Out with a Bang
Black Square (Maria Giudici)

고시원 엿보기
블랙스퀘어(마리아 지우디치)

도시의 경기장
NP2F

집합도시: 현대 인도 현대 건축의 집합적 수행의 형태
사미프 파도라 건축연구소(sP+a/ sPare) 기획 / 아키텍쳐 레드 RED, 아비지트 무쿨 키쇼어 & 로한 시브쿠마르, 반드라 콜렉티브, 버스라이드 디자인 스튜디오, St+art 인디아 재단, 아빈 디자인 스튜디오, 피케이 다스 앤 어소시에이츠, 공간적 대안을 위한 모임, 아다르카르 어소시에이츠

타인과 어울리는 방법
제너럴 아키텍처 콜라보레이티브 (GAC)(유타카 쇼, 제임스 세츨러, 레이튼 비먼)

공간적 가치의 창조
CBC(차이나 빌딩 센터)(펑 리샤오)

일곱가지 서적을 올린 제단
바쿠

매니
켈러 이스터링

약속의 땅, 저가형 주거지와 건축에 관하여
도그마(피에르 비토리오 아우렐리, 마르티노 타타라) + 뉴 아카데미 (레오나르드 마, 투오마스 토이보넨)

PARK
신경섭

세 도시의 현장조사
볼스 + 윌슨(줄리아 볼스-윌슨, 피터 윌슨)

전환기 카이로의 공공 공간
클러스터(카이로도시환경연구소)

세상이 이토록 거대할 줄이야!
스틸스.언리미티드(안나 조코치, 마크 닐렌), 예레 쿠즈마니치, 프레드라호 밀리치

다카: 백만가지 이야기
마리나 타바시움 + 벵갈 인스티튜트 포 아키텍쳐, 랜드스케이프 앤 세틀먼츠

집 없는 문명
아미드.세로9(크리스티나 디아즈 모레노 + 에프렌 가르시아 그린다 / 팀: 프란시스코 크라베이루, 제레비 쉬퍼, 헤주스 빌라, 유탄 선)

생산의 장소, 알루미늄
누라 알 사예, 안네 홀트롭

국경 접경지대 공유정류장
에스투디오 테디 크루즈 + 포나 포르만

집합 공간의 변신: 공공 공간, 민주주의의 반영
아르키우르바노 + 타부(존 오르티스, 이반 아세베도)

푸른 혁명: 초투명 유리의 환경 파괴적 도시화에 대한 개입 조치
안드레스 하케 [오피스 포 폴리티컬 이노베이션(OFFPOLINN)]
연구팀: 안드레스 하케, 에노 첸, 마르코스 가르시아, 이새 매코믹

생각할(먹) 거리
페르난도 드 멜로 프랑코, 안나 카이저, 카롤리나 파소스, 리안드로 리마, 마르셀라 페레이라, 마르타 보게아

기록의 재정립
오픈워크숍(니라지 바티야)

숲의 모양들
LCLA 오피스(루이스 까예하스, 샬로트 한손, 데일 위브)

리오 세코: 그린 네트워크
빌딩 소사이어티 포 아키텍처(베라미노 산토스 "키우라")

밤섬 당인리 라이브
매스스터디스(조민석)

저항의 지도
포렌식 아키텍처

프로젝트 발자취
사미프 파도라 건축연구소

이미지와 건축 #11: 팔만대장경
바스 프린센

도시와 농촌의 교류
아틀리에 바우와우(요시하루 츠카모토, 모모야 카이지마, 요이치 타마이)

독창성의 도구
알렉산더 아이젠슈미트(비져너리 시티스 프로젝트)

도시전

큐레이터
임동우, 라파엘 루나

협력큐레이터
김유빈

코디네이터
조웅희

관람경험 디자인
NOLGONG

참여자

1. 21세기 산업도시

Urban Enclosure
NP2F

Collective Cities: Notes on Forms Of Collective – Practices In Contemporary Indian Architecture
Curated by Sameep Padora Architecture and Research (sP+a/ sPare): Architecture RED, Avijit Mukul Kishore & Rohan Shivkumar, Bandra collective, Anthill design, The Burside design studio, St+art India Foundation, Abin Design Studio, PK Das & Associates, Collective for Spatial Alternatives (CSA), Adarkar Associates

Playing well with Others
General Architecture Collaborative (Yutaka Sho, James Setzler, Leighton Beaman)

Creation of Spatial Value
CBC (China Building Centre) Curated by Peng Lixiao

Altar of the Seven Books
Baukuh

MANY
Keller Easterling

Promised Land, Rethinking Typology and Construction of Affordable Housing
Dogma (Pier Vittorio Aureli and Martino Tattara) and New Academy (Leonard Ma and Tuomas Toivonen)

PARK
Kyungsub Shin

3 Urban Field Researches
BOLLES + WILSON (Julia Bolles-Wilson and Peter Wilson)

Negotiating Public Space in Cairo During a Time of Transition
CLUSTER (Cairo Lab for Urban Studies, Training and Environmental Research)

I Did Not Know The World Is So Big
STEALTH.unlimited (Ana Džokic, Marc Neelen) and Jere Kuzmanic, with contribution from Predrag Milic

Dhaka: A Million Stories
Marina Tabaussum + The Bengal Institute for Architecture, Landscapes and Settlements

A Civilization without Homes
amid.cero9 (Cristina Diaz Moreno + Efren Garcia Grinda / team: Francisco Craveiro, Jeremy Shipper, Jesus M. Villar and Yutan Sun)

Places of Production, Aluminium
Noura Al-Sayeh and Anne Holtrop

Cross-Border Community Stations
Estudio Teddy Cruz + Fonna Forman

Transformation of Collective Spaces: Public Spaces, The Reflection of Democracy
ARQUIURBANO + TABUÚ (John O. Ortíz and Iván D. Acevedo)

Blue Rebellion: An Intervention on the Toxic Urbanism of Ultra-Clear Glass
Andrés Jaque [Office for Political Innovation (OFFPOLINN)]
Research team: Andrés Jaque, Eno Chen, Marcos García Mouronte, Jesse McCormick

Food for Thought
Fernando de Mello Franco, with the collaboration of Anna Kaiser, Carolina Passos, Leandro Lima, Marcela Ferreira and Marta Bogea

Re-Assembling the Archive
The Open Workshop (Neeraj Bhatia)

Shapes of the Forest
LCLA Office (Luis Callejas and Charlotte Hansson with Dale Wiebe)

Rio Seco: Green Network
Building Society for Architecture (Belarmino Santos "Kiwla")

Bamseom Danginri Live
Mass Studies / Minsuk Cho

Maps of Defiance
Forensic Architecture

Projective Histories
Sameep Padora Architecture and Research (sP+a/ sPare)

Image and Architecture #11: Tripitaka Koreana
Bas Princen

Urban Rural Exchange
Atelier Bow-Wow (Yoshiharu Tsukamoto, Momoyo Kaijima and Yoichi Tamai)

City of Urban Inventions
Alexander Eisenschmidt (Visionary Cities Project)

Cities Exhibition

Curator
Dongwoo Yim, Rafael Luna

Associate Curator
YouBeen Kim

Coordinator
Tony Woonghee Cho

Visitor Experience Design
NOLGONG

Participants

1. INDUSTRY IN THE 21ST CENTURY

잔지바르, 탄자니아
응암보: 잔지바르 타운의 새로운 중심지
잔지바르 토지위원회, 아프리칸
아키텍처 매터스

코펜하겐, 덴마크
작은 행성의 거대한 건축
씬 그린 라인 프로덕션

밀란, 이탈리아
밀란: 건축적 맥락
스튜디오 디 마우리치오 카로네스
(마우리치오 카로네스, 조항준, 루카
스칼린지)

상트페테르부르크, 러시아
상트페테르부르크를 통해 보는 구소련
도시의 다층 구조
MLA+

글로벌 스튜디오

큐레이터
최상기

협력큐레이터
이희원

보조큐레이터
최영민

전시 디자인
(주)건축사사무소오드투에이

참여자

베이루트 아메리칸대학교
지도교수: 카를라 아라모니, 니콜라스
파야드, 라나 사마라, 크리스토스
마르코폴로스

아키텍처럴 어쏘시에이션 건축대학교
지도교수: 샘 자코비

어썸션대학교 + 서울시립대학교 +
호치민 건축대학교
지도교수: 윤정원, 한부디홍,
응우엔호광, 헝레디투, 프리마
비리야바다나

바틀렛 건축대학교 + 한양대학교 에리카
지도교수: 사빈 스토프, 패트릭 웨버,
김소영

캘리포니아 예술대학교
지도교수: 니라지 바티야

홍콩 중문대학교
지도교수: 피터 페레토

컬럼비아대학교
지도교수: 데이비드 유진 문

동아대학교
지도교수: 차윤석

이화여자대학교 + 라드바우드대학교
지도교수: 클라스 크레세, 에르빈 판
데르 크라벤

한양대학교
지도교수: 라파엘 루나

하버드대학교
지도교수: 안드레스 세브츠크

홍익대학교
지도교수: 김주원, 임동우

출랄롱코른대학교 INDA
지도교수: 알리샤 라차로니

게이오대학교
지도교수: 호르헤 알마산

국민대학교
지도교수: 최혜정, 봉일범, 김우일,
이규환

쿠웨이트대학교
지도교수: 샤이카 알 무바라키

싱가포르국립대학교
지도교수: 에릭 루뢰

영국왕립예술대학교
지도교수: 다비드 사코니, 잔프랑코
봄바치, 마테오 코스탄초, 프란체스카
로마나 델랄리오

서울대학교
지도교수: 존 홍

싱가포르 기술디자인대학교
지도교수: 캘빈 추아

성균관대학교 + 계명대학교
+ 영남대학교 + 카를스루에 공과대학교
+ 카를스루에 응용과학대학교 +
슈투트가르트 응용과학대학교
지도교수: 토르스텐 슈처

시라큐스대학교
지도교수: 페이 왕

베를린 공과대학교
지도교수: 도미니크 바르트만스키,
김선주, 에밀리 켈링, 마르티나 뢰브,
세브린 마르갱, 티머시 파프, 다그마르
펠거, 요르그 슈톨만

빈 공과대학교
지도교수: 블라덴 야드릭

텍사스 테크대학교
지도교수: 박건, 임리사

케이프타운대학교
지도교수: 파들리 아이작스, 멜린다
실버먼

홍콩대학교
지도교수: 제럴딘 보리오

Global Studio

Curator
Sanki Choe

Associate Curator
Heewon Lee

Assistant Curator
Youngmin Choi

Exhibition design
ODETO.A

Global Studio Participant

American University of Beirut
Professor: Carla Aramouny
Nicolas Fayad
Rana Samara
Christos Marcopoulos

Architectural Association School of
Architecture
Professor: Sam Jacoby

Assumption University
Professor: Hương Le Thi Thu
Viriyavadhana Prima
Leelapattanaputi Veera
Chokchaiyakul Laddaphan
Tantilertanant Nuttee

University of Seoul
Professor: Jungwon Yoon

University of Architecture Ho Chi
Minh City
Professor: Hanh Vu Thi Hong
Nguyên Trần Phạm Sĩ
Quang Lê Hồng
Dũng Đinh Xuân
Phương Tô Thanh
An Trần Duy An

Bartlett School of Architecture
Professor: Sabine Storp
Patrick Weber

Hanyang University Erica
Professor: So Young Kim

California College of The Arts
Professor: Neeraj Bhatia

Chinese University of Hong Kong
Professor: Peter Ferretto
Haoran Howard Wang
(Exhibition Designer)
Ling Cai

Columbia University
Professor: Nayun Hwang

Dong-A University
Professor: Youn Suk Cha

Ewha Womans University
Professor: Klaas Kresse

Radboud University
Professor: Erwin Van Der Krabben

Hanyang University
Professor: Rafael Luna

Harvard University
Professor: Andres Sevtsuk

Hongik University
Professor: Juwon Kim
Dongwoo Yim

INDA – Chulalongkorn University
Professor: Alicia Lazzaroni

Keio University
Professor: Jorge Almazan

Kookmin University
Professor: Helen Hejung Choi
Ilburm Bong
Wooil Kim
Kyu Hwan Lee

Kuwait University
Professor: Shaikha Al Mubaraki

National University of Singapore
Professor: Erik L'heureux

Royal College of Art
Professor: Davide Sacconi
Gianfranco Bombaci
Matteo Costanzo
Francesca Dell'aglio

Seoul National University
Professor: John Hong

Singapore University of Technology
And Design
Professor: Calvin Chua

Sungkyunkwan University
Professor: Thorsten Schuetze

Yeungnam University
Professor: Emilien Gohaud
Hyunhak Do

Keimyung University
Professor: Hansoo Kim

Karlsruhe Institute of Technology
Professor: Markus Kaltenbach

Karlsruhe University of Applied
Science
Professor: Jan Riel

University of Applied Science
Stuttgart
Professor: Philipp Dechow

Syracuse University
Professor: Fei Wang

Technische Universität Berlin
Professor: Dominik Bartmanski,
Seonju Kim, Emily Kelling, Martina
Löw, Séverine Marguin, Timothy
Pape, Dagmar Pelger, Jörg
Stollmann

Techische Universität Wien
Professor: Mladen Jadric
Federica Rizzo

Texas Tech University
Professor: Kuhn Park
Lisa Lim

University of Cape Town
Professor: Fadly Isaacs, Melinda
Silverman

매니토바대학교 + 칼턴대학교 + 토론토대학교
지도교수: 전재성

펜실베이니아대학교
지도교수: 사이먼 김

서울시립대학교
지도교수: 마르크 브로사, 변효진

시드니 공과대학교
지도교수: 앤드루 벤저민, 제라드 라인무스

텍사스대학교알링턴
지도교수: 조슈아 네이슨

EAFIT 대학교 URBAM
지도교수: 훌리아나 퀸테로 마린

연세대학교
지도교수: 이상윤

현장 프로젝트

큐레이터
장영철

협력큐레이터
유아람, 최주연, 홍주석

집합도시장

서울 2045
오영욱

2019 display_02
토마즈 히폴리토

동대문시장과 배후기지
OOO간

시장의 초상
노경

무엇이 가만히 스치는 소리
오재우

데이터스케이프: 서울장의 형태
방정인, 스튜디오 둘 셋

서울도시장

을地:공존
윤현상재

도시상회
어반플레이

파빌리온 프로젝트

감각 場
UAUS: 국민대학교

풍경재생
UAUS: 연세대학교

가설.가설.가설.
서승모

컵플라워
UAUS: 서울시립대학교

받히다, 바치다
UAUS: 가천대학교

내가 만드는 키오스크
UAUS: 중앙대학교

리:커버
UAUS: 선문대학교

플로트폼
UAUS: 한양대학교

서울시장산책
어반플레이(이희준)

서울마당

총괄
임재용

협력 큐레이터
강민선

보조 큐레이터
정진우

시민참여프로그램

총괄
김나연

오픈하우스 서울 특별프로그램
정혜린

영화영상프로그램
천주현

시민참여프로그램(교육, 투어)
최예지

교육프로그램

2019 비엔날레 주제강연
프란시스코 사닌, 홍은주와 김형재, 홍주석, 최상기, 조민석, 유아람, 장영철, 임동우

특별강연
이이남, 김일현, 서현석, 김소현, 박정현, 최호철, 박인하

전시연계체험프로그램
권현정(아키에듀)

서울시 건축학교
권현정(아키에듀)

투어프로그램

University of Hong Kong
Professor: Géraldine Borio

University of Manitoba
Professor: Jaesung Chon

Carleton University
Professor: Ozayr Saloojee

University of Toronto
Professor: Adrian Phiffer

University of Pennsylvania
Professor: Simon Kim

University of Seoul
Professor: Marc Brossa
Hyojin Byun (Tutor)

University of Technology Sydney
Professor: Gerald Reinmuth, Andrew Benjamin

University of Texas at Arlington
Professor: Joshua Nason

URBAM - Universidad Eafit
Professor: Juliana Quintero Marin

Yonsei University
Professor: Sang Yun Lee

Live Projects

Curator
Young Chul Jang

Associate Curator
Aram You, Jooyeon Choi, Jooseok Hong

Collective Market City

Collective Market City
Aram You

Seoul 2045
Young Wook Oh

2019 display_02
Tomaz Hipólito

Dongdaemun Market and the Backstreets
OOOgan

Portraits from the Market
Kyung Roh

The Sound of gently brushing by
Jaewoo Oh

Datascape: Form of Seoul Market
Jeongin Bang, Studio Twothree

Seoul City Market

Eulji: Coexistence
Younhyun Trading

City Markets
Urban Play

Pavilion Project

Weaving Dome
UAUS: Kookmin University

Play-scape
UAUS: Yonsei University

Perhaps. Perhaps. Perhaps
Seungmo Seo

CUPLOWER
UAUS: University of Seoul

Crater: Create With Crate
UAUS: Gachon University

Ikio
UAUS: Chung-Ang University

RE: COVER
UAUS: Sun Moon University

Floatform
UAUS: Hanyang University

Discovery Seoul Market
Urbanplay (Heejun Lee)

Seoul Madang

Director
Jaeyong Lim

Associate Curator
Minsun Kang

Assistant Curator
Jinwoo Jung

Public Program

General Manager
Nayeon Kim

Open House Seoul Special Program
Hye Rin Jeong

Film & Video Program
Ju Hyun Cheon

Public Program (Education, Tour)
Ye Ji Choe

Education Program

2019 SBAU Lectures
Francisco Sanin, Eunjoo Hong and Hyungjae Kim, Jooseok Hong, Sanki Choe, Minsuk Cho, Aram You, Young Jang, Dongwoo Yim,

Special Lectures
Leenam Lee, Ilhyun Kim, Hyunseok Seo, Sohyun Kim, Junghyun Park, Hochul Choi, Inha Park

Hands-on Exhibition Program
Hyun-jeong Kwon (ArchiEdu)

Seoul Metropolitan City Architecture School Program
Hyun-jeong Kwon (ArchiEdu)

Tour Program

Seoul History Tour
Changmo Ahn, Junggoo Cho, Hyoungnam Lim

서울 역사투어
안창모, 조정구, 임형남

서울 테마투어
심영규, 이희준, 김선재, 정소익

국외 총감독 지원
강민선

국외 총감독 연구 보조
무니라 알라비 셰리프 파리드

2019 서울도시건축비엔날레
전시 아이덴티티 및 그래픽디자인
홍은주, 김형재

전시 대행사
(주)리쉬이야기(양희석, 서지영,
이기행, 허필, 손보라, 박현정)

운영 대행사
(주)스튜디오 블룸(류제원, 전평재,
이해존, 강장원, 이병연, 김윤주, 김정원,
이경진, 윤정애)

홍보 대행사
(주)피알하우스(손혜경, 이다겸,
이유진, 조주현, 조아라, 조수빈, 김종빈,
박규희, 권해진, 조민정)

출판

편집
임여진
김형재

번역
한국문화예술번역원

영문 교정
앨리스 김

사진
김태윤(주제전), 진효숙(도시전),
김용순(글로벌 스튜디오, 현장프로젝트,
시민참여프로그램), 윤현(서울도시장),
어반플레이(서울도시장),
최연정(서울시장산책),
스튜디오블룸(시민참여프로그램),
아키에듀(시민참여프로그램)

디자인
홍은주와 김형재

인쇄 및 제책
(주)으뜸프로세스

후원 및 협찬

협찬
삼성전자(주)
코오롱글로텍(주)
더플라자호텔
신라스테이

후원
국립현대미술관
네덜란드 대사관 및 건축미술관
네덜란드창조산업기금
서울역사박물관
서울주택공사
세운협업지원센터
주한독일문화원
주한 이탈리아 문화원
주한 프랑스 문화원
암스테르담 시
베를린 도시개발 및 주택부 상원위원회
브뤼셀시 공간 계획 기관

협력
서울도시건축전시관
(사)한국건축가협회
(사)대한건축사협회
AA 건축대학교
EAFIT 대학교
게이오대학교
국민대학교
동아대학교
매니토바대학교
바틀렛건축대학교
베를린공과대학교
베이루트 아메리칸대학교
빈 공과대학교
서울대학교
서울시립대학교
서울역사박물관
성균관대학교
시드니공과대학교
시라큐스대학교
싱가포르과학기술대학교
싱가포르국립대학교
연세대학교
영국왕립예술학교
영남대학교
이화여자대학교
출랄롱코른대학교 INDA
캘리포니아예술대학교
케이프타운 대학교
콜럼비아대학교
쿠웨이트 대학교
텍사스공과대학교
텍사스알링턴대학교
펜실베이니아대학교
하버드대학교
한양대학교
한양대학교 에리카
홍익대학교
홍콩대학교
홍콩중문대학교
로드 아일랜드 스쿨 오브 디자인
카탈루냐 자치정부, 바르셀로나
　　광역행정청
태국디지털경제홍보처,
　　쇼우헝디자인기술주식회사
아랍사회과학협의회
비판적 방송을 위한 MIT 연구소
하버드 건축대학원, 앤드류 멜론 재단
퀸즐랜드 대학교
뉴욕 주립 대학교 버펄로 건축대학
취리히 연방 공과 대학교
뉴욕 공과 대학교, 시드니 공과 대학교
평등주의와 도시를 위한 미시간 멜론
　　프로젝트, 미시간 대학교 터브먼

Seoul Themed Tour
Youngkyu Shim, Heejun Lee, Sun
Jae Kim, Soik Jung

Foreign Directorial Assistant
Minsun Kang

Foreign Directorial Research
Assistants
Muneerah Alrabe, Cherif Farid

2019 SBAU Identity Design &
Graphic Design
Eunjoo Hong and Hyungjae Kim

Exhibit Agency
RISH IYAGI Co., Ltd.
Richard Yang, JiYoung Seo,
KiHaeng Lee, Philip Heo, Mirae
Park, Bora Son, Hyunjung Park

Managing Agency
Studio Bloom Co., Ltd.
Je-Won Lew, Pyeong-jae Jeon,
Hae-John Lee, Jang-won Kang,
Byung-yeon Lee, Yunju Kim,
Jeongwon Kim, Kyeong-jin Lee,
Jeong-ae Yoon

PR Agency
PR House Co., Ltd.
Haekyung Son, Dagyeom Lee,
Eujene Lee, Juhyun Cho, Ara Jo,
Soobin Jo, Jongbin Kim, Gyuhee
Park, Haejin Kwon, Minjoung Cho

Publications

Edited by
Jean Im
Hyungjae Kim

Translation
Korea Institute of Culture and Arts
Translation

English Proofreading
Alice S Kim

Photography
Taeyoon Kim (Thematic
Exhibition), Hyosook Jin (Cities
Exhibition), Yongsun Kim (Global
Studio, Live Projects, Public
Program), Younhyun (Seoul
City Market), Urbanplay (Seoul
City Market), Yeongjung Choi
(Discovery Seoul Market), Studio
Bloom (Public Program), ArchiEdu
(Public Program)

Design
Eunjoo Hong and Hyungjae Kim

Print
Top Process., Ltd.

Sponsors and Partners

Sponsored by
Samsung Electronics Co., Ltd.
Kolon Glotech, Inc.
The Plaza Seoul, Autograph
Collection
Shilla Stay

Projects Sponsored by
National Museum of Modern and
　　Contemporary Art
Netherlands Embassy and Het
　　Nieuwe Instituut
Creative Industries Fund NL
Seoul Museum of History
Seoul Housing & Communities
　　Corporation
Sewoon Collaboration Support
　　Center
Goethe-Institut Korea
Istituto Italiano di Cultura - Seoul
Institut français de Corée du Sud
City of Amsterdam
Senate Department for Urban
　　Development and Housing
　　Berlin (Senatsverwaltung für
　　Stadtentwicklung und Wohnen
　　Berlin)
Perspective, Spatial Planning
　　Agency of the Brussels Capital
　　Region

Partners Institutes
Seoul Hall of Urbanism &
　　Architecture
Korean Institute of Architects
Korean Institute of Registered
　　Architects
Architectural Association School of
　　Architecture
URBAM - Universidad de EAFIT
Keio University
Kookmin University
Donga University
University of Manitoba
Bartlett School of Architecture
Technische Universität Berlin
American University of Beirut
Technische Universität Wien
Seoul National University
University of Seoul
Seoul Museum of History
Sungkyunkwan University
University of Technology Sydney
Syracuse University
Singapore University of Technology
　　and Design
National University of Singapore
Yonsei University
Royal College of Art
Yeungnam University
Ewha Womans University
Chulalongkorn University INDA
California College of the Arts
University of Cape Town
Columbia University
Kuwait University
Texas Tech University
University of Texas at Arlington
University of Pennsylvania
Harvard University
Hanyang University
Hanyang University ERICA
Hongik University
University of Hong Kong
Chinese University of Hong Kong
RISD Architecture & Canon
　　Foundation
Generalitat de Catalunya,
　　Area Metropolitana de
　　Barcelona
"depa X SHDT" (Digital Economy
　　Promotion Agency of Thailand
　　and Shouheng Design and

건축대학
덴마크 예술 재단, 주한 덴마크 대사관
SAC, 하인츠와 지젤라 프리드리히 재단
　/ 인스부르크 대학교
개성공단재단
서매틱 콜라보레이티브, 예일대학교
건축대학
선전 인택트 스튜디오(영상 후반 작업:
　선전 위롱 테크놀로지)
자카르타특별지구, 인도네시아
　창조경제부, 카르손 & 사갈라
　건축디자인, noMaden & Pupla
　프로젝트
런던대학교 바틀릿 건축대학,
　포스터+파트너스, 웨스턴
　윌리엄슨+파트너스
르노-닛산 자동차, 포드 자동차
메데인 시
밀라노 시
그레이엄 예술고등교육재단, 컬럼비아
　대학교 건축대학원
주한 코스타리카 대사관
하다드 재단
로열 멜버른 공과대학교
　건축&도시설계대학
오토데스크(Autodesk), 펜실베이니아
　대학교
울산발전연구원
빈 비지니스 에이전시, 빈 마케팅,
　BKA(오스트리아 문화예술부),
　오타크링거 양조장 GmbH, 요제프
　마너 &Comp. AG
뉴욕 공과 대학교, 웰링턴 빅토리아
　대학교
영주시
암스테르담 시
프로 헬베티아
서던캘리포니아 대학교 프라이스
　행정대학 공간분석 연구소
이음피움 봉제역사관
뮤지스땅스
배화여자고등학교
서소문성지역사박물관
서울시립미술관
서울도시건축센터
서울시 경제정책실
서울시 도시공간개선단
서울시 도시재생실
서울시 역사도심재생과
서울시 푸른도시국
서울시 행정국

Technology Inc)
ACSS (Arab Council for Social
　Sciences)
MIT Critical Broadcasting Lab (Ana
　Miljacki, Faculty lead)
Harvard University Graduate
　School of Design, Andrew W.
　Mellon Foundation
The University of Queensland
University at Buffalo School of
　Architecture and Planning
ETH Zürich (Eidgenössische
　Technische Hochschule Zürich)
New York Institute of Technology
　(NYIT), University of Technology
　Sydney (UTS)
Michigan Mellon Project for
　Egalitarianism and the
　Metropolis, Taubman College
　University of Michigan
Danish Arts Foundation, Embassy
　of Denmark in Korea
SAC & Heinz und Gisela Friedrichs
　Stiftung / University of
　Innsbruck
Gaesong Industrial District
　Foundation
Somatic Collaborative/ Yale School
　of Architecture
Intact Studio, Shenzhen (video post
　production: Yulong Technology,
　Shenzhen
Capital Special Region of Jakarta,
　Badan Ekonomi Kreatif
　(BEKRAF), Kalson Sagala
　Architecture & Design (KSAD),
　noMADen & Pupla Project
The Bartlett School of Architecture,
　Foster + Partners, Weston
　Williamson + Partner
Renault Nissan; Ford
Alcaldía de Medellín
City of Milano
Graham Foundation for Advanced
　Studies in the Fine Arts, and a
　Columbia University Graduate
　school of Architecture, Planning
　and Preservation
Embajada de Costa Rica en Corea
Haddad Foundation
RMIT School of Architecture &
　Urban Design
Autodesk, University of
　Pennsylvania
Ulsan Development Institute
Vienna Business Agency, Wien
　Marketing, BKA (The Arts
　and Culture Division of the
　Federal Chancellery of Austria);
　Ottakringer Brauerei GmbH,
　Josef Manner & Comp. AG
　(inquired)
New York Institute of Technology
　(NYIT) & Victoria University of
　Wellington (VUW)
City of Yeongju
City of Amsterdam
Pro Helvetia
SLAB: Spatial Analysis Lab of USC
　Price School of Public Policy
Iumpium Sawing History Center
Musistance
Paiwha Girls' High School
Seosomun Shrine History Museum
Seoul Museum of Art
Seoul Center for Architecture &
　Urbanism
Seoul Economic Policy Office
Seoul Urban Improvement Bureau

Seoul Urban Regeneration Office
Seoul Historic City Center
　Regeneration Division
Seoul Green Seoul Bureau
Seoul Administrative Services

2019 서울도시건축비엔날레
집합도시

초판 1쇄 발행
2019. 11. 7

발행처
서울특별시

발행인
서울특별시장 박원순

제작부서
도시공간개선단

주소
서울특별시 중구 세종대로 110

기획
서울디자인재단 서울도시건축비엔날레 사무국

ISBN
979-11-6161-513-4 (03540)

2019 Seoul Biennale of Architecture and Urbanism
COLLECTIVE CITY

First Edition Printed on
Novemeber 7th, 2019

Published by
Seoul Metropolitan Government

Publisher
Won-soon Park, Mayor of Seoul

Department
Seoul Metropolitan Government
Urban Improvement Bureau

Address
Sejongdaero 110, Junggu, Seoul

Planning and Management
Seoul Design Foundation Seoul Biennale of
Architecture and Urbanism Division

ISBN
979-11-6161-513-4 (03540)